LINEAR RELATIONS AND FUNCTIONS (2-4)

$f(x) = mx + c$ Linear function
$(y - y_1) = m(x - x_1)$ Point–slope form
$y = C$ Horizontal line
$x = C$ Vertical line

POLYNOMIAL AND RATIONAL FUNCTIONS (2-5, 2-6, 3-1–3-7)

$f(x) = a_n x^n + a_{n-1} x^{n-1} + \cdots + a_1 x + a_0,$ Polynomial
     $a_n \neq 0$, n a nonnegative integer   function

$f(x) = \dfrac{p(x)}{q(x)},$ p and q polynomial   Rational
           functions, $q(x) \neq 0$ function

EXPONENTIAL AND LOGARITHMIC FUNCTIONS (4-1–4-5)

$f(x) = b^x$, $b > 0, b \neq 1$ Exponential function
$f(x) = \log_b x$, $b > 0, b \neq 1$ Logarithmic function
$y = \log_b x$ if and only if $x = b^y$, $b > 0, b \neq 1$

EQUATIONS OF A CIRCLE (2-1)

$(x - h)^2 + (y - k)^2 = r^2$ Center at (h, k), radius r
$x^2 + y^2 = r^2$ Center at $(0, 0)$, radius r

VARIATION (2-8)

$y = kx$, $y \neq 0$ Direct
$y = k/x$, $k \neq 0$ Inverse
$w = kxy$, $k \neq 0$ Joint

MATRICES AND DETERMINANTS (5-2, 6-1–6-5)

$\begin{bmatrix} a & b & c \\ d & e & f \end{bmatrix}$ Matrix

$\begin{vmatrix} a & b & c \\ d & e & f \\ g & h & i \end{vmatrix}$ Determinant

ARITHMETIC SEQUENCE

$a_1, a_2, \ldots, a_n, \ldots$
$a_n - a_{n-1} = d$ Common
$a_n = a_1 + (n - 1)d$ nth
$S_n = a_1 + \cdots + a_n = \dfrac{n}{2}[2a_1 + (n - 1)d]$ Sum of n terms
$S_n = \dfrac{n}{2}(a_1 + a_n)$

GEOMETRIC SEQUENCE (7-4)

$a_1, a_2, \ldots, a_n, \ldots$
$\dfrac{a_n}{a_{n-1}} = r$ Common ratio
$a_n = a_1 r^{n-1}$ nth-term formula
$S_n = a_1 + \cdots + a_n = \dfrac{a_1 - a_1 r^n}{1 - r}$, $r \neq 1$ Sum of n terms
$S_n = \dfrac{a_1 - r a_n}{1 - r}$, $r \neq 1$
$S_\infty = a_1 + a_2 + \cdots = \dfrac{a_1}{1 - r}$, $|r| < 1$ Sum of infinitely many terms

FACTORIAL AND BINOMIAL FORMULAS (7-6)

$n! = n(n - 1) \cdots 2 \cdot 1$, $n \in N$ n factorial
$0! = 1$
$\dbinom{n}{r} = \dfrac{n!}{r!(n - r)!}$
$(a + b)^n = \displaystyle\sum_{k=0}^{n} \dbinom{n}{k} a^{n-k} b^k$, $n \geq 1$ Binomial formula

PERMUTATIONS, COMBINATIONS, AND SET PARTITIONING (8-3)

$P_{n,r} = \dfrac{n!}{(n - r)!}$ Permutation

$C_{n,r} = \dbinom{n}{r} = \dfrac{n!}{r!(n - r)!}$ Combination

$\dbinom{n}{r_1, r_2, \ldots, r_k} = \dfrac{n!}{r_1! r_2! \cdots r_k!}$ Partition

COLLEGE ALGEBRA

COLLEGE ALGEBRA

Raymond A. Barnett
Merritt College

THIRD EDITION

McGRAW-HILL BOOK COMPANY

New York   St. Louis   San Francisco   Auckland   Bogotá   Hamburg
Johannesburg   London   Madrid   Mexico   Montreal   New Delhi   Panama
Paris   São Paulo   Singapore   Sydney   Tokyo   Toronto

COLLEGE ALGEBRA

1 2 3 4 5 6 7 8 9 0 D O C D O C 8 9 8 7 6 5 4 3

ISBN 0-07-003861-9

This book was set in Melior.
The editor was Peter R. Devine;
the designer was Janet Bollow;
the cover was designed by Anne Canevari Green;
the production supervisor was Leroy A. Young.
The drawings were done by Carl Brown.
Project supervision was done by Phyllis Niklas.
R.R. Donnelley & Sons Company was printer and binder.

Chapter-Opening Photo Credits

Photo for Chapter 0 © by P. J. Bryant, University of California, Irvine/BPS.
Photos for Chapters 1, 6, and 8 © by Peter Pearce. Photos for Chapters 2, 5,
and 7 © by Anne Monk. Photo for Chapter 3 © by William C. Ferguson.
Photo for Chapter 4 © by S. K. Webster, Monterey Bay Aquarium/BPS.

Cover Photo Credit

This colonial organism is a bryozoan (*Spirobranchus grandis*), commonly
called the Christmas tree worm. Photo © by S. K. Webster, Monterey Bay
Aquarium/BPS.

Library of Congress Cataloging in Publication Data

Barnett, Raymond A.
    College algebra.

    Includes index.
    1. Algebra.   I. Title.
QA154.2.B35 1984   512.9   83-12076
ISBN 0-07-003861-9

Contents

*Note to Instructor: May be omitted, depending on time, need, and interest. (Some topics are review topics and some are new topics.)

Tables A3

Preface

This third edition of *College Algebra* reflects experience and feedback from a large number of users of the earlier edition.

■ Principal Changes from the Second Edition

1. The most significant change is in Chapter 2 on **graphs and functions,** which was completely rewritten. The treatment of function, composite function, and inverse function is more comprehensive and better motivated. The discussion of graphing techniques and the graphing of special functions such as polynomial functions and rational functions has been substantially expanded.

2. Chapter 0 provides a comprehensive **review of basic algebraic operations** that are usually covered in intermediate algebra. This material may be treated systematically if a class is weak in basics; it may be briefly reviewed and the class tested to screen out students who need a more elementary course before this one; or it may be omitted altogether if time is a concern (students can then refer to the material as needed).

3. The material in Chapter 1 on **equations and inequalities** has been arranged to increase comprehension and motivation. For example, the real number line is introduced in conjunction with solving and graphing linear inequalities in one variable, and complex numbers are introduced just prior to solving quadratic equations.

4. Chapter 3 on **polynomial functions and the theory of equations** has been streamlined, and there is now a better focus on key topics and processes.

5. Much greater **use of scientific calculators** is made throughout the text both in examples and in exercise sets. For example, calculators are used in graphing; solving equations, including exponential and logarithmic equations (discussion of table use is included but is optional); evaluating permutations and combinations; and determining probabilities. The author believes every student in the class should have a scientific calculator with a user's book for that calculator.

6. Chapter 5 on **systems of equations and inequalities** now includes solutions of linear systems using Gauss–Jordan elimination. The presentation is gradual, well motivated, and carefully done. (See Sections 5-1, 5-2, and 5-3.)

7. Chapter 6 on **matrices and determinants** has been reorganized and the material on matrices has been expanded. Inverses of square matrices and matrix equations are now treated in some detail. (See Sections 6-1, 6-2, and 6-3.)

8. Functional **use of a second color** is made throughout the text for increased clarity in topic presentation and graphing.

9. To **increase student support** and understanding, additional discussion, examples, and exercises have been added to almost every section, and some sections have been completely rewritten. These additions and changes have increased the length of the book, but have not added substantially to the number of topics covered. With this added student support, an instructor should be able to cover the same amount of material as before and with greater student comprehension.

10. There are now separate **chapter review sections** that include concise summaries of terms, formulas, and symbols discussed in each section; chapter review exercise sets keyed to specific sections through the answers in the back of the book; and practice tests, also keyed to specific sections through the answers in the back of the book.

■ Important Features Retained from the Second Edition

1. The text is still **written for student comprehension.** Each concept is illustrated with an example, followed by a parallel problem with an answer so that a student can immediately check his or her understanding of the concept. These followup problems also encourage active rather than passive reading of the text.

2. An **informal style** is used for exposition, statements of definitions, and proofs of theorems.

3. The text includes **more than 3,300 carefully selected and graded problems.** The exercises are divided into A, B, and C groupings, with the A problems easy and routine, the B problems more challenging but still emphasizing mechanics, and the C problems a mixture of theoretical and difficult mechanics. In short, the text is designed so that an average or below-average student will be able to experience success and a very capable student will be challenged.

4. The subject matter is related to the real world through many carefully selected **realistic applications** from the physical

sciences, business and economics, life sciences, and social sciences. Thus, the text is equally suited for students interested in any of these areas.

5. Following the recommendations of many national and international mathematical organizations and the author's own convictions, the **function concept is used as a unifying notion** from the second chapter onward.

6. **Answers** to all chapter review exercises and practice tests and to all odd-numbered problems from the other exercises are in the back of the book.

■ Student Aids

1. **Common student errors** are clearly identified at places where they naturally occur (see Sections 0-2 and 1-2).

2. **Think boxes** (dashed boxes) are used to enclose steps that are usually performed mentally (see Sections 1-2 and 1-4).

3. **Annotation** of examples and developments is found throughout the text to help students through critical stages (see Sections 1-2 and 1-4).

4. **Functional use of a second color** guides students through critical steps (see Sections 1-2 and 1-4).

5. **Chapter review sections** include a review of all important terms and symbols, a comprehensive review exercise, and a practice test. Answers to all review exercises and practice test problems are included in the back of the book and are keyed (with numbers in italics) to the corresponding text sections.

6. **Summaries** of formulas and symbols (keyed to the sections in which they are introduced) and the metric system are inside the front and back covers of the text for convenient reference.

7. A **solutions manual** is available at a nominal cost through a bookstore. The manual includes detailed solutions to all odd-numbered problems, all chapter review exercises, and all practice test problems.

■ Instructor Aids

1. A **comprehensive test battery** is included in the instructor's manual, which can be obtained from the publisher at no cost. The test battery includes two regular and two multiple-choice tests for each chapter. All tests have easy-to-grade solution keys and sample student solution sheets. The format is $8\frac{1}{2}$ by 11 inches for ease of reproduction.

2. **Answers to even-numbered problems,** which are not included in the text, are given in the instructor's manual.
3. A **solutions manual** (see student aids) is available to instructors at no cost from the publisher.

■ Error Check

Because of the careful checking and proofing by a number of very competent people (acting independently), the author and publisher believe this book to be substantially error-free. For any errors remaining, the author would be grateful if they were sent to: Mathematics Editor, College Division, 27th floor, McGraw-Hill Book Company, 1221 Avenue of the Americas, New York, New York 10020.

■ Acknowledgments

The preparation of a book requires the effort and skills of many people in addition to an author. I wish to thank the reviewers for their many helpful suggestions and comments. (It is this process of use, feedback, and adjustment that produces an increasingly effective book for both students and instructors.) In particular I wish to thank Thomas A. Atchison, Stephen F. Austin State University; Martin Broadwell, Jr., Florida College; Eddie J. Brown, St. Clair County Community College; Leland Fry, Kirkwood Community College; Lotus Hershberger, Illinois State University; Sidney Katoni, New York City Technical College; Peter Lindstrom, North Lake College; Gary Ling, Golden Gate University; Stanley Lukawecki, Clemson University; Francis E. Masat, Glassboro State College; Ronald Prielipp, Bethany College; Bruce Reed, Virginia Polytechnic Institute; Donald G. Spencer, Northeast Louisiana University; and George N. Trytten, Luther College. Special thanks go to Margaret Barnett-Burnette, Fred Safier (City College of San Francisco), and Ward A. Soper (Walla Walla College) for their careful checking of all examples, matched problems, and exercise sets.

Raymond A. Barnett

To the Student

The following suggestions are made to help you get the most out of this book and your efforts.

As you study the text we suggest a five-step process. For each section:

1. Read a mathematical development. ⎫ Repeat the 1-2-3
2. Work through the illustrative example. ⎬ cycle until the
3. Work the matched problem. ⎭ section is finished.
4. Review the main ideas in the section.
5. Work the assigned exercise at the end of the section.

All of this should be done with plenty of paper, pencils, and a wastebasket at hand. In fact, no mathematics text should be read without pencil and paper in hand; mathematics is not a spectator sport. Just as you cannot learn to swim by watching someone else swim, you cannot learn mathematics by simply reading worked examples — you must work problems, lots of them.

If you have difficulty with the course, then, in addition to doing the regular assignments, spend more time on the examples and matched problems and work more A exercises, even if they are not assigned. If the A exercises continue to be difficult for you, you probably should take an intermediate algebra course before attempting this one. If you find the course too easy, then work more C exercises, even if they are not assigned. If the C exercises are consistently easy for you, you are probably ready to start the calculus sequence.

Raymond A. Barnett

Possible Courses

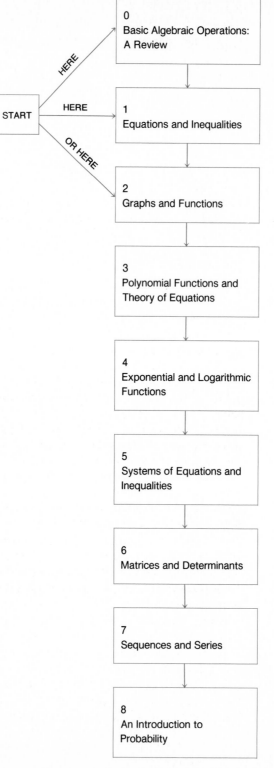

Remarks on Calculator Use

Hand calculators are of two basic types relative to their internal logic (the way they compute): algebraic and reverse Polish notation (RPN). Throughout the book we will identify algebraic calculator steps with "A" and reverse Polish notation calculator steps with "P." Let us see how each type of calculator would compute

$$\frac{(5)(3)(2) - (7)(6)}{2(11)}$$

PRESS   DISPLAY

Some people prefer the algebraic logic and others prefer the Polish. Which is better is still being debated. The answer seems to rest with the type of problems one encounters and with individual preferences. The author owns both types and uses the one with Polish logic most frequently. However, he knows people who prefer the algebraic type, and they seem quite happy with their choice.

In any case, irrespective of the type of calculator you own, it is essential that you read the user's manual for your own calculator. A large variety of calculators are on the market, and each is slightly different from the others. Therefore, it is important that you take the time to read the manual. Do not try to read and understand everything the calculator can do; this will only tend to confuse you. Read only those sections that pertain to the operations you are or will be using; then return to the manual as necessary when you encounter new operations.

In many places in the text calculator steps for new types of calculations will be shown (similar to those steps shown here). These are only aids. Try the calculation without the aid; then use the aid only if you get stuck.

It is important to remember that *a calculator is not a substitute for thinking*. It can save you a great deal of time in certain types of problems, but you still must know how and when to use it.

Raymond A. Barnett

Basic Algebraic Operations: A Review

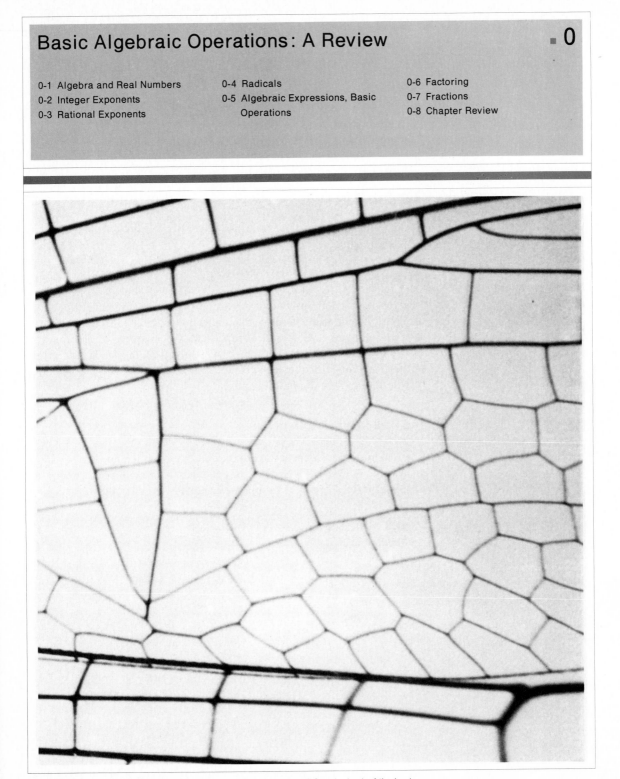

A natural design of mathematical interest. Can you guess the source? See the back of the book.

Chapter 0 ■ Basic Algebraic Operations: A Review

Section 0-1 Algebra and Real Numbers

- The Real Number System
- Basic Properties
- Further Properties
- Fraction Properties

In algebra we are interested in manipulating symbols in order to change or simplify algebraic expressions and to solve algebraic equations. Because many of these symbols represent real numbers, it is important to briefly review the real number system and some of its important properties. These properties provide the basic rules for much of the manipulation of symbols in algebra.

■ The Real Number System

The real number system is the number system in which you have worked most of your life. Table 1 describes the set* of real numbers and some of the important types of numbers within the set of real numbers.

TABLE 1 The Set of Real Numbers	SYMBOL	NUMBER SYSTEM	DESCRIPTION	EXAMPLES
	N	Natural numbers	Counting numbers (also called positive integers)	$1, 2, 3, \ldots$
	Z	Integers	Set of natural numbers, their negatives, and 0	$\ldots, -2, -1, 0, 1, 2, \ldots$
	Q	Rationals	Any number that can be represented as a/b, where a and b are integers and $b \neq 0$	$-4; \frac{-3}{5}; 0; 1; \frac{2}{3}; 3.67$
	R	Reals	Set of all rational and irrational numbers (the irrational numbers are all the real numbers that are not rational)	$-4; \frac{-3}{5}; 0; 1; \frac{2}{3}; 3.67; \sqrt{2}; \pi; \sqrt[3]{5}$

* A **set** is a collection of objects. More will be said about sets in Chapter 1.

Figure 1 illustrates how these sets of numbers are related to one another.

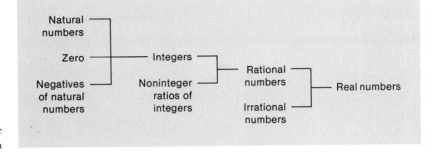

FIGURE 1   The real number
system

The set of integers contains all the natural numbers and something else (their negatives and 0). The set of rational numbers contains all the integers and something else (noninteger ratios of integers). And the set of real numbers contains all the rational numbers and something else (irrational numbers).

Rational numbers have repeating decimal representations, whereas irrational numbers have infinite nonrepeating decimal representations. For example, the decimal representations of the rational numbers 2, $\frac{4}{3}$, and $\frac{5}{11}$ are, respectively,

$$2 = 2.0000 \ldots \qquad \frac{4}{3} = 1.333 \ldots \qquad \frac{5}{11} = 0.454545 \ldots$$

whereas those of the irrational numbers $\sqrt{2}$ and π are, respectively,

$$\sqrt{2} = 1.41421356 \ldots \qquad \pi = 3.14159265 \ldots$$

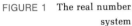 Basic Properties

We now review (informally) a few basic real number properties. Other real number properties will be discussed as needed in other parts of the text. These properties of the real numbers become operational rules in the algebra of real numbers.

Real numbers can be added, subtracted, multiplied, and divided (except for division by 0). Does it matter in which order we perform addition, subtraction, multiplication, or division? In general, we may add in any order or multiply in any order [but this is not true for subtraction or division (for example, $4 - 2 \neq 2 - 4$ and $4 \div 2 \neq 2 \div 4$)]. This property is referred to as the **commutative property** (for addition and multiplication) for real numbers.

Commutative Property

For all real numbers a and b

Addition	Multiplication
$a + b = b + a$	$ab = ba$
$7 + 2 = 2 + 7$	$3 \cdot 5 = 5 \cdot 3$

When computing

$12 + 6 + 2$

$12 - 6 - 2$

$12 \cdot 6 \cdot 2$

$12 \div 6 \div 2$

does it matter how the numbers are grouped? That is, which of the following are true?

$(12 + 6) + 2 = 12 + (6 + 2)$

$(12 - 6) - 2 = 12 - (6 - 2)$

$(12 \cdot 6) \cdot 2 = 12 \cdot (6 \cdot 2)$

$(12 \div 6) \div 2 = 12 \div (6 \div 2)$

We see that the first and third are true, but the second and fourth are false. In general, we may group terms in addition and factors in multiplication in any way we please (but we are not free to group as we please in subtraction and division). This property is referred to as the **associative property** (for addition and multiplication) for real numbers.

Associative Property

For all real numbers a, b, and c,

Addition	Multiplication
$(a + b) + c = a + (b + c)$	$(a \cdot b) \cdot c = a \cdot (b \cdot c)$
$(5 + 2) + 3 = 5 + (2 + 3)$	$(4 \cdot 3) \cdot 2 = 4 \cdot (3 \cdot 2)$

Conclusion

In addition, commutativity and associativity permit us to change the order at will and insert or remove parentheses as we please. The same is true for multiplication, but not for subtraction and division.

What number added to a given number will give that number back again? What number times a given number will give that number back again? The answers are 0 and 1, respectively; thus, 0 and 1 are called the **identity elements** for the real numbers.

Identities

For each real number a,

$$a + 0 = a \qquad 1 \cdot a = a$$

$$(-3) + 0 = -3 \qquad 1 \cdot 7 = 7$$

0 is the additive identity.
1 is the multiplicative identity.

We now consider inverses. For each real number a, is there a real number that when added to a produces 0? For each real number $a \neq 0$, is there a real number that when multiplied times a produces 1? The answer in both cases is yes.

Inverses

For each real number a, there is a unique real number $-a$ such that

$$a + (-a) = 0$$

$$3 + (-3) = 0$$

For each real number $a \neq 0$, there is a unique real number $1/a$ such that

$$a\left(\frac{1}{a}\right) = 1$$

$$3\left(\frac{1}{3}\right) = 1$$

$-a$ is called the **additive inverse** of a or **negative** of a.*

$1/a$ is called the **multiplicative inverse** of a or **reciprocal** of a.

Zero has no multiplicative inverse.   What number times 0 is 1?

---

* $-a$ is not necessarily a negative number; it is positive if a is negative and negative if a is positive.

We now turn to an important real number property that involves both multiplication and addition. Consider the two computations:

$$3(4 + 2) = 3 \cdot 6 \qquad 3 \cdot 4 + 3 \cdot 2 = 12 + 6$$
$$= 18 \qquad\qquad\qquad = 18$$

Thus,

$$3(4 + 2) = 3 \cdot 4 + 3 \cdot 2$$

and we say that the factor 3 *distributes* over the sum $(4 + 2)$. In general, in the real number system multiplication always **distributes** over addition.

Distributive Property

For all real numbers a, b, and c,

$$a(b + c) = ab + ac \qquad (b + c)a = ba + ca$$
$$2(x + y) = 2x + 2y \qquad (3 + 5)x = 3x + 5x$$

All the properties listed in the preceding boxes are axioms (mathematical statements that are assumed true without proof). Many other real number properties (theorems) can be proved using these few axioms.

Throughout the rest of the book all variables represent real numbers unless stated to the contrary, and the properties given here will be assumed.

EXAMPLE 1   STATEMENT   PROPERTY ILLUSTRATED

(A) $(7x)y = 7(xy)$ Associative $(\times)$
(B) $a(b + c) = (b + c)a$ Commutative $(\times)$
(C) $(2x + 3y) + 5y = 2x + (3y + 5y)$ Associative $(+)$
(D) $(x + y)(a + b) = (x + y)a + (x + y)b$ Distributive
(E) If $a + b = 0$, then $b = -a$.   Inverse

PROBLEM 1*   Which real number property justifies the indicated statement?

(A) $4 + (2 + x) = (4 + 2) + x$
(B) $(a + b) + c = c + (a + b)$
(C) $3x + 7x = (3 + 7)x$
(D) $(2x + 3y) + 0 = 2x + 3y$
(E) If $ab = 1$ and $a \neq 0$, then $b = 1/a$.

* Answers to matched problems in a given section are found near the end of the section, before the exercise set.

■ Further Properties

Subtraction and division can be defined in terms of addition and multiplication, respectively:

SUBTRACTION: $a - b = a + (-b)$ $(-5) - (-3) = (-5) + (3) = -2$

DIVISION: $b\overline{)a} = a \div b = \dfrac{a}{b} = a\left(\dfrac{1}{b}\right)$ $b \neq 0$ $3 \div 2 = 3\left(\dfrac{1}{2}\right)$

Thus, to subtract b from a, add the negative (the additive inverse) of b to a. To divide a by b, multiply a by the reciprocal (the multiplicative inverse) of b. Note that division by 0 is not defined (one cannot divide by 0 ever!), since 0 does not have a reciprocal.

The following properties of negatives (called *theorems*) can be proved using the preceding axioms and definitions.

Properties of Negatives

For all real numbers a and b,

$-(-a) = a$

$(-a)b = -(ab) = a(-b)$

$(-a)(-b) = ab$

$(-1)a = -a$

$\dfrac{-a}{b} = -\dfrac{a}{b} = \dfrac{a}{-b}$ $b \neq 0$

$\dfrac{-a}{-b} = -\dfrac{-a}{b} = -\dfrac{a}{-b} = \dfrac{a}{b}$ $b \neq 0$

We now state two important properties (theorems) involving 0.

Zero Properties

For all real numbers a and b,

$a \cdot 0 = 0$

$ab = 0$ if and only if $a = 0$ or $b = 0$ or both.

EXAMPLE 2   STATEMENT   PROPERTY OR DEFINITION ILLUSTRATED

(A) $3 - (-2) = 3 + [-(-2)]$ Subtraction

(B) $-(-2) = 2$ Negatives

(C) $-\dfrac{-3}{2} = \dfrac{3}{2}$ Negatives

(D) $\dfrac{5}{-2} = -\dfrac{5}{2}$ Negatives

(E) If $(x - 3)(x + 5) = 0$, then either   Zero
     $x - 3 = 0$ or $x + 5 = 0$.

PROBLEM 2   Which real number property or definition justifies each statement?

(A) $\dfrac{3}{5} = 3\left(\dfrac{1}{5}\right)$ (B) $(-5)(2) = -(5 \cdot 2)$ (C) $(-1)3 = -3$

(D) $\dfrac{-7}{9} = -\dfrac{7}{9}$ (E) If $(x + 5) = 0$, then $(x - 3)(x + 5) = 0$.

■   **Fraction Properties**

Recall that the quotient $a \div b$, $b \neq 0$, written in the form a/b is called a **fraction**. The quantity a is called the **numerator** and the quantity b the **denominator**.

Fraction Properties

For all real numbers a, b, c, d, and k (division by 0 excluded),

$$\frac{a}{b} = \frac{c}{d} \quad \text{if and only if } ad = bc$$

$$\frac{4}{6} = \frac{6}{9} \quad \text{since } 4 \cdot 9 = 6 \cdot 6$$

$$\frac{ka}{kb} = \frac{a}{b} \qquad \frac{a}{b} \cdot \frac{c}{d} = \frac{ac}{bd} \qquad \frac{a}{b} \div \frac{c}{d} = \frac{a}{b} \cdot \frac{d}{c}$$

$$\frac{7 \cdot 3}{7 \cdot 5} = \frac{3}{5} \qquad \frac{3}{5} \cdot \frac{7}{8} = \frac{3 \cdot 7}{5 \cdot 8} \qquad \frac{2}{3} \div \frac{5}{7} = \frac{2}{3} \cdot \frac{7}{5}$$

$$\frac{a}{b} + \frac{c}{b} = \frac{a + c}{b} \qquad \frac{a}{b} - \frac{c}{b} = \frac{a - c}{b} \qquad \frac{a}{b} + \frac{c}{d} = \frac{ad + bc}{bd}$$

$$\frac{3}{6} + \frac{5}{6} = \frac{3 + 5}{6} \qquad \frac{7}{8} - \frac{3}{8} = \frac{7 - 3}{8} \qquad \frac{2}{3} + \frac{3}{5} = \frac{2 \cdot 5 + 3 \cdot 3}{3 \cdot 5}$$

Answers to Matched Problems

1. (A) Associative   (B) Commutative   (C) Distributive
   (D) Identity   (E) Inverse

2. (A) Division   (B) Negatives   (C) Negatives
   (D) Negatives   (E) Zero

Exercise 0-1 ■ A

In Problems 1–34 each statement illustrates the use of one of the following properties or definitions. Indicate which.

Commutative   Identity   Division
Associative   Inverse   Zero
Distributive   Subtraction   Negatives

1. $5 + 2x = 2x + 5$

2. $x + ym = x + my$

3. $7(3m) = (7 \cdot 3)m$

4. $(2w + 8) + 3 = 2w + (8 + 3)$

5. $x(y + z) = xy + xz$

6. $5(u + v) = 5u + 5v$

7. $-(-12) = 12$

8. $-\dfrac{3}{-5} = \dfrac{3}{5}$

9. $(-5) - (-2) = (-5) + [-(-2)]$

10. $8 - 12 = 8 + (-12)$

11. $\dfrac{-7}{9} = -\dfrac{7}{9}$

12. $\dfrac{-5}{-8} = \dfrac{5}{8}$

13. $3xyz + 0 = 3xyz$

14. $1 \cdot \left(-\dfrac{2}{3}\right) = -\dfrac{2}{3}$

15. $5 \div (-6) = 5\left(\dfrac{1}{-6}\right)$

16. $7 \div 9 = 7\left(\dfrac{1}{9}\right)$

17. $w + (-w) = 0$

18. $(-u) + [-(-u)] = 0$

B
19. $(7 + 12)x = 7x + 12x$

20. $8m + 5m = (8 + 5)m$

21. $4uv + 7uv = (4 + 7)uv$

22. $7x + 7y = 7(x + y)$

23. $(2x - 3)(x + 5) = 0$ if either $2x - 3 = 0$ or $x + 5 = 0$

24. $(u - 4)(3u - 7) = 0$ if either $u - 4 = 0$ or $3u - 7 = 0$

25. $(3x + 5) + 7 = 7 + (3x + 5)$

26. $(mn)p = p(mn)$

27. $(3x + 2) + (x + 5) = 3x + [2 + (x + 5)]$

28. $(5x)(7y) = 5[x(7y)]$

29. $(x + 3)(x + 5) = (x + 3)x + (x + 3)5$

30. $(m + n)(u + v) = m(u + v) + n(u + v)$

31. $x(x - y) + y(x - y) = (x + y)(x - y)$

32. $2x(x + 4) + 3(x + 4) = (2x + 3)(x + 4)$

33. $\dfrac{5}{-(x - 3)} = -\dfrac{5}{x - 3}$

34. $\dfrac{-7}{-(m + n)} = \dfrac{7}{m + n}$

35. If $ab = 0$, does either a or b have to be 0?

36. If $ab = 1$, does either a or b have to be 1?

37. Indicate which of the following are true.
(A) All natural numbers are integers.
(B) All real numbers are irrational.
(C) All rational numbers are real numbers.

38. Indicate which of the following are true.
(A) All integers are natural numbers.
(B) All rational numbers are real numbers.
(C) All natural numbers are rational numbers.

39. Give an example of a rational number that is not an integer.

40. Give an example of a real number that is not a rational number.

41. Given the sets of numbers: N (natural numbers), Z (integers), Q (rational numbers), and R (real numbers). Indicate to which set(s) each of the following numbers belong(s).
(A) -3 (B) 3.14 (C) π (D) $\frac{2}{3}$

42. Given the sets of numbers N, Z, Q, and R (see Problem 41), indicate to which set(s) each of the following numbers belong(s).
(A) 8 (B) $\sqrt{2}$ (C) -1.414 (D) $\dfrac{-5}{2}$

43. Indicate true (T) or false (F), and for each false statement find real number replacements for a and b that will illustrate its falseness. For all real numbers a and b,
(A) $a + b = b + a$ (B) $a - b = b - a$
(C) $ab = ba$ (D) $a \div b = b \div a$

44. Indicate true (T) or false (F), and for each false statement find real number replacements for a, b, and c that will illustrate its falseness. For all real numbers a, b, and c,
(A) $(a + b) + c = a + (b + c)$ (B) $(a - b) - c = a - (b - c)$
(C) $a(bc) = (ab)c$ (D) $(a \div b) \div c = a \div (b \div c)$

C **45.** If $c = 0.151515\ldots$, then $100c = 15.1515\ldots$ and

$$100c - c = (15.1515\ldots) - (0.151515\ldots)$$
$$99c = 15$$
$$c = \tfrac{15}{99} = \tfrac{5}{33}$$

$100c = 9.0909\ldots$

$100c - c = 9.0909\ldots - .090909$

$99c = 9 \qquad \dfrac{9}{99} = \dfrac{1}{11}$

Proceeding similarly, convert the repeating decimal $0.090909\ldots$ into a fraction. (All repeating decimals are rational numbers, and all rational numbers have repeating decimal representations.)

46. Repeat Problem 45 for $0.181818.\ldots$.

47. To see how the distributive property is behind the mechanics of long multiplication, compute each of the following and compare.

LONG MULTIPLICATION $\qquad$ USE OF THE DISTRIBUTIVE PROPERTY

$$\begin{array}{r} 23 \\ \times\ 12 \\ \hline \end{array} \qquad\qquad \begin{aligned} 23 \cdot 12 &= 23(2 + 10) \\ &= \end{aligned}$$

48. For a and b real numbers, justify each step using a property in this section.

STATEMENT	REASON
1. $[a + b] + (-a) = (-a) + [a + b]$	**1.**
2. $ = [(-a) + a] + b$	**2.**
3. $ = 0 + b$	**3.**
4. $ = b$	**4.**

CALCULATOR PROBLEMS

Express each number as a decimal fraction to the capacity of your calculator. Observe the repeating decimal representation of the rational numbers and the apparent nonrepeating decimal representation of the irrational numbers.

49. (A) $\dfrac{8}{9}$ $\quad$ (B) $\dfrac{3}{11}$ $\quad$ (C) $\sqrt{5}$ $\quad$ (D) $\dfrac{11}{8}$

50. (A) $\dfrac{13}{6}$ $\quad$ (B) $\sqrt{21}$ $\quad$ (C) $\dfrac{7}{16}$ $\quad$ (D) $\dfrac{29}{111}$

Section 0-2 Integer Exponents

■ Integer Exponents
■ Scientific Notation

■ Integer Exponents

We now turn to the exponent form a^n, where the **exponent** n is an integer and the **base** a is a real number. Tables 2 and 3 summarize the definition and properties of exponents.

TABLE 2   Definition of a^n, n an Integer and a Real

1. *For n a positive integer:*

$$a^n = a \cdot a \cdot \cdots \cdot a$$

n factors of a

$$3^5 = 3 \cdot 3 \cdot 3 \cdot 3 \cdot 3$$

2. *For n = 0:*

$$a^0 = 1 \qquad a \neq 0$$
0^0 is not defined

$$132^0 = 1$$

3. *For n a negative integer:*

$$a^n = \frac{1}{a^{-n}} \qquad a \neq 0$$

$$7^{-3} \boxed{= \frac{1}{7^{-(-3)}}}^* = \frac{1}{7^3}$$

[*Note:* In general, it can be shown that

$$a^{-n} = \frac{1}{a^n}$$

$$a^{-5} = \frac{1}{a^5}; \quad a^{-(-3)} = \frac{1}{a^{-3}}$$

for *all* integers n.]

* Dashed boxes are used throughout the text to indicate steps that are usually assumed or done mentally.

TABLE 3   Properties of Exponents

For n and m integers and a and b real numbers:

1. $a^m a^n = a^{m+n}$

$$a^5 a^{-7} \boxed{= a^{5+(-7)}} = a^{-2}$$

2. $(a^n)^m = a^{mn}$

$$(a^3)^{-2} \boxed{= a^{(-2)3}} = a^{-6}$$

3. $(ab)^m = a^m b^m$

$$(ab)^3 = a^3 b^3$$

4. $\left(\dfrac{a}{b}\right)^m = \dfrac{a^m}{b^m} \qquad b \neq 0$

$$\left(\frac{a}{b}\right)^4 = \frac{a^4}{b^4}$$

5. $\dfrac{a^m}{a^n} = \begin{cases} a^{m-n} \\[2mm] \dfrac{1}{a^{n-m}} \end{cases} \qquad a \neq 0$

$$\frac{a^3}{a^{-2}} = a^{3-(-2)} = a^5$$

$$\frac{a^3}{a^{-2}} = \frac{1}{a^{-2-3}} = \frac{1}{a^{-5}}$$

Many students are reasonably satisfied when we define 2^4 to be $2 \cdot 2 \cdot 2 \cdot 2$ but find it harder to accept 2^0 defined as 1 and 2^{-3} defined as $1/2^3$. If we want all five laws of exponents to continue to hold even if some of the exponents are 0 or negative integers, we do not have much choice as to how 2^0 or 2^{-3} must be defined. For example, if the exponent laws are to hold for *all* integer exponents, then, in particular,

$$2^0 \cdot 2^3 = 2^{0+3} = 2^3$$

Thus, 2^0 must be 1, the multiplicative identity. (The number 1 is the only real number that has the property $1 \cdot a = a$ for all real numbers a.)

What about 0^0? Proceeding as above,

$$0^0 \cdot 0^2 = 0^{0+2} = 0^2 = 0 \cdot 0 = 0$$

Thus, 0^0 could be any real number (since $0^2 = 0$ and the product of 0 and any number is 0); hence, 0^0 is not uniquely determined. For this reason we choose not to define the symbol 0^0.

Now let us turn to 2^{-3}. Again assuming the exponent laws hold for *all* integer exponents, then, in particular,

$$2^{-3} \cdot 2^3 = 2^{-3+3} = 2^0 = 1$$

Thus, 2^{-3} must be the multiplicative inverse (reciprocal) of 2^3; that is, 2^{-3} must be $1/2^3$.

This discussion is included to show you that the definitions of integer exponents on page 12 are reasonable. We have not given proofs of the definitions—definitions do not require proofs.

In all examples and discussion that follows we assume without statement that all variables are restricted to avoid division by 0.

EXAMPLE 3 (A) $(u^3v^2)^0 = 1$ $u \neq 0,$ $v \neq 0$ (B) $10^{-3} = \dfrac{1}{10^3} = \dfrac{1}{1{,}000} = 0.001$

(C) $x^{-8} = \dfrac{1}{x^8}$ (D) $\dfrac{x^{-3}}{y^{-5}} \;\left[= \dfrac{x^{-3}}{1} \cdot \dfrac{1}{y^{-5}} = \dfrac{1}{x^3} \cdot \dfrac{y^5}{1} \right] = \dfrac{y^5}{x^3}$

PROBLEM 3 Write (A) to (D) as decimal fractions and (E) and (F) with positive exponents.

(A) 635^0 (B) $(x^2)^0,$ $x \neq 0$ (C) 10^{-5}
(D) $1/10^{-3}$ (E) $1/x^{-4}$ (F) u^{-7}/v^{-3}

From the definition of negative exponents and the five laws of exponents, we can easily establish the following properties that are used very frequently when dealing with exponent forms.

Further Exponent Properties

For a and b any real numbers and m, n, and p any integers, then (division by 0 excluded)

1. $(a^m b^n)^p = a^{pm} b^{pn}$ **2.** $\left(\dfrac{a^m}{b^n}\right)^p = \dfrac{a^{pm}}{b^{pn}}$

3. $\dfrac{a^{-n}}{b^{-m}} = \dfrac{b^m}{a^n}$ **4.** $\left(\dfrac{a}{b}\right)^{-n} = \left(\dfrac{b}{a}\right)^n$

We prove properties 1 and 4 and leave 2 and 3 to the reader.

1. $(a^m b^n)^p = (a^m)^p (b^n)^p$ Exponent law 3

$\qquad\quad = a^{pm} b^{pn}$ Exponent law 2

4. $\left(\dfrac{a}{b}\right)^{-n} = \dfrac{a^{-n}}{b^{-n}}$ Exponent law 4

$\qquad\quad = \dfrac{b^n}{a^n}$ Property 3

$\qquad\quad = \left(\dfrac{b}{a}\right)^n$ Exponent law 4

EXAMPLE 4   Simplify using exponent properties, and express answers using positive exponents only.*

(A) $(3a^5)(2a^{-3}) \boxed{= (3 \cdot 2)(a^5 a^{-3})} = 6a^2$

(B) $\dfrac{6x^{-2}}{8x^{-5}} = \dfrac{3x^{-2-(-5)}}{4} = \dfrac{3x^3}{4}$

(C) $(2a^{-3}b^2)^{-2} = 2^{-2}a^6 b^{-4} = \dfrac{a^6}{4b^4}$

(D) $\left(\dfrac{a^3}{b^5}\right)^{-2} = \left(\dfrac{b^5}{a^3}\right)^2 = \dfrac{b^{10}}{a^6}$

[*Note:* In simplifying exponent forms there is often more than one sequence of steps that will lead to the same result. For example, in Example 4D we can also proceed as follows:

$$\left(\frac{a^3}{b^5}\right)^{-2} = \frac{a^{-6}}{b^{-10}} = \frac{b^{10}}{a^6}$$

Use whichever sequence of steps makes sense to you; however, do not create new rules unless they can be justified.]

---

* By "simplify" we mean eliminate common factors from numerators and denominators and reduce to a minimum the number of times a given constant or variable appears in an expression. We ask that answers be expressed using positive exponents only in order to have a definite form for an answer. Later (in this section and elsewhere) we will encounter situations where we will want negative exponents in a final answer.

PROBLEM 4   Simplify using exponent properties, and express answers using positive exponents only.

(A) $(5x^{-3})(3x^4)$ (B) $\dfrac{9y^{-7}}{6y^{-4}}$

(C) $(3x^4y^{-3})^{-2}$ (D) $\left(\dfrac{x^2}{y^4}\right)^{-3}$

EXAMPLE 5   Simplify using exponent properties, and express answers using positive exponents only.

(A) $\dfrac{4x^{-3}y^{-5}}{6x^{-4}y^3} = \dfrac{2x^{-3-(-4)}}{3y^{3-(-5)}} = \dfrac{2x}{3y^8}$

(B) $\left(\dfrac{m^{-3}m^3}{n^{-2}}\right)^{-2} = \left(\dfrac{m^{-3+3}}{n^{-2}}\right)^{-2} = \left(\dfrac{m^0}{n^{-2}}\right)^{-2} = \left(\dfrac{1}{n^{-2}}\right)^{-2} = \dfrac{1}{n^4}$

(C) $\left[\left(\dfrac{12x^{-3}}{8y^4}\right)^{-1}\right]^{-2} = \left(\dfrac{3x^{-3}}{2y^4}\right)^2 = \dfrac{9x^{-6}}{4y^8} = \dfrac{9}{4x^6y^8}$

(D) $(x+y)^{-3} = \dfrac{1}{(x+y)^3}$ Common error:
$(x+y)^{-3} = x^{-3} + y^{-3}$
$= \dfrac{1}{x^3} + \dfrac{1}{y^3}$

[Note: Exponent laws deal primarily with products and quotients. Some students cannot resist making up their own new "laws" (usually invalid) for sums and differences (see Example 5D).]

PROBLEM 5   Simplify using exponent properties, and express answers using positive exponents only.

(A) $\dfrac{6m^{-2}n^3}{15m^{-1}n^{-2}}$ (B) $\left(\dfrac{x^{-3}}{y^4y^{-4}}\right)^{-3}$

(C) $\left[\left(\dfrac{4a^2}{2b^5}\right)^2\right]^{-1}$ (D) $\dfrac{1}{(a-b)^{-2}}$

■ Scientific Notation

Scientific work often involves the use of very large numbers or very small numbers. For example, the average cell contains about 200,000,000,000,000 molecules, and the diameter of an electron is about 0.000 000 000 0004 centimeters. It is generally troublesome to write and work with numbers of this type in standard decimal form. The two numbers written here cannot even be entered into most hand calculators as they are written. With exponents now defined for all integers, however, it is possible to express any decimal fraction as the product of a number between 1 and 10 and an integer power of 10— that is, in the form

$$a \times 10^n \qquad 1 \leq a < 10, \quad n \text{ an integer}$$

A number expressed in this form is said to be in **scientific notation**.

EXAMPLE 6

$$7 = 7 \times 10^0$$
$$720 = 7.2 \times 10^2$$
$$6{,}430 = 6.43 \times 10^3$$
$$5{,}350{,}000 = 5.35 \times 10^6$$

$$0.5 = 5 \times 10^{-1}$$
$$0.08 = 8 \times 10^{-2}$$
$$0.000\ 32 = 3.2 \times 10^{-4}$$
$$0.000\ 000\ 0738 = 7.38 \times 10^{-8}$$

Can you discover a rule relating the number of decimal places that the decimal is moved to the power of 10 that is used?

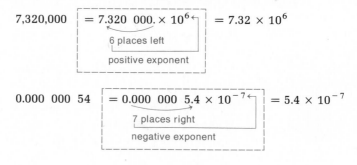

PROBLEM 6   (A) Write each number in scientific notation: 430; 23,000;
            345,000,000; 0.3; 0.0031; 0.000 000 683
           (B) Write as decimal fractions: 4×10^3, 5.3×10^5, 2.53×10^{-2},
            7.42×10^{-6}

Most scientific and business hand calculators express very large and very small numbers in scientific notation. Read the instruction manual for your calculator to see how numbers in scientific notation are entered into your calculator. Numbers in scientific notation are displayed in most calculators as follows.

CALCULATOR DISPLAY   NUMBER REPRESENTED

| 5.427493 -17 | $5.427\ 493 \times 10^{-17}$

| 2.359779 12 | $2.359\ 779 \times 10^{12}$

EXAMPLE 7   Write each number that cannot be entered directly into your calculator
              in decimal form in scientific notation; then carry out the computation
              using your calculator. (Refer to the user's manual accompanying your
              calculator for the procedure.) Express the answer to three significant
              digits* in scientific notation.

$$\frac{(25.32)(325,100,000,000)}{(0.08)(0.000\ 000\ 000\ 000\ 0871)} = \frac{(25.32)(3.251 \times 10^{11})}{(0.08)(8.71 \times 10^{-14})}$$

$$= \boxed{1.181333\ \ 27}\quad \text{Calculator display}$$

$$= 1.18 \times 10^{27}\quad \text{To three significant digits}$$

CALCULATOR OPERATIONS

A: $\boxed{25.32}\ \boxed{\times}\ \boxed{3.251}\ \boxed{\text{EE}}\ \boxed{11}\ \boxed{=}\ \boxed{\div}\ \boxed{0.08}\ \boxed{\div}\ \boxed{8.71}\ \boxed{\text{EE}}\ \boxed{14}\ \boxed{+/-}\ \boxed{=}$

P: $\boxed{25.32}\ \boxed{\text{ENTER}}\ \boxed{3.251}\ \boxed{\text{EE}}\ \boxed{11}\ \boxed{\times}\ \boxed{0.08}\ \boxed{\div}\ \boxed{8.71}\ \boxed{\text{EE}}\ \boxed{14}\ \boxed{+/-}\ \boxed{\div}$

PROBLEM 7   Repeat Example 7 for

$$\frac{(0.371)(0.000\ 000\ 006\ 932)}{(532)(62,600,000,000)}$$

Answers to Matched Problems   **3.** (A) 1   (B) 1   (C) 0.000 01   (D) 1,000   (E) x^4
                                      (F) v^3/u^7

                                4. (A) $15x$ (B) $3/2y^3$ (C) $y^6/9x^8$ (D) y^{12}/x^6

                                5. (A) $2n^5/5m$ (B) x^9 (C) $b^{10}/4a^4$ (D) $(a-b)^2$

                                6. (A) $4.3 \times 10^2, 2.3 \times 10^4, 3.45 \times 10^8, 3 \times 10^{-1}, 3.1 \times 10^{-3},$
                                          6.83×10^{-7}

                                     (B) 4,000; 530,000; 0.0253; 0.000 007 42

                                7. 7.72×10^{-23}

Exercise 0-2 ■ A   *Simplify Problems 1–16 and write the answers using positive exponents only.*

1. $y^{-5}y^5$ **2.** x^3x^{-3} **3.** $(2x^2)(3x^3)(x^4)$

4. $(2x^5)(3x^7)(4x^2)$ **5.** $(3x^3y^{-2})^2$ **6.** $(2cd^2)^{-3}$

7. $\left(\dfrac{ab^3}{c^2d}\right)^4$ **8.** $\left(\dfrac{x^2y}{2w^2}\right)^3$ **9.** $\dfrac{10^{23} \cdot 10^{-11}}{10^{-3} \cdot 10^{-2}}$

---

* For those not familiar with the meaning of "significant digits," see Appendix A for a brief
discussion of this concept.

10. $\dfrac{10^{-13} \cdot 10^{-4}}{10^{-21} \cdot 10^{3}}$ **11.** $\dfrac{4x^{-2}y^{-3}}{2x^{-3}y^{-1}}$ **12.** $\dfrac{2a^{6}b^{-2}}{16a^{-3}b^{2}}$

13. $\left(\dfrac{n^{-3}}{n^{-2}}\right)^{-2}$ **14.** $\left(\dfrac{x^{-1}}{x^{-8}}\right)^{-1}$ **15.** $\dfrac{8 \times 10^{3}}{2 \times 10^{-5}}$

16. $\dfrac{18 \times 10^{12}}{6 \times 10^{-4}}$

Write the numbers in Problems 17–22 in scientific notation.

17. 32,250,000   **18.** 4,930   **19.** 0.085

20. 0.017   **21.** 0.000 000 0729   **22.** 0.000 592

In Problems 23–28 write each number as a decimal fraction.

23. 5×10^{-3} **24.** 4×10^{-4} **25.** 2.69×10^{7}

26. 6.5×10^{9} **27.** 5.9×10^{-10} **28.** 6.3×10^{-6}

B *Simplify Problems 29–44 and write answers using positive exponents only.*

29. $\dfrac{27x^{-5}x^{5}}{18y^{-6}y^{2}}$ **30.** $\dfrac{32n^{5}n^{-8}}{24m^{-7}m^{7}}$

31. $\left(\dfrac{x^{4}y^{-1}}{x^{-2}y^{3}}\right)^{2}$ **32.** $\left(\dfrac{m^{-2}n^{3}}{m^{4}n^{-1}}\right)^{2}$

33. $\left(\dfrac{2x^{-3}y^{2}}{4xy^{-1}}\right)^{-2}$ **34.** $\left(\dfrac{6mn^{-2}}{3m^{-1}n^{2}}\right)^{-3}$

35. $\left[\left(\dfrac{u^{3}v^{-1}w^{-2}}{u^{-2}v^{-2}w}\right)^{-2}\right]^{2}$ **36.** $\left[\left(\dfrac{x^{-2}y^{3}t}{x^{-3}y^{-2}t^{2}}\right)^{2}\right]^{-1}$

37. $\left(\dfrac{3^{3}x^{0}y^{-2}}{2^{3}x^{3}y^{-5}}\right)^{-1}\left(\dfrac{3^{3}x^{-1}y}{2^{2}x^{2}y^{-2}}\right)^{2}$ **38.** $\left(\dfrac{2^{2}x^{2}y^{0}}{8x^{-1}}\right)^{-2}\left(\dfrac{x^{-3}}{x^{-5}}\right)^{3}$

39. $(x + y)^{-2}$ **40.** $(a^{2} - b^{2})^{-1}$

C **41.** $\dfrac{12(a + 2b)^{-3}}{6(a + 2b)^{-8}}$ **42.** $\dfrac{4(x - 3)^{-4}}{8(x - 3)^{-2}}$

43. $\dfrac{5(u - v + w)^{8}}{(u - v + w)^{11}}$ **44.** $\dfrac{3(x + y)^{3}(x - y)^{4}}{(x - y)^{2}6(x + y)^{5}}$

APPLICATIONS   **45.** *Earth science.* If the mass of the earth is approximately 6.1×10^{27} grams and each gram is 2.2×10^{-3} pound, what is the mass of the earth in pounds?

       46. *Biology.* In 1929 Vernadsky, a biologist, estimated that all the free oxygen of the earth weighs 1.5×10^{21} grams and that it is produced by life alone. If 1 gram is approximately 2.2×10^{-3} pound, what is the weight of the free oxygen in pounds?

47. *Computer science.* Today's fastest computers can perform a single addition in 10^{-8} second, and the next generation of computers working with tiny superconducting devices are expected to be able to perform a single addition in 10^{-10} second. How many additions will both types of computers be able to perform in 1 second? In 1 minute?

48. *Computer science.* If electricity travels in a computer circuit at the speed of light (1.86×10^5 miles/second), how far will electricity travel in the superconducting computer (see Problem 47) in the time it takes it to perform one addition? (Size of circuits is a critical problem in computer design.) Give the answer in miles, feet, and inches (1 mile = 5,280 feet). Compute answers to two significant digits.

CALCULATOR PROBLEMS   *Evaluate Problems 49–52 to three significant digits using scientific*
                        notation where appropriate and a calculator.

49. $\dfrac{(32.7)(0.000\ 000\ 008\ 42)}{(0.0513)(80,700,000,000)}$ **50.** $\dfrac{(4,320)(0.000\ 000\ 000\ 704)}{(835)(635,000,000,000)}$

51. $\dfrac{(5,760,000,000)}{(527)(0.000\ 007\ 09)}$ **52.** $\dfrac{0.000\ 000\ 007\ 23}{(0.0933)(43,700,000,000)}$

Use y^x or a comparable key on your calculator to evaluate each of the following problems to five significant digits. (Read the instruction book accompanying your calculator.)

53. $(23.8)^8$ **54.** $(-302)^7$

55. $(-302)^{-7}$ **56.** $(23.8)^{-8}$

57. $(9,820,000,000)^3$ **58.** $(0.000\ 000\ 000\ 482)^{-4}$

Section 0-3 Rational Exponents

■ Roots of Real Numbers
■ Rational Exponents

■ **Roots of Real Numbers**

A **square root** of a number b is a number r such that $r^2 = b$, and a **cube root** of a number b is a number r such that $r^3 = b$.

2 is a square root of 4, since $2^2 = 4$.

-2 is a square root of 4, since $(-2)^2 = 4$.

2 is a cube root of 8, since $2^3 = 8$.

-2 is a cube root of -8, since $(-2)^3 = 8$.

In general, for n a natural number,

r is an **nth root** of b if $r^n = b$

How many real square roots of 16 exist? Of 7? Of -4? How many real fourth roots of 7 exist? Of -7? How many real cube roots of -8 exist? Of 11? Theorem 1 (which we state without proof) answers these questions completely.

THEOREM 1

Number of Real nth Roots of a Real Number b		
	n even	n odd
b positive	Two real nth roots	One real nth root
	-2 and 2 are both 4th roots of 16	2 is the only real cube root of 8
b negative	No real nth root	One real nth root
	-4 has no real square roots	-2 is the only real cube root of -8

Thus:

7 has two real square roots, two real 4th roots, and so on.

10 has one real cube root, one real 5th root, and so on.

-13 has one real cube root, one real 5th root, and so on.

-8 has no real square roots, no real 4th roots, and so on.

What symbols do we use to represent the various kinds of real nth roots? We turn to this question now.

■ Rational Exponents

If all exponent laws are to continue to hold even if some of the exponents are not integers, then, in particular,

$$(6^{1/2})^2 = 6^{2/2} = 6 \quad \text{and} \quad (5^{1/3})^3 = 5^{3/3} = 5$$

Hence, $6^{1/2}$ must name a square root of 6, since $(6^{1/2})^2 = 6$. Similarly, $5^{1/3}$ must name a cube root of 5, since $(5^{1/3})^3 = 5$.

In general, for n a positive integer and b not negative when n is even,

$$(b^{1/n})^n = b^{n/n} = b$$

Thus, $b^{1/n}$ must name an nth root of b. Which of the two real nth roots of b does $b^{1/n}$ represent if n is even and b is positive? We answer this question in the definition of $b^{1/n}$.

Definition of $b^{1/n}$

For n a positive integer,

$b^{1/n}$ is an nth root of b

If n is even and b is positive, then $b^{1/n}$ represents the positive real nth root of b (sometimes called the **principal nth root** of b), $-b^{1/n}$ represents the negative real nth root of b, and $(-b)^{1/n}$ does not represent a real number. If n is odd and b is either positive or negative, then $b^{1/n}$ represents the real nth root of b. $0^{1/n} = 0$ for all positive integers n.

$25^{1/2} = 5$	$-25^{1/2} = -5$	$(-25)^{1/2}$ is not real
	$[-25^{1/2}$ and $(-25)^{1/2}$	
	are not the same]	
$32^{1/5} = 2$	$(-32)^{1/5} = -2$	$0^{1/4} = 0$

EXAMPLE 8   (A) $4^{1/2} = 2$ (B) $-4^{1/2} = -2$ (C) $(-4)^{1/2}$ is not a real number
(D) $(-8)^{1/3} = -2$ (E) $0^{1/5} = 0$

PROBLEM 8   Find integer representations of each of the following, if they exist.

(A) $9^{1/2}$ (B) $-9^{1/2}$ (C) $(-9)^{1/2}$
(D) $27^{1/3}$ (E) $(-27)^{1/3}$ (F) $0^{1/4}$

How should a symbol such as $7^{2/3}$ be defined? If the properties of exponents are to hold for rational exponents, then $7^{2/3} = (7^{1/3})^2$; that is, $7^{2/3}$ must represent the square of the cube root of 7. We are thus led to the general definition of $b^{m/n}$ and $b^{-m/n}$.

Definition of $b^{m/n}$ and $b^{-m/n}$

For m and n positive integers and b any real number, except b cannot be negative when n is even,

$$b^{m/n} = (b^{1/n})^m \quad \text{and} \quad b^{-m/n} = \frac{1}{b^{m/n}}$$

$4^{3/2} = (4^{1/2})^3 = 2^3 = 8 \qquad (-32)^{3/5} = [(-32)^{1/5}]^3 = (-2)^3 = -8$

$4^{-3/2} = \dfrac{1}{4^{3/2}} = \dfrac{1}{8}$

$(-4)^{3/2}$ is not a real number

We have discussed $b^{m/n}$ for all rational numbers m/n and real numbers b. It can be shown that all five laws of exponents discussed in Section 0-2 continue to hold for rational number exponents as long as we avoid even roots of negative numbers. With the latter restriction in effect, the following useful relationship is an immediate consequence of the exponent properties:

$$b^{m/n} = (b^{1/n})^m = (b^m)^{1/n}$$

To see why we require b to be positive when n is even, consider what happens when we relax that restriction. Can you resolve the following contradiction?

$$-1 = (-1)^{2/2} = [(-1)^2]^{1/2} = 1^{1/2} = 1$$

The second member of the equality chain, $(-1)^{2/2}$, involves an even root of a negative number, which is not real. Thus, we see that the properties of exponents do not necessarily hold when we are dealing with nonreal quantities unless further restrictions are imposed. One such restriction is to require all rational exponents be reduced to lowest terms.

EXAMPLE 9   Assume all letters represent positive real numbers.

(A) $8^{2/3} = (8^{1/3})^2 = 2^2 = 4 \quad \text{or} \quad 8^{2/3} = (8^2)^{1/3} = 64^{1/3} = 4$

(B) $(-8)^{5/3} = [(-8)^{1/3}]^5 = (-2)^5 = -32$

(C) $(3x^{1/3})(2x^{1/2}) = 6x^{1/3 + 1/2} = 6x^{5/6}$

(D) $\left(\dfrac{4x^{1/3}}{x^{1/2}}\right)^{1/2} = \dfrac{4^{1/2}x^{1/6}}{x^{1/4}} = \dfrac{2}{x^{1/4-1/6}} = \dfrac{2}{x^{1/12}}$

(E) $(x+2)^{3/4}(x+2)^{2/3} = (x+2)^{3/4+2/3} = (x+2)^{17/12}$

PROBLEM 9 Simplify, and express answers using positive exponents only. All letters represent positive real numbers.

(A) $9^{3/2}$ (B) $(-27)^{4/3}$

(C) $(5y^{3/4})(2y^{1/3})$ (D) $(2x^{-3/4}y^{1/4})^4$

(E) $\left(\dfrac{8x^{1/2}}{x^{2/3}}\right)^{1/3}$ (F) $\dfrac{6(a+b)^{2/3}}{8(a+b)^{1/2}}$

EXAMPLE 10 Evaluate to four significant digits using a hand calculator. (Refer to the instruction book for your particular calculator to see how exponential forms are evaluated.)

(A) $11^{3/4}$ (B) $3.1046^{-2/3}$ (C) $(0.000\ 000\ 008\ 437)^{3/11}$

Solution (A) First change $\frac{3}{4}$ to the decimal fraction 0.75; then evaluate $11^{0.75}$ using y^x (or a comparable key) on your calculator.

$$11^{3/4} = 6.040$$

A: $\boxed{11}\ \boxed{y^x}\ \boxed{(}\ \boxed{3}\ \boxed{\div}\ \boxed{4}\ \boxed{)}\ \boxed{=}$

P: $\boxed{11}\ \boxed{\text{ENTER}}\ \boxed{3}\ \boxed{\text{ENTER}}\ \boxed{4}\ \boxed{\div}\ \boxed{y^x}$

(B) $3.1046^{-2/3} = 0.4699$

A: $\boxed{3.1046}\ \boxed{y^x}\ \boxed{(}\ \boxed{2}\ \boxed{\div}\ \boxed{3}\ \boxed{+/-}\ \boxed{)}\ \boxed{=}$

P: $\boxed{3.1046}\ \boxed{\text{ENTER}}\ \boxed{2}\ \boxed{\text{ENTER}}\ \boxed{3}\ \boxed{\div}\ \boxed{+/-}\ \boxed{y^x}$

(C) $(0.000\ 000\ 008\ 437)^{3/11} = (8.437 \times 10^{-9})^{3/11}$

$$= 0.006\ 281$$

PROBLEM 10 Repeat Example 10 for:

(A) $2^{3/8}$ (B) $57.28^{-5/6}$ (C) $(83,240,000,000)^{5/3}$

Answers to Matched Problems **8.** (A) 3   (B) -3 (C) Not a real number   (D) 3
      (E) -3 (F) 0

9. (A) 27   (B) 81   (C) $10y^{13/12}$ (D) $16y/x^3$
     (E) $2/x^{1/18}$ (F) $[3(a+b)^{1/6}]/4$

10. (A) 1.297   (B) 0.034 28   (C) 1.587×10^{18}

Exercise 0-3 ■

All letters represent positive real numbers unless otherwise stated.

A   *In Problems 1–12 find rational representations for each, if they exist.*

1. $16^{1/2}$ **2.** $64^{1/3}$ **3.** $16^{3/2}$

$= \sqrt{36} \quad -6$

4. $16^{3/4}$ **5.** $-36^{1/2}$ **6.** $32^{3/5}$

7. $(-36)^{1/2}$ **8.** $(-32)^{3/5}$ **9.** $(\frac{4}{25})^{3/2}$

10. $(\frac{8}{27})^{2/3}$ **11.** $9^{-3/2}$ **12.** $8^{-2/3}$

Simplify Problems 13–36 and express answers using positive exponents only.

13. $y^{1/5}y^{2/5}$ **14.** $x^{1/4}x^{3/4}$

15. $d^{2/3}d^{-1/3}$ **16.** $x^{1/4}x^{-3/4}$

17. $(y^{-8})^{1/16}$ **18.** $(x^{-2/3})^{-6}$

19. $(8x^3y^{-6})^{1/3}$ **20.** $(4u^{-2}v^4)^{1/2}$

B 21. $\left(\dfrac{a^{-3}}{b^4}\right)^{1/12}$ **22.** $\left(\dfrac{m^{-2/3}}{n^{-1/2}}\right)^{-6}$

23. $\left(\dfrac{4x^{-2}}{y^4}\right)^{-1/2}$ **24.** $\left(\dfrac{w^4}{9x^{-2}}\right)^{-1/2}$

25. $\left(\dfrac{8a^{-4}b^3}{27a^2b^{-3}}\right)^{1/3}$ **26.** $\left(\dfrac{25x^5y^{-1}}{16x^{-3}y^{-5}}\right)^{1/2}$

27. $\left(\dfrac{9}{25}x^2y^{-1/3}\right)^{-3/2}$ **28.** $\left(\dfrac{27}{8}x^{-3}y^{1/2}\right)^{-4/3}$

29. $\dfrac{8x^{-1/3}}{12x^{1/4}}$ **30.** $\dfrac{6a^{3/4}}{15a^{-1/3}}$

31. $\left(\dfrac{a^{2/3}b^{-1/2}}{a^{1/2}b^{1/2}}\right)^2$ **32.** $\left(\dfrac{x^{-1/3}y^{1/2}}{x^{-1/4}y^{1/3}}\right)^6$

33. $\left(\dfrac{9x^{1/3}x^{1/2}}{x^{-1/6}}\right)^{1/2}$ **34.** $\left(\dfrac{8y^{1/3}y^{-1/4}}{y^{-1/12}}\right)^2$

35. $(125x^{1/2}y^{-1/3})^{-2/3}(x^{1/3}y^{-2/3})$ **36.** $(9x^{1/3}y^{-1/2})^{3/2}(x^{-1/3}y^{1/4})$

C   *In Problems 37–40, m and n represent positive integers. Simplify and express answers using positive exponents.*

37. $(a^{3/n}b^{3/m})^{1/3}$ **38.** $(a^{n/2}b^{n/3})^{1/n}$

39. $(x^{m/4}y^{n/3})^{-12}$ **40.** $(a^{m/3}b^{n/2})^{-6}$

41. Find a real value of x such that
    (A) $(x^2)^{1/2} \neq x$ (B) $(x^2)^{1/2} = x$

42. Find a real value of x such that
    (A) $(x^2)^{1/2} \neq -x$ (B) $(x^2)^{1/2} = -x$

CALCULATOR PROBLEMS   *In Problems 43–50 evaluate to four significant digits using a hand calculator. (Refer to the instruction book for your calculator to see how exponential forms are evaluated.)*

43. $15^{5/4}$ **44.** $22^{3/2}$

45. $103^{-3/4}$ **46.** $827^{-3/8}$

47. $2.876^{8/5}$ **48.** $37.09^{7/3}$

49. $(0.000\ 000\ 077\ 35)^{-2/7}$ **50.** $(491,300,000,000)^{7/4}$

Section 0-4 Radicals

- From Rational Exponents to Radicals
- Properties of Radicals
- Changing Radical Forms
- Simplifying $\sqrt[n]{x^n}$ for All Real x

■ **From Rational Exponents to Radicals**

For n a natural number greater than 1 and b any real number except b negative when n is even, we define

nTH ROOT OF b: $\sqrt[n]{b} = b^{1/n}$

The symbol $\sqrt{}$ is called a **radical**, n is called the **index**, and b is called the **radicand**. (Note that, if $n = 2$, we write $\sqrt{b}$ in place of $\sqrt[2]{b}$.) There are occasions when it is more convenient to work with radicals than with rational exponents, or vice versa. It is often an advantage to be able to shift back and forth between the two forms. The following relationships are useful in this regard:

$$b^{m/n} = (b^m)^{1/n} = \sqrt[n]{b^m}$$
$$b^{m/n} = (b^{1/n})^m = (\sqrt[n]{b})^m$$

where b is not negative for n even.

[*Note:* Unless stated to the contrary, all variables in the rest of the discussion represent positive real numbers.]

EXAMPLE 11   From rational exponent form to radical form.

(A) $x^{1/7} = \sqrt[7]{x}$

(B) $(3u^2v^3)^{3/5} = \sqrt[5]{(3u^2v^3)^3}$ or $(\sqrt[5]{3u^2v^3})^3$ The first is usually preferred.

(C) $y^{-2/3} = \dfrac{1}{y^{2/3}} = \dfrac{1}{\sqrt[3]{y^2}}$ or $\sqrt[3]{y^{-2}}$ or $\sqrt[3]{\dfrac{1}{y^2}}$

From radical form to rational exponent form.

(D) $\sqrt[5]{6} = 6^{1/5}$ (E) $-\sqrt[3]{x^2} = -x^{2/3}$ (F) $\sqrt{x^2+y^2} = (x^2+y^2)^{1/2}$

[Note: $(x^2+y^2)^{1/2} \neq x+y$ (Why?)]

PROBLEM 11   Convert to radical form.

(A) $u^{1/5}$ (B) $(6x^2y^5)^{2/9}$ (C) $(3xy)^{-3/5}$

Convert to rational exponent form.

(D) $\sqrt[4]{9u}$ (E) $-\sqrt[7]{(2x)^4}$ (F) $\sqrt[3]{x^3+y^3}$

■   Properties of Radicals

Changing and simplifying radical expressions are aided by the introduction of several properties of radicals that follow directly from exponent properties considered earlier.

Properties of Radicals

For k, n, and m natural numbers 2 or larger, and x and y positive real numbers:

1. $\sqrt[n]{x^n} = x$ $\sqrt[3]{x^3} = x$

2. $\sqrt[n]{xy} = \sqrt[n]{x}\,\sqrt[n]{y}$ $\sqrt[5]{xy} = \sqrt[5]{x}\,\sqrt[5]{y}$

3. $\sqrt[n]{\dfrac{x}{y}} = \dfrac{\sqrt[n]{x}}{\sqrt[n]{y}}$ $\sqrt[4]{\dfrac{x}{y}} = \dfrac{\sqrt[4]{x}}{\sqrt[4]{y}}$

4. $\sqrt[kn]{x^{km}} = \sqrt[n]{x^m}$ $\sqrt[12]{x^8} = \sqrt[4\cdot3]{x^{4\cdot2}} = \sqrt[3]{x^2}$

5. $\sqrt[m]{\sqrt[n]{x}} = \sqrt[mn]{x}$ $\sqrt[4]{\sqrt[3]{x}} = \sqrt[12]{x}$

    Can you supply the reasons for the proofs of properties 1, 2, and 4?
(Proofs of properties 3 and 5 are asked for in Exercise 0-4, Problems 71
and 72.)

Proofs of 1, 2, and 4   **1.** $\sqrt[n]{x^n} = (x^n)^{1/n} = x^{n/n} = x$

2. $\sqrt[n]{xy} = (xy)^{1/n} = x^{1/n}y^{1/n} = \sqrt[n]{x}\sqrt[n]{y}$

4. $\sqrt[kn]{x^{km}} = (x^{km})^{1/kn} = x^{km/kn} = x^{m/n} = (x^m)^{1/n} = \sqrt[n]{x^m}$

EXAMPLE 12   (A) $\sqrt[5]{(3x^2y)^5} = 3x^2y$

(B) $\sqrt{10}\sqrt{5} = \sqrt{50} = \sqrt{25 \cdot 2} = \sqrt{25}\sqrt{2} = 5\sqrt{2}$

(C) $\sqrt[3]{\dfrac{x}{27}} = \dfrac{\sqrt[3]{x}}{\sqrt[3]{27}} = \dfrac{\sqrt[3]{x}}{3}$ or $\dfrac{1}{3}\sqrt[3]{x}$

(D) $\sqrt[6]{x^4}$ $\;\boxed{= \sqrt[2 \cdot 3]{x^{2 \cdot 2}}}\; = \sqrt[3]{x^2}$

(E) $\sqrt[3]{\sqrt[4]{x}}$ $\;\boxed{= \sqrt[3 \cdot 4]{x}}\; = \sqrt[12]{x}$

PROBLEM 12   Simplify as in Example 12.

(A) $\sqrt[7]{(u^2 + v^2)^7}$ (B) $\sqrt{6}\sqrt{2}$ (C) $\sqrt[3]{\dfrac{x^2}{8}}$

(D) $\sqrt[8]{y^6}$ (E) $\sqrt[5]{\sqrt{x}}$

■ **Changing Radical Forms**

The properties of radicals provide us with the means of changing algebraic expressions containing radicals to a variety of equivalent forms. One form that is often useful is the simplest radical form. An algebraic expression that contains radicals is said to be in the **simplest radical form** if all four of the conditions listed in the box are satisfied.

Simplest Radical Form

1. A radicand (the expression within the radical sign) contains no factor to a power greater than or equal to the index of the radical. ($\sqrt{x^5}$ violates this condition.)
2. The power of the radicand and the index of the radical have no common factor other than 1. ($\sqrt[6]{x^4}$ violates this condition.)
3. No radical appears in a denominator. ($y/\sqrt{x}$ violates this condition.)
4. No fraction appears within a radical. ($\sqrt{\frac{3}{5}}$ violates this condition.)

EXAMPLE 13   Express radicals in simplest radical form.

(A) $\sqrt[3]{54} = \sqrt[3]{3^3 \cdot 2}$ $54 = 27 \cdot 2 = 3^3 \cdot 2$

$\quad\quad = \sqrt[3]{3^3} \sqrt[3]{2}$ Condition 1 is not met.

$\quad\quad = 3\sqrt[3]{2}$

(B) $\sqrt{12x^3y^5z^2} = \sqrt{(4x^2y^4z^2)(3xy)}$ Condition 1 is not met.

$\quad\quad = \sqrt{(2xy^2z^2)^2(3xy)}$

$\quad\quad = \sqrt{(2xy^2z^2)^2}\sqrt{3xy}$

$\quad\quad = 2xy^2z\sqrt{3xy}$

(C) $\sqrt[6]{16x^4y^2} = \sqrt[6]{(4x^2y)^2}$ Condition 2 is not met.

$\quad\quad = \sqrt[3]{(4x^2y)^{2 \cdot 1}}$

$\quad\quad = \sqrt[3]{4x^2y}$

(D) $\sqrt[3]{\sqrt{27}} = \sqrt[6]{27}$ Condition 2 is not met.

$\quad\quad = \sqrt[3 \cdot 2]{3^{3 \cdot 1}} = \sqrt{3}$

PROBLEM 13   Express in simplest radical form.

(A) $\sqrt[3]{16}$ (B) $\sqrt{18x^5y^2z^3}$ (C) $\sqrt[9]{8x^6y^3}$ (D) $\sqrt{\sqrt[3]{4}}$

Eliminating a radical from a denominator or a fraction from within a radical is referred to as **rationalizing denominators**.

EXAMPLE 14   Rationalize denominators.

(A) $\dfrac{3}{\sqrt{5}} = \dfrac{3}{\sqrt{5}} \cdot \dfrac{\sqrt{5}}{\sqrt{5}}$ Condition 3 is not met.

$\quad\quad = \dfrac{3\sqrt{5}}{\sqrt{5^2}}$

$\quad\quad = \dfrac{3\sqrt{5}}{5}$ or $\tfrac{3}{5}\sqrt{5}$

(B) $\dfrac{6x^2}{\sqrt[3]{9x}} = \dfrac{6x^2}{\sqrt[3]{9x}} \cdot \dfrac{\sqrt[3]{3x^2}}{\sqrt[3]{3x^2}}$ Condition 3 is not met.

$\quad\quad = \dfrac{6x^2\sqrt[3]{3x^2}}{\sqrt[3]{3^3x^3}}$

$\quad\quad = \dfrac{6x^2\sqrt[3]{3x^2}}{3x} = 2x\sqrt[3]{3x^2}$

(C) $\sqrt[3]{\dfrac{2a^2}{3b^2}} = \sqrt[3]{\dfrac{2a^2}{3b^2} \cdot \dfrac{3^2 b}{3^2 b}}$ Condition 4 is not met.

$= \sqrt[3]{\dfrac{18a^2 b}{3^3 b^3}}$

$= \dfrac{\sqrt[3]{18a^2 b}}{\sqrt[3]{3^3 b^3}} = \dfrac{\sqrt[3]{18a^2 b}}{3b}$

PROBLEM 14   Write in simplest radical form by rationalizing denominators.

(A) $\dfrac{6}{\sqrt{2x}}$ (B) $\dfrac{10x^3}{\sqrt[3]{2x^2}}$ (C) $\sqrt[3]{\dfrac{3y^2}{2x^4}}$

Simplest radical forms are often useful in simplifying more involved algebraic expressions that contain radicals. However, there are many situations in which the simplest radical form may not be the most useful form. For example, in finding a decimal approximation for $\sqrt{\tfrac{2}{5}}$ using a hand calculator, it is probably easier to evaluate $\sqrt{\tfrac{2}{5}}$ directly rather than expressing it in the simplest radical form $\sqrt{10}/5$ first.

EXAMPLE 15   (A) $\sqrt[3]{6x^2 y}\,\sqrt[3]{4x^5 y^2}$ $\boxed{= \sqrt[3]{(6x^2 y)(4x^5 y^2)}}$

$= \sqrt[3]{24x^7 y^3}$

$= \sqrt[3]{(8x^6 y^3)(3x)}$

$\boxed{\begin{array}{l} = \sqrt[3]{(2x^2 y)^3 (3x)} \\ = \sqrt[3]{(2x^2 y)^3}\,\sqrt[3]{3x} \end{array}}$

$= 2x^2 y \sqrt[3]{3x}$

(B) $\sqrt{2u^5 v}\,\sqrt[3]{4u^4 v^2} = \sqrt[3\cdot 2]{(2u^5 v)^3}\,\sqrt[2\cdot 3]{(4u^4 v^2)^2}$

$= \sqrt[6]{2^3 u^{15} v^3}\,\sqrt[6]{2^4 u^8 v^4}$

$= \sqrt[6]{2^7 u^{23} v^7}$

$= \sqrt[6]{(2^6 u^{18} v^6)(2u^5 v)}$

$\boxed{= \sqrt[6]{(2u^3 v)^6 (2u^5 v)}}$

$= 2u^3 v \sqrt[6]{2u^5 v}$

[*Note:*   Index numbers must be the same for both radicals before they can be multiplied (see property 2 on page 26).]

PROBLEM 15   Multiply (as in Example 15) and express the answer in simplest radical form.

(A) $\sqrt[4]{27a^3b^3}\,\sqrt[4]{3a^5b^3}$ (B) $\sqrt[4]{8x^7y^3}\,\sqrt{2x^3y}$

■ Simplifying $\sqrt[n]{x^n}$ for All Real x

We digress for a moment to recall that the absolute value of a real number is defined as follows:

Absolute Value

For all real numbers a, the **absolute value** of a, denoted by $|a|$, is given by

$$|a| = \begin{cases} a & \text{if } a \text{ is positive} \\ 0 & \\ -a & \text{if } a \text{ is negative} \end{cases}$$
$$\begin{aligned} |3| &= 3 \\ |0| &= 0 \\ |-3| &= -(-3) = 3 \end{aligned}$$

Now let us return to the first property of radicals:

$$\sqrt[n]{x^n} = x \quad \text{for } x \text{ a positive number}$$

What happens if we do not restrict x to positive numbers? For example, does

$$\sqrt{x^2} = x$$

for *all* real numbers x? The answer is no. Testing the equation for $x = 2$ and $x = -2$, we see that

$$\begin{array}{cc} x = 2 & x = -2 \\ \sqrt{2^2} \overset{?}{=} 2 & \sqrt{(-2)^2} \overset{?}{=} -2 \\ \sqrt{4} \overset{?}{=} 2 & \sqrt{4} \overset{?}{=} -2 \\ 2 \overset{\checkmark}{=} 2 & 2 \neq -2 \end{array}$$

Thus, if x is negative, we must write

$$\sqrt{x^2} = -x$$

then both sides will represent the same positive number.

In summary, for x any real number,

$$\sqrt{x^2} = \begin{cases} x & \text{if } x \text{ is positive} \\ 0 & \text{if } x \text{ is } 0 \\ -x & \text{if } x \text{ is negative} \end{cases}$$

And we see that $\sqrt{x^2}$ and $|x|$ are the same; thus,

For any real number x,

$$\sqrt{x^2} = |x|^*$$

Now let us turn to $\sqrt[3]{x^3}$. Here we do not have the same problem as earlier.

For any real number x,

$$\sqrt[3]{x^3} = x$$

Evaluate both sides for $x = 2$ and for $x = -2$.

If asked to simplify $\sqrt[3]{x^3} + \sqrt{x^2}$, many students would write

$$\sqrt[3]{x^3} + \sqrt{x^2} = x + x \;\;\boxed{= (1 + 1)x} \;\; = 2x$$

and not think any more about it. But if we evaluate both sides for $x = -2$, we find that

$$\sqrt[3]{(-2)^3} + \sqrt{(-2)^2} = \sqrt[3]{-8} + \sqrt{4} = -2 + 2 = 0 \quad \text{Left side}$$

and

$$2(-2) = -4 \quad \text{Right side}$$

Both sides are not equal! What is wrong? When x is not restricted to positive real values—that is, when x is allowed to take on any real value—then we should write

$$\sqrt[3]{x^3} + \sqrt{x^2} = x + |x|$$

Then the left side will equal the right side for all real numbers.

EXAMPLE 16 For x a positive number,

$$\sqrt[3]{x^3} + \sqrt{x^2} = x + |x| = x + x \;\;\boxed{= (1 + 1)x} \;\; = 2x \quad |x| = x \text{ if } x \text{ is positive.}$$

---

* This relationship is also of use when a programmable calculator without an $|x|$ function key is used.

For x a negative number,

$$\sqrt[3]{x^3} + \sqrt{x^2} = x + |x| = x + (-x) = 0 \qquad |x| = -x \text{ if }$$

$$x \text{ is negative.}$$

PROBLEM 16   Given $4\sqrt[3]{x^3} + \sqrt{x^2}$, simplify.

(A) For x a positive number   (B) For x a negative number

Following the same kind of reasoning, we can obtain the more general result:

For x any real number and n a positive integer 2 or larger,

$$\sqrt[n]{x^n} = \begin{cases} |x| & \text{if } n \text{ is even} \\ x & \text{if } n \text{ is odd} \end{cases}$$

Answers to Matched Problems
11. (A) $\sqrt[5]{u}$ (B) $\sqrt[9]{(6x^2y^5)^2}$ or $\left(\sqrt[9]{6x^2y^5}\right)^2$ (C) $1/\sqrt[5]{(3xy)^3}$
    (D) $(9u)^{1/4}$ (E) $-(2x)^{4/7}$ (F) $(x^3 + y^3)^{1/3}$ (not x + y)
12. (A) $u^2 + v^2$ (B) $2\sqrt{3}$ (C) $(\sqrt[3]{x^2})/2$ or $\frac{1}{2}\sqrt[3]{x^2}$
    (D) $\sqrt[4]{y^3}$ (E) $\sqrt[10]{x}$
13. (A) $2\sqrt[3]{2}$ (B) $3x^2yz\sqrt{2xz}$ (C) $\sqrt[3]{2x^2y}$ (D) $\sqrt[3]{2}$
14. (A) $(3\sqrt{2x})/x$ (B) $5x^2\sqrt[3]{4x}$ (C) $(\sqrt[3]{12x^2y^2})/2x^2$
15. (A) $3a^2b\sqrt[4]{b^2} = 3a^2b\sqrt{b}$ (B) $2x^3y\sqrt[4]{2xy}$
16. (A) 5x   (B) 3x

Exercise 0-4

■ Unless stated to the contrary, all letters and radicands represent positive real numbers.

A   *Change to radical form. Do not simplify.*

1. $m^{2/3}$
2. $n^{4/5}$
3. $6x^{3/5}$
4. $7y^{2/5}$
5. $(4xy^3)^{2/5}$
6. $(7x^2y)^{5/7}$
7. $(x + y)^{1/2}$
8. $x^{1/2} + y^{1/2}$

Change to rational exponent form. Do not simplify.

9. $\sqrt[5]{b}$
10. $\sqrt{c}$
11. $5\sqrt[4]{x^3}$
12. $7m\sqrt[5]{n^2}$
13. $\sqrt[5]{(2x^2y)^3}$
14. $\sqrt[9]{(3m^4n)^2}$
15. $\sqrt[3]{x} + \sqrt[3]{y}$
16. $\sqrt[3]{x + y}$

Simplify and write in simplest radical form.

17. $\sqrt[3]{-8}$
18. $\sqrt[3]{-27}$
19. $\sqrt{9x^8y^4}$
20. $\sqrt{16m^4y^8}$
21. $\sqrt[4]{16m^4n^8}$
22. $\sqrt[5]{32a^{15}b^{10}}$

23. $\sqrt{8a^3b^5}$ 24. $\sqrt{27m^2n^7}$ 25. $\sqrt[3]{2^4x^4y^7}$

26. $\sqrt[4]{2^4x^5y^8}$ 27. $\sqrt[4]{m^2}$ 28. $\sqrt[10]{n^6}$

29. $\sqrt[5]{\sqrt[3]{xy}}$ 30. $\sqrt{\sqrt[4]{5x}}$ 31. $\sqrt[3]{9x^2}\sqrt[3]{9x}$

32. $\sqrt{2x}\sqrt{8xy}$ 33. $1/\sqrt{5}$ 34. $1/\sqrt{7}$

35. $6/\sqrt[3]{3}$ 36. $2/\sqrt[3]{2}$ 37. $\sqrt{6x/7y}$

38. $\sqrt{3a/2b}$

B Simplify and write in simplest radical form.

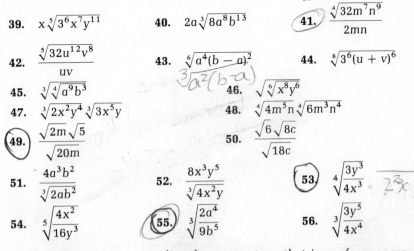

39. $x\sqrt[5]{3^6x^7y^{11}}$ 40. $2a\sqrt[3]{8a^8b^{13}}$ 41. $\dfrac{\sqrt[4]{32m^7n^9}}{2mn}$

42. $\dfrac{\sqrt[5]{32u^{12}v^8}}{uv}$ 43. $\sqrt[6]{a^4(b-a)^2}$ 44. $\sqrt[8]{3^6(u+v)^6}$

45. $\sqrt[3]{\sqrt[4]{a^9b^3}}$ 46. $\sqrt{\sqrt[6]{x^8y^6}}$

47. $\sqrt[3]{2x^2y^4}\sqrt[3]{3x^5y}$ 48. $\sqrt[4]{4m^5n}\sqrt[4]{6m^3n^4}$

49. $\dfrac{\sqrt{2m}\sqrt{5}}{\sqrt{20m}}$ 50. $\dfrac{\sqrt{6}\sqrt{8c}}{\sqrt{18c}}$

51. $\dfrac{4a^3b^2}{\sqrt[3]{2ab^2}}$ 52. $\dfrac{8x^3y^5}{\sqrt[3]{4x^2y}}$ 53. $\sqrt[4]{\dfrac{3y^3}{4x^3}}$

54. $\sqrt[5]{\dfrac{4x^2}{16y^3}}$ 55. $\sqrt[3]{\dfrac{2a^4}{9b^5}}$ 56. $\sqrt[3]{\dfrac{3y^5}{4x^4}}$

In Problems 57–62 rationalize the numerators; that is, perform operations on the algebraic expression that eliminate radicals from the numerator. (This is a particularly useful operation in some problems in calculus.)

57. $\dfrac{\sqrt{2x}}{4}$ 58. $\dfrac{\sqrt{3y}}{\sqrt{2x}}$ 59. $\dfrac{\sqrt[3]{4m}}{2m^2}$

60. $\dfrac{\sqrt[3]{3x^2}}{6xy}$ 61. $\sqrt[4]{8x^3}$ 62. $\sqrt[5]{8y^3}$

C Simplify Problems 63–70 and express answers in simplest radical form.

63. $\sqrt[3]{\sqrt[5]{\sqrt{32a^{15}b^{10}}}}$ 64. $\sqrt[4]{\sqrt[3]{\sqrt{8x^6y^9}}}$

65. $\sqrt[3]{8\sqrt{16x^6y^4}}$ 66. $\sqrt[4]{16x^4\sqrt[3]{16x^{24}y^4}}$

67. $\sqrt{2a^5b^3}\sqrt[3]{16a^7b^7}$ 68. $\sqrt[3]{3x^2y^2}\sqrt[4]{3x^3y^2}$

69. $\dfrac{\sqrt[3]{4a^2b^2}}{\sqrt{2ab}}$ 70. $\dfrac{\sqrt{2x}}{\sqrt[3]{x}}$

71. Show that: $\sqrt[n]{\dfrac{x}{y}} = \dfrac{\sqrt[n]{x}}{\sqrt[n]{y}}$ 72. Show that: $\sqrt[m]{\sqrt[n]{x}} = \sqrt[mn]{x}$

Simplify Problems 73–78 for (A) x a positive number and (B) x a negative number.

73. $4\sqrt[3]{x^3} + 2\sqrt{x^2}$ **74.** $7\sqrt[3]{x^3} + 2\sqrt{x^2}$ **75.** $\sqrt[5]{x^5} + \sqrt[4]{x^4}$

76. $\sqrt[7]{x^7} + \sqrt[8]{x^8}$ **77.** $2\sqrt[4]{x^4} + 3\sqrt[5]{x^5}$ **78.** $3\sqrt[6]{x^6} + 4\sqrt[9]{x^9}$

CALCULATOR PROBLEMS

Evaluate to four significant digits using a hand calculator. (Read the instruction booklet accompanying your calculator for the process required to evaluate $\sqrt[n]{x}$.)

79. $\sqrt{0.049\ 375}$ **80.** $\sqrt{306.721}$

81. $\sqrt[5]{27.0635}$ **82.** $\sqrt[8]{0.070\ 144}$

83. $\sqrt[7]{0.000\ 000\ 008\ 066}$ **84.** $\sqrt[12]{6,423,000,000,000}$

85. $\sqrt[3]{7} + \sqrt[3]{7}$ **86.** $\sqrt[5]{4} + \sqrt[5]{4}$

87. $\sqrt[3]{\sqrt[4]{2}}$ and $\sqrt[12]{2}$ **88.** $\sqrt[3]{\sqrt{5}}$ and $\sqrt[6]{5}$

89. $\dfrac{1}{\sqrt[3]{4}}$ and $\dfrac{\sqrt[3]{2}}{2}$ **90.** $\dfrac{1}{\sqrt[3]{5}}$ and $\dfrac{\sqrt[3]{25}}{5}$

Section 0-5   Algebraic Expressions, Basic Operations

- Algebraic Expressions and Polynomials
- Addition and Subtraction
- Multiplication
- Combined Operations

In this section we will review the basic operations of addition, subtraction, and multiplication on algebraic expressions. We will also introduce a special algebraic form called a *polynomial*.

- Algebraic Expressions and Polynomials

Algebraic expressions are formed by using constants and variables* and the algebraic operations of addition, subtraction, multiplication, division, and the taking of roots. For example,

$$\sqrt[3]{x^3 - 2x + 1} \qquad \frac{x - 5}{x^2 + 2x - 5} \qquad (3x^{-5} - 2x^{-3})^{2/3}$$

---

* A **constant** is any symbol that is used to name exactly one thing; a **variable** is any symbol used as a placeholder for constants from a set of two or more constants.

are all algebraic expressions. An algebraic expression involving only the operations of addition, subtraction, and multiplication on variables and constants, such as $x^3 - 2x^2 + 5x - 1$, is called a **polynomial**.

Polynomial in x

A **polynomial in x** is an algebraic expression of the form

$$a_n x^n + a_{n-1} x^{n-1} + \cdots + a_1 x + a_0$$

where the coefficients $a_0, a_1, \ldots, a_n$ are real numbers and n is a nonnegative integer.

Of course, we may consider polynomials in more than one variable. A polynomial in the two variables x and y is an algebraic expression formed by adding terms of the form $ax^m y^n$, where a is a real number and m and n are nonnegative integers. For example,

$$3x^3 - \sqrt{2}x^2 y + xy - \tfrac{1}{2}xy^2 + y^3 + 2x - 3$$

is a polynomial in two variables. Polynomials in three and more variables are defined in a similar way.

Polynomial forms are encountered frequently in mathematics, and for their more efficient study it is useful to classify them according to their degree. If a term in a polynomial has only one variable as a factor, then the **degree of that term** is the power of the variable. If two or more variables are present in a term as factors, then the **degree of the term** is the sum of the powers of the variables. The **degree of a polynomial** is the degree of the nonzero term with the highest degree in the polynomial. Any nonzero constant is defined to be a **polynomial of degree 0**. The number 0 is also a polynomial but is not assigned a degree.

EXAMPLE 17   (A) Polynomials in one variable:

$$x^2 - 3x + 2 \qquad 6x^3 - \sqrt{2}x - \tfrac{1}{3}$$

(B) Polynomials in several variables:

$$3x^2 - 2xy + y^2 \qquad 4x^3 y^2 - \sqrt{3}xy^2 z^5$$

(C) Nonpolynomials:

$$\sqrt{2x} - \frac{3}{x} + 5 \qquad \frac{x^2 - 3x + 2}{x - 3} \qquad \sqrt{x^2 - 3x + 1}$$

(D) The degree of the first term in $6x^3 - \sqrt{2}x - \tfrac{1}{3}$ is 3, the second term 1, the third term 0, and the whole polynomial 3.

(E) The degree of the first term in $4x^3y^2 - \sqrt{3}xy^2$ is 5, the second 3, and the whole polynomial 5.

PROBLEM 17   (A) Which of the following are polynomials?

$$3x^2 - 2x + 1 \qquad \sqrt{x - 3} \qquad x^2 - 2xy + y^2 \qquad \frac{x - 1}{x^2 + 2}$$

(B) Given the polynomial $3x^5 - 6x^3 + 5$, what is the degree of the first term? The second term? The whole polynomial?

(C) Given the polynomial $6x^4y^2 - 3xy^3$, what is the degree of the first term? The second term? The whole polynomial?

In addition to classifying polynomials by degree, we also call a single-term polynomial a **monomial**, a two-term polynomial a **binomial**, and a three-term polynomial a **trinomial**.

■ Addition and Subtraction

We now turn to the addition and subtraction of polynomials and other algebraic expressions. All letters in the following discussion and examples represent real numbers; hence, all the properties of real numbers we have discussed apply. We now add one more property that we will find useful—namely, that multiplication distributes over subtraction:

$$a(b - c) = (b - c)a = ab - ac$$

This is easy to see as follows:

$$a(b - c) = a[b + (-c)]$$ Can you supply the reasons
$$= ab + a(-c)$$ for each of these steps?
$$= ab + [-(ac)]$$
$$= ab - ac$$

EXAMPLE 18   (A) Add $5x^3 - 2x^2 + x - 3$ and $7x^3 + 5x^2 + 9$.

Solution $(5x^3 - 2x^2 + x - 3) + (7x^3 + 5x^2 + 9)$ By placing 1's in
front of the
$$= 1(5x^3 - 2x^2 + x - 3) + 1(7x^3 + 5x^2 + 9)$$ parentheses, we
can use the
$$= 5x^3 - 2x^2 + x - 3 + 7x^3 + 5x^2 + 9$$ distributive property
to "clear" the
$$= 5x^3 + 7x^3 + 5x^2 - 2x^2 + x + 6$$ parentheses. Notice
$$= (5 + 7)x^3 + (5 - 2)x^2 + x + 6$$ how the
commutative and
$$= 12x^3 + 3x^2 + x + 6$$ associative
properties are also
(B) Subtract $5x^3 - 2x^2 + x - 3$ from $7x^3 + 5x^2 + 9$.   used in this example.

Solution $(7x^3 + 5x^2 + 9) - (5x^3 - 2x^2 + x - 3)$

$$= 1(7x^3 + 5x^2 + 9) + (-1)(5x^3 - 2x^2 + x - 3)$$

$$= 7x^3 + 5x^2 + 9 - 5x^3 + 2x^2 - x + 3$$

$$= 7x^3 - 5x^3 + 5x^2 + 2x^2 - x + 12$$
$$= (7 - 5)x^3 + (5 + 2)x^2 - x + 12$$

$$= 2x^3 + 7x^2 - x + 12$$

Subtracting a number is the same as adding its negative. The negative of a number can be obtained by multiplying it by -1.

PROBLEM 18   Given the polynomials $x^3 - 7x + 2$ and $4x^3 - x^2 + x - 1$.

(A) Add them.   (B) Subtract the first from the second.

EXAMPLE 19   (A) Add $3\sqrt{x} + 5\sqrt{y} + 2$ and $\sqrt{x} + \sqrt[3]{y} - 4$.

Solution $\left(3\sqrt{x} + 5\sqrt{y} + 2\right) + \left(\sqrt{x} + \sqrt[3]{y} - 4\right)$ These are not polynomials.

$$= 3\sqrt{x} + 5\sqrt{y} + 2 + \sqrt{x} + \sqrt[3]{y} - 4$$

$$= 3\sqrt{x} + \sqrt{x} + 5\sqrt{y} + \sqrt[3]{y} - 2$$
$$= (3 + 1)\sqrt{x} + 5\sqrt{y} + \sqrt[3]{y} - 2$$

$$= 4\sqrt{x} + 5\sqrt{y} + \sqrt[3]{y} - 2$$

(B) Subtract $4x^{2/3} - x^{1/3} + 2$ from $3x^{2/3} + 2x^{1/3} - 8$.

Solution $(3x^{2/3} + 2x^{1/3} - 8) - (4x^{2/3} - x^{1/3} + 2)$ These are not polynomials.

$$= 3x^{2/3} + 2x^{1/3} - 8 - 4x^{2/3} + x^{1/3} - 2$$

$$= 3x^{2/3} - 4x^{2/3} + 2x^{1/3} + x^{1/3} - 10$$
$$= (3 - 4)x^{2/3} + (2 + 1)x^{1/3} - 10$$

$$= -x^{2/3} + 3x^{1/3} - 10$$

PROBLEM 19   (A) Add $\left(2\sqrt[4]{m} + 3\sqrt{pq} - 3\right)$ and $\left(\sqrt[3]{m} - \sqrt{pq} + 5\right)$.
(B) Subtract the first algebraic expression from the second in part (A).

■ Multiplication

Multiplication of algebraic expressions involves the extensive use of distributive properties for real numbers, as well as other real number properties.

EXAMPLE 20   Multiply $(2x - 3)(3x^2 - 2x + 3)$.

Solution $(2x - 3)(3x^2 - 2x + 3)$

$$= 2x(3x^2 - 2x + 3) - 3(3x^2 - 2x + 3)$$

$$= 6x^3 - 4x^2 + 6x - 9x^2 + 6x - 9$$
$$= 6x^3 - 13x^2 + 12x - 9$$

or

$$3x^2 - 2x + 3$$
$$\underline{2x - 3 }$$
$$6x^3 - 4x^2 + 6x $$
$$\underline{ - 9x^2 + 6x - 9}$$
$$6x^3 - 13x^2 + 12x - 9$$

PROBLEM 20   Multiply $(2x - 3)(2x^2 + 3x - 2)$.

Certain types of binomial products occur so frequently that it is useful to note formulas for them.

Special Products

$$(ax + b)(cx + d) = acx^2 + (ad + bc)x + bd$$
$$(A - B)(A + B) = A^2 - B^2$$
$$(A + B)^2 = A^2 + 2AB + B^2$$
$$(A - B)^2 = A^2 - 2AB + B^2$$

EXAMPLE 21   (A) $(2x - 3y)(5x + 2y) = 10x^2 - 11xy - 6y^2$

(B) $\left(\sqrt{2} - \sqrt{3}\right)\left(\sqrt{2} + \sqrt{3}\right) = \left(\sqrt{2}\right)^2 - \left(\sqrt{3}\right)^2$

$$= 2 - 3 = -1$$

(C) $(3x^{1/2} - 2y^{1/2})^2 = 9x - 12x^{1/2}y^{1/2} + 4y$

(D) $\left(3\sqrt{x} + 2\sqrt{y}\right)\left(2\sqrt{x} - \sqrt{y}\right) = 6x + \sqrt{xy} - 2y$

PROBLEM 21   Multiply and simplify.

(A) $(5u + 3v)(4u - v)$ (B) $\left(\sqrt{x} - \sqrt{y}\right)\left(\sqrt{x} + \sqrt{y}\right)$

(C) $(2^{1/2} - 3^{1/2})^2$ (D) $\left(4\sqrt{a} - \sqrt{b}\right)\left(3\sqrt{a} + 2\sqrt{b}\right)$

■ Combined Operations

We complete this section by considering several examples that combine operations discussed in this and preceding sections.

EXAMPLE 22

(A) $\sqrt{2}(\sqrt{10} - 3) - 4(\sqrt{45} + \sqrt{2})$

Multiply, transform terms to simplest radical form, and combine terms where possible.

$$= \sqrt{20} - 3\sqrt{2} - 4\sqrt{45} - 4\sqrt{2}$$
$$= \sqrt{4 \cdot 5} - 3\sqrt{2} - 4\sqrt{9 \cdot 5} - 4\sqrt{2}$$
$$= 2\sqrt{5} - 3\sqrt{2} - 12\sqrt{5} - 4\sqrt{2}$$
$$= -10\sqrt{5} - 7\sqrt{2}$$

(B) $3x - \{5 - 3[x - x(3 - x)]\} = 3x - \{5 - 3[x - 3x + x^2]\}$
$$= 3x - \{5 - 3x + 9x - 3x^2\}$$
$$= 3x - 5 + 3x - 9x + 3x^2$$
$$= 3x^2 - 3x - 5$$

(C) $\left(\sqrt[3]{m} + \sqrt[3]{n^2}\right)\left(\sqrt[3]{m^2} - \sqrt[3]{n}\right) = \sqrt[3]{m^3} - \sqrt[3]{mn} + \sqrt[3]{m^2n^2} - \sqrt[3]{n^3}$
$$= m - \sqrt[3]{mn} + \sqrt[3]{m^2n^2} - n$$

PROBLEM 22

Multiply and simplify.

(A) $\sqrt{10}(\sqrt{5} + \sqrt{2}) + \sqrt{6}(\sqrt{3} - \sqrt{2})$ (B) $2t - \{7 - 2[t - t(4 + t)]\}$

(C) $\left(\sqrt[3]{x^2} - \sqrt[3]{y^2}\right)\left(\sqrt[3]{x} + \sqrt[3]{y}\right)$

Answers to Matched Problems

17. (A) $3x^2 - 2x + 1, x^2 - 2xy + y^2$ (B) 5, 3, 5
(C) 6, 4, 6

18. (A) $5x^3 - x^2 - 6x + 1$ (B) $3x^3 - x^2 + 8x - 3$

19. (A) $2\sqrt[4]{m} + \sqrt[3]{m} + 2\sqrt{pq} + 2$ (B) $-2\sqrt[4]{m} + \sqrt[3]{m} - 4\sqrt{pq} + 8$

20. $4x^3 - 13x + 6$

21. (A) $20u^2 + 7uv - 3v^2$ (B) $x - y$
(C) $5 - 2 \cdot 2^{1/2} \cdot 3^{1/2}$ or $5 - 2(6)^{1/2}$ (D) $12a + 5\sqrt{ab} - 2b$

22. (A) $8\sqrt{2} + 2\sqrt{5} - 2\sqrt{3}$ (B) $-2t^2 - 4t - 7$
(C) $x + \sqrt[3]{x^2y} - \sqrt[3]{xy^2} - y$

Exercise 0-5 ■ Unless stated to the contrary, express all answers in simplest radical form with positive exponents. All variables are positive real numbers.

A   Given the polynomials: $2x^3 - 3x^2 + x + 5$, $2x^2 + x - 1$, and $3x - 2$.

1. What is the degree of the first?

2. What is the degree of the second?

3. Add the first and second.

4. Add the second and third.

5. Subtract the second from the first.

6. Subtract the third from the second.

7. Multiply the first and third.

8. Multiply the second and third.

In Problems 9–28 perform the indicated operations and simplify.

9. $2(x-1)+3(2x-3)-(4x-5)$

10. $2(u-1)-(3u+2)-2(2u-3)$

11. $2y - 3y[4 - 2(y - 1)]$

12. $4a - 2a[5 - 3(a + 2)]$

13. $\sqrt{5} - 2\sqrt{3} + 3\sqrt{5}$

14. $3\sqrt{2} - 2\sqrt{3} - \sqrt{2}$

15. $2\sqrt[3]{a} + 3\sqrt[3]{a} - \sqrt[4]{a}$

16. $4\sqrt[3]{y} - \sqrt[3]{y} + 2\sqrt{y}$

17. $(3x - 2y)(3x + 2y)$

18. $(4m + 3n)(4m - 3n)$

19. $(\sqrt{c} - \sqrt{d})(\sqrt{c} + \sqrt{d})$

20. $(x^{1/2} + y^{1/2})(x^{1/2} - y^{1/2})$

21. $(4x - y)^2$

22. $(3u + 4v)^2$

23. $(x^{1/2} + y^{1/2})^2$

24. $(\sqrt{y} + \sqrt{z})^2$

25. $(\sqrt{c} - \sqrt{d})^2$

26. $(x^{1/2} - y^{1/2})^2$

27. $(a + b)(a^2 - ab + b^2)$

28. $(a - b)(a^2 + ab + b^2)$

B *In Problems 29–54 perform the indicated operations and simplify (or vice versa).*

29. $2x - 3\{x + 2[x - (x + 5)] + 1\}$

30. $m - \{m - [m - (m - 1)]\}$

31. $(2x^2 + x - 2)(x^2 - 3x + 5)$

32. $(x^2 - 2xy + y^2)(x^2 + 2xy + y^2)$

33. $(2x - 1)^2 - (3x + 2)(3x - 2)$

34. $(3a - b)(3a + b) - (2a - 3b)^2$

35. $(2m - n)^3$

36. $(x - 2y)^3$

37. $\sqrt{8} - \sqrt{20} + 4\sqrt{2}$

38. $3\sqrt{3} - \sqrt{12} + \sqrt{24}$

39. $2\sqrt{12x} - 3\sqrt{27x} + \sqrt{3x}$

40. $\sqrt[3]{8a} - 2\sqrt[3]{27a} + \sqrt{4a}$

41. $(a\sqrt{2} - b\sqrt{5})(a\sqrt{2} + b\sqrt{5})$

42. $(x + y\sqrt{3})(x - y\sqrt{3})$

43. $(3\sqrt{x} - \sqrt{y})^2$

44. $(5\sqrt{x} + 2)(2\sqrt{x} - 3)$

45. $(2x^{1/2} + y^{1/2})(x^{1/2} + y^{1/2})$

46. $(3u^{1/2} - 2)(2u^{1/2} + 4)$

47. $2x^{1/3}(3x^{2/3} - x^6)$

48. $3m^{3/4}(4m^{1/4} - 2m^8)$

49. $\sqrt{3 - \sqrt{8}}\sqrt{3 + \sqrt{8}}$

50. $\sqrt{4 - \sqrt{10}}\sqrt{4 + \sqrt{10}}$

51. $\sqrt{ab}\left(\sqrt{\dfrac{a}{b}} + \sqrt{\dfrac{b}{a}}\right)$

52. $(xy)^{1/3}\left[\left(\dfrac{x^2}{y}\right)^{1/3} - \left(\dfrac{y^2}{x}\right)^{1/3}\right]$

53. $(\sqrt{x + h} - \sqrt{x})(\sqrt{x + h} + \sqrt{x})$

54. $[(u + k)^{1/2} - u^{1/2}][(u + k)^{1/2} + u^{1/2}]$

In Problems 55–58 evaluate each polynomial for the indicated value.

55. $x^2 + 2x - 5$ for $x = -1 + \sqrt{6}$ **56.** $m^2 - 4m + 1$ for $m = 2 + \sqrt{3}$

57. $u^2 - 6u + 3$ for $u = 3 - \sqrt{2}$ **58.** $y^2 - 3y - 2$ for $y = 1 - \sqrt{3}$

Simplify.

59. $(x^{-1/2} - y^{-1/2})^2$

60. $(a^{-1/2} + 3b^{-1/2})(2a^{-1/2} - b^{-1/2})$

61. $(\sqrt[3]{t} - \sqrt[3]{x})(\sqrt[3]{t^2} + \sqrt[3]{tx} + \sqrt[3]{x^2})$

62. $(\sqrt[3]{t} + \sqrt[3]{x})(\sqrt[3]{t^2} - \sqrt[3]{tx} + \sqrt[3]{x^2})$

63. $[3(x + 3)^{1/2} + 2][2(x + 3)^{1/2} - 3]$

64. $[2(x - 2)^{1/2} - 3][(x - 2)^{1/2} + 3]$

65. $(\sqrt[3]{x} - \sqrt[3]{y^2})(\sqrt[3]{x^2} + 2\sqrt[3]{y})$ **66.** $(\sqrt[5]{y^2} - \sqrt[5]{z^3})(\sqrt[5]{y^3} + \sqrt[5]{z^2})$

APPLICATIONS **67.** *Geometric.*   The width of a rectangle is 5 centimeters less than its length. If x represents the length, write an algebraic expression in terms of x that represents the perimeter of the rectangle. Simplify the expression.

68. *Geometric.*   The length of a rectangle is 8 meters more than its width. If x represents the width of the rectangle, write an algebraic expression in terms of x that represents its area. Change the expression to a form without parentheses.

69. *Coin problem.*   A parking meter contains nickels, dimes, and quarters. There are five fewer dimes than nickels, and two more quarters than dimes. If x represents the number of nickels, write an algebraic expression in terms of x that represents the value of all the coins in the meter in cents. Simplify the expression.

70. *Coin problem.*   A vending machine contains dimes and quarters only. There are four more dimes than quarters. If x represents the number of quarters, write an algebraic expression in terms of x that represents the value of all the coins in the vending machine. Simplify the expression.

CALCULATOR PROBLEMS   *Evaluate each to four significant digits using a hand calculator.*

71. $(2\sqrt{5} - \sqrt{2})^2 - (24 - 4\sqrt{10})$ **72.** $(\sqrt{3} - \sqrt{5})^2 - (8 - 2\sqrt{15})$

73. $\sqrt{28 - 10\sqrt{3}} - (5 - \sqrt{3})$ **74.** $(\sqrt{2} + \sqrt{3}) - \sqrt{5 + 2\sqrt{6}}$

Section 0-6 Factoring

- Factoring—What Does It Mean?
- Common Factors

- Factoring by Grouping
- Special Factoring Formulas

- **Factoring—What Does It Mean?**

If a number is written as the product of other numbers, then each number in the product is called a **factor** of the original number. Similarly, if an algebraic expression is written as the product of other algebraic expressions, then each algebraic expression in the product is called a **factor** of the original algebraic expression.

$$30 = 2 \cdot 3 \cdot 5 \qquad \text{2, 3, and 5 are factors of 30.}$$
$$x^2 - 4 = (x - 2)(x + 2) \quad \text{$(x - 2)$ and $(x + 2)$ are factors of $x^2 - 4$.}$$

The process of writing a number or algebraic expression as the product of other numbers or algebraic expressions is called **factoring**. We start our discussion of factoring with the positive integers.

An integer such as 30 can be represented in a factored form in many ways. The products

$$6 \cdot 5 \qquad (\tfrac{1}{2})(10)(6) \qquad 15 \cdot 2 \qquad \sqrt{15} \cdot \sqrt{60}$$

all yield 30. A particularly useful way of factoring positive integers greater than 1 is in terms of prime numbers.

Prime and Composite Numbers

A positive integer greater than 1 is **prime** if its only positive integer factors are itself and 1. A positive integer greater than 1 that is not prime is called a **composite number**. The integer 1 is neither prime nor composite.

PRIME NUMBERS: 2, 3, 5, 7, 11, 13, . . .

COMPOSITE NUMBERS: 4, 6, 8, 9, 10, 12, . . .

A composite integer greater than 1 is said to be **factored completely** if it is represented as a product of prime factors. The only factoring of 30 given that meets this condition is $30 = 2 \cdot 3 \cdot 5$.

EXAMPLE 23   Write 60 in a completely factored form.

Solution $60 = 6 \cdot 10 = 2 \cdot 3 \cdot 2 \cdot 5 = 2^2 \cdot 3 \cdot 5$

or

$$60 = 5 \cdot 12 = 5 \cdot 4 \cdot 3 = 2^2 \cdot 3 \cdot 5$$

or

$$60 = 2 \cdot 30 = 2 \cdot 2 \cdot 15 = 2^2 \cdot 3 \cdot 5$$

Notice in Example 23 that we end up with the same prime factors for 60 irrespective of how we progress through the factoring process. This illustrates a basic property of integers:

Fundamental Theorem of Arithmetic

Each positive integer greater than 1 is either prime or has, except for the order of factors, a unique set of prime factors.

PROBLEM 23   Write 180 in a completely factored form.

We can also talk about writing polynomials in a completely factored form. The following polynomials are written in a factored form:

$$x^2 - 9 = (x - 3)(x + 3)$$
$$2x^3 - 4x = 2x\left(x - \sqrt{2}\right)\left(x + \sqrt{2}\right)$$
$$2x^4 - 15x^2 - 27 = (x^2 - 9)(2x^2 + 3)$$
$$x^2 + 3x + \tfrac{9}{4} = (x + \tfrac{3}{2})^2$$

But which are in a completely factored form? Paralleling our discussion with prime numbers, we define a **prime polynomial** as follows:

Prime Polynomials

A polynomial is said to be **prime** relative to a given set of numbers if: (1) it has coefficients from that set, and (2) it cannot be written as a product of two polynomials of positive degree having coefficients from that set.

For example, $x^2 - 2$ is prime relative to the integers but is not prime relative to the real numbers [since $x^2 - 2 = (x - \sqrt{2})(x + \sqrt{2})$]. A nonprime polynomial is said to be **factored completely relative to a given set of numbers** if it is represented as a product of prime polynomials relative to that set of numbers.

Writing polynomials in a completely factored form is often a difficult task. But accomplishing it can lead to the simplification of certain algebraic expressions and to the solutions of certain types of equations.

Our objective in this section is to review some of the standard factoring techniques; then in Chapter 3 we will go into the topic in detail from a more advanced point of view.

■ Common Factors

In the preceding sections a number of problems involved factoring out common factors. The distributive property for real numbers is, of course, behind the process.

EXAMPLE 24   Take out all factors common to all terms.

(A) $2x^3y - 8x^2y^2 - 6xy^3$ (B) $2x(3x - 2) - 7(3x - 2)$

Solutions   (A) $2x^3y - 8x^2y^2 - 6xy^3$ $= (2xy)x^2 - (2xy)4xy - (2xy)3y^2$

$= 2xy(x^2 - 4xy - 3y^2)$

(B) $2x(3x - 2) - 7(3x - 2)$ $= 2x(3x - 2) - 7(3x - 2)$

$= (2x - 7)(3x - 2)$

PROBLEM 24   Take out all factors common to all terms.

(A) $3x^3y - 6x^2y^2 - 3xy^3$ (B) $3y(2y + 5) + 2(2y + 5)$

■ Factoring by Grouping

Some polynomials can be factored by first grouping terms in such a way that we obtain an algebraic expression that looks something like Example 18B. We can then complete the factoring by the method used in that example.

EXAMPLE 25   Factor completely, relative to the integers, by grouping.

(A) $3x^2 - 6x + 4x - 8$ (B) $2y^2 - 2y - y + 1$
(C) $4x^2 + 8xy - xy - 2y^2$ (D) $2ac - 2ad - bc + bd$

Solutions   (A) $3x^2 - 6x + 4x - 8$ Group the first two and last two terms.

$= (3x^2 - 6x) + (4x - 8)$ Remove common factors from each group.

$= 3x(x - 2) + 4(x - 2)$ The common factor $(x - 2)$ can be taken out.

$= (3x + 4)(x - 2)$ The factoring is complete.

(B) $2y^2 - 2y - y + 1$

$= (2y^2 - 2y) - (y - 1)$ Be careful of sign errors here.

$= 2y(y - 1) - 1(y - 1)$

$= (2y - 1)(y - 1)$

(C) $4x^2 + 8xy - xy - 2y^2 = (4x^2 + 8xy) - (xy + 2y^2)$
$$= 4x(x + 2y) - y(x + 2y)$$
$$= (4x - y)(x + 2y)$$

(D) $2ac - 2ad - bc + bd = (2ac - 2ad) - (bc - bd)$
$$= 2a(c - d) - b(c - d)$$
$$= (2a - b)(c - d)$$

PROBLEM 25 Factor completely, relative to the integers, by grouping.

(A) $2x^2 + 6x + 5x + 15$ (B) $3m^2 + 3m - m - 1$
(C) $6u^2 - 3uv - 2uv + v^2$ (D) $6wy + 3wz - 2xy - xz$

■ Special Factoring Formulas

Several factoring formulas are worthwhile observing, since they show us how to factor certain polynomial forms that occur often.

Special Factoring Formulas

1. $u^2 + (a + b)u + ab = (u + a)(u + b)$

2. $acu^2 + (ad + bc)u + bd = (au + b)(cu + d)$

3. $a^2u^2 + 2abuv + b^2v^2 = (au + bv)^2$ PERFECT SQUARE

4. $u^2 - v^2 = (u - v)(u + v)$ DIFFERENCE OF TWO SQUARES

5. $u^3 - v^3 = (u - v)(u^2 + uv + v^2)$ DIFFERENCE OF TWO CUBES

6. $u^3 + v^3 = (u + v)(u^2 - uv + v^2)$ SUM OF TWO CUBES

The formulas in the box can be established by multiplying the factors on the right.

EXAMPLE 26 Factor completely relative to the integers.

(A) $x^2 - 5x - 6$ (B) $6x^2 - 5x - 4$ (C) $x^2 + 6xy + 9y^2$
(D) $9x^2 - 4y^2$ (E) $8m^3 - 1$ (F) $x^3 + y^3z^3$

Solutions (A) $x^2 - 5x - 6 = (x - 6)(x + 1)$
(B) $6x^2 - 5x - 4 = (3x - 4)(2x + 1)$
(C) $x^2 + 6xy + 9y^2 = (x + 3y)^2$
(D) $9x^2 - 4y^2 = (3x - 2y)(3x + 2y)$

(E) $8m^3 - 1$ $\boxed{\begin{aligned} &= (2m)^3 - 1^3 \\ &= (2m - 1)[(2m)^2 + (2m)(1) + 1^2] \end{aligned}}$

$= (2m - 1)(4m^2 + 2m + 1)$

(F) $x^3 + y^3z^3$ $\boxed{= x^3 + (yz)^3}$

$= (x + yz)(x^2 - xyz + y^2z^2)$

PROBLEM 26   Factor completely relative to the integers.

(A) $x^2 + 7x - 8$ (B) $4m^2 - 4mn - 3n^2$
(C) $4m^2 - 12mn + 9n^2$ (D) $x^2 - 16y^2$
(E) $z^3 - 1$ (F) $m^3 + n^3$

We complete this section by considering factoring that involves combinations of the preceding techniques as well as a few additional ones. Generally speaking, **when asked to factor a polynomial, we first take out all factors common to all terms, if they are present, and then proceed as above until all factors are prime.**

EXAMPLE 27   Factor completely relative to the integers.

(A) $18x^3 - 8x$ (B) $x^2 - 6x + 9 - y^2$
(C) $4m^3n - 2m^2n^2 + 2mn^3$ (D) $2t^4 - 16t$
(E) $2y^4 - 5y^2 - 12$ (F) $x^4 + x^2 + 1$

Solutions   (A) $18x^3 - 8x = 2x(9x^2 - 4)$

$= 2x(3x - 2)(3x + 2)$

(B) $x^2 - 6x + 9 - y^2$ Group the first three terms.

$= (x^2 - 6x + 9) - y^2$ Factor $x^2 - 6x + 9$.

$= (x - 3)^2 - y^2$ Difference of two squares

$= [(x - 3) - y][(x - 3) + y]$

$= (x - 3 - y)(x - 3 + y)$

(C) $4m^3n - 2m^2n^2 + 2mn^3 = 2mn(2m^2 - mn + n^2)$

(D) $2t^4 - 16t = 2t(t^3 - 8)$

$= 2t(t - 2)(t^2 + 2t + 4)$

(E) $2y^4 - 5y^2 - 12 = (2y^2 + 3)(y^2 - 4)$

$= (2y^2 + 3)(y - 2)(y + 2)$

(F) $x^4 + x^2 + 1$

$\qquad = (x^4 + 2x^2 + 1) - x^2$

$\qquad = (x^2 + 1)^2 - x^2$

$\qquad = [(x^2 + 1) - x][(x^2 + 1) + x]$

$\qquad = (x^2 - x + 1)(x^2 + x + 1)$

Proceeding as in (E) does not work; this involves a "trick." Adding and subtracting x^2 lead to a difference of two squares form.

PROBLEM 27   Factor completely relative to the integers.

(A) $3x^3 - 48x$ (B) $x^2 - y^2 - 4y - 4$

(C) $3u^4 - 3u^3v - 9u^2v^2$ (D) $3m^4 - 24mn^3$

(E) $3x^4 - 5x^2 + 2$ (F) $x^4 + x^2 + 25$

Remark: It should be noted that if one writes a polynomial with integer coefficients at random, then the resulting polynomial is more likely to be prime than not prime; that is, it most likely will not have polynomial factors of positive degree relative to the integers. If it does, however, the results may be very useful, as was pointed out earlier.

Answers to Matched Problems

23. $2^2 \cdot 3^2 \cdot 5$

24. (A) $3xy(x^2 - 2xy - y^2)$ (B) $(3y + 2)(2y + 5)$

25. (A) $(2x + 5)(x + 3)$ (B) $(3m - 1)(m + 1)$

       (C) $(3u - v)(2u - v)$ (D) $(3w - x)(2y + z)$

26. (A) $(x + 8)(x - 1)$ (B) $(2m - 3n)(2m + n)$

       (C) $(2m - 3n)^2$ (D) $(x - 4y)(x + 4y)$

       (E) $(z - 1)(z^2 + z + 1)$ (F) $(m + n)(m^2 - mn + n^2)$

27. (A) $3x(x - 4)(x + 4)$ (B) $(x - y - 2)(x + y + 2)$

       (C) $3u^2(u^2 - uv - 3v^2)$ (D) $3m(m - 2n)(m^2 + 2mn + 4n^2)$

       (E) $(3x^2 - 2)(x - 1)(x + 1)$ (F) $(x^2 - 3x + 5)(x^2 + 3x + 5)$

Exercise 0-6 ■ A   *Factor out all factors common to all terms.*

1. $6x^4 - 8x^3 - 2x^2$ **2.** $6m^4 - 9m^3 - 3m^2$

3. $10x^3y + 20x^2y^2 - 15xy^3$ **4.** $8u^3v - 6u^2v^2 + 4uv^3$

5. $5x(x + 1) - 3(x + 1)$ **6.** $7m(2m - 3) + 5(2m - 3)$

7. $2w(y - 2z) - x(y - 2z)$ **8.** $a(3c + d) - 4b(3c + d)$

Factor completely relative to the integers. Start by grouping the first two and last two terms.

9. $x^2 - 2x + 3x - 6$ **10.** $2y^2 - 6y + 5y - 15$

11. $6m^2 + 10m - 3m - 5$ **12.** $5x^2 - 40x - x + 8$

13. $2x^2 - 4xy - 3xy + 6y^2$ **14.** $3a^2 - 12ab - 2ab + 8b^2$

15. $8ac - 4ad - 6bc + 3bd$ **16.** $3pr + 6ps - qr - 2qs$

Factor completely relative to the integers. If a polynomial is prime relative to the integers, say so.

17. $2x^2 + 5x - 3$ **18.** $3y^2 - y - 2$

19. $x^2 - 4xy - 12y^2$ **20.** $u^2 - 2uv - 15v^2$

21. $x^2 + x - 4$ **22.** $m^2 - 6m - 3$

23. $25m^2 - 16n^2$ **24.** $w^2x^2 - y^2$

25. $x^2 + 10xy + 25y^2$ **26.** $9m^2 - 6mn + n^2$

27. $u^2 + 81$ **28.** $y^2 + 16$

29. $6x^2 + 48x + 72$ **30.** $4z^2 - 28z + 48$

31. $2y^3 - 22y^2 + 48y$ **32.** $2x^4 - 24x^3 + 40x^2$

33. $16x^2y - 8xy + y$ **34.** $4xy^2 - 12xy + 9x$

35. $6s^2 + 7st - 3t^2$ **36.** $6m^2 - mn - 12n^2$

37. $x^3y - 9xy^3$ **38.** $4u^3v - uv^3$

39. $3m^3 - 6m^2 + 15m$ **40.** $2x^3 - 2x^2 + 8x$

41. $m^3 + n^3$ **42.** $r^3 - t^3$

43. $c^3 - 1$ **44.** $a^3 + 1$

B *Factor completely relative to the integers. In polynomials involving more than three terms, try grouping the terms in various combinations as a first step. If a polynomial is prime relative to the integers, say so.*

45. $(a - b)^2 - 4(c - d)^2$ **46.** $(x + 2)^2 - 9y^2$

47. $2am - 3an + 2bm - 3bn$ **48.** $15ac - 20ad + 3bc - 4bd$

49. $3x^2 - 2xy - 4y^2$ **50.** $5u^2 + 4uv - 2v^2$

51. $x^3 - 3x^2 - 9x + 27$ **52.** $x^3 - x^2 - x + 1$

53. $a^3 - 2a^2 - a + 2$ **54.** $t^3 - 2t^2 + t - 2$

55. $4(A + B)^2 - 5(A + B) - 6$ **56.** $6(x - y)^2 + 23(x - y) - 4$

57. $m^4 - n^4$ **58.** $y^4 - 3y^2 - 4$

59. $s^4t^4 - 8st$ **60.** $27a^2 + a^5b^3$

61. $m^2 + 2mn + n^2 - m - n$ **62.** $y^2 - 2xy + x^2 - y + x$

Factor the following algebraic expressions by using the factoring formulas discussed in this section.

63. $\dfrac{4}{x^2} - \dfrac{9}{y^2}$ **64.** $u^2 - \dfrac{4}{v^2}$

65. $x^2 + 3x + \dfrac{9}{4}$ **66.** $x^2 - 5x + \dfrac{25}{4}$

67. $\dfrac{1}{x^3} - 8$

68. $27 + \dfrac{8}{t^3}$

69. $x^{2/3} - 3x^{1/3} + 2$

70. $9x^{2/3} - 4y^{2/5}$

71. $4x^{-2} - y^{-4}$

72. $2y^{-2} - y^{-1} - 3$

C *Factor completely to the integers. In polynomials that involve more than three terms, try grouping the terms in various combinations as a first step. Also try the "trick" illustrated in Example 27F where appropriate.*

73. $18a^3 - 8a(x^2 + 8x + 16)$

74. $25(4x^2 - 12xy + 9y^2) - 9a^2b^2$

75. $x^4 + 2x^2 + 1 - x^2$

76. $a^4 + 2a^2b^2 + b^4 - a^2b^2$

77. $16x^4 + 4x^2 + 1$

78. $a^4 + a^2b^2 + b^4$

79. $x^6 - 1$

80. $a^6 - b^6$

81. $x^4 + 6x^2y^2 + 25y^4$

82. $4a^4 + 8a^2b^2 + 9b^4$

Section 0-7 Fractions

■ Multiplication and Division
■ Addition and Subtraction
■ Complex Fractions

Algebraic fractions represent quotients, and for those replacements of the variables by real numbers that result in the quotient of real numbers, division by 0 excluded, the properties of real fractions discussed in Section 0-1 apply. In particular, we will very frequently be using the **fundamental principle of fractions**:

$$\frac{ak}{bk} = \frac{a}{b} \qquad b, k \neq 0$$

Using this principle from left to right to eliminate all common factors from a numerator and a denominator of a given fraction is referred to as **reducing a fraction to lowest terms**. We are actually dividing the numerator and denominator by the same nonzero common factor.

Using the principle from right to left—that is, multiplying a numerator and a denominator by the same nonzero factor—is referred to as **raising a fraction to higher terms**. We will use the principle in both directions in the material that follows.

A particular type of algebraic fraction, the quotient of two polynomials, is called a **rational expression**. We say that a rational expression is reduced to lowest terms if the numerator and denominator do not have any prime factors in common. (Unless stated to the contrary, *prime* will mean relative to the integers.)

EXAMPLE 28   Reduce each rational expression to lowest terms.

(A) $\dfrac{x^2 - 6x + 9}{x^2 - 9} = \dfrac{(x-3)^2}{(x-3)(x+3)}$ Factor numerator and denominator completely. Divide numerator and denominator by $(x-3)$—a valid operation as long as $x \neq 3$ and $x \neq -3$ as well.

$= \dfrac{x-3}{x+3}$

(B) $\dfrac{x^3 - 1}{x^2 - 1} = \dfrac{(x-1)(x^2 + x + 1)}{(x-1)(x+1)}$

$= \dfrac{x^2 + x + 1}{x + 1}$

[*Note:* Throughout our work on fractions, we will always assume without specific statement that variables are restricted to avoid division by 0.]

PROBLEM 28   Reduce to lowest terms.

(A) $\dfrac{6x^2 + x - 2}{2x^2 + x - 1}$ (B) $\dfrac{x^4 - 8x}{3x^3 - 2x^2 - 8x}$

■ Multiplication and Division

Since in each algebraic fraction we restrict variable replacements to real numbers that produce real fractions, multiplication and division of algebraic fractions follow the rules for multiplying and dividing fractions in real numbers; that is (excluding division by 0),

$$\frac{a}{b} \cdot \frac{c}{d} = \frac{ac}{bd} \qquad \frac{a}{b} \div \frac{c}{d} = \frac{a}{b} \cdot \frac{d}{c}$$

EXAMPLE 29   (A) $\dfrac{10x^3 y}{3xy + 9y} \cdot \dfrac{x^2 - 9}{4x^2 - 12x}$ Factor numerators and denominators; then divide any numerator and any denominator with a like common factor (this can be done either before or after multiplying numerators or denominators).

$= \dfrac{\overset{5x^2}{\cancel{10x^3 y}}}{\underset{3 \cdot 1}{\cancel{3y(x+3)}}} \cdot \dfrac{\overset{1 \cdot 1}{\cancel{(x-3)(x+3)}}}{\underset{2 \cdot 1}{\cancel{4x(x-3)}}}$

$= \dfrac{5x^2}{6}$

(B) $\dfrac{4 - 2x}{4} \div (x - 2) = \dfrac{\overset{1}{2(2 - x)}}{\underset{2}{\cancel{4}}} \cdot \dfrac{1}{x - 2}$ $x - 2$ is the same as $\dfrac{x - 2}{1}$.

$= \dfrac{2 - x}{2(x - 2)} = \dfrac{\overset{-1}{-\cancel{(x - 2)}}}{\underset{1}{2\cancel{(x - 2)}}}$ $b - a = -(a - b)$, a useful change in some problems.

$= -\frac{1}{2}$

(C) $\dfrac{2x^3 - 2x^2y + 2xy^2}{x^3y - xy^3} \div \dfrac{x^3 + y^3}{x^2 + 2xy + y^2}$

$= \dfrac{\overset{2}{2x}(\overset{1}{x^2 - xy + y^2})}{\underset{y}{xy}\underset{1}{(x + y)}(x - y)} \cdot \dfrac{\overset{1}{(x + y)^2}}{\underset{1}{(x + y)}\underset{1}{(x^2 - xy + y^2)}}$

$= \dfrac{2}{y(x - y)}$

PROBLEM 29   Perform the indicated operations and reduce to lowest terms.

(A) $\dfrac{12x^2y^3}{2xy^2 + 6xy} \cdot \dfrac{y^2 + 6y + 9}{3y^3 + 9y^2}$

(B) $(4 - x) \div \dfrac{x^2 - 16}{5}$

(C) $\dfrac{m^3 + n^3}{2m^2 + mn - n^2} \div \dfrac{m^3n - m^2n^2 + mn^3}{2m^3n^2 - m^2n^3}$

We will now use the fundamental principle of fractions to rationalize denominators and numerators in fractional expressions that involve radicals.

EXAMPLE 30   (A) Rationalize the denominator: $\dfrac{\sqrt{x} - \sqrt{y}}{\sqrt{x} + \sqrt{y}}$

Solution $\dfrac{\sqrt{x} - \sqrt{y}}{\sqrt{x} + \sqrt{y}} = \dfrac{\sqrt{x} - \sqrt{y}}{\sqrt{x} + \sqrt{y}} \cdot \dfrac{\sqrt{x} - \sqrt{y}}{\sqrt{x} - \sqrt{y}}$ Multiplying numerator and denominator by $(\sqrt{x} - \sqrt{y})$ eliminates radicals from the denominator, since $(a + b)(a - b) = a^2 - b^2$.

$= \dfrac{x - 2\sqrt{xy} + y}{x - y}$

(B) Rationalize the numerator: $\dfrac{\sqrt{x + h} - \sqrt{x}}{h}$

Solution

$$\frac{\sqrt{x + h} - \sqrt{x}}{h} = \frac{\sqrt{x + h} - \sqrt{x}}{h} \cdot \frac{\sqrt{x + h} + \sqrt{x}}{\sqrt{x + h} + \sqrt{x}}$$

$$= \frac{x \times h - x}{h(\sqrt{x + h} + \sqrt{x})}$$

$$= \frac{h}{h(\sqrt{x + h} + \sqrt{x})} = \frac{1}{\sqrt{x + h} + \sqrt{x}}$$

PROBLEM 30   (A) Rationalize the denominator: $\dfrac{\sqrt{m} + \sqrt{n}}{\sqrt{m} - \sqrt{n}}$

(B) Rationalize the numerator: $\dfrac{\sqrt{3 + h} - \sqrt{3}}{h}$

■   Addition and Subtraction

Because in each algebraic fraction we restrict variable replacements to real numbers that produce real fractions, addition and subtraction of algebraic fractions follow the rules for adding and subtracting fractions in real numbers; that is (excluding division by 0),

$$\frac{a}{b} + \frac{c}{b} = \frac{a + c}{b} \qquad \frac{a}{b} - \frac{c}{b} = \frac{a - c}{b}$$

Thus, we add algebraic fractions, if their denominators are the same, by adding or subtracting their numerators and placing the result over the common denominator. If the denominators are not the same, we raise the fractions to higher terms, using the fundamental principle of fractions to obtain common denominators, and then proceed as described.

Even though any common denominator will do, the problem will generally become less involved if the least common denominator (LCD)

The Least Common Denominator

The LCD of two or more rational expressions is found as follows:

1. Factor each denominator completely.
2. Form a product that contains each different factor from all denominators to the highest power it occurs in any one denominator. This product is the LCD.

is used. Often, the LCD is obvious, but if it is not, proceed as described in the box to find it.

EXAMPLE 31   Combine into a single fraction and reduce to lowest terms.

(A) $\dfrac{3}{10} + \dfrac{5}{6} - \dfrac{11}{45}$ (B) $\dfrac{4}{9x} - \dfrac{5x}{6y^2} + 1$

(C) $\dfrac{x + 3}{x^2 - 6x + 9} - \dfrac{x + 2}{x^2 - 9} - \dfrac{5}{3 - x}$

Solution   (A) To find the LCD, factor each denominator completely.

$$\left.\begin{array}{l} 10 = 2 \cdot 5 \\ 6 = 2 \cdot 3 \\ 45 = 3^2 \cdot 5 \end{array}\right\} \text{LCD} = 2 \cdot 3^2 \cdot 5 = 90$$

Now use the fundamental principle of fractions to make each denominator 90.

$$\frac{3}{10} + \frac{5}{6} - \frac{11}{45} = \frac{9 \cdot 3}{9 \cdot 10} + \frac{15 \cdot 5}{15 \cdot 6} - \frac{2 \cdot 11}{2 \cdot 45}$$

$$\boxed{= \frac{27}{90} + \frac{75}{90} - \frac{22}{90}}$$

$$= \frac{27 + 75 - 22}{90} = \frac{80}{90} = \frac{8}{9}$$

(B) $\left.\begin{array}{l} 9x = 3^2 x \\ 6y^2 = 2 \cdot 3y^2 \end{array}\right\} \text{LCD} = 2 \cdot 3^2 xy^2 = 18xy^2$

$$\frac{4}{9x} - \frac{5x}{6y^2} + 1 = \frac{2y^2 \cdot 4}{2y^2 \cdot 9x} - \frac{3x \cdot 5x}{3x \cdot 6y^2} + \frac{18xy^2}{18xy^2}$$

$$= \frac{8y^2 - 15x^2 + 18xy^2}{18xy^2}$$

(C) $\dfrac{x + 3}{x^2 - 6x + 9} - \dfrac{x + 2}{x^2 - 9} - \dfrac{5}{3 - x} = \dfrac{x + 3}{(x - 3)^2} - \dfrac{x + 2}{(x - 3)(x + 3)} + \dfrac{5}{x - 3}$

Note: $-\dfrac{5}{3 - x} = -\dfrac{5}{-(x - 3)} = \dfrac{5}{x - 3}$ We have again used the fact that $(a - b) = -(b - a)$.

The LCD $= (x - 3)^2(x + 3)$. Thus,

$$\frac{(x + 3)^2}{(x - 3)^2(x + 3)} - \frac{(x - 3)(x + 2)}{(x - 3)^2(x + 3)} + \frac{5(x - 3)(x + 3)}{(x - 3)^2(x + 3)}$$

$$= \frac{(x^2 + 6x + 9) - (x^2 - x - 6) + 5(x^2 - 9)}{(x - 3)^2(x + 3)}$$

$$= \frac{x^2 + 6x + 9 - x^2 + x + 6 + 5x^2 - 45}{(x - 3)^2(x + 3)}$$

$$= \frac{5x^2 + 7x - 30}{(x - 3)^2(x + 3)}$$

PROBLEM 31   Combine into a single fraction and reduce to lowest terms.

(A) $\dfrac{5}{28} - \dfrac{1}{10} + \dfrac{6}{35}$

(B) $\dfrac{1}{4x^2} - \dfrac{2x + 1}{3x^3} + \dfrac{3}{12x}$

(C) $\dfrac{y - 3}{y^2 - 4} - \dfrac{y + 2}{y^2 - 4y + 4} - \dfrac{2}{2 - y}$

■ Complex Fractions

A fractional form with fractions in its numerator, denominator, or both is called a **complex fraction**. It is often necessary to represent a complex fraction as a **simple fraction**—that is (in all cases we will consider), as the quotient of two polynomials. The process does not involve any new concepts. It is a matter of applying old concepts and processes in the right sequence. We will illustrate two approaches to the problem, each with its own merits, depending on the particular problem under consideration. One of the methods makes very effective use of the fundamental principle of fractions:

$$\frac{a}{b} = \frac{ka}{kb} \qquad b, k \neq 0$$

EXAMPLE 32   Express as a simple fraction: $\dfrac{\dfrac{y}{x^2} - \dfrac{x}{y^2}}{\dfrac{y}{x} - \dfrac{x}{y}}$

Solution   *Method 1:* Multiply the numerator and denominator by the LCD of all fractions in the numerator and denominator—in this case, x^2y^2.

$$\frac{x^2 y^2 \left(\dfrac{y}{x^2} - \dfrac{x}{y^2}\right)}{x^2 y^2 \left(\dfrac{y}{x} - \dfrac{x}{y}\right)} = \frac{y^3 - x^3}{xy^3 - x^3 y} = \frac{\overset{1}{\cancel{(y - x)}}(y^2 + xy + x^2)}{xy\underset{1}{\cancel{(y - x)}}(y + x)}$$

$$= \frac{y^2 + xy + x^2}{xy(y + x)} \quad \text{or} \quad \frac{x^2 + xy + y^2}{xy(x + y)}$$

Method 2: Write the numerator and denominator as single fractions. Then treat as a quotient.

$$\frac{\dfrac{y}{x^2} - \dfrac{x}{y^2}}{\dfrac{y}{x} - \dfrac{x}{y}} = \frac{\dfrac{y^3 - x^3}{x^2 y^2}}{\dfrac{y^2 - x^2}{xy}} = \frac{y^3 - x^3}{x^2 y^2} \div \frac{y^2 - x^2}{xy}$$

$$= \frac{\overset{1}{\cancel{(y - x)}}(y^2 + xy + x^2)}{\underset{xy}{x^2 y^2}} \cdot \frac{\overset{xy}{}}{\underset{1}{\cancel{(y - x)}}(y + x)}$$

$$= \frac{x^2 + xy + y^2}{xy(x + y)}$$

PROBLEM 32 Express as a simple fraction reduced to lowest terms. Use the two methods described in Example 32.

$$\frac{\dfrac{a}{b} - \dfrac{b}{a}}{\dfrac{a}{b} + 2 + \dfrac{b}{a}}$$

EXAMPLE 33 Express as simple fractions reduced to lowest terms.

$$\text{(A)} \quad \frac{x^{-2} - y^{-2}}{x^{-1} + y^{-1}} = \frac{\dfrac{1}{x^2} - \dfrac{1}{y^2}}{\dfrac{1}{x} + \dfrac{1}{y}} = \frac{x^2 y^2 \left(\dfrac{1}{x^2} - \dfrac{1}{y^2}\right)}{x^2 y^2 \left(\dfrac{1}{x} + \dfrac{1}{y}\right)}$$

$$= \frac{y^2 - x^2}{xy^2 + x^2 y} = \frac{(y - x)\overset{1}{\cancel{(y + x)}}}{xy\underset{1}{\cancel{(y + x)}}}$$

$$= \frac{y - x}{xy}$$

(B) $2 - \cfrac{1}{2 - \cfrac{2}{2 + \cfrac{1}{x}}} = 2 - \cfrac{1}{2 - \cfrac{x \cdot 2}{x\left(2 + \cfrac{1}{x}\right)}}$ Write $\cfrac{2}{2 + \cfrac{1}{x}}$ as a simple fraction first.

$$= 2 - \cfrac{1}{2 - \cfrac{2x}{2x + 1}}$$

$$= 2 - \cfrac{(2x + 1) \cdot 1}{(2x + 1)\left(2 - \cfrac{2x}{2x + 1}\right)}$$

$$= 2 - \frac{2x + 1}{4x + 2 - 2x}$$

$$= 2 - \frac{2x + 1}{2x + 2} = \frac{4x + 4 - 2x - 1}{2x + 2}$$

$$= \frac{2x + 3}{2x + 2}$$

PROBLEM 33   Express as simple fractions reduced to lowest terms.

(A) $\dfrac{x - x^{-1}}{1 - x^{-2}}$ (B) $2 - \cfrac{1}{2 - \cfrac{1}{1 - \cfrac{1}{x}}}$

Answers to Matched Problems

28. (A) $(3x + 2)/(x + 1)$ (B) $(x^2 + 2x + 4)/(3x + 4)$
29. (A) $2x$ (B) $-5/(x + 4)$ (C) mn
30. (A) $(m + 2\sqrt{mn} + n)/(m - n)$ (B) $1/(\sqrt{3 + h} + \sqrt{3})$
31. (A) $\frac{1}{4}$ (B) $(3x^2 - 5x - 4)/12x^3$
        (C) $(2y^2 - 9y - 6)/[(y - 2)^2(y + 2)]$
32. $(a - b)/(a + b)$
33. (A) x (B) $(x - 3)/(x - 2)$

Exercise 0-7 ■ **A** *Perform the indicated operations and reduce to lowest terms. Represent all complex fractions as simple fractions reduced to lowest terms.*

1. $\left(\dfrac{d^5}{3a} \div \dfrac{d^2}{6a^2}\right) \cdot \dfrac{a}{4d^3}$ **2.** $\dfrac{d^5}{3a} \div \left(\dfrac{d^2}{6a^2} \cdot \dfrac{a}{4d^3}\right)$

3. $\dfrac{2y}{18} - \dfrac{-1}{28} - \dfrac{y}{42}$

4. $\dfrac{x^2}{12} + \dfrac{x}{18} - \dfrac{1}{30}$

5. $\dfrac{3x + 8}{4x^2} - \dfrac{2x - 1}{x^3} - \dfrac{5}{8x}$

6. $\dfrac{4m - 3}{18m^3} + \dfrac{3}{4m} - \dfrac{2m - 1}{6m^2}$

7. $\dfrac{2x^2 + 7x + 3}{4x^2 - 1} \div (x + 3)$

8. $\dfrac{x^2 - 9}{x^2 - 3x} \div (x^2 - x - 12)$

9. $\dfrac{m + n}{m^2 - n^2} \div \dfrac{m^2 - mn}{m^2 - 2mn + n^2}$

10. $\dfrac{x^2 - 6x + 9}{x^2 - x - 6} \div \dfrac{x^2 + 2x - 15}{x^2 + 2x}$

11. $\dfrac{1}{a^2 - b^2} + \dfrac{1}{a^2 + 2ab + b^2}$

12. $\dfrac{3}{x^2 - 1} - \dfrac{2}{x^2 - 2x + 1}$

13. $m - 3 - \dfrac{m - 1}{m - 2}$

14. $\dfrac{x + 1}{x - 1} - 1$

15. $\dfrac{5}{x - 3} - \dfrac{2}{3 - x}$

16. $\dfrac{3}{a - 1} - \dfrac{2}{1 - a}$

17. $\dfrac{2}{y + 3} - \dfrac{1}{y - 3} + \dfrac{2y}{y^2 - 9}$

18. $\dfrac{2x}{x^2 - y^2} + \dfrac{1}{x + y} - \dfrac{1}{x - y}$

19. $\dfrac{1 - \dfrac{y^2}{x^2}}{1 - \dfrac{y}{x}}$

20. $\dfrac{1 + \dfrac{3}{x}}{x - \dfrac{9}{x}}$

21. $\dfrac{m^{-1} + 1}{m + 1}$

22. $\dfrac{m^{-2} - 1}{m^{-1} + 1}$

B

23. $\dfrac{y}{y^2 - y - 2} - \dfrac{1}{y^2 + 5y - 14} - \dfrac{2}{y^2 + 8y + 7}$

24. $\dfrac{x^2}{x^2 + 2x + 1} + \dfrac{x - 1}{3x + 3} - \dfrac{1}{6}$

25. $\dfrac{9 - m^2}{m^2 + 5m + 6} \cdot \dfrac{m + 2}{m - 3}$

26. $\dfrac{2 - x}{2x + x^2} \cdot \dfrac{x^2 + 4x + 4}{x^2 - 4}$

27. $\dfrac{x + 7}{ax - bx} + \dfrac{y + 9}{by - ay}$

28. $\dfrac{c + 2}{5c - 5} - \dfrac{c - 2}{3c - 3} + \dfrac{c}{1 - c}$

29. $\dfrac{x^2 - 16}{2x^2 + 10x + 8} \div \dfrac{x^2 - 13x + 36}{x^3 + 1}$

30. $\left(\dfrac{x^3 - y^3}{y^3} \cdot \dfrac{y}{x - y} \right) \div \dfrac{x^2 + xy + y^2}{y^2}$

31. $\dfrac{x^2 - xy}{xy + y^2} \div \left(\dfrac{x^2 - y^2}{x^2 + 2xy + y^2} \div \dfrac{x^2 - 2xy + y^2}{x^2y + xy^2} \right)$

32. $\left(\dfrac{x^2 - xy}{xy + y^2} \div \dfrac{x^2 - y^2}{x^2 + 2xy + y^2}\right) \div \dfrac{x^2 - 2xy + y^2}{x^2y + xy^2}$

33. $\left(\dfrac{x}{x^2 - 16} - \dfrac{1}{x + 4}\right) \div \dfrac{4}{x + 4}$

34. $\left(\dfrac{3}{x - 2} - \dfrac{1}{x + 1}\right) \div \dfrac{x + 4}{x - 2}$

35. $\dfrac{x^{-1} + y^{-1}}{x + y}$

36. $\dfrac{c - d}{c^{-1} - d^{-1}}$

37. $\dfrac{1 + \dfrac{2}{x} - \dfrac{15}{x^2}}{1 + \dfrac{4}{x} - \dfrac{5}{x^2}}$

38. $\dfrac{\dfrac{x}{y} - 2 + \dfrac{y}{x}}{\dfrac{x}{y} - \dfrac{y}{x}}$

39. $\dfrac{xy^{-2} - yx^{-2}}{y^{-1} - x^{-1}}$

40. $\dfrac{b^{-2} - c^{-2}}{b^{-3} - c^{-3}}$

41. $\dfrac{y - \dfrac{y^2}{y - x}}{1 + \dfrac{x^2}{y^2 - x^2}}$

42. $\dfrac{\dfrac{s^2}{s - t} - s}{\dfrac{t^2}{s - t} + t}$

43. $2 - \dfrac{1}{1 - \dfrac{2}{a + 2}}$

44. $1 - \dfrac{1}{1 - \dfrac{1}{1 - \dfrac{1}{x}}}$

45. $\left(\dfrac{-b + \sqrt{b^2 - 4ac}}{2a}\right)\left(\dfrac{-b - \sqrt{b^2 - 4ac}}{2a}\right)$

46. $\dfrac{-b + \sqrt{b^2 - 4ac}}{2a} + \dfrac{-b - \sqrt{b^2 - 4ac}}{2a}$

Rationalize denominators. [Hint: $(a - b)(a + b) = a^2 - b^2$.]

47. $\dfrac{3 - \sqrt{a}}{\sqrt{a} - 2}$

48. $\dfrac{2 + \sqrt{x}}{\sqrt{x} - 3}$

49. $\dfrac{2\sqrt{5} - 3\sqrt{2}}{5\sqrt{5} + 2\sqrt{2}}$

50. $\dfrac{3\sqrt{2} + 2\sqrt{3}}{2\sqrt{2} - 3\sqrt{3}}$

51. $\dfrac{x^2}{\sqrt{x^2 + 9} - 3}$

52. $\dfrac{-y^2}{2 - \sqrt{y^2 + 4}}$

Rationalize numerators.

53. $\dfrac{\sqrt{t} - \sqrt{x}}{t - x}$

54. $\dfrac{\sqrt{x} - \sqrt{y}}{\sqrt{x} + \sqrt{y}}$

55. $\dfrac{\sqrt{x + h} - \sqrt{x}}{h}$

56. $\dfrac{\sqrt{2 + h} + \sqrt{2}}{h}$

C *Combine into single terms. First write each radical in simplest radical form.*

57. $\sqrt{\dfrac{3xy}{2}} + \sqrt{\dfrac{2xy}{3}}$

58. $\sqrt{\dfrac{5a}{8}} - \sqrt{\dfrac{2a}{5}}$

59. $\sqrt{50} - \dfrac{2}{\sqrt{2}} + \sqrt{\dfrac{9}{2}}$

60. $\sqrt{\dfrac{1}{3}} - \dfrac{2}{\sqrt{3}} + \sqrt{12}$

Write as simple fractions.

61. $\left(\dfrac{x^{-1}}{x^{-1} - y^{-1}}\right)^{-1}$

62. $\left[\dfrac{u^{-2} - v^{-2}}{(u^{-1} - v^{-1})^2}\right]^{-1}$

63. $1 - \cfrac{1}{1 - \cfrac{1}{1 - \cfrac{1}{1 - \cfrac{1}{x}}}}$

64. $1 + \cfrac{1}{1 + \cfrac{1}{1 + \cfrac{1}{1 + x}}}$

Rationalize the numerators. [*Hint:* $(a - b)(a^2 + ab + b^2) = a^3 - b^3$.]

65. $\dfrac{\sqrt[3]{t} - \sqrt[3]{x}}{t - x}$

66. $\dfrac{\sqrt[3]{x + h} - \sqrt[3]{x}}{h}$

Section 0-8   Chapter Review

Important Terms
and Symbols

0-1 Algebra and Real Numbers.   The real number system, natural numbers, integers, rational numbers, irrational numbers, commutative property, associative property, identities, inverses, distributive property, subtraction, division, properties of negatives, zero properties, fraction properties, numerator, denominator, N, Z, Q, R

0-2 Integer Exponents.   Exponent, base, positive integer exponent, zero exponent, negative integer exponent, five exponent properties, scientific notation, b^n, b^{-n}, b^0

0-3 Rational Exponents.   Square root, cube root, nth root, principal nth root, rational exponent, properties of rational exponents, $b^{1/n}$, $b^{m/n}$, $b^{-m/n}$

0-4 Radicals.   Radical, index, radicand, properties of radicals, simplest radical form, rationalizing denominators, absolute value, $\sqrt[n]{b}$, $|a|$, $\sqrt[n]{x^n}$

0-5 Algebraic Expressions, Basic Operations.   Algebraic expressions; polynomials; degree of a term; degree of a polynomial; polynomial of degree 0; monomial; binomial; trinomial; adding, subtracting, and multiplying polynomials

0-6 Factoring.   Factor, prime number, composite number, positive integer, factored completely, fundamental theorem of arithmetic, prime polynomial, polynomial factored completely, common factors, factoring by grouping, special factoring formulas

0-7 Fractions.   Fundamental principle of fractions, lowest terms, canceling, higher terms, rational expression, multiplication, division, rationalize denominator, rationalize numerator, addition and subtraction, least common denominator, LCD, complex fraction, simple fraction

Exercise 0-8   Chapter Review

Work through all the problems in this chapter review and check answers in the back of the book. (Answers to all problems are there, and following each answer is a number in italics indicating the section in which that type of problem is discussed.) Where weaknesses show up, review appropriate sections in the text. When you are satisfied that you know the material, take the practice test following this review.

A   *Simplify Problems 1–6, and write answers using positive exponents only. All variables represent positive real numbers.*

1. $6(xy^3)^5$ **2.** $\dfrac{9u^8v^6}{3u^4v^8}$ **3.** $(2 \times 10^5)(3 \times 10^{-3})$

4. $(x^{-3}y^2)^{-2}$ **5.** $u^{5/3}u^{2/3}$ **6.** $(9a^4b^{-2})^{1/2}$

7. Change to radical form: $3x^{2/5}$
8. Change to rational exponent form: $-3\sqrt[3]{(xy)^2}$

Simplify and express in simplest radical form. All variables represent positive real numbers.

9. $3x\sqrt[3]{x^5y^4}$ **10.** $\sqrt{2x^2y^5}\sqrt{18x^3y^2}$ **11.** $\dfrac{6ab}{\sqrt{3a}}$

12. $\sqrt{7} + 2\sqrt{3} - 4\sqrt{3}$ **13.** $\dfrac{\sqrt{5}}{3 - \sqrt{5}}$ **14.** $\sqrt[8]{y^6}$

Given the polynomials $3x - 4$, $x + 2$, $3x^2 + x - 8$, and $x^3 + 8$.

15. Add all four.
16. Subtract the sum of the first and third from the sum of the second and fourth.
17. Multiply the third and fourth.

Write each polynomial in a completely factored form relative to the integers. If the polynomial is prime relative to the integers, say so.

18. $9x^2 - 12x + 4$ **19.** $t^2 - 4t - 6$ **20.** $6n^3 - 9n^2 - 15n$

Perform the indicated operations and reduce to lowest terms. Represent all complex fractions as simple fractions reduced to lowest terms.

21. $\dfrac{2}{5b} - \dfrac{4}{3a^3} - \dfrac{1}{6a^2b^2}$ **22.** $\dfrac{3x}{3x^2 - 12x} + \dfrac{1}{6x}$

23. $\dfrac{y - 2}{y^2 - 4y + 4} \div \dfrac{y^2 + 2y}{y^2 + 4y + 4}$ **24.** $\dfrac{u - \dfrac{1}{u}}{1 - \dfrac{1}{u^2}}$

B   In Problems 25–30, each statement illustrates the use of one of the following real number properties or definitions. Indicate which.

Commutative	Identity	Division
Associative	Inverse	Zero
Distributive	Subtraction	Negatives

25. $(-3) - (-2) = (-3) + [-(-2)]$

26. $3y + (2x + 5) = (2x + 5) + 3y$

27. $(2x + 3)(3x + 5) = (2x + 3)3x + (2x + 3)5$

28. $3 \cdot (5x) = (3 \cdot 5)x$

29. $\dfrac{a}{-(b - c)} = -\dfrac{a}{b - c}$ **30.** $3xy + 0 = 3xy$

31. Indicate true (T) or false (F):

(A) An integer is a rational number and a real number.

(B) An irrational number has a repeating decimal representation.

32. Give an example of an integer that is not a natural number.

33. Given the algebraic expressions

(a) $2x^2 - 3x + 5$ (b) $x^2 - \sqrt{x - 3}$

(c) $x^{-3} + x^{-2} - 3x^{-1}$ (d) $x^2 - 3xy - y^2$

(A) Identify all second-degree polynomials.

(B) Identify all third-degree polynomials.

34. Simplify: $-2x\{(x^2 + 2)(x - 3) - x[x - x(3 - x)]\}$

Write in a completely factored form relative to the integers.

35. $(4x - y)^2 - 9x^2$ **36.** $2x^2 + 4xy - 5y^2$

37. $6x^3y + 12x^2y^2 - 15xy^3$ **38.** $(y - b)^2 - y + b$

39. $3x^3 + 24y^3$ **40.** $y^3 + 2y^2 - 4y - 8$

Perform the indicated operations and reduce to lowest terms. Represent all complex fractions as simple fractions reduced to lowest terms.

41. $\dfrac{m-1}{m^2-4m+4}+\dfrac{m+3}{m^2-4}+\dfrac{2}{2-m}$

42. $\dfrac{y}{x^2}\div\left(\dfrac{x^2+3x}{2x^2+5x-3}\div\dfrac{x^3y-x^2y}{2x^2-3x+1}\right)$

43. $\dfrac{1-\dfrac{1}{1+\dfrac{x}{y}}}{1-\dfrac{1}{1-\dfrac{x}{y}}}$ $\dfrac{x}{x+y}$ $\dfrac{-x}{y-x}$

44. $\dfrac{a^{-1}-b^{-1}}{ab^{-2}-ba^{-2}}$

Perform the indicated operations, simplify, and write answers using positive exponents only. All variables represent positive real numbers.

45. $\left(\dfrac{8u^{-1}}{2^2u^2v^0}\right)^{-2}\left(\dfrac{u^{-5}}{u^{-3}}\right)^3$

46. $\dfrac{5^0}{3^2}+\dfrac{3^{-2}}{2^{-2}}$

47. $\left(\dfrac{27x^2y^{-3}}{8x^{-4}y^3}\right)^{1/3}$

48. $(a^{-1/3}b^{1/4})(9a^{1/3}b^{-1/2})^{3/2}$

49. $(x^{1/2}+y^{1/2})^2$

50. $(x^2y^3)^{1/5}[(x^3y^{-3})^{1/5}-(x^{-2}y^2)^{1/5}]$

51. Convert to scientific notation and simplify:

$$\dfrac{0.000\ 000\ 000\ 52}{(1,300)(0.000\ 002)}$$

Perform the indicated operations and express answers in simplest radical form. All radicands represent positive real numbers.

52. $-2x\sqrt[5]{3^6x^7y^{11}}$

53. $\dfrac{2x^2}{\sqrt[3]{4x}}$

54. $\sqrt[5]{\dfrac{3y^2}{8x^2}}$

55. $\sqrt[9]{8x^6y^{12}}$

56. $\sqrt[3]{3}-\dfrac{6}{\sqrt[3]{9}}+\sqrt[3]{\dfrac{1}{9}}$

57. $(2\sqrt{x}-5\sqrt{y})(\sqrt{x}+\sqrt{y})$

58. $\dfrac{y^2}{\sqrt{y^2+4}-2}$

59. $\dfrac{3\sqrt{x}}{2\sqrt{x}-\sqrt{y}}$

60. $\sqrt{\sqrt[3]{4x^4}}$

61. Evaluate x^2-4x+1 for $x=2-\sqrt{3}$.

C **62.** Write 0.545 454 54 . . . in the form a/b, reduced to lowest terms, where a and b are positive integers.

63. Simplify: $\left(x - \dfrac{1}{1 - \dfrac{1}{x}}\right) \div \left(\dfrac{x}{x + 1} - \dfrac{x}{1 - x}\right)$

Simplify and express answers using positive exponents only. (m is an integer greater than 1.)

64. $\dfrac{8(x - 2)^{-3}(x + 3)^2}{12(x - 2)^{-4}(x + 3)^{-2}}$

65. $\left(\dfrac{a^{-2}}{b^{-1}} + \dfrac{b^{-2}}{a^{-1}}\right)^{-1}$

66. $(x^{1/3} - y^{1/3})(x^{2/3} + x^{1/3}y^{1/3} + y^{2/3})$

67. $\left(\dfrac{x^{m^2}}{x^{2m-1}}\right)^{1/(m-1)}, \quad m > 1$

Perform the indicated operations and express the answer in simplest radical form. All radicands represent positive real numbers.

68. $\sqrt{2xy}\,\sqrt[3]{4x^2y^2}$

69. $\dfrac{\sqrt{2xy}}{\sqrt[3]{2xy}}$

70. $\sqrt[3]{8\sqrt{64\sqrt{xy}}}$

71. $\sqrt[(n+1)]{x^{n^2}x^{2n+1}}, \quad n > 0$

72. Simplify $4\sqrt[5]{x^5} - 2\sqrt[4]{x^4}$
   (A) For x positive   (B) For x negative

73. Rationalize the numerator: $\dfrac{\sqrt{5 + h} - \sqrt{5}}{h}$

Factor using integer coefficients.

74. $4x^{-2/5} - 25y^{2/3}$

75. $25x^2y^4 - 4(9x^2 + 12xy + 4y^2)$

Practice Test Chapter 0

Take this practice test as if it were a graded test. Allow yourself up to 50 minutes. Work the problems without looking back in the chapter. Correct your work using the answers (keyed to appropriate sections) in the back of the book.

1. Each statement illustrates the use of one of the following real number properties or definitions. Indicate which.

Commutative	Identity	Division
Associative	Inverse	Zero
Distributive	Subtraction	Negatives

(A) $(xy)z = z(xy)$
(B) $(a + b)(x + y) = (a + b)x + (a + b)y$
(C) $(-3) - (-7) = (-3) + [-(-7)]$
(D) $1(7x + 3y) = 7x + 3y$
(E) $(2x + 3) + (3x + 2) = 2x + [3 + (3x + 2)]$

2.   Indicate true (T) or false (F).
(A)   A natural number is an integer and a real number.
(B)   A rational number has a nonrepeating decimal representation.

Perform the indicated operations and simplify.

3. $(3x - 2)(x + 3) - 2[3x^2 - 2x(4 - 3x)]$

4. $\dfrac{2}{3 - x} - \dfrac{5 - 6x}{x^2 + x - 12}$

5. $\dfrac{x^4 + 4x^3 + 4x^2}{x^2 - x - 6} \div \dfrac{7x^4 + 56x}{x^5 - 2x^4 + 4x^3}$

6. $\dfrac{4 + \dfrac{6}{x + 3}}{\dfrac{5}{x - 3} - \dfrac{1}{x + 3}}$

Factor completely relative to the integers.

7. $8ax - 2ay - 12bx + 3by$ 8. $16x^4 - 8x^2y^2 + y^4$

9. $x^3 + 5x^2 - 25x - 125$

Simplify and write answers using positive exponents only.

10. $\left(\dfrac{64x^{-5}y^2}{27x^4y^{-3}}\right)^{1/3}$ 11. $(x^{1/2}y^{-1/3})(8x^{-1/2}y^2)^{2/3}$

12. $\left(\dfrac{y^{-1} - x^{-1}}{xy^{-1} + yx^{-1}}\right)^{-1}$

Simplify and write in simplest radical form.

13. $\sqrt[3]{12x^5y^4}\,\sqrt[3]{2x^3y^5}$ 14. $\dfrac{21a^4b^3}{\sqrt[3]{7ab^2}}$

15. $\dfrac{3\sqrt{a}}{\sqrt{a} - 2\sqrt{b}}$ 16. $\dfrac{1}{\sqrt[3]{4}} + \sqrt[3]{16}$

17. $\sqrt[3]{\sqrt{8x^3}}$ 18. $\sqrt[3]{4x^5y^2}\,\sqrt{2x^3y^3}$

19. Convert to scientific notation and simplify:

$$\frac{(0.000\ 000\ 0009)(500{,}000)}{0.000\ 015}$$

20. Simplify $5 \sqrt[3]{x^3} - 3 \sqrt{x^2}$
(A) For x positive   (B) For x negative

Equations and Inequalities

1

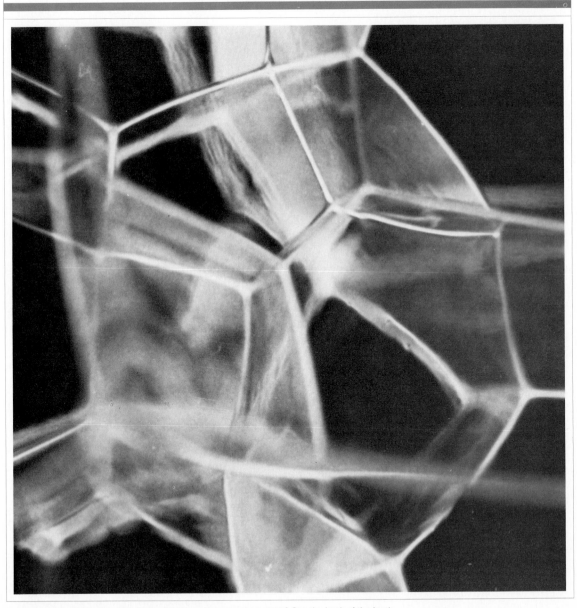

A natural design of mathematical interest. Can you guess the source? See the back of the book.

Chapter 1 ■ Equations and Inequalities

Section 1-1   Sets

- ■ Set Notation
- ■ Subsets and Equality
- ■ Set Operations and Venn Diagrams

George Cantor (1845–1918), when about thirty, created a new mathematical concept, the set, and subsequently developed a theory of sets. This new theory, an outgrowth of his studies on infinity, has become a milestone in the development of mathematics. We will use only a few key ideas and symbols from this theory, ideas and symbols that will help us discuss certain mathematical developments with increased clarity and precision.

■ Set Notation

We can think of a **set** as any collection of objects that is **well defined**; that is, the collection is specified in such a way that we can tell whether any given object is or is not in the collection. In this course we will usually be interested in certain sets of numbers. Capital letters, such as A, B, and C, are often used to represent sets. For example,

$$A = \{1, 3, 5\} \qquad B = \{2, 5, 6\}$$

specify sets A and B. Each object in a set is called an **element** or **member** of the set. Symbolically:

$a \in A$ means   "a is an element of set A."

$a \notin A$ means   "a is not an element of set A."

Referring to sets A and B, we see that

$3 \in A$ 3 is an element of set A.

$4 \notin A$ 4 is not an element of set B.

A set without any elements is called the **empty** or **null set**. For example, the set of all solutions to the equation $x + 8 = x + 2$ is the empty set. Symbolically:

$\varnothing$ represents the empty or null set

A set is usually described in one of two ways:

1. By **listing** the elements between braces $\{\ \}$: $\{1, 3, 5\}$.
2. By enclosing a **rule** within braces $\{\ \}$ that determines the elements of the set:

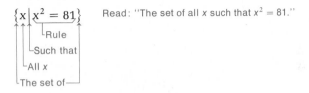

Read: "The set of all x such that $x^2 = 81$."

EXAMPLE 1   Let $D =$ The set of all numbers x such that $x^2 = 4$. Set D may be described by either the listing or the rule method:

Listing method: $D = \{-2, 2\}$
Rule method: $D = \{x \mid x^2 = 4\}$

PROBLEM 1   Let $G =$ The set of all numbers x such that $x^2 = 64$.

(A) Write G using the listing method.
(B) Write G using the rule method.
(C) Indicate true (T) or false (F).

$\qquad$ $4 \in G$ $\qquad$ $8 \in G$ $\qquad$ $16 \notin G$

■ Subsets and Equality

If each element of set A is also an element of set B, we say that A is a **subset** of set B. The set of all women in a class is a subset of the whole class. (Note that the definition of a subset allows a set to be a subset of itself.) If two sets have exactly the same elements (the order of listing does not matter), the sets are said to be **equal**. Set A is equal to set B if and only if A is a subset of B and B is a subset of A.

Symbolically,

Subsets
$A \subset B$ means "A is a subset of B." $\{3, 5\} \subset \{3, 5, 7\}$ $A = B$ means "A is equal to B." $\{4, 6\} = \{6, 4\}$

It is useful and interesting to note that

$\varnothing$ is a subset of every set.

It is certainly true that every element of $\varnothing$ is an element of any given set, since $\varnothing$ has no elements.

EXAMPLE 2   Let $A = \{-3, 0, 5\}$, $B = \{0, 5, -3\}$, and $C = \{0, 5\}$. Then each of the following statements is true.

$$C \subset A \qquad C \subset B \qquad A = B$$
$$A \subset B \qquad \varnothing \subset A \qquad A \neq C$$

PROBLEM 2   Let $M = \{-4, 6\}$, $N = \{6, -4\}$, and $P = \{-4\}$. Indicate true (T) or false (F).

(A) $M = N$ (B) $P \subset N$ (C) $N \neq P$
(D) $N \subset M$ (E) $\varnothing \subset P$ (F) $M \subset P$

■   Set Operations and Venn Diagrams

The **union** of sets A and B, denoted by $A \cup B$, is the set of all elements formed by placing all the elements of A and all the elements of B into one set (the same element is not repeated). Symbolically,

Union
$A \cup B = \{x \mid x \in A \quad \textbf{or} \quad x \in B\}$

Here we use the word *or* in the way it is most frequently used in mathematics; that is, x may be an element of set A or set B or both.

Venn diagrams are useful aids in visualizing set relationships. The union of two sets can be illustrated as shown in Figure 1.

The **intersection** of sets A and B, denoted by $A \cap B$, is the set of elements in set A that are also in set B. Symbolically,

Intersection
$A \cap B = \{x \mid x \in A \ \textbf{and} \ x \in B\}$

This relationship is easily visualized in the Venn diagram shown in Figure 2.

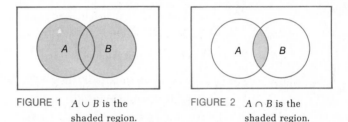

FIGURE 1 $A \cup B$ is the shaded region.

FIGURE 2 $A \cap B$ is the shaded region.

If $A \cap B = \varnothing$, then the sets A and B are said to be **disjoint**; this is illustrated in Figure 3.

The set of all elements under consideration is called the **universal set** U. Once the universal set is determined for a particular discussion, all other sets in that discussion must be subsets of U. We now define one more operation on sets called the *complement*. The **complement** of A (relative to U), denoted by A', is the set of elements in U that are not in A (see Fig. 4). Symbolically,

Complement
$A' = \{x \in U \mid x \notin A\}$

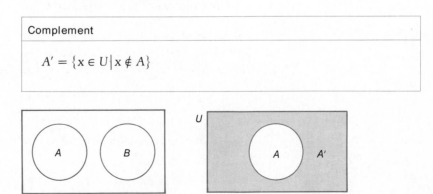

FIGURE 3 $A \cap B = \varnothing$

FIGURE 4   The complement of A is A'.

EXAMPLE 3   If $A = \{4, 5, 7\}$, $B = \{3, 6, 9\}$, and $C = \{3, 4, 5, 6, 7\}$, then

$B \cup C = \{3, 4, 5, 6, 7, 9\}$ Common elements are listed only once.

$B \cap C = \{3, 6\}$

$A \cap B = \varnothing$ A and B are disjoint.

A' relative to C is $\{3, 6\}$

PROBLEM 3   If $M = \{1, 2, 3, 4\}$, $N = \{1, 3, 5, 7\}$, and $Q = \{2, 4\}$, find:

(A) $M \cup N$ (B) $M \cap N$ (C) $N \cap Q$ (D) Q' relative to M

EXAMPLE 4   From a survey involving 100 college students, a marketing research company found that seventy-three students owned stereos, fifty-four owned bicycles, and forty-one owned bicycles and stereos.

(A) How many students owned either a stereo or a bicycle?
(B) How many students owned neither a bicycle nor a stereo?

Solution   Venn diagrams are very useful for this type of problem. If we let

U = Set of students in the sample (100)

S = Set of students who own stereos (73)

B = Set of students who own bicycles (54)

$B \cap S$ = Set of students who own bicycles and stereos (41)

then

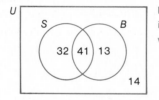

Place the number in the intersection first, then work outward.

(A) The number of students who own either a bicycle or a stereo is the number of students in the set $B \cup S$. You might be tempted to say that this is just the number of students in B plus the number of students in S, $54 + 73 = 127$, but this sum is larger than the sample we started with! What is wrong? We have actually counted the number in the intersection (41) twice. The correct answer, as seen in the Venn diagram, is

$$13 + 41 + 32 = 86 \quad \text{or} \quad 54 + 73 - 41 = 86$$

(B) The number of students who own neither a bicycle nor a stereo is the number of students in the set $(B \cup S)'$—that is, 14.

PROBLEM 4   Refer to Example 4.

(A) How many students own a bicycle but not a stereo?

(B) How many students do not own both a bicycle and a stereo?

Answers to Matched Problems   **1.** (A) $\{-8, 8\}$ (B) $\{x \mid x^2 = 64\}$ (C) F, T, T

2. (A) T   (B) T   (C) T   (D) T   (E) T   (F) F

3. (A) $\{1, 2, 3, 4, 5, 7\}$ (B) $\{1, 3\}$ (C) $\varnothing$ (D) $\{1, 3\}$

4. (A) 13   (B) 59

Exercise 1-1 ■ A   *Indicate true (T) or false (F)*

1. $5 \in \{2, 3, 5\}$ **2.** $7 \notin \{2, 3, 5\}$

3. $\{3, 5\} \subset \{2, 3, 5\}$ **4.** $\{2, 3, 5\} = \{5, 2, 3\}$

5. $\{2, 3, 5\} \subset \{5, 2, 3\}$ **6.** $\{3, 5\} \in \{2, 3, 5\}$

7. $\varnothing \subset \{2, 3, 5\}$ **8.** $\varnothing = \{0\}$

In Problems 9–14 write the resulting sets using the listing method.

9. $\{3, 5, 7\} \cup \{4, 5, 6\}$ **10.** $\{1, 3, 5\} \cup \{2, 3, 4\}$

11. $\{3, 5, 7\} \cap \{4, 5, 6\}$ **12.** $\{1, 3, 5\} \cap \{2, 3, 4\}$

13. $\{3, 5, 7\} \cap \{4, 6, 8\}$ **14.** $\{3, 4, 7, 8\} \cap \{5, 6, 9\}$

B   *In Problems 15–20 write the resulting sets using the listing method.*

15. $\{x \mid x - 8 = 0\}$ **16.** $\{x \mid x + 3 = 0\}$

17. $\{x \mid x^2 = 49\}$ **18.** $\{x \mid x^2 = 1\}$

19. $\{x \mid x$ is a prime number between 1 and 9, inclusive$\}$

20. $\{x \mid x$ is a composite number between 1 and 9, inclusive$\}$

21. For $U = \{-2, -1, 0, 1, 2\}$ and $M = \{-1, 0\}$, find M'.

22. For $U = \{1, 2, 3, 4, 5, 6, 7\}$ and $N = \{2, 4, 6\}$, find N'.

Problems 23–34 refer to the accompanying Venn diagram. How many elements are in the indicated sets?

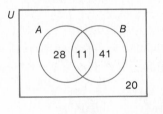

23. U **24.** B **25.** A' **26.** B'

27. $A \cup B$ **28.** $A \cap B$ **29.** $A' \cap B$ **30.** $A \cap B'$

31. $(A \cap B)'$ **32.** $(A \cup B)'$ **33.** $A' \cap B'$ **34.** U'

35. If $M = \{2, 3, 4, 5\}$ and $N = \{4, 5, 6, 7\}$, find:
(A) $\{x \mid x \in M \quad \text{or} \quad x \in N\}$ (B) $M \cup N$

36. For sets M and N in Problem 35, find:
(A) $\{x \mid x \in M \quad \text{and} \quad x \in N\}$ (B) $M \cap N$

37. For $A = \{1, 2, 3, 4\}$, $B = \{2, 4, 6\}$, and $C = \{3, 4, 5, 6\}$, find $A \cup (B \cap C)$.

38. For sets A, B, and C in Problem 37, find $A \cap (B \cup C)$.

C   *Venn diagrams may be of help in Problems 39–44.*

39. If $M \cap N = N$, can we always conclude that $N \subset M$?

40. If $M \cup N = N$, can we always conclude that $M \subset N$?

41. If $M \cap N = \varnothing$, can we always conclude that $M = \varnothing$ or $N = \varnothing$?

42. If M and N are arbitrary sets, can we always conclude that $M \cap N \subset N$?

43. If $M \subset N$ and $x \in N$, can we always conclude that $x \in M$?

44. If $M \subset N$ and $x \in M$, can we always conclude that $x \in N$?

45. How do the sets $\varnothing$, $\{0\}$, and $\{\varnothing\}$ differ from one another?

46. How many subsets does each of the following sets have?
(A) $\{a\}$ (B) $\{a, b\}$ (C) $\{a, b, c\}$
(D) A set with n elements

APPLICATIONS

Business and Economics

Problems 47–58 refer to the following marketing survey: From a random sample of 1,000 students on a college campus, it was found that 720 owned tape cassettes, 670 owned records, and 540 owned both. Let

$C = $ *Set of students in the sample who owned cassettes*
$R = $ *Set of students in the sample who owned records*

Following the procedures in Example 4, find the number of students in each set.

47. $C \cup R$ **48.** $C \cap R$ **49.** $(C \cup R)'$

50. $(C \cap R)'$ **51.** $C' \cap R$ **52.** $C \cap R'$

53. Set of students who owned either cassettes or records

54. Set of students who owned both cassettes and records

55. Set of students who owned neither cassettes nor records

56. Set of students who did not own both cassettes and records

57. Set of students who owned records but no cassettes

58. Set of students who owned cassettes but no records

Medicine—Blood Types

Problems 59–66 refer to the following breakdown in blood types: When receiving a blood transfusion, a recipient must have all the antigens of the donor. A person may have one or more of the three antigens A, B, and Rh, or none at all. Eight blood types are possible, as indicated in the Venn diagram in the figure, where U is the set of all people under consideration. An A− person has A antigens but no B or Rh; an O+ person has Rh but neither A nor B; an AB− person has A and B antigens but no Rh; and so on. Using the figure, indicate which of the eight blood types are included in each set.

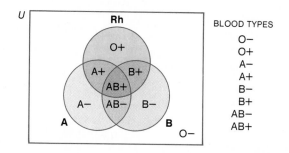

59. $A \cap B$ **60.** $A \cap Rh$ **61.** $A \cup B$

62. $A \cup Rh$ **63.** $(A \cup B \cup Rh)'$ **64.** $(A \cup B)'$

65. $Rh' \cap A$ **66.** $A' \cap B$

Section 1-2   Linear Equations

- Equality
- Solving Linear Equations
- A Common Error
- Equations Reducible to Linear Forms

- Equality

The equal sign, $=$, is used to join two expressions if the two expressions are names or descriptions of exactly the same object. Thus,

$$a = b$$

means a and b are names for the same object. Of course, $a \neq b$ means **a is not equal to b**. It is interesting to note that the equality sign did not appear until rather late in history—the sixteenth century. It was introduced by the English mathematician Robert Recorde (1510–1558).

If two algebraic expressions involving at least one variable are joined with an equal sign, the resulting form is called an **algebraic equation**. Since a variable is a placeholder for constants from a given replacement set, an equation is neither true nor false as it stands; it does not become so until the variables have been replaced by constants.

Several important properties of the equality symbol, $=$, follow directly from its logical meaning. These properties must hold any time the symbol is used.

Basic Properties of Equality

If a, b, and c are names of objects, then:

1. $a = a$ REFLEXIVE PROPERTY
2. If $a = b$, then $b = a$.   SYMMETRIC PROPERTY
3. If $a = b$ and $b = c$, then $a = c$.   TRANSITIVE PROPERTY
4. If $a = b$, then either may replace   SUBSTITUTION PRINCIPLE
   the other in any statement
   without changing the truth or falsity
   of the statement.

The properties of equality are used extensively throughout mathematics. For example, using the symmetric property, we may reverse the left and right members of an equation any time we wish. That is,

if $A = P + Prt$,   then $P + Prt = A$

Using the transitive property, we find that if

$$2x + 3x = (2 + 3)x \quad \text{and} \quad (2 + 3)x = 5x$$

then

$$2x + 3x = 5x$$

And, finally, if we know that

$$C = \pi D \quad \text{and} \quad D = 2R$$

then, using the substitution principle, D in the first formula may be replaced by $2R$ from the second formula to obtain

$$C = \pi(2R) = 2\pi R$$

■ Solving Linear Equations

We now turn our attention to methods of solving **first-degree** or **linear equations** in one variable—that is, to solving any equation that can be written in the following form:

LINEAR EQUATION $ax + b = 0$ $a \neq 0$

---

where a and b are real constants and x is a variable.

The **replacement set** for a variable is defined to be the set of constants that are permitted to replace the variable, and the **solution set** for an equation is defined to be the set of elements from its replacement set that makes the equation a true statement. Any element of the solution set is called a **solution** or **root** of the equation. To **solve an equation** is to find the solution set for the equation.

For example, if the replacement set for the variable x is the set of integers, then the solution set for the equation

$$x^2 - \tfrac{1}{9} = 0$$

is the empty set. If, on the other hand, the replacement set is the set of rational numbers, then the solution set is $\left\{ -\tfrac{1}{3}, \tfrac{1}{3} \right\}$, and we call $-\tfrac{1}{3}$ and $\tfrac{1}{3}$ *solutions* or *roots* of the equation. What is the solution set for the equation $x^2 - 4 = 0$ if the replacement set for the variable x is the set of natural numbers? The set of integers? [*Answer:* $\{2\}, \{-2, 2\}$.]

Knowing what we mean by the solution set of an equation is one thing; finding it is another. To this end we introduce the idea of equivalent equations. Two equations are said to be **equivalent** if they both have the same solution set. A basic technique for solving equations is to perform operations on equations that produce simpler equivalent equations, and to continue the process until an equation is reached whose solution is obvious.

The properties of equality given in Theorem 1 produce equivalent equations when applied. These further properties follow directly from the basic properties of equality stated earlier.

THEOREM 1 | Further Properties of Equality

For a, b, and c any real numbers,

1. If $a = b$, then $a + c = b + c$.   ADDITION PROPERTY
2. If $a = b$, then $a - c = b - c$.   SUBTRACTION PROPERTY
3. If $a = b$, then $ca = cb$, $c \neq 0$.   MULTIPLICATION PROPERTY
4. If $a = b$, then $\dfrac{a}{c} = \dfrac{b}{c}$, $c \neq 0$.   DIVISION PROPERTY

EXAMPLE 5   Solve $3x - 2(2x - 5) = 2(x + 3) - 8$ and check.

Solution

$3x - 2(2x - 5) = 2(x + 3) - 8$ Clear parentheses.

$3x - 4x + 10 = 2x + 6 - 8$ Combine like terms.

$-x + 10 = 2x - 2$ Use subtraction property.

$-3x = -12$ Use division property.

$x = 4$

The solution set to this last equation is obvious:

Solution set $= \{4\}$

And since $x = 4$ is equivalent to the first equation, $\{4\}$ is also the solution set for the first equation.

[*Note:* If an equation has only one element in its solution set, we generally use the last equation (in this case $x = 4$) rather than set notation to represent the solution.]

Check $3x - 2(2x - 5) = 2(x + 3) - 8$

$3(4) - 2[2(4) - 5] \overset{?}{=} 2[(4) + 3] - 8$

$6 \overset{\vee}{=} 6$

PROBLEM 5   Solve and check: $2(3 - x) - (3x + 1) = 8 - 2(x + 2)$

EXAMPLE 6   Solve: $\dfrac{x + 1}{3} - \dfrac{x}{4} = \dfrac{1}{2}$

Solution   If we could find a number that is exactly divisible by each denominator, then we could use the multiplication property in Theorem 1 to clear the equation of fractions. The smallest positive integer that is exactly divisible by each element in a set of positive integers is called the **least common multiple** (LCM) of the set. In this case, the LCM of 3, 4, and 2 is 12. Thus,

$$\frac{x + 1}{3} - \frac{x}{4} = \frac{1}{2}$$ Multiply both sides by 12.

$$12\left(\frac{x + 1}{3} - \frac{x}{4}\right) = 12 \cdot \frac{1}{2}$$ These steps can usually be done mentally.

$$12 \cdot \frac{x + 1}{3} - 12 \cdot \frac{x}{4} = 6$$

$4(x + 1) - 3x = 6$ The equation is now free of fractions.

$4x + 4 - 3x = 6$

$x = 2$ Solution set $= \{2\}$

The check is left to the reader.

PROBLEM 6   Solve: $\dfrac{x}{5} - \dfrac{x-2}{2} = \dfrac{3}{4}$

■ **A Common Error**

A very common error occurs about now—students tend to confuse *algebraic expressions* involving fractions with *algebraic equations* involving fractions.

   Consider the two problems:

(A)   Solve: $\dfrac{x}{2} + \dfrac{x}{3} = 10$ (B)   Add: $\dfrac{x}{2} + \dfrac{x}{3} + 10$

The problems look very much alike but are actually very different. To solve the equation in (A) we multiply both sides by 6 (the LCM of 2 and 3) to clear the fractions. This works so well for equations that students want to do the same thing for problems like (B). The only catch is that (B) is not an equation and the multiplication property of equality does not apply. If we multiply (B) by 6, we obtain an expression six times as large as the original.

   Compare the following:

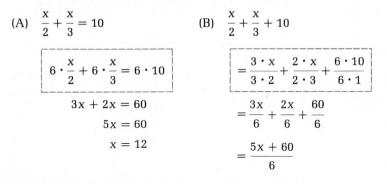

(A) $\dfrac{x}{2} + \dfrac{x}{3} = 10$ (B) $\dfrac{x}{2} + \dfrac{x}{3} + 10$

$$6 \cdot \frac{x}{2} + 6 \cdot \frac{x}{3} = 6 \cdot 10$$ $$= \frac{3 \cdot x}{3 \cdot 2} + \frac{2 \cdot x}{2 \cdot 3} + \frac{6 \cdot 10}{6 \cdot 1}$$

$$3x + 2x = 60$$ $$= \frac{3x}{6} + \frac{2x}{6} + \frac{60}{6}$$

$$5x = 60$$

$$x = 12$$ $$= \frac{5x + 60}{6}$$

■ **Equations Reducible to Linear Forms**

Some equations involving variables in a denominator can be transformed into linear equations. We may proceed in essentially the same way as in the preceding example; however, the replacement set of the equation must exclude any value of the variable that will make a denominator 0. As long as we use the replacement set of the original equation, we may multiply through by the LCM of the denominators even though it contains a variable, and, according to Theorem 1, the new equation will be equivalent to the old.

EXAMPLE 7   Solve and check: $\dfrac{x}{2x-4} - \dfrac{2}{3} = \dfrac{7-2x}{3x-6}$

Solution $\dfrac{x}{2x-4} - \dfrac{2}{3} = \dfrac{7-2x}{3x-6}$ Factor denominators.

$\dfrac{x}{2(x-2)} - \dfrac{2}{3} = \dfrac{7-2x}{3(x-2)}$ $x \neq 2$ (Why?)

Multiply both sides by the LCM of the denominators, $6(x-2)$, to obtain

$$6(x-2)\,\dfrac{x}{2(x-2)} - 6(x-2)\,\dfrac{2}{3} = 6(x-2)\,\dfrac{7-2x}{3(x-2)}$$

$$3x - 4x + 8 = 14 - 4x$$

$$3x = 6$$

$$x = 2 \quad \text{No solution}$$

Since 2 is not in the replacement set of the original equation, the original equation has no solution. (Note that when $x = 2$ is replaced in the left and right members of the original equation, neither is defined.)

PROBLEM 7   Solve and check: $\dfrac{x-3}{2x-2} = \dfrac{1}{6} - \dfrac{1-x}{3x-3}$

EXAMPLE 8   Solve $A = P + Prt$ for P.

Solution $A = P + Prt$ Think of P as a variable and A, r, and t as constants.

$P + Prt = A$ Symmetric property of equality

$P(1 + rt) = A$ Distributive property

$P = \dfrac{A}{1 + rt}$ Division property of equality ($rt \neq -1$)

PROBLEM 8   Solve $C = \tfrac{5}{9}(F - 32)$ for F.

It appears that any equation that can be written in the form

$$ax + b = 0 \qquad a \neq 0 \tag{1}$$

has exactly one solution. That this is true in general can be seen by solving equation (1) for x in terms of a and b.

$ax + b = 0$

$ax = -b$ Subtraction property of equality

$x = \dfrac{-b}{a}$ Division property of equality

Requiring $a \neq 0$ in equation (1) is an important restriction because without it we are able to write equations with first-degree members that have no solutions or have infinitely many solutions. For example,

$$2x - 3 = 2x + 5$$

has no solution. (Why?) And

$$3x - 4 = 5 + 3(x - 3)$$

has infinitely many solutions. (Why?) Try to solve each equation to see what happens.

Answers to Matched Problems

5. $x = \frac{1}{3}$

6. $x = \frac{5}{6}$

7. No solution

8. $F = \frac{9}{5}C + 32$

Exercise 1-2 ■ A Solve each equation, if possible.

1. $3(x + 2) = 5(x - 6)$

2. $5x + 10(x - 2) = 40$

3. $5 + 4(t - 2) = 2(t + 7) + 1$

4. $5x - (7x - 4) - 2 = 5 - (3x + 2)$

5. $x(x + 2) = x(x + 4) - 12$

6. $x(x + 4) - 2 = x^2 - 4(x + 3)$

7. $3 - \dfrac{2x - 3}{3} = \dfrac{5 - x}{2}$

8. $\dfrac{x - 2}{3} + 1 = \dfrac{x}{7}$

9. $5 - \dfrac{2x - 1}{4} = \dfrac{x + 2}{3}$

10. $\dfrac{x + 3}{4} - \dfrac{x - 4}{2} = \dfrac{3}{8}$

11. $0.1(x - 7) + 0.05x = 0.8$

12. $0.4(x + 5) - 0.3x = 17$

13. $0.3x - 0.04(x + 1) = 2.04$

14. $0.02x - 0.5(x - 2) = 5.32$

15. $\dfrac{3x}{24} - \dfrac{2 - x}{10} = \dfrac{5 + x}{40} - \dfrac{1}{15}$

16. $\dfrac{2x - 3}{9} - \dfrac{x + 5}{6} = \dfrac{3 - x}{2} + 1$

17. $\dfrac{1}{m} - \dfrac{1}{9} = \dfrac{4}{9} - \dfrac{2}{3m}$

18. $\dfrac{2}{3x} + \dfrac{1}{2} = \dfrac{4}{x} + \dfrac{4}{3}$

19. $\dfrac{5x}{x + 5} = 2 - \dfrac{25}{x + 5}$

20. $\dfrac{3}{2x - 1} + 4 = \dfrac{6x}{2x - 1}$

B Solve each equation, if possible.

21. $\dfrac{2x}{10} - \dfrac{3 - x}{14} = \dfrac{2 + x}{5} - \dfrac{1}{2}$

22. $\dfrac{3x}{24} - \dfrac{2 - x}{10} = \dfrac{5 + x}{40} - \dfrac{1}{15}$

23. $\dfrac{1}{3} - \dfrac{s - 2}{2s + 4} = \dfrac{s + 2}{3s + 6}$

24. $\dfrac{n - 5}{6n - 6} = \dfrac{1}{9} - \dfrac{n - 3}{4n - 4}$

25. $\dfrac{3x}{2-x} + \dfrac{6}{x-2} = 3$

26. $5 - \dfrac{2x}{3-x} = \dfrac{6}{x-3}$

27. $\dfrac{5t-22}{t^2-6t+9} - \dfrac{11}{t^2-3t} - \dfrac{5}{t} = 0$

28. $\dfrac{5}{x-3} = \dfrac{33-x}{x^2-6x+9}$

29. $\dfrac{1}{x^2-x-2} - \dfrac{3}{x^2-2x-3} = \dfrac{1}{x^2-5x+6}$

30. $\dfrac{10}{x} - \dfrac{22}{3x-x^2} = \dfrac{10x-44}{x^2-6x+9}$

Solve for the indicated letter.

31. $a_n = a_1 + (n-1)d$ for d (arithmetic progressions)

32. $F = \frac{5}{9}C + 32$ for C (temperature scale)

33. $1/f = (1/d_1) + (1/d_2)$ for f (simple lens formula)

34. $1/R = (1/R_1) + (1/R_2)$ for R_1 (electric circuit)

35. $A = 2ab + 2ac + 2bc$ for a (area of a rectangular solid)

36. $A = 2ab + 2ac + 2bc$ for c

37. $y = \dfrac{2x-3}{3x+5}$ for x

38. $x = \dfrac{3y+2}{y-3}$ for y

C **39.** Solve for x: $\dfrac{x - \dfrac{1}{x}}{x + \dfrac{1}{x}} = 3$

40. Solve $\dfrac{y}{1-y} = \left(\dfrac{x}{1-x}\right)^3$ for y in terms of x.

41. Solve $y = \dfrac{a}{1 + \dfrac{b}{x+c}}$ for x in terms of y.

42. Let m and n be real numbers with m larger than n. Then there exists a positive real number p such that $m = n + p$. Find the fallacy in the following argument:

$$m = n + p$$
$$(m - n)m = (m - n)(n + p)$$
$$m^2 - mn = mn + mp - n^2 - np$$
$$m^2 - mn - mp = mn - n^2 - np$$
$$m(m - n - p) = n(m - n - p)$$
$$m = n$$

43. Prove the subtraction property of equality in Theorem 1.

44. Prove the multiplication property of equality in Theorem 1.

Section 1-3 Applications

- A Strategy for Solving Word Problems
- Number and Geometric Problems
- Rate-Time Problems
- Mixture Problems

■ A Strategy for Solving Word Problems

A great many practical problems can be solved using algebraic techniques—so many, in fact, there is no one method of attack that will work for all. However, we can formulate a strategy that will help you organize your approach.

Strategy for Solving Word Problems

1. Read the problem carefully—several times if necessary; that is, until you understand the problem, know what is to be found, and know what is given.

2. Draw figures or diagrams and label known and unknown parts.

3. Look for formulas connecting the known quantities with the unknown quantities.

4. Let one of the unknown quantities be represented by a variable, say x, and try to represent all other unknown quantities in terms of x. This is an important step and must be done carefully.

5. Form an equation relating the unknown quantities with the known quantities.

6. Solve the equation and write answers to *all* parts of the problem requested.

7. Check and interpret all solutions in terms of the original problem and not just the equation found in step 5 (a mistake might have been made in setting up the equation in step 5).

The examples in this section contain worked-out solutions to a variety of word problems. It is suggested that you cover up a solution, try solving the problem yourself, and uncover just enough of a solution to get going again in case you get stuck. After successfully completing an example, try the matched problem. Completing the section in this way, you will be ready to attack the fairly large variety of applications in Exercise 1-3.

■   Number and Geometric Problems

Let us start with a couple of fairly simple examples that will introduce the process of setting up and solving word problems in a simple context. Following these, we will consider several examples of a more substantive nature.

EXAMPLE 9   Find four consecutive even integers such that the sum of the first three exceeds the fourth by eight.

Solution   Let $x =$ The first even integer, then

x Remember, even integers

$x + 2$ Increase two at a time.

$x + 4$

$x + 6$

represent four consecutive even integers starting with the even integer x. The phrase "the sum of the first three exceeds the fourth by eight" translates into an equation:

$$x + (x + 2) + (x + 4) = (x + 6) + 8$$
$$3x + 6 = x + 14$$
$$2x = 8$$
$$x = 4$$

The four consecutive integers are 4, 6, 8, and 10.

Check
$$\begin{array}{rl} (4 + 6 + 8) = & 18 \quad \text{Sum of first three} \\ & -10 \quad \text{Fourth} \\ \hline & 8 \end{array}$$

PROBLEM 9   Find three consecutive odd numbers such that three times their sum is five more than eight times the middle one.

EXAMPLE 10   If one side of a triangle is one-third the perimeter, the second side is 7 meters, and the third side is one-fifth the perimeter, what is the perimeter of the triangle?

Solution   Let $P = $ The perimeter. Draw a triangle and label the sides.

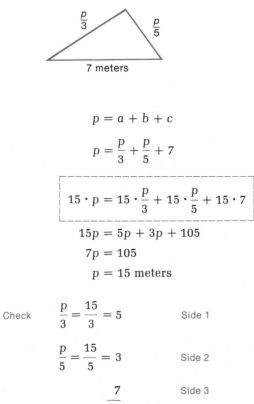

$$p = a + b + c$$

$$p = \frac{p}{3} + \frac{p}{5} + 7$$

$$15 \cdot p = 15 \cdot \frac{p}{3} + 15 \cdot \frac{p}{5} + 15 \cdot 7$$

$$15p = 5p + 3p + 105$$

$$7p = 105$$

$$p = 15 \text{ meters}$$

Check $\dfrac{p}{3} = \dfrac{15}{3} = 5$ Side 1

$\dfrac{p}{5} = \dfrac{15}{5} = 3$ Side 2

$\dfrac{7}{15}$ meters   Side 3

Perimeter

PROBLEM 10   If one side of a triangle is one-fourth the perimeter, the second side is 7 centimeters, and the third side is two-fifths the perimeter, what is the perimeter?

■ Rate-Time Problems

There are many types of rate-time problems in addition to distance-rate-time problems. In general, if Q is the quantity of something produced (kilometers, words, parts, and so on) in T units of time (hours, years, minutes, seconds, and so on), then the formulas given in the box are relevant.

Quantity-Rate-Time Formulas

$R = \dfrac{Q}{T}$ $Rate = \dfrac{Quantity}{Time}$

$Q = RT$ $Quantity = (Rate)(Time)$

$T = \dfrac{Q}{R}$ $Time = \dfrac{Quantity}{Rate}$

If Q is distance D, then

$R = \dfrac{D}{T}$ $D = RT$ $T = \dfrac{D}{R}$

[*Note:*   R is an average rate.]

EXAMPLE 11   The distance along a shipping route between San Francisco and Honolulu is 2,100 nautical miles. If one ship leaves San Francisco at the same time another leaves Honolulu, and if the former travels at 15 knots* and the latter at 20 knots, how long will it take the two ships to rendezvous? How far will they be from Honolulu and San Francisco at that time?

Solution   Let T = Number of hours until both ships meet. Draw a diagram and label known and unknown parts. Both ships will have traveled the same amount of time when they meet.

$$\begin{pmatrix} \text{Distance ship} \\ \text{from Honolulu} \\ \text{travels to} \\ \text{meeting point} \end{pmatrix} + \begin{pmatrix} \text{Distance ship} \\ \text{from San Francisco} \\ \text{travels to meeting} \\ \text{point} \end{pmatrix} = \begin{pmatrix} \text{Total distance} \\ \text{from Honolulu} \\ \text{to San Francisco} \end{pmatrix}$$

D_1	$+$	D_2	$= 2{,}100$
$20T$	$+$	$15T$	$= 2{,}100$
		$35T$	$= 2{,}100$
		T	$= 60$ hours (or 2.5 days)

* 15 knots means 15 nautical miles/hour. There are 6,076.1 feet in 1 nautical mile.

Distance from Honolulu = 20 · 60 = 1,200 nautical miles

Distance from San Francisco = 15 · 60 = 900 nautical miles

Check   1,200 + 900 = 2,100 nautical miles

PROBLEM 11   An older piece of equipment can print, stuff, and label thirty-eight mailing pieces per minute. A newer model can handle eighty-two per minute. How long will it take for both pieces of equipment to prepare a mailing of 6,000 pieces? [*Note:* Mathematically Problem 11 and Example 11 are the same.]

EXAMPLE 12   An excursion boat takes 1.5 times as long to go 360 miles up a river than to return. If the boat cruises at 15 miles/hour in still water, what is the rate of the current?

Solution   Let

$$x = \text{Rate of current}$$

$$15 - x = \text{Rate of boat upstream}$$

$$15 + x = \text{Rate of boat downstream}$$

$$\text{Time upstream} = (1.5)(\text{Time downstream})$$

$$\frac{\text{Distance upstream}}{\text{Rate upstream}} = (1.5)\frac{\text{Distance downstream}}{\text{Rate downstream}}$$

$$\frac{360}{15 - x} = (1.5)\frac{360}{15 + x}$$

$$\frac{360}{15 - x} = \frac{540}{15 + x}$$

Multiply both sides by $(15 - x)(15 + x)$ to clear fractions.

$$360(15 + x) = 540(15 - x)$$

$$5,400 + 360x = 8,100 - 540x$$

$$900x = 2,700$$

$$x = 3 \text{ miles/hour} \quad \text{Rate of current}$$

PROBLEM 12   A jet airliner takes 1.2 hours longer to fly from Paris to New York (3,600 miles) than to return. If the jet cruises at 550 miles/hour in still air, what is the average rate of the wind blowing in the direction of Paris from New York?

EXAMPLE 13   In an advertising firm an older machine can prepare a whole mailing in 6 hours. With the help of a newer machine the job is complete in 2 hours. How long would it take the newer machine to do the job alone?

Solution   Let x = Time for the newer machine to do the whole job alone.

$$\left(\begin{array}{l}\text{Part of job completed} \\ \text{in a given length of time}\end{array}\right) = (\text{Rate})(\text{Time})$$

$$\text{Rate of old machine} = \frac{1}{6} \text{ job per hour}$$

$$\text{Rate of new machine} = \frac{1}{x} \text{ job per hour}$$

$$\left(\begin{array}{l}\text{Part of job} \\ \text{completed by} \\ \text{old machine} \\ \text{in 2 hours}\end{array}\right) + \left(\begin{array}{l}\text{Part of job} \\ \text{completed by} \\ \text{new machine} \\ \text{in 2 hours}\end{array}\right) = 1 \text{ whole job}$$

$$(\text{Rate})(\text{Time}) \quad + \quad (\text{Rate})(\text{Time}) \quad = 1$$

$$\frac{1}{6}(2) \quad\quad + \quad\quad \frac{1}{x}(2) \quad\quad = 1$$

$$\frac{1}{3} + \frac{2}{x} = 1$$

$$x + 6 = 3x$$

$$-2x = -6$$

$$x = 3 \text{ hours}$$

Check $$\frac{1}{6}(2) + \frac{1}{3}(2) = \frac{1}{3} + \frac{2}{3} = 1 \text{ whole job}$$

PROBLEM 13   Two pumps are used to fill a water storage tank at a resort. One pump can fill the tank by itself in 9 hours, and the other can fill it in 6 hours. How long will it take both pumps operating together to fill the tank?

■ Mixture Problems

A variety of applications can be classified as mixture problems. Even though the problems come from different areas, their mathematical treatment is essentially the same.

EXAMPLE 14   A specialty coffee shop wishes to blend a $4-per-pound coffee with a $7-per-pound coffee to produce a blend that will sell for $5 per pound. How much of each should be used to produce 300 pounds of the new blend?

Solution   Let x = Amount of $4-per-pound coffee used, then

300 − x = Amount of $7-per-pound coffee used

VALUE BEFORE BLENDING   VALUE AFTER BLENDING

$$\begin{pmatrix} \text{Value of} \\ \text{\$4-per-pound} \\ \text{coffee used} \end{pmatrix} + \begin{pmatrix} \text{Value of} \\ \text{\$7-per-pound} \\ \text{coffee used} \end{pmatrix} = \begin{pmatrix} \text{Total value} \\ \text{of 300 pounds} \\ \text{of the blend} \end{pmatrix}$$

$$4x \quad + \quad 7(300 - x) \quad = \quad 5(300)$$

$$4x + 2{,}100 - 7x = 1500$$

$$-3x = -600$$

$$x = 200 \text{ pounds of \$4 coffee}$$

$$300 - x = 100 \text{ pounds of \$7 coffee}$$

PROBLEM 14   A concert brought in $41,080 on the sale of 5,000 tickets. If tickets sold for $6 and $10, how many of each were sold?

EXAMPLE 15   How many liters of a mixture containing 80% alcohol should be added to 5 liters of a 20% solution to yield a 30% solution?

Solution   Let x = Amount of 80% solution used.

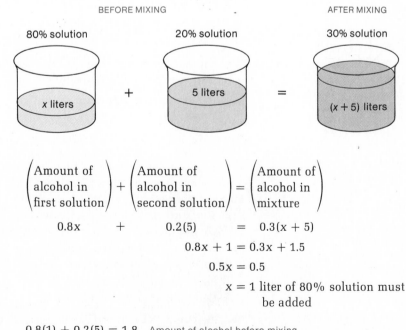

BEFORE MIXING   AFTER MIXING

80% solution   20% solution   30% solution

x liters   +   5 liters   =   (x + 5) liters

$$\begin{pmatrix} \text{Amount of} \\ \text{alcohol in} \\ \text{first solution} \end{pmatrix} + \begin{pmatrix} \text{Amount of} \\ \text{alcohol in} \\ \text{second solution} \end{pmatrix} = \begin{pmatrix} \text{Amount of} \\ \text{alcohol in} \\ \text{mixture} \end{pmatrix}$$

$$0.8x \quad + \quad 0.2(5) \quad = \quad 0.3(x + 5)$$

$$0.8x + 1 = 0.3x + 1.5$$

$$0.5x = 0.5$$

$$x = 1 \text{ liter of 80\% solution must be added}$$

Check $0.8(1) + 0.2(5) = 1.8$ Amount of alcohol before mixing

$0.3(1 + 5) = 1.8$ Amount of alcohol after mixing

PROBLEM 15   A chemical storeroom has a 90% acid solution and a 40% acid solution. How many centiliters must be taken from each to obtain 25 centiliters of a 50% acid solution?

Answers to Matched Problems

9. 3, 5, 7   **10.** 20 centimeters

11. 50 minutes   **12.** 50 miles/hour

13. 3.6 hours

14. 2,230 $6 tickets; 2,770 $10 tickets

15. 5 centiliters of 90% solution, 20 centiliters of 40% solution

Exercise 1-3 ■

These problems are not grouped from easy (A) to difficult or theoretical (C). They are grouped according to type. However, the most difficult problems are double-starred (★★), moderately difficult problems are single-starred (★), and the easier problems are not marked.

APPLICATIONS

Numbers

1. Find four consecutive even numbers so that the sum of the first three is two more than twice the fourth.

4, 6, 8

2. Find three consecutive even numbers so that the first plus twice the second is twice the third.

Geometry

3. Find the dimensions of a rectangle with a perimeter of 54 meters, if its length is 3 meters less than twice its width.

24 m x 6 m

4. A rectangle 24 meters long has the same area as a square that is 12 meters on a side. What are the dimensions of the rectangle?

Business and Economics

5. The sale price on a camera after a 20% discount is $72. What was the price before the discount?

$90

6. A stereo store marks up each item it sells 60% above wholesale price. What is the wholesale price on a record player that retails at $144?

7. It costs a record company $5,600 to prepare a record album. This is a one-time *fixed cost* that covers recording, album design, and so on. *Variable costs*, including such things as manufacturing, marketing, and royalties, are $2.60 per record. If the album is sold to record shops for $4 each, how many must be sold for the company to *break even*?

★**8.** Three individuals decided to share equally the cost of a sailboat. They found, however, that if they took in one more partner, the cost of the boat for each of the original three would be reduced by $300. What is the cost of the boat?

★★**9.** A person got a 10% raise one month and a 10% cut in salary the next month. What was the original monthly salary if the salary after the raise and cut was $1,980 per month?

Earth Science   **10.**   In case you did not know it, continents drift. The idea was vividly described by Thomas McEvilly in a lecture series at the University of California in Berkeley, when he said, "We can see the Farallon Islands [off San Francisco] sailing north at two and a half inches a year. If you stand on Mount Diablo, you can watch them sailing by. There they go!" How long will it take the islands to move 1 mile (5,280 feet) north? Set up an equation and solve.

11.   Pressure in seawater increases by 1 atmosphere (14.7 pounds/square inch) for each 33 feet of depth; it is 14.7 pounds/square inch at the surface. Thus, $p = 14.7 + 14.7(d/33)$, where p is the pressure in pounds per square inch at a depth of d feet below the surface. How deep (to the nearest foot) is a diver if she observes that the pressure is 165 pounds/square inch?

*12.   An earthquake emits a primary wave and a secondary wave. Near the surface of the earth the primary wave travels at about 5 miles/second, and the secondary wave at about 3 miles/second. From the time lag between the two waves arriving at a given seismic station, it is possible to estimate the distance to the quake. (The *epicenter* can be located by obtaining distance bearings at three or more stations.) Suppose a station measures a time difference of 12 seconds between the arrival of the two waves. How far is the earthquake from the station?

*13.   A ship using sound-sensing devices above and below water recorded a surface explosion 39 seconds sooner on its underwater device than on its above-water device. If sound travels in air at about 1,100 feet/second, and in water at about 5,000 feet/second, how far away was the explosion?

Life Science   **14.**   In biology there is an approximate rule, called the bioclimatic rule for temperate climates, which states that in spring and early summer periodic phenomena, such as blossoming for a given species, appearance of certain insects, and ripening of fruit, usually occur about 4 days later for each 500 feet of altitude increase or 1 degree latitude increase from any given base. In formulas we have

$$d = 4\left(\frac{h}{500}\right) \quad \text{and} \quad d = 4L$$

where d = Days, h = Change in altitude in feet, and L = Change in latitude in degrees. What change in altitude would delay pear trees from blossoming for 16 days? What change in latitude would accomplish the same thing?

*15.   A wildlife manager estimated the total number of deer in a national forest by using the popular capture-mark-recapture technique. She

captured, marked, and released 100 deer. A month later, allowing time for thorough mixing, she captured 100 more deer and found four marked ones among them. Assuming that the proportion of marked deer in the second sample is the same as the proportion of all marked deer in the total population, estimate the number of deer in the forest.

2 gal.

Chemistry ★16. How many gallons of hydrochloric acid must be added to 12 gallons of a 30% solution to obtain a 40% solution? 30

★17. A chemist has two solutions of sulfuric acid—a 20% solution and an 80% solution. How much of each should be used to obtain 100 liters of a 62% solution?

Rate-Time 18. In a computer center two electronic card sorters are used to sort 52,000 computer cards. If the first sorter operates at 225 cards per minute and the second sorter operates at 175 cards per minute, how long will it take both sorters together to sort all the cards?

★19. A skydiver free-falls (because of air resistance) at about 176 feet/second, or 120 miles/hour; with his parachute open he falls at about 22 feet/second, or 15 miles/hour. If the skydiver opened his chute halfway down and the total time for the descent was 6 minutes, how high was the plane when he jumped?

★20. You are at a river resort and rent a motor boat for 5 hours starting at 7 AM. You are told that the boat will travel at 8 miles/hour upstream and 12 miles/hour returning. You decide that you would like to go as far up the river as you can and still be back at noon. At what time should you turn back, and how far from the resort will you be at that time?

★★21. The cruising speed of an airplane is 150 miles/hour (relative to ground). You wish to hire the plane for a 3-hour sightseeing trip. You instruct the pilot to fly north as far as he can and still return to the airport at the end of the allotted time.
(A)   How far north should the pilot fly if there is a 30 miles/hour wind blowing from the north?
(B)   How far north should the pilot fly if there is no wind blowing?

Music 22. The three major chords in music are composed of notes whose frequencies are in the ratio 4:5:6. If the first note of a chord has a frequency of 264 hertz (middle C on the piano), find the frequencies of the other two notes. [Hint: Set up two proportions using 4:5 and 4:6.]

23. The three minor chords are composed of notes whose frequencies are in the ratio 10:12:15. If the first note of a minor chord is A, with a frequency of 220 hertz, what are the frequencies of the other two notes?

Psychology **24.** In 1948, Professor Brown, a psychologist, trained a group of rats (in an experiment on motivation) to run down a narrow passage in a cage to receive food in a goal box. He then put a harness on each rat and connected it to an overhead wire attached to a scale. In this way he could place the rat different distances from the food and measure the pull (in grams) of the rat toward the food. He found that the relation between motivation (pull) and position was given approximately by the equation

$$p = -\tfrac{1}{5}d + 70 \qquad 30 \le d \le 170$$

where pull p is measured in grams and distance d in centimeters. When the pull registered was 40 grams, how far was the rat from the goal box?

25. Professor Brown performed the same kind of experiment as described in Problem 24, except that he replaced the food in the goal box with a mild electric shock. With the same kind of apparatus, he was able to measure the avoidance strength relative to the distance from the object to be avoided. He found that the avoidance strength a (measured in grams) was related to the distance d that the rat was from the shock (measured in centimeters) approximately by the equation

$$a = -\tfrac{4}{3}d + 230 \qquad 30 \le d \le 170$$

If the same rat were trained as described in this problem and in Problem 24, at what distance (to one decimal place) from the goal box would the approach and avoidance strengths be the same? (What do you think the rat would do at this point?)

Puzzle **26.** A person has \$3.35 in coins in a change purse. If there are three times as many dimes as quarters and three fewer nickels than dimes, how many of each type of coins are in the purse?

★**27.** An oil-drilling rig in the Gulf of Mexico stands so that one-fifth of it is in sand, 20 feet of it is in water, and two-thirds of it is in the air. What is the total height of the rig?

★**28.** During a camping trip in the North Woods in Canada, a couple went one-third of the way by boat, 10 miles by foot, and one-sixth of the way by horse. How long was the trip?

★★**29.** After exactly 12 o'clock noon, what time will the hands of a clock be together again?

Section 1-4   Real Number Line; Linear Inequalities

- ■ The Real Number Line
- ■ Inequality Relations
- ■ Solving Linear Inequalities

Figure 5 illustrates the set of real numbers and some of its important
subsets using a Venn diagram and a "tree" diagram.

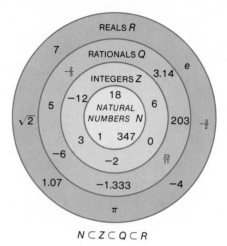

$$N \subset Z \subset Q \subset R$$

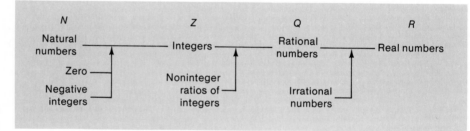

FIGURE 5   The set of
real numbers

■   The Real Number Line

A one-to-one correspondence exists between the set of real numbers
and the set of points on a line; that is, each real number corresponds
to exactly one point, and each point to exactly one real number. A line
with a real number associated with each point, and vice versa, as in
Figure 6, is called a **real number line**, or simply a **real line**. Each number
associated with a point is called the **coordinate** of the point. The point
with coordinate 0 is called the **origin**. The arrow indicates a positive

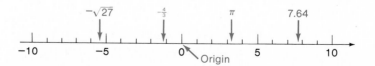

FIGURE 6   A real number line

direction; the coordinates of all points to the right of the origin are called **positive real numbers** and those to the left of the origin are called **negative real numbers**.

■ Inequality Relations

We now define "less than" and "greater than" for the set of real numbers.

Definition of $a < b$ and $b > a$

For a and b real numbers, we say that **a is less than b** or **b is greater than a** and write

$a < b$ or $b > a$

if there exists a positive real number p such that $a + p = b$ (or equivalently $b - a = p$).

We would certainly expect that if a positive number was added to *any* real number, the sum would be larger than the original. That is essentially what the definition states. When we write

a ≤ b

we mean **a is less than or equal to b**, and when we write

a ≥ b

we mean that **a is greater than or equal to b**.

The inequality symbols $<$ and $>$ have a very clear geometric interpretation on the real number line. If $a < b$, then a is to the left of b; if $c > d$, then c is to the right of d (Fig. 7).

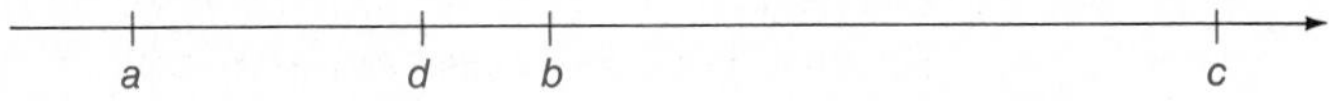

FIGURE 7 $a < b, c > d$

It is an interesting and useful fact that for any two real numbers a and b, $a < b$, $a > b$, or $a = b$. This property (called the **trichotomy** property) is not shared with all number systems, as we will see later in this chapter when we extend the set of real numbers to the set of complex numbers.

The double inequality $a < x \le b$ means that $a < x$ and $x \le b$; that is, x is between a and b, including b but not including a. Other variations on the theme, as well as a useful **interval notation**, are shown in Table 1.

TABLE 1	INTERVAL NOTATION	INEQUALITY NOTATION	LINE GRAPH
	$[a, b]$	$a \le x \le b$	
	$[a, b)$	$a \le x < b$	
	$(a, b]$	$a < x \le b$	
	(a, b)	$a < x < b$	
	$[b, \infty)^*$	$x \ge b$	
	(b, ∞)	$x > b$	
	$(-\infty, a]$	$x \le a$	
	$(-\infty, a)$	$x < a$	

* The symbol ∞ (read "infinity") is not a number. When we write $[b, \infty)$, we are simply referring to the interval starting at b and continuing indefinitely to the right. We would never write $[b, \infty]$.

EXAMPLE 16   Write each of the following in inequality notation and graph on a real number line.

(A) $[-2, 3)$ (B) $(-4, 2)$ (C) $[-2, \infty)$ (D) $(-\infty, 3)$

Solution   (A) $-2 \le x < 3$

(B) $-4 < x < 2$

(C) $x \ge -2$

(D) $x < 3$

PROBLEM 16   Write each of the following in interval notation and graph on a real number line.

(A) $-3 < x \le 3$ (B) $-1 \le x \le 2$
(C) $x > 1$ (D) $x \le 2$

■ Solving Linear Inequalities

We now turn to the problem of solving linear inequalities in one variable, such as

$$2(2x + 3) < 6(x - 2) + 10$$

and

$$-3 < 2x + 3 \leq 9$$

The **solution set** for an inequality is the set of elements from its replacement set that make the inequality a true statement. Any element of the solution set is called a **solution** of the inequality. To **solve an inequality** is to find its solution set. Two inequalities are **equivalent** if they have the same solution set. Just as with equations, we try to perform operations on inequalities that produce simpler equivalent inequalities, and to continue the process until an inequality is reached whose solution is obvious. The properties of inequalities given in Theorem 2 produce equivalent inequalities when applied.

THEOREM 2

Inequality Properties

For a, b, and c any real numbers:

1. If $a < b$, then $a + c < b + c$.   ADDITION PROPERTY

$$-2 < 4 \qquad -2 + 3 < 4 + 3$$

2. If $a < b$, then $a - c < b - c$.   SUBTRACTION PROPERTY

$$-2 < 4 \qquad -2 - 3 < 4 - 3$$

3. If $a < b$ and c is positive, then $ca < ca$.   MULTIPLICATION PROPERTY

$$-2 < 4 \qquad 3(-2) < 3(4)$$

4. If $a < b$ and c is negative, then $ca > cb$.   (NOTE DIFFERENCE BETWEEN 3 AND 4)

$$-2 < 4 \qquad (-3)(-2) > (-3)(4)$$

5. If $a < b$ and c is positive, then $\dfrac{a}{c} < \dfrac{b}{c}$.

$$-2 < 4 \qquad \frac{-2}{2} < \frac{4}{2}$$

DIVISION PROPERTY (NOTE DIFFERENCE BETWEEN 5 AND 6)

6. If $a < b$ and c is negative, then $\dfrac{a}{c} > \dfrac{b}{c}$.

$$-2 < 4 \qquad \frac{-2}{-2} > \frac{4}{-2}$$

Similar properties hold if each inequality sign is reversed, or if $<$ is replaced with $\leq$ and $>$ is replaced with $\geq$. Thus, we find that we can perform essentially the same operations on inequalities that we perform on equations. When working with inequalities, we have to be particularly careful of the use of the multiplication and division properties.

The sense of the inequality reverses if we multiply or divide both sides of an inequality statement by a negative number.

Let us sketch a proof of the multiplication property: If $a < b$, then by definition of $<$, there exists a positive number p such that $a + p = b$. If we multiply both sides of $a + p = b$ by a positive number c, we obtain $ca + cp = cb$, where cp is positive. (Why?) Thus, by definition of $<$, we see that $ca < cb$. Now if we multiply both sides of $a + p = b$ by a negative number c, we obtain $ca + cp = cb$ or $ca = cb - cp$ where cp is negative (Why?) and $-cp$ is positive. Hence, by definition of $<$, we see that $cb < ca$ or $ca > cb$.

Now let us see how the inequality properties are used to solve linear inequalities. Several examples will illustrate the process.

EXAMPLE 17   Solve and graph: $2(2x + 3) - 10 < 6(x - 2)$

Solution

$2(2x + 3) - 10 < 6(x - 2)$ Simplify left and right sides.

$4x + 6 - 10 < 6x - 12$

$4x - 4 < 6x - 12$

$4x - 4 + 4 \leq 6x - 12 + 4$ Addition property

$4x < 6x - 8$

$4x - 6x < 6x - 8 - 6x$ Subtraction property

$-2x < -8$

$\dfrac{-2x}{-2} > \dfrac{-8}{-2}$ Division property— note that sense reverses. (Why?)

$x > 4$ or $(4, \infty)$

PROBLEM 17   Solve and graph: $3(x - 1) \geq 5(x + 2) - 5$

EXAMPLE 18   Solve and graph: $\dfrac{2x - 3}{4} + 6 \geq 2 + \dfrac{4x}{3}$

Solution

$$\frac{2x - 3}{4} + 6 \geq 2 + \frac{4x}{3}$$

Multiply both sides by 12, the LCM of 4 and 3.

$$12 \cdot \frac{2x - 3}{4} + 12 \cdot 6 \geq 12 \cdot 2 + 12 \cdot \frac{4x}{3}$$

$$3(2x - 3) + 72 \geq 24 + 4 \cdot 4x$$

$$6x - 9 + 72 \geq 24 + 16x$$

$$6x + 63 \geq 24 + 16x$$

$$-10x \geq -39$$

$$x \leq 3.9 \quad \text{or} \quad (-\infty, 3.9] \quad \text{Sense reverses. (Why?)}$$

PROBLEM 18   Solve and graph: $\dfrac{4x - 3}{3} + 8 < 6 + \dfrac{3x}{2}$

EXAMPLE 19   Solve and graph: $-3 \leq 4 - 7x < 18$

Solution   We proceed as in the preceding examples, except we try to isolate x in the middle with a coefficient of 1.

$$-3 \leq 4 - 7x < 18$$

Subtract 4 from each member.

$$-3 - 4 \leq 4 - 7x - 4 < 18 - 4$$

$$-7 \leq -7x < 14$$

Divide each member by -7.

$$\frac{-7}{-7} \geq \frac{-7x}{-7} > \frac{14}{-7}$$

Sense reverses. (Why?)

$$1 \geq x > -2 \quad \text{or} \quad -2 < x \leq 1 \quad \text{or} \quad (-2, 1]$$

PROBLEM 19   Solve and graph: $-3 < 7 - 2x \leq 7$

EXAMPLE 20   In a chemistry experiment a solution of hydrochloric acid is to be kept between 30° and 35° Celsius—that is, $30 \leq C \leq 35$. What is the range in temperature in degrees Fahrenheit? $[C = \frac{5}{9}(F - 32)]$

Solution $30 \leq C \leq 35$ Replace C with $\frac{5}{9}(F - 32)$.

$$30 \leq \frac{5}{9}(F - 32) \leq 35$$ Multiply each member by $\frac{9}{5}$. (Why?)

$$\frac{9}{5} \cdot 30 \leq \frac{9}{5} \cdot \frac{5}{9}(F - 32) \leq \frac{9}{5} \cdot 35$$

$$54 \leq F - 32 \leq 63$$ Add 32 to each member.

$$54 + 32 \leq F - 32 + 32 \leq 63 + 32$$

$$86 \leq F \leq 95 \quad \text{or} \quad [86, 95]$$

PROBLEM 20   A film developer is to be kept between 68° and 77° Fahrenheit—that is, $68 \leq F \leq 77$. What is the range in temperature in degrees Celsius? $(F = \frac{9}{5}C + 32)$

Answers to Matched Problems   **16.** (A) $(-3, 3]$

(B) $[-1, 2]$

(C) $(1, \infty)$

(D) $(-\infty, 2]$

17. $x \leq -4 \quad \text{or} \quad (-\infty, -4]$

18. $x > 6 \quad \text{or} \quad (6, \infty)$

19. $5 > x \geq 0 \quad \text{or} \quad 0 \leq x < 5 \text{ or } [0, 5)$

20. $20 \leq C \leq 25 \quad \text{or} \quad [20, 25]$

Exercise 1-4 ■ *Unless otherwise stated, the replacement set for all variables is the set of real numbers.*

A   Write in inequality notation and graph on a real number line.

1. $[-8, 7]$ **2.** $(-4, 8)$ **3.** $[-6, 6)$

4. $(-3, 3]$ **5.** $[-6, \infty)$ **6.** $(-\infty, 7)$

Write in interval notation and graph on a real number line.

7. $-2 < x \leq 6$ **8.** $-5 \leq x \leq 5$ **9.** $-7 < x < 8$

10. $-4 \leq x < 5$ **11.** $x \leq -2$ **12.** $x > 3$

Write in interval and inequality notation.

13.

14.

15.

16.

Solve and graph.

17. $7x - 8 < 4x + 7$ **18.** $4x + 8 \geq x - 1$

19. $3 - x \geq 5(3 - x)$ **20.** $2(x - 3) + 5 < 5 - x$

21. $\dfrac{N}{-2} > 4$ **22.** $\dfrac{M}{-3} \leq -2$

23. $-5t < -10$ **24.** $-7n \geq 21$

25. $3 - m < 4(m - 3)$ **26.** $2(1 - u) \geq 5u$

27. $-2 - \dfrac{B}{4} \leq \dfrac{1 + B}{3}$ **28.** $\dfrac{y - 3}{4} - 1 > \dfrac{y}{2}$

29. $-4 < 5t + 6 \leq 21$ **30.** $2 \leq 3m - 7 < 14$

B Write each set by the listing method assuming x is an integer. For example, $\{x \mid -2 < x \leq 2\} = \{-1, 0, 1, 2\}$.

31. $\{x \mid -4 \leq x \leq 2\}$ **32.** $\{x \mid -1 < x \leq 1\}$

33. $\{x \mid -4 < x < 2\}$ **34.** $\{x \mid -1 \leq x < 1\}$

Solve and graph.

35. $\dfrac{q}{7} - 3 > \dfrac{q - 4}{3} + 1$ **36.** $\dfrac{p}{3} - \dfrac{p - 2}{2} \leq \dfrac{p}{4} - 4$

37. $\dfrac{2x}{5} - \dfrac{1}{2}(x - 3) \leq \dfrac{2x}{3} - \dfrac{3}{10}(x + 2)$

38. $\dfrac{2}{3}(x + 7) - \dfrac{x}{4} > \dfrac{1}{2}(3 - x) + \dfrac{x}{6}$

39. $-4 \leq \frac{9}{5}x + 32 \leq 68$ **40.** $-1 \leq \frac{2}{3}A + 5 \leq 11$

41. $-12 < \frac{3}{4}(2 - x) \leq 24$ **42.** $24 \leq \frac{2}{3}(x - 5) < 36$

43. $16 < 7 - 3x \leq 31$ **44.** $-1 \leq 9 - 2x < 5$

45. $-6 < -\frac{2}{5}(1 - x) \leq 4$ **46.** $15 \leq 7 - \frac{2}{5}x \leq 21$

STOP   C   (47.)   If both a and b are negative numbers and b/a is greater than 1, then is $a - b$ positive or negative?

48.   If both a and b are positive numbers and b/a is greater than 1, then is $a - b$ positive or negative?

49.   Indicate true (T) or false (F).

(A)   If $p > q$ and $m > 0$, then $mp < mq$.

(B)   If $p < q$ and $m < 0$, then $mp > mq$.

(C)   If $p > 0$ and $q < 0$, then $p + q > q$.

50.   Assume that $m > n > 0$; then

$$mn > n^2$$
$$mn - m^2 > n^2 - m^2$$
$$m(n - m) > (n + m)(n - m)$$
$$m > n + m$$
$$0 > n$$

But it was assumed $n > 0$. Can you find the error?

APPLICATIONS   51.   *Earth science.*   As dry air moves upward it expands, and in so doing cools at a rate of about 5.5°F for each 1,000-foot rise up to about 40,000 feet. If the ground temperature is 70°F, then the temperature T at height h is given approximately by $T = 70 - 0.0055h$. For what range in altitude will the temperature be between 26° and -40°F?

52.   *Energy.*   If the power demands in a 110-volt electric circuit in a home vary between 220 and 2,750 watts, what is the range of current flowing through the circuit? ($W = EI$, where $W =$ Power in watts, $E =$ Pressure in volts, and $I =$ Current in amperes.)

53.   *Business and economics.*   For a business to make a profit it is clear that revenue R must be greater than cost C; in short, a profit will result only if $R > C$. If a company manufactures records and its cost equation for a week is $C = 300 + 1.5x$ and its revenue equation is $R = 2x$, where x is the number of records sold in a week, how many records must be sold for the company to realize a profit?

54.   *Psychology.*   IQ is given by the formula

$$IQ = \frac{MA}{CA} 100$$

where MA is mental age and CA is chronological age. If

$$80 \leq IQ \leq 140$$

for a group of 12-year-old children, find the range of their mental ages.

Section 1-5   Absolute Value in Equations and Inequalities

- ■ Absolute Value and Distance
- ■ Absolute Value in Equations and Inequalities

■ **Absolute Value and Distance**

We start with a geometric definition of absolute value. If a is the co-ordinate of a point on a real number line, then the (nondirected) distance from the origin to a, a nonnegative quantity, is represented by $|a|$ and is referred to as the **absolute value** of a (Fig. 8). Thus, if $|x| = 5$, then x can be either -5 or 5.

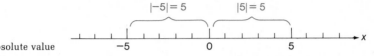

FIGURE 8   Absolute value

Symbolically, and more formally, recall (see Section 0-4) that we defined absolute value as follows:

Absolute Value
$$|x| = \begin{cases} x & \text{if } x \text{ is positive} \\ 0 & \text{if } x \text{ is 0} \\ -x & \text{if } x \text{ is negative} \end{cases}$$
[*Note:* $-x$ is positive if x is negative.]

Both the geometric and nongeometric definitions of absolute value are useful, as will be seen in the material that follows. Remember:

The absolute value of a number is never negative.

EXAMPLE 21
(A) $|7| = 7$

(B) $|\pi - 3| = \pi - 3$ Since $\pi - 3$ is nonnegative

(C) $|-7| = -(-7) = 7$

(D) $|3 - \pi| = -(3 - \pi) = \pi - 3$ Since $3 - \pi$ is negative

PROBLEM 21
Write without the absolute value sign.

(A) $|8|$ (B) $\left|\sqrt[3]{9} - 2\right|$ (C) $\left|-\sqrt{2}\right|$ (D) $\left|2 - \sqrt[3]{9}\right|$

Following the same reasoning used in Example 21B and D, it can be shown (see Problem 61 in Exercise 1-5) that

---

For all real numbers a and b,

$$|b - a| = |a - b|$$

We use this result in defining the distance between two points on a real number line.

Distance between Points A and B

Let A and B be two points on a real number line with coordinates a and b, respectively. The **distance between A and B** (also called the **length of the line segment** joining A and B) is given by

$$d(A, B) = |b - a|$$

EXAMPLE 22   Find the distance between points A and B with coordinates a and b, respectively, as given.

(A) $a = 4, \quad b = 9$ (B) $a = 9, \quad b = 4$
(C) $a = 0, \quad b = 6$ (D) $a = -3, \quad b = 5$

Solution

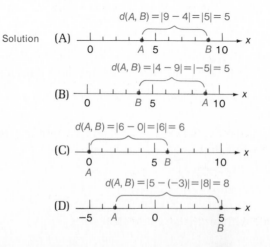

$d(A, B) = |9 - 4| = |5| = 5$

(A)

$d(A, B) = |4 - 9| = |-5| = 5$

(B)

$d(A, B) = |6 - 0| = |6| = 6$

(C)

$d(A, B) = |5 - (-3)| = |8| = 8$

(D)

It is clear, since $|b - a| = |a - b|$, that

d(A, B) = d(B, A)

Hence, in computing the distance between two points on a real number line, it does not matter how the two points are labeled—point A can be to the left or to the right of point B. Note also that if A is at the origin O, then

d(O, B) = |b - 0| = |b|

PROBLEM 22   Find the indicated distances given.

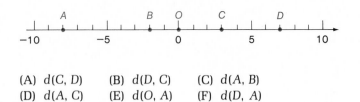

(A) $d(C, D)$ (B) $d(D, C)$ (C) $d(A, B)$
(D) $d(A, C)$ (E) $d(O, A)$ (F) $d(D, A)$

■   **Absolute Value in Equations and Inequalities**

Absolute value is frequently encountered in equations and inequalities. Some of these forms have immediate geometric interpretation.

EXAMPLE 23   Solve geometrically and graph.

(A) $|x - 3| = 5$ (B) $|x - 3| < 5$
(C) $0 < |x - 3| < 5$ (D) $|x - 3| > 5$

Solution   (A) Geometrically, $|x - 3|$ represents the distance between x and 3; thus, in $|x - 3| = 5$, x is a number whose distance from 3 is 5. That is,

x = 3 ± 5 = −2   or   8

(B) Geometrically, in $|x - 3| < 5$, x is a number whose distance from 3 is less than 5; that is,

$-2 < x < 8$ or $(-2, 8)$

(C) The form $0 < |x - 3| < 5$ is encountered in calculus and more advanced mathematics. Geometrically, x is a number whose distance from 3 is less than 5, but x cannot equal 3. Thus,

$$-2 < x < 8 \qquad x \neq 3$$

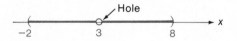

(D) Geometrically, in $|x - 3| > 5$, x is a number whose distance from 3 is greater than 5; that is,

$$x < -2 \quad \text{or} \quad x > 8$$

Note: This cannot be written as a double inequality.

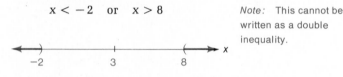

We summarize the preceding results in Table 2.

TABLE 2

FORM ($d > 0$)	GEOMETRIC INTERPRETATION	GRAPH
$|x - c| = d$	Distance between x and c is equal to d.	⊢—d—⊣⊢—d—⊣ •————+————•→ x $c - d$ c $c + d$
$|x - c| < d$	Distance between x and c is less than d.	(————+————)→ x $c - d$ c $c + d$
$0 < |x - c| < d$	Distance between x and c is less than c, but $x \neq c$.	(————○————)→ x $c - d$ c $c + d$
$|x - c| > d$	Distance between x and c is greater than d.	←——)———+———(——→ x $c - d$ c $c + d$

PROBLEM 23   Solve geometrically and graph.

(A) $|x + 2| = 6$ (B) $|x + 2| < 6$

(C) $0 < |x + 2| < 6$ (D) $|x + 2| > 6$

$\left[\textit{Hint:} \quad |x + 2| = |x - (-2)|. \right]$

Reasoning geometrically as before (noting that $|x| = |x - 0|$), we can establish Theorem 3.

THEOREM 3

For $p > 0$,

1. $|x| = p$ is equivalent to $x = \pm p$.
2. $|x| < p$ is equivalent to $-p < x < p$.
3. $|x| > p$ is equivalent to $x < -p$ or $x > p$.

If we replace x in Theorem 3 with $ax + b$, we obtain the more general Theorem 4.

THEOREM 4

For $p > 0$,

1. $|ax + b| = p$ is equivalent to $ax + b = \pm p$.
2. $|ax + b| < p$ is equivalent to $-p < ax + b < p$.
3. $|ax + b| > p$ is equivalent to $ax + b < -p$ or $ax + b > p$.

EXAMPLE 24   Solve.

(A) $|3x + 5| = 4$ (B) $|x| < 5$
(C) $|2x - 1| < 3$ (D) $|7 - 3x| \leq 2$

Solution   (A) $|3x + 5| = 4$ (B) $|x| < 5$

$$3x + 5 = \pm 4 \qquad\qquad\qquad -5 < x < 5$$
$$3x = -5 \pm 4$$
$$x = \frac{-5 \pm 4}{3}$$
$$x = -3, \ -\tfrac{1}{3}$$

(C) $|2x - 1| < 3$ (D) $|7 - 3x| \leq 2$

$$-3 < 2x - 1 < 3 \qquad\qquad -2 \leq 7 - 3x \leq 2$$
$$-2 < 2x < 4 \qquad\qquad\qquad -9 \leq -3x \leq -5$$
$$-1 < x < 2 \qquad\qquad\qquad\quad 3 \geq x \geq \tfrac{5}{3}$$
$$\tfrac{5}{3} \leq x \leq 3$$

PROBLEM 24   Solve.

(A) $|2x - 1| = 8$ (B) $|x| \leq 7$
(C) $|3x + 3| \leq 9$ (D) $|5 - 2x| < 9$

EXAMPLE 25   Solve.

(A) $|x| > 3$ (B) $|2x - 1| \geq 3$ (C) $|7 - 3x| > 2$

Solution   (A) $|x| > 3$

$x < -3$ or $x > 3$

(B) $|2x - 1| \geq 3$

$2x - 1 < -3$ or $2x - 1 > 3$

$2x < -2$ or $2x > 4$

$x < -1$ or $x > 2$

(C) $|7 - 3x| > 2$

$7 - 3x < -2$ or $7 - 3x > 2$

$-3x < -9$ or $-3x > -5$

$x > 3$ or $x < \frac{5}{3}$

PROBLEM 25   Solve.

(A) $|x| \geq 5$ (B) $|4x - 3| > 5$ (C) $|6 - 5x| > 16$

Answers to Matched Problems   **21.**   (A)   8   (B) $\sqrt[3]{9} - 2$ (C) $\sqrt{2}$ (D) $\sqrt[3]{9} - 2$

22.   (A)   4   (B)   4   (C)   6   (D)   11   (E)   8   (F)   15

23.   (A) $x = -8, 4$

(B) $-8 < x < 4$ or $(-8, 4)$

(C) $-8 < x < 4$, $x \neq -2$

(D) $x < -8$ or $x > 4$

24.   (A) $-\frac{7}{2}, \frac{9}{2}$ (B) $-7 \leq x \leq 7$ (C) $-4 \leq x \leq 2$
(D) $-2 < x < 7$

25.   (A) $x \leq -5$ or $x \geq 5$ (B) $x < -\frac{1}{2}$ or $x > 2$
(C) $x < -2$ or $x > \frac{22}{5}$

Exercise 1-5 ■   A   *Simplify, and write without absolute value signs. Leave radicals in simplest radical form.*

1. $|\sqrt{5}|$ **2.** $|-\frac{3}{4}|$ **3.** $|(-6) - (-2)|$

4. $|(-2) - (-6)|$ **5.** $|5 - \sqrt{5}|$ **6.** $|\sqrt{7} - 2|$

7. $|\sqrt{5} - 5|$ **8.** $|2 - \sqrt{7}|$

Find the distance between points A and B with coordinates a and b, respectively, as given.

9. $a = -7, \quad b = 5$

10. $a = 3, \quad b = 12$

11. $a = 5, \quad b = -7$

12. $a = 12, \quad b = 3$

13. $a = -16, \quad b = -25$

14. $a = -9, \quad b = -17$

Find the indicated distances, given

15. $d(B, O)$

16. $d(A, B)$

17. $d(O, B)$

18. $d(B, A)$

19. $d(B, C)$

20. $d(D, C)$

Solve and graph.

21. $|x| = 7$

22. $|x| = 5$

23. $|x| \le 7$

24. $|t| \le 5$

25. $|x| \ge 7$

26. $|x| \ge 5$

27. $|y - 5| = 3$

28. $|t - 3| = 4$

29. $|y - 5| < 3$

30. $|t - 3| < 4$

31. $|y - 5| > 3$

32. $|t - 3| > 4$

33. $|u + 8| = 3$

34. $|x + 1| = 5$

35. $|u + 8| \le 3$

36. $|x + 1| \le 5$

37. $|u + 8| \ge 3$

38. $|x + 1| \ge 5$

B *Solve.*

39. $|3x + 4| = 8$

40. $|2x - 3| = 5$

41. $|5x - 3| \le 12$

42. $|2x - 3| \le 5$

43. $|2y - 8| > 2$

44. $|3u + 4| > 3$

45. $|5t - 7| = 11$

46. $|6m + 9| = 13$

47. $|9 - 7u| < 14$

48. $|7 - 9M| < 15$

49. $|1 - \frac{2}{3}x| \ge 5$

50. $|\frac{3}{4}x + 3| \ge 9$

51. $|\frac{9}{5}C + 32| < 31$

52. $|\frac{5}{9}(F - 32)| < 40$

C *For what values of x does each of the following hold?*

53. $|x - 5| = x - 5$

54. $|x + 7| = x + 7$

55. $|x + 8| = -(x + 8)$

56. $|x - 11| = -(x - 11)$

57. $|4x + 3| = 4x + 3$

58. $|5x - 9| = (5x - 9)$

59. $|5x - 2| = -(5x - 2)$

60. $|3x + 7| = -(3x + 7)$

61. Show that $|b - a| = |a - b|$ for all real numbers a and b.

62. Prove that $|x|^2 = x^2$ for all real numbers x.

Section 1-6 Nonlinear Inequalities

■ Polynomial Inequalities
■ Rational Inequalities

■ **Polynomial Inequalities**

You now know how to solve first-degree (linear) inequalities such as

$$3x - 7 \geq 5(x - 2) + 3$$

But how do we solve second-degree (quadratic) inequalities such as

$$x^2 - 12 < x$$

If, after collecting all nonzero terms on the left, we find that we are able to factor the left side in terms of first-degree factors, then we will be able to solve the inequality.

$$x^2 - 12 < x \quad \text{Move all nonzero terms to the left side.}$$
$$x^2 - x - 12 < 0 \quad \text{Factor left side.}$$
$$(x + 3)(x - 4) < 0$$

We are looking for values of x that will make the left side less than 0—that is, negative. What will the signs of $(x + 3)$ and $(x - 4)$ have to be so that their product is negative? They must have opposite signs.

Let us see whether we can determine where each of the factors is positive, negative, and 0. The point at which either factor is 0 is called a **critical point**. We will see why in a moment.

Sign analysis for $(x + 3)$:

CRITICAL POINT	$(x + 3)$ IS POSITIVE WHEN	$(x + 3)$ IS NEGATIVE WHEN
$x + 3 = 0$	$x + 3 > 0$	$x + 3 < 0$
$x = -3$	$x > -3$	$x < -3$

It is useful to summarize these results on a real number line, as shown in Figure 9:

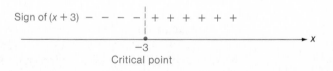

FIGURE 9

Thus, $(x + 3)$ is negative for values of x to the left of -3 and is positive for values of x to the right of -3.

Sign analysis for $(x - 4)$:

CRITICAL POINT	$(x - 4)$ IS POSITIVE WHEN	$(x - 4)$ IS NEGATIVE WHEN
$x - 4 = 0$	$x - 4 > 0$	$x - 4 < 0$
$x = 4$	$x > 4$	$x < 4$

This is illustrated geometrically in Figure 10:

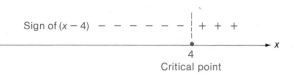

FIGURE 10

Thus, $(x - 4)$ is negative for values of x to the left of 4 and is positive for values of x to the right of 4.

Combining the results on a single real number line (Fig. 11) leads to a simple solution to the original problem.

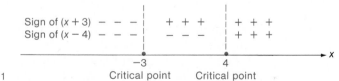

FIGURE 11

We see that the factors have opposite signs (thus their product is negative) for x between -3 and 4. We can now give the solution and graph for $x^2 - 12 \leq x$:

$-3 < x < 4$ Inequality notation

$(-3, 4)$ Interval notation

Proceeding as in the above example, we can easily prove Theorem 5, which is behind the sign-analysis method of solving second- and higher-degree inequalities as well as other types of inequalities.

THEOREM 5

The value of x at which $(ax + b)$ is 0 is called the **critical point** for $ax + b$. To the left of this critical point on a real number line $(ax + b)$ has one sign and to the right of this critical point $(ax + b)$ has the opposite sign $(a \neq 0)$.

EXAMPLE 26

Solve and graph: $3x^2 + 10x \geq 8$

Solution

$$3x^2 + 10x \geq 8 \qquad \text{Move all nonzero terms to the left side.}$$

$$3x^2 + 10x - 8 \geq 0 \qquad \text{Factor the left side (if possible).}$$

$$(3x - 2)(x + 4) \geq 0 \qquad \text{Find critical points.}$$

Critical points: $-4, \frac{2}{3}$

Locate the critical points on a real number line and determine the sign of each linear factor to the left and right of its critical point.

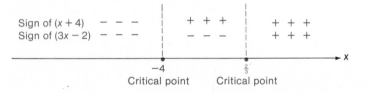

Note that the equality part of the inequality statement is satisfied at the critical points. The inequality part is satisfied when the product of the factors is positive—that is, when the factors have the same sign. From the figure we see that this happens to the left of -4 or to the right of $\frac{2}{3}$. We can now give the solution and graph.

$$x \leq -4 \quad \text{or} \quad x \geq \tfrac{2}{3} \qquad \text{Inequality notation}$$

$$(-\infty, -4] \cup [\tfrac{2}{3}, \infty) \qquad \text{Interval notation}$$

PROBLEM 26

Solve and graph: $2x^2 \geq 3x + 9$

EXAMPLE 27

Solve and graph: $x^3 - 4 \leq x - 4x^2$

Solution

$$x^3 - 4 \leq x - 4x^2 \qquad \text{Move all nonzero terms to left side.}$$

$$x^3 + 4x^2 - x - 4 \leq 0 \qquad \text{Factor left side by grouping.}$$

$$(x^3 + 4x^2) - (x + 4) \leq 0$$

$$x^2(x + 4) - (x + 4) \leq 0$$

$$(x^2 - 1)(x + 4) \leq 0$$

$$(x - 1)(x + 1)(x + 4) \leq 0 \qquad \text{Find critical points.}$$

Critical points: $-4, -1, 1$

Equality holds at all the critical points. The inequality holds when the left side is less than 0—that is, when the left side is negative. The left side is negative when $(x - 1)$, $(x + 1)$, and $(x + 4)$ are all negative or when one is negative and two are positive. We chart the sign of each factor on a real number line.

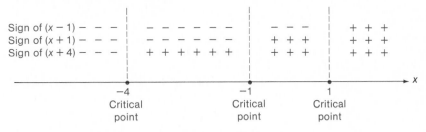

The solution is easily determined from the figure.

$$x \leq -4 \quad \text{or} \quad -1 \leq x \leq 1 \quad \text{Inequality notation}$$

$$(-\infty, -4] \cup [-1, 1] \qquad \text{Interval notation}$$

PROBLEM 27 Solve and graph: $x^3 + 12 > 3x^2 + 4x$

Remark: The key to solving polynomial inequalities is factoring. At this point we are able to factor only a few very special types of polynomials. In Chapter 3 procedures will be developed to help us factor polynomials of higher degree. We will then be able to apply the factoring technique of solving polynomial inequalities to a much wider class of problems.

■ Rational Inequalities

The sign-analysis technique described for solving polynomial inequalities can also be used to solve inequalities involving rational forms such as

$$\frac{x - 3}{x + 5} > 0 \qquad \frac{x^2 + 5x - 6}{5 - x} \leq 1$$

EXAMPLE 28 Solve and graph: $\dfrac{x^2 - x + 1}{2 - x} \geq 1$

Solution   We might be tempted to start by multiplying both sides by $2 - x$ (as we would do if the inequality were an equation). However, since we do not know whether $2 - x$ is positive or negative, we do not know whether the sense of the inequality is to be changed.

We proceed instead as follows:

$$\frac{x^2 - x + 1}{2 - x} \geq 1 \quad \text{Move all nonzero terms to left side.}$$

$$\frac{x^2 - x + 1}{2 - x} - 1 \geq 0 \quad \text{Combine left side into a single fraction.}$$

$$\frac{x^2 - x + 1 - (2 - x)}{2 - x} \geq 0$$

$$\frac{x^2 - 1}{2 - x} \geq 0 \quad \text{Factor numerator.}$$

$$\frac{(x - 1)(x + 1)}{2 - x} \geq 0$$

Critical points: $-1, 1, 2$. Equality holds when $x = \pm 1$. The left side is not defined when $x = 2$.

The inequality holds when $(x - 1)$, $(x + 1)$, and $(2 - x)$ are all positive or two are negative and one is positive. We chart the sign of each on a real number line:

```
Sign of (x − 1)  − − −  |  − − − − − −  |  + + +  |  + + +
Sign of (x + 1)  − − −  |  + + + + + +  |  + + +  |  + + +
Sign of (2 − x)  + + +  |  + + + + + +  |  + + +  |  − − −
─────────────────●──────────────────────●────────○───────→ x
               −1                        1        2
             Critical                 Critical  Critical
              point                    point     point
```

Note the sign pattern for $(2 - x)$. It is positive to the left of its critical point and negative to the right. From the figure it is easy to write the solution.

$x \leq -1 \quad \text{or} \quad 1 \leq x < 2 \quad$ Inequality notation

$(-\infty, -1] \cup [1, 2) \qquad$ Interval notation

```
◄────┤        ┌────)───→ x
    −1   0    1    2
```

PROBLEM 28   Solve and graph: $\dfrac{3}{2 - x} \leq \dfrac{1}{x + 4}$

Answers to Matched Problems   **26.** $x \le -\frac{3}{2}$ or $x \ge 3$

$(-\infty, -\frac{3}{2}] \cup [3, \infty)$

27. $-2 < x < 2$ or $x > 3$

$(-2, 2) \cup (3, \infty)$

28. $-4 < x \le -\frac{5}{2}$ or $x > 2$

$(-4, -\frac{5}{2}] \cup (2, \infty)$

Exercise 1-6 ■ *Solve and graph. Express answers in both inequality and interval notation.*

A **1.** $x^2 - x - 12 < 0$ **2.** $x^2 - 2x - 8 < 0$

3. $x^2 - x - 12 \ge 0$ **4.** $x^2 - 2x - 8 \ge 0$

5. $x^2 < 10 - 3x$ **6.** $x^2 + x < 12$

7. $x^2 + 21 > 10x$ **8.** $x^2 + 7x + 10 > 0$

9. $x^2 \le 8x$ **10.** $x^2 + 6x \ge 0$

11. $x^2 + 5x \le 0$ **12.** $x^2 \le 4x$

13. $x^2 > 4$ **14.** $x^2 \le 9$

B **15.** $x^2 + 9 \ge 6x$ **16.** $x^2 + 4 \ge 4x$

17. $x^3 + 5 \ge 5x^2 + x$ **18.** $x^3 + x^2 < 9x + 9$

19. $x^3 + 75 < 3x^2 + 25x$ **20.** $x^3 + 4x^2 \ge 4x + 16$

21. $\dfrac{x-2}{x+4} \le 0$ **22.** $\dfrac{x+3}{x-1} \ge 0$

23. $\dfrac{x^2 + 5x}{x-3} \ge 0$ **24.** $\dfrac{x-4}{x^2 + 2x} \le 0$

25. $\dfrac{x+4}{1-x} \le 0$ **26.** $\dfrac{3-x}{x+5} \le 0$

27. $\dfrac{1}{x} < 4$ **28.** $\dfrac{5}{x} > 3$

29. $\dfrac{2x}{x+3} \ge 1$ **30.** $\dfrac{2}{x-3} \le -2$

31. $\dfrac{3x+1}{x+4} \le 1$ **32.** $\dfrac{5x-8}{x-5} \ge 2$

33. $\dfrac{2}{x+1} \ge \dfrac{1}{x-2}$ **34.** $\dfrac{3}{x-3} \le \dfrac{2}{x+2}$

C   **35.** $x^2 + 1 < 2x$ **36.** $x^2 + 25 < 10x$

   37. $x^3 + 5x > 4x^2 + 20$ **38.** $x^3 + 3x^2 + x + 3 < 0$

   39. $4x^4 + 4 \leq 17x^2$ **40.** $x^4 + 36 \geq 13x^2$

   41. $\left|x^2 - 1\right| \leq 3$ **42.** $\left|\dfrac{x + 1}{x}\right| > 2$

Section 1-7   Complex Numbers

- Introductory Remarks
- The Complex Number System
- Complex Numbers and Radicals

■ **Introductory Remarks**

The Pythagoreans (500–275 BC) found that the simple equation

$$x^2 = 2 \tag{1}$$

had no rational number solutions. If equation (1) were to have a solution, then a new kind of number had to be invented—the irrational numbers. The irrational numbers $\sqrt{2}$ and $-\sqrt{2}$ are both solutions to (1). Irrational numbers were not put on a firm mathematical foundation until the last century. The rational and irrational numbers together constitute the real number system.

Is there any need to extend the real number system further? Yes, since we find that another simple equation

$$x^2 = -1$$

has no real solutions (what real number squared is negative?). Once again, we are forced to invent a new kind of number, a number that has the possibility of being negative when it is squared. These new numbers are called **complex numbers**. The complex numbers evolved over a long period of time,* but, like the real numbers, it was not until the last century that they were placed on a firm mathematical foundation.

---

* BRIEF HISTORY OF COMPLEX NUMBERS

Approximate Date	Person	Event
50	Heron of Alexandria	First recorded encounter of a square root of a negative number
850	Mahavira of India	Said that a negative has no square root, since it is not a square
1545	Cardano of Italy	Solutions to cubic equations involved square roots of negative numbers.
1637	Descartes of France	Introduced the terms *real* and *imaginary*
1748	Euler of Switzerland	Used *i* for $\sqrt{-1}$
1832	Gauss of Germany	Introduced the term *complex number*

■ The Complex Number System

A **complex number** is a number of the form

$$a + bi$$

where a and b are real numbers and i is called the **imaginary unit**. Thus,

$$3 - 2i \qquad \tfrac{1}{2} + 5i \qquad 2 - \tfrac{1}{3}i$$
$$0 + 3i \qquad 5 + 0i \qquad 0 + 0i$$

are all complex numbers. Particular kinds of complex numbers are given special names as follows:

REAL NUMBERS:	$a + 0i = a$
PURE IMAGINARY NUMBERS:	$0 + bi = bi$
ZERO:	$0 + 0i = 0$
IMAGINARY UNIT:	$1i = i$
CONJUGATE OF $a + bi$:	$a - bi$

Thus, we see that just as every integer is a rational number, every real number is a complex number; that is, the real numbers form a subset of the set of complex numbers. The complex number system is related to the other number systems that we have studied as shown in Figure 12.

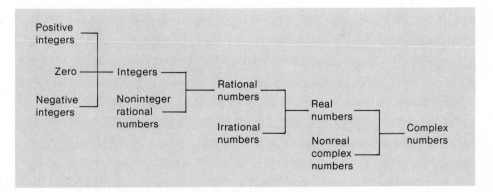

FIGURE 12

[*Note*: $a + bi$ is a real number if $b = 0$; $a + bi$ is a nonreal complex number if $b \neq 0$.]

To use complex numbers we must know how to add, subtract, multiply, and divide them. We start by defining equality, addition, and multiplication.

EQUALITY: $a + bi = c + di$ if and only if $a = c$ and $b = d$

ADDITION: $(a + bi) + (c + di) = (a + c) + (b + d)i$

MULTIPLICATION: $(a + bi)(c + di) = (ac - bd) + (ad + bc)i$

The definitions, particularly the one for multiplication, may seem a little strange to you. But it turns out that if we want many of the properties for real numbers (commutative, associative, distributive, etc.) to continue to hold for complex numbers, and if we also want the possibility of having the square of a number negative, then we must define addition and multiplication in this way. Let us use the definition of multiplication to see what happens to i when it is squared:

$$\overset{\displaystyle a \quad\; b \;\; c \quad\; d}{i^2 = (0 + 1i)(0 + 1i)}$$

$$\overset{\displaystyle a \;\; c \quad b \;\; d \quad\;\; a \;\; d \quad b \;\; c}{= (0 \cdot 0 - 1 \cdot 1) + (0 \cdot 1 + 1 \cdot 0)i}$$

$$= -1 + 0i$$

$$= -1$$

Thus,

$$i^2 = -1$$

---

and we have a number whose square is negative (and a solution to $x^2 = -1$). We choose to let

---

$$i = \sqrt{-1} \quad \text{and} \quad -i = -\sqrt{-1}$$

Fortunately, you do not have to memorize the definitions of addition and multiplication. We can show that the complex numbers under these definitions are associative and commutative and that multiplication distributes over addition. As a consequence, we can manipulate complex numbers as if they were binomial forms in real number algebra, with the exception that i^2 is to be replaced with -1. The following example illustrates the mechanics of carrying out addition, subtraction, multiplication, and division.

EXAMPLE 29   Carry out the following operations and write each answer in the form $a + bi$.

(A) $(2 - 3i) + (6 + 2i)$ (B) $(7 - 3i) - (6 + 2i)$
(C) $(2 - 3i)(6 + 2i)$ (D) $(3 + 2i)/(5 + 3i)$

Solution   (A) $(2 - 3i) + (6 + 2i) = 2 - 3i + 6 + 2i$ Remove parentheses and
                                 $= 8 - i$ combine like terms.

(B) $(7 - 3i) - (6 + 2i) = 7 - 3i - 6 - 2i$ Remove parentheses and
                                $= 1 - 5i$ combine like terms.

(C) $(2 - 3i)(6 + 2i) = 12 - 14i - 6i^2$ Multiply; then replace i^2
                          $= 12 - 14i - 6(-1)$ with -1.
                          $= 12 - 14i + 6$
                          $= 18 - 14i$

(D) To eliminate i from the denominator, we multiply the numerator and denominator by the conjugate of $5 + 3i$—that is, by $5 - 3i$:

$$\frac{3 + 2i}{5 + 3i} \cdot \frac{5 - 3i}{5 - 3i} = \frac{15 + i - 6i^2}{25 - 9i^2}$$

$$= \frac{15 + i + 6}{25 + 9}$$

$$= \frac{21 + i}{34} = \frac{21}{34} + \frac{1}{34}i \quad \text{Form } a + bi.$$

PROBLEM 29   Carry out the indicated operations and write each answer in the $a + bi$ form.

(A) $(3 + 2i) + (6 - 4i)$ (B) $(3 - 5i) - (1 - 3i)$

(C) $(2 - 4i)(3 + 2i)$ (D) $\dfrac{2 + 4i}{3 + 2i}$

EXAMPLE 30   Carry out the indicated operations and write each answer in the form $a + bi$.

(A) $(3 - 2i)^2 - 6(3 - 2i) + 13$ (B) $\dfrac{2 - 3i}{2i}$

Solution (A) $(3 - 2i)^2 - 6(3 - 2i) + 13 = 9 - 12i + 4i^2 - 18 + 12i + 13$

$$= 9 - 12i - 4 - 18 + 12i + 13$$

$$= 0 + 0i \quad \text{or} \quad 0$$

(B) $\dfrac{2 - 3i}{2i} \cdot \dfrac{i}{i} = \dfrac{2i - 3i^2}{2i^2} = \dfrac{2i + 3}{-2} = -\dfrac{3}{2} - i$

PROBLEM 30 Carry out the indicated operations and write each answer in the form $a + bi$.

(A) $(3 + 2i)^2 - 6(3 + 2i) + 13$ (B) $\dfrac{4 - i}{3i}$

■ Complex Numbers and Radicals

Recall that we say that a is a square root of b if $a^2 = b$. It can be shown that if x is a positive or negative real number, then x has two square roots. We denote one by $\sqrt{x}$ and the other by $-\sqrt{x}$. If x is negative, the square roots of x are complex numbers. If we let $x = -a, a > 0$, then

$$\sqrt{-a} = i\sqrt{a} \qquad a > 0$$

This is readily verified by noting that $(i\sqrt{a})^2 = i^2 a = -a$. Hence, a square root of any negative real number can be written as the product of i and a square root of a positive real number. Thus,

$$\sqrt{-3} = i\sqrt{3} \qquad \sqrt{-4} = i\sqrt{4} = 2i$$

What are $\sqrt{-7}$ and $\sqrt{-9}$? [Answer: $i\sqrt{7}$ and $3i$]

EXAMPLE 31 Write in the form $a + bi$.

(A) $\sqrt{-4}$ (B) $4 + \sqrt{-5}$

(C) $\dfrac{-3 - \sqrt{-5}}{2}$ (D) $\dfrac{1}{1 - \sqrt{-9}}$

Solution (A) $\sqrt{-4} = i\sqrt{4} = 2i$

(B) $4 + \sqrt{-5} = 4 + i\sqrt{5}$

(C) $\dfrac{-3 - \sqrt{-5}}{2} = \dfrac{-3 - i\sqrt{5}}{2} = -\dfrac{3}{2} - \dfrac{\sqrt{5}}{2}i$

(D) $\dfrac{1}{1 - \sqrt{-9}} = \dfrac{1}{1 - 3i} = \dfrac{1}{(1 - 3i)} \dfrac{(1 + 3i)}{(1 + 3i)}$

$= \dfrac{1 + 3i}{1 - 9i^2} = \dfrac{1 + 3i}{10} = \dfrac{1}{10} + \dfrac{3}{10}i$

PROBLEM 31   Write in the form $a + bi$.

(A) $\sqrt{-16}$ (B) $5 + \sqrt{-7}$ (C) $\dfrac{-5 - \sqrt{-2}}{2}$ (D) $\dfrac{1}{3 - \sqrt{-4}}$

Early resistance to these new numbers is suggested by the words used
to name them: *complex* and *imaginary*. In spite of this early resistance,
complex numbers have come into widespread use in both pure and
applied mathematics. They are used extensively, for example, in electrical
engineering, physics, chemistry, statistics, and aeronautical engineering.
Our first use of them will be in connection with solutions of second-
degree equations in the next section.

Answers to Matched Problems   **29.** (A) $9 - 2i$ (B) $2 - 2i$ (C) $14 - 8i$ (D) $\frac{14}{13} + \frac{8}{13}i$

30. (A) 0 (B) $-\frac{1}{3} - \frac{4}{3}i$

31. (A) $4i$ (B) $5 + i\sqrt{7}$ (C) $-\frac{5}{2} - (\sqrt{2}/2)i$ (D) $\frac{3}{13} + \frac{2}{13}i$

Exercise 1-7 ■ A

Perform the indicated operations and write each answer in the form
$a + bi$.

1. $(2 + 4i) + (5 + i)$ **2.** $(3 + i) + (4 + 2i)$

3. $(-2 + 6i) + (7 - 3i)$ **4.** $(6 - 2i) + (8 - 3i)$

5. $(6 + 7i) - (4 + 3i)$ **6.** $(9 + 8i) - (5 + 6i)$

7. $(3 + 5i) - (-2 - 4i)$ **8.** $(8 - 4i) - (11 - 2i)$

9. $(4 - 5i) + 2i$ **10.** $6 + (3 - 4i)$

11. $(4i)(6i)$ **12.** $(3i)(8i)$

13. $-3i(2 - 4i)$ **14.** $-2i(5 - 3i)$

15. $(3 + 3i)(2 - 3i)$ **16.** $(-2 - 3i)(3 - 5i)$

17. $(2 - 3i)(7 - 6i)$ **18.** $(3 + 2i)(2 - i)$

19. $(7 + 4i)(7 - 4i)$ **20.** $(5 + 3i)(5 - 3i)$

21. $\dfrac{1}{2 + i}$ **22.** $\dfrac{1}{3 - i}$

23. $\dfrac{3 + i}{2 - 3i}$ **24.** $\dfrac{2 - i}{3 + 2i}$

25. $\dfrac{13 + i}{2 - i}$ **26.** $\dfrac{15 - 3i}{2 - 3i}$

B Convert square roots of negative numbers to complex form, perform the indicated operations, and express answers in the form $a + bi$.

27. $(2 - \sqrt{-4}) + (5 - \sqrt{-9})$ **28.** $(3 - \sqrt{-4}) + (-8 + \sqrt{-25})$

29. $(9 - \sqrt{-9}) - (12 - \sqrt{-25})$ **30.** $(-2 - \sqrt{-36}) - (4 + \sqrt{-49})$

31. $(3 - \sqrt{-4})(-2 + \sqrt{-49}$ **32.** $(2 - \sqrt{-1})(5 + \sqrt{-9})$

33. $\dfrac{5 - \sqrt{-4}}{7}$ **34.** $\dfrac{6 - \sqrt{-64}}{2}$

35. $\dfrac{1}{2 - \sqrt{-9}}$ **36.** $\dfrac{1}{3 - \sqrt{-16}}$

Write in the form $a + bi$.

37. $\dfrac{2}{5i}$ **38.** $\dfrac{1}{3i}$

39. $\dfrac{1 + 3i}{2i}$ **40.** $\dfrac{2 - i}{3i}$

41. $(2 - 3i)^2 - 2(2 - 3i) + 9$ **42.** $(2 - i)^2 + 3(2 - i) - 5$

43. Evaluate $x^2 - 2x + 2$ for $x = 1 - i$.

44. Evaluate $x^2 - 2x + 2$ for $x = 1 + i$.

45. Simplify: $i^2,\ i^3,\ i^4,\ i^5,\ i^6,\ i^7,\ i^8$

46. Simplify: $i^{12},\ i^{13},\ i^{14},\ i^{15},\ i^{16}$

stop _____

47. For what real values of x and y will $(2x - 1) + (3y + 2)i = 5 - 4i$?

48. For what real values of x and y will $3x + (y - 2)i = (5 - 2x) + (3y - 8)i$?

C Perform the indicated operations and write each answer in the form $a + bi$.

49. $(a + bi) + (c + di)$ **50.** $(a + bi) - (c + di)$

51. $(a + bi)(a - bi)$ **52.** $(u - vi)(u + vi)$

53. $(a + bi)(c + di)$ **54.** $\dfrac{a + bi}{c + di}$

55. Show that $i^{4k} = 1, k \in N$.   **56.** Show that $i^{4k+1} = i, k \in N$.

Supply the reasons in the proofs for the following two theorems.

57. Theorem: The complex numbers are commutative under addition.
   Proof: Let $a + bi$ and $c + di$ be two arbitrary complex numbers; then

STATEMENT	REASON
1. $(a + bi) + (c + di) = (a + c) + (b + d)i$	**1.**
2. $\qquad\qquad\quad = (c + a) + (d + b)i$	**2.**
3. $\qquad\qquad\quad = (c + di) + (a + bi)$	**3.**

58. *Theorem:* The complex numbers are commutative under multi-plication.

Proof: Let $a + bi$ and $c + di$ be two arbitrary complex numbers; then

STATEMENT	REASON
1. $(a \cdot bi) \cdot (c + di) = (ac - bd) + (ad + bc)i$	**1.**
2. $\qquad\qquad\quad = (ca - db) + (da + cb)i$	**2.**
3. $\qquad\qquad\quad = (c + di)(a + bi)$	**3.**

Section 1-8 Quadratic Equations

- Solution by Factoring
- Solution by Square Root
- Solution by Completing the Square
- Solution by Quadratic Formula

The next class of equations we will consider are the second-degree poly-nomial equations in one variable, called *quadratic equations. A* **quadratic equation** in one variable is any equation that can be written in the following form:

$$\text{QUADRATIC EQUATION, STANDARD FORM:} \quad ax^2 + bx + c = 0 \qquad a \neq 0$$

where x is a variable and a, b, and c are constants. We will refer to this form as the **standard form** for the quadratic equation.

- Solution by Factoring

If the coefficients a, b, and c are integers and are such that $ax^2 + bx + c$ can be written as the product of two first-degree factors with integral coefficients, then the quadratic equation can be quickly and easily solved. The method of solution by factoring rests on the zero property of real numbers.

Zero Property

If m and n are real numbers, then

$mn = 0$ if and only if $m = 0$ or $n = 0$ (or both).

EXAMPLE 32   Solve by factoring if possible.

(A) $6x^2 - 19x - 7 = 0$ (B) $2x^2 - 8x + 3 = 0$ (C) $2x^2 - 3x$

Solution   (A) $6x^2 - 19x - 7 = 0$

$(2x - 7)(3x + 1) = 0$

$2x - 7 = 0$ or $3x + 1 = 0$

$x = \frac{7}{2}$ $x = -\frac{1}{3}$

(B) $2x^2 - 8x + 3$ cannot be factored using integers as coefficients. Another method must be used to solve this equation.

(C) $2x^2 = 3x$ Why shouldn't both members be divided by x?

$2x^2 - 3x = 0$

$x(2x - 3) = 0$

$x = 0$ or $2x - 3 = 0$

$2x = 3$

$x = \frac{3}{2}$

PROBLEM 32   Solve by factoring if possible.

(A) $3x^2 + 7x - 20 = 0$ (B) $2x^2 - 3x - 3 = 0$ (C) $4x^2 = 5x$

■ Solution by Square Root

We now turn our attention to quadratic equations of the form

$$ax^2 + c = 0 \qquad a \neq 0$$

that is, quadratic equations that have the first-degree term missing. The method of solution makes direct use of the definition of a square root of a number. The process is illustrated in the following example.

EXAMPLE 33   Solve by the square root method.

(A) $2x^2 - 3 = 0$ (B) $3x^2 + 27 = 0$ (C) $(x + \frac{1}{2})^2 = \frac{5}{4}$

Solution   (A) $2x^2 - 3 = 0$

$x^2 = \frac{3}{2}$ What number squared is $\frac{3}{2}$?

$x = \pm\sqrt{\frac{3}{2}}$ or $\pm\dfrac{\sqrt{6}}{2}$

(B) $3x^2 + 27 = 0$

$\qquad x^2 = -9$ What number squared is -9?

$\qquad x = \pm\sqrt{-9}$ or $\pm 3i$

(C) $(x + \frac{1}{2})^2 = \frac{5}{4}$

$\qquad x + \frac{1}{2} = \pm\sqrt{\frac{5}{4}}$

$\qquad x = -\frac{1}{2} \pm \dfrac{\sqrt{5}}{2}$

$\qquad x = \dfrac{-1 \pm \sqrt{5}}{2}$

PROBLEM 33   Solve by the square root method.

(A) $3x^2 - 5 = 0$ (B) $2x^2 + 8 = 0$ (C) $(x + \frac{1}{3})^2 = \frac{2}{9}$

■ Solution by Completing the Square

The methods of square root and factoring are generally fast when they apply; however, there are equations, such as $2x^2 - 8x + 3 = 0$ (see Example 32B), that cannot be solved by these methods. A more general method must be developed to take care of this type of equation. The method of completing the square is such a method. This method is based on the process of transforming the standard quadratic equation

$\qquad ax^2 + bx + c = 0$

into the form

$\qquad (x + A)^2 = B$

where A and B are constants. The last equation can be easily solved by the square root method just discussed. But how do we transform the first equation into the second? The following brief discussion provides the key to the process.

What number must be added to $x^2 + bx$ so that the result is the square of a first-degree polynomial? There is an easy mechanical rule for finding this number, based on the square of the following binomials:

$\qquad (x + m)^2 = x^2 + 2mx + m^2$

$\qquad (x - m)^2 = x^2 - 2mx + m^2$

In either case, we see that the third term on the right is the square of one-half of the coefficient of x in the second term on the right. This observation leads directly to the rule for completing the square.

Completing the Square

To complete the square of a quadratic of the form $x^2 + bx$, add the square of one-half the coefficient of x; that is, add $(b/2)^2$. Thus,

$$x^2 + bx + \left(\frac{b}{2}\right)^2 = \left(x + \frac{b}{2}\right)^2$$

EXAMPLE 34   Complete the square for each of the following.

(A) $x^2 + 6x$ (B) $x^2 - 3x$ (C) $x^2 + bx$

Solution   (A) $x^2 + 6x$ Add $\left(\frac{6}{2}\right)^2$—that is, 9.

$x^2 + 6x + 9 = (x + 3)^2$

(B) $x^2 - 3x$ Add $\left(\frac{-3}{2}\right)^2$—that is, $\frac{9}{4}$.

$x^2 - 3x + \frac{9}{4} = (x - \frac{3}{2})^2$

(C) $x^2 + bx$ Add $\left(\frac{b}{2}\right)^2$—that is, $\frac{b^2}{4}$.

$x^2 + bx + \frac{b^2}{4} = \left(x + \frac{b}{2}\right)^2$

PROBLEM 34   Complete the square for each of the following.

(A) $x^2 + 10x$ (B) $x^2 - 5x$ (C) $x^2 + mx$

It is important to note that the rule for completing the square applies only to quadratic forms in which the coefficient of the second-degree term is 1. This causes little trouble, however, as you will see. We now solve two problems by the method of completing the square.

EXAMPLE 35   Solve by completing the square.

(A) $x^2 + 6x - 2 = 0$ (B) $2x^2 - 4x + 3 = 0$

Solution   (A) $x^2 + 6x - 2 = 0$

$x^2 + 6x = 2$ Complete the square on the left by adding 9 to both members of the equation.

$x^2 + 6x + 9 = 2 + 9$

$(x + 3)^2 = 11$

$x + 3 = \pm\sqrt{11}$

$x = -3 \pm \sqrt{11}$

(B) $2x^2 - 4x + 3 = 0$ Make the leading coefficient 1 by dividing by 2.

$x^2 - 2x + \frac{3}{2} = 0$

$x^2 - 2x = -\frac{3}{2}$ Complete the square.

$x^2 - 2x + 1 = -\frac{3}{2} + 1$

$(x - 1)^2 = -\frac{1}{2}$

$x - 1 = \pm\sqrt{-\frac{1}{2}}$

$x = 1 \pm \frac{\sqrt{2}}{2}i$

PROBLEM 35   Solve by completing the square.

(A) $x^2 + 8x - 3 = 0$ (B) $3x^2 - 12x + 13 = 0$

■   **Solution by Quadratic Formula**

Let us now consider the general quadratic equation, with unspecified coefficients,

$$ax^2 + bx + c = 0 \qquad a \neq 0$$

and solve it by completing the square exactly as we did in the preceding examples in which the coefficients were specified. To make the leading coefficient 1, multiply both members of the equation by $1/a$. Thus,

$$x^2 + \frac{b}{a}x + \frac{c}{a} = 0$$

Adding $-c/a$ to both members and then completing the square of the left member, we have

$$x^2 + \frac{b}{a}x + \frac{b^2}{4a^2} = \frac{b^2}{4a^2} - \frac{c}{a}$$

We now factor the left member and solve by the square root method.

$$\left(x + \frac{b}{2a}\right)^2 = \frac{b^2 - 4ac}{4a^2}$$

$$x + \frac{b}{2a} = \pm\sqrt{\frac{b^2 - 4ac}{4a^2}}$$

$$x = -\frac{b}{2a} \pm \frac{\sqrt{b^2 - 4ac}}{2a} \qquad \text{See Problem 69 in Exercise 1-8.}$$

$$x = \frac{-b \pm \sqrt{b^2 - 4ac}}{2a} \qquad a \neq 0 \quad \text{QUADRATIC FORMULA}$$

The last equation is called the **quadratic formula**. It should be memorized and used to solve quadratic equations when all other methods fail.

EXAMPLE 36   Solve $2x + \frac{3}{2} = x^2$ by use of the quadratic formula. Leave the answer in simplest radical form.

Solution

$$2x + \frac{3}{2} = x^2 \qquad \text{Multiply both sides by 2.}$$

$$4x + 3 = 2x^2 \qquad \text{Write in standard form.}$$

$$2x^2 - 4x - 3 = 0 \qquad \begin{array}{l}\text{Cannot be solved by factoring, so go}\\ \text{directly to the quadratic formula.}\end{array}$$

$$x = \frac{-b \pm \sqrt{b^2 - 4ac}}{2a} \qquad a = 2, \quad b = -4, \quad c = -3$$

$$x = \frac{-(-4) \pm \sqrt{(-4)^2 - 4(2)(-3)}}{2(2)}$$

$$x = \frac{4 \pm \sqrt{40}}{4} = \frac{4 \pm 2\sqrt{10}}{4} = \frac{2 \pm \sqrt{10}}{2}$$

$\left\{ \dfrac{2 + \sqrt{10}}{2}, \dfrac{2 - \sqrt{10}}{2} \right\}$ is the solution set for the equation.

$\Bigg[$ *Note:*   A common mistake is to cancel the 4's in $\dfrac{\cancel{4} \pm 2\sqrt{10}}{\cancel{4}}$.   Also, some students carelessly write $2 \pm \dfrac{\sqrt{10}}{2}$ for $\dfrac{2 \pm \sqrt{10}}{2}$. $\Bigg]$

PROBLEM 36   Solve $x^2 - \frac{5}{2} = -3x$ by use of the quadratic formula. Leave the answer in simplest radical form.

EXAMPLE 37   Solve $5.37x^2 - 6.03x + 1.17 = 0$ to two decimal places using a hand calculator.

Solution $5.37x^2 - 6.03x + 1.17 = 0$

$$x = \frac{6.03 \pm \sqrt{(-6.03)^2 - 4(5.37)(1.17)}}{2(5.37)}$$

$$x = 0.25, \, 0.87$$

CALCULATOR STEPS FOR THE SOLUTION INVOLVING THE NEGATIVE RADICAL

PROBLEM 37   Solve $2.79x^2 + 5.07x - 7.69 = 0$ to two decimal places using a hand calculator.

We conclude this section by noting that $b^2 - 4ac$ in the quadratic formula is called the **discriminant** and gives us useful information about the corresponding roots as shown in Table 3.

TABLE 3   DISCRIMINANT AND ROOTS

Discriminant $b^2 - 4ac$	Roots of $ax^2 + bx + c = 0$ $a \neq 0$; a, b, and c real
Positive	Two distinct real roots
0	One real root
Negative	Two complex roots, one the conjugate of the other

Answers to Matched Problems
32. (A) $-4, \frac{5}{3}$ (B) Does not factor using integer coefficients
     (C) $0, \frac{5}{4}$
33. (A) $\pm\sqrt{\frac{5}{3}}$ or $\pm\sqrt{15}/3$ (B) $\pm 2i$ (C) $(-1 \pm \sqrt{2})/3$
34. (A) $x^2 + 10x + 25 = (x + 5)^2$ (B) $x^2 - 5x + \frac{25}{4} = (x - \frac{5}{2})^2$
     (C) $x^2 + mx + (m^2/4) = [x + (m/2)]^2$
35. (A) $-4 \pm \sqrt{19}$ (B) $(6 \pm i\sqrt{3})/3$ or $2 \pm (\sqrt{3}/3)i$
36. $(-3 \pm \sqrt{19})/2$ 37. $-2.80, 0.98$

Exercise 1-8 ■   *Leave all answers involving radicals in simplest radical form unless otherwise stated.*

A   *Solve by factoring.*

1. $4u^2 = 8u$ 2. $3A^2 = -12A$
3. $2d^2 + 15d = 8$ 4. $3x^2 = 10x + 8$
5. $11x = 2x^2 + 12$ 6. $8 - 10x = 3x^2$
7. $6x^2 + 5x = 4$ 8. $6x^2 = 47x + 8$

Solve by the square root method.

9. $x^2 - 25 = 0$ **10.** $x^2 - 16 = 0$ **11.** $x^2 + 25 = 0$

12. $x^2 + 16 = 0$ **13.** $m^2 - 12 = 0$ **14.** $y^2 - 45 = 0$

15. $9y^2 - 16 = 0$ **16.** $4x^2 - 9 = 0$ **17.** $4x^2 + 25 = 0$

18. $16a^2 + 9 = 0$ **19.** $(n + 5)^2 = 9$ **20.** $(m - 3)^2 = 25$

21. $(d - 3)^2 = -4$ **22.** $(t + 1)^2 = -9$

Solve using the quadratic formula.

23. $x^2 - 10x - 3 = 0$ **24.** $x^2 - 6x - 3 = 0$

25. $t^2 = 1 - t$ **26.** $u^2 = 1 - 3u$

27. $x^2 + 8 = 4x$ **28.** $y^2 + 3 = 2y$

29. $2x^2 + 1 = 4x$ **30.** $2m^2 + 3 = 6m$

31. $3q + 2q^2 = 1$ **32.** $p = 1 - 3p^2$

33. $5x^2 + 2 = 2x$ **34.** $7x^2 + 6x + 4 = 0$

B   *Solve by completing the square.*

35. $x^2 - 6x - 3 = 0$ **36.** $y^2 - 10y - 3 = 0$

37. $2y^2 - 6y + 3 = 0$ **38.** $2d^2 - 4d + 1 = 0$

39. $3x^2 - 2x - 2 = 0$ **40.** $3x^2 + 5x - 4 = 0$

41. $x^2 + mx + n = 0$ **42.** $ax^2 + bx + c = 0, \quad a \neq 0$

Solve by any method.

43. $12x^2 + 7x = 10$ **44.** $9x^2 + 9x = 4$

45. $(2y - 3)^2 = 5$ **46.** $(3m + 2)^2 = -4$

47. $x^2 = 3x + 1$ **48.** $x^2 + 2x = 2$

49. $7n^2 = -4n$ **50.** $8u^2 + 3u = 0$

51. $2y = \dfrac{2}{y} + 3$ **52.** $L = \dfrac{15}{L - 2}$

53. $1 + \dfrac{8}{x^2} = \dfrac{4}{x}$ **54.** $\dfrac{2}{u} = \dfrac{3}{u^2} + 1$

55. $\dfrac{24}{10 + m} + 1 = \dfrac{24}{10 - m}$ **56.** $\dfrac{1.2}{y - 1} + \dfrac{1.2}{y} = 1$

57. $\dfrac{2}{x - 2} = \dfrac{4}{x - 3} - \dfrac{1}{x + 1}$ **58.** $\dfrac{3}{x - 1} - \dfrac{2}{x + 3} = \dfrac{4}{x - 2}$

59. $\dfrac{x + 2}{x + 3} - \dfrac{x^2}{x^2 - 9} = 1 - \dfrac{x - 1}{3 - x}$ **60.** $\dfrac{11}{x^2 - 4} + \dfrac{x + 3}{2 - x} = \dfrac{2x - 3}{x + 2}$

In Problems 61–64, solve for the indicated letters in terms of the other letters. Use positive square roots only.

61. $s = \frac{1}{2}gt^2$ for t **62.** $a^2 + b^2 = c^2$ for a

63. $P = EI - RI^2$ for I **64.** $A = P(1 + r)^2$ for r

C 65. Show that if r_1 and r_2 are the two roots of $ax^2 + bx + c = 0$, then $r_1 r_2 = c/a$.

66. For r_1 and r_2 in Problem 65, show that $r_1 + r_2 = -b/a$.

67. Show that if r_1 and r_2 are any nonzero numbers such that $r_1 r_2 = c/a$ and $r_1 + r_2 = -b/a$, then they are roots of $ax^2 + bx + c = 0$.

68. Use the results of Problems 65, 66, and 67 to check which of the following are roots of $2x^2 - 2x + 5 = 0$.
   (A) $-1, \quad 2$ (B) $2 + \sqrt{3}, \quad 2 - \sqrt{3}$ (C) $\frac{1}{2} - \frac{3}{2}i, \quad \frac{1}{2} + \frac{3}{2}i$

69. In one stage of the derivation of the quadratic formula, we replaced $\pm\sqrt{(b^2 - 4ac)/4a^2}$ with $\pm\sqrt{b^2 - 4ac}/2a$. What justifies using $2a$ in place of $|2a|$?

70. Find the fallacy.

$$(n + 1)^2 = n^2 + 2n + 1$$

$$(n + 1)^2 - (2n + 1) = n^2$$

$$(n + 1)^2 - (2n + 1) - n(2n + 1) = n^2 - n(2n + 1)$$

$$(n + 1)^2 - (2n + 1) - n(2n + 1) + \frac{(2n + 1)^2}{4} = n^2 - n(2n + 1)$$

$$+ \frac{(2n + 1)^2}{4}$$

$$\left[(n + 1) - \left(\frac{2n + 1}{2} \right) \right]^2 = \left(n - \frac{2n + 1}{2} \right)^2$$

$$(n + 1) - \frac{2n + 1}{2} = n - \frac{2n + 1}{2}$$

$$n + 1 = n$$

CALCULATOR PROBLEMS

Solve to two decimal places using a hand calculator.

71. $2.07x^2 - 3.79x + 1.34 = 0$ **72.** $0.61x^2 - 4.28x + 2.93 = 0$

73. $4.83x^2 + 2.04x - 3.18 = 0$ **74.** $5.13x^2 + 7.27x - 4.32 = 0$

Use the discriminant to determine which equations have real solutions.

75. $0.0134x^2 + 0.0414x + 0.0304 = 0$

76. $0.543x^2 - 0.182x + 0.003\ 12 = 0$

77. $0.0134x^2 + 0.0214x + 0.0304 = 0$

78. $0.543x^2 - 0.182x + 0.0312 = 0$

Section 1-9   Applications

We will now consider a variety of applications that use quadratic equations in their solution. We restate the strategy for solving word problems presented in Section 1-3.

Strategy for Solving Word Problems

---

1. Read the problem carefully—several times if necessary; that is, until you understand the problem, know what is to be found, and know what is given.
2. Draw figures or diagrams and label known and unknown parts.
3. Look for formulas connecting the known quantities with the unknown quantities.
4. Let one of the unknown quantities be represented by a variable, say x, and try to represent all other unknown quantities in terms of x. This is an important step and must be done carefully.
5. Form an equation relating the unknown quantities with the known quantities.
6. Solve the equation and write answers to *all* parts of the problem requested.
7. Check and interpret all solutions in terms of the original problem and not just the equation found in step 5 (a mistake might have been made in setting up the equation in step 5).

EXAMPLE 38   The sum of a number and its reciprocal is $\frac{13}{6}$. Find all such numbers.

Solution   Let x = The number, then

$$x + \frac{1}{x} = \frac{13}{6} \qquad \text{Multiply both sides by 6x. [\textit{Note:}   } x \neq 0]$$

$$(6x)x + (6x)\frac{1}{x} = (6x)\frac{13}{6}$$

$$6x^2 + 6 = 13x$$

$$6x^2 - 13x + 6 = 0$$

$$(2x - 3)(3x - 2) = 0$$

$$2x - 3 = 0 \quad \text{or} \quad 3x - 2 = 0$$

$$x = \tfrac{3}{2} \qquad\qquad x = \tfrac{2}{3}$$

Check $\tfrac{3}{2} + \tfrac{2}{3} = \tfrac{13}{6}$ $\tfrac{2}{3} + \tfrac{3}{2} = \tfrac{13}{6}$

PROBLEM 38   The sum of two numbers is 23 and their product is 132. Find the two numbers. [*Hint:*   If one number is x, then the other number is 23 − x.]

EXAMPLE 39   An excursion boat takes 1.6 hours longer to go up a river than to return. If the rate of the current is 4 miles/hour, what is the rate of the boat in still water?

Solution   Let x = Rate of boat in still water, then

x + 4 = Rate downstream

x − 4 = Rate upstream

$$\left(\begin{array}{c}\text{Time}\\\text{upstream}\end{array}\right) - \left(\begin{array}{c}\text{Time}\\\text{downstream}\end{array}\right) = 1.6$$

$$\frac{36}{x - 4} \quad - \quad \frac{36}{x + 4} \quad = 1.6 \qquad\qquad T = \frac{D}{R}$$

$$36(x + 4) \quad - \quad 36(x - 4) \quad = 1.6(x - 4)(x + 4)$$

$$36x + 144 - 36x + 144 = 1.6x^2 - 25.6$$

$$1.6x^2 = 313.6$$

$$x^2 = 196$$

$$x = \sqrt{196} = 14 \text{ miles/hour} \quad \begin{array}{l}\text{Rate in}\\\text{still water}\end{array}$$

[*Note:* $-\sqrt{196} = -14$ must be discarded, since it does not make sense in the problem.]

Check   Time upstream $= \dfrac{D}{R} = \dfrac{36}{14 - 4} = 3.6$

        Time downstream $= \dfrac{D}{R} = \dfrac{36}{14 + 4} = \underset{\underline{1.6}}{2} \quad$ Difference of times

PROBLEM 39   Two boats travel at right angles to each other after leaving a dock at the same time. One hour later they are 25 miles apart. If one boat travels 5 miles/hour faster than the other, what is the rate of each? [*Hint:*   Use the Pythagorean theorem,* remembering that distance equals rate times time.]

* *Pythagorean theorem.* A triangle is a right triangle if and only if the square of the longest side is equal to the sum of the squares of the two shorter sides.

$$c^2 = a^2 + b^2$$

EXAMPLE 40 A tank can be filled in 4 hours by two pipes when both are used. How many hours are required for each pipe to fill the tank alone if the smaller pipe requires 3 hours more than the larger one? Compute answers to two decimal places.

Solution Let

$$x = \text{Time for larger pipe to fill tank alone}$$
$$x + 3 = \text{Time for smaller pipe to fill tank alone}$$
$$4 = \text{Time for both pipes to fill tank together}$$

Then

$$\frac{1}{x} = \text{Rate for larger pipe} \left(\text{fills } \frac{1}{x} \text{ of the tank per hour} \right)$$

$$\frac{1}{x + 3} = \text{Rate for smaller pipe} \left(\text{fills } \frac{1}{x + 3} \text{ of the tank per hour} \right)$$

$$\frac{1}{4} = \text{Rate together} \left(\text{fills } \frac{1}{4} \text{ of the tank per hour} \right)$$

$$\left(\begin{array}{c} \text{Rate of} \\ \text{larger pipe} \end{array} \right) + \left(\begin{array}{c} \text{Rate of} \\ \text{smaller pipe} \end{array} \right) = \left(\begin{array}{c} \text{Rate} \\ \text{together} \end{array} \right)$$

$$\frac{1}{x} \quad + \quad \frac{1}{x + 3} \quad = \quad \frac{1}{4}$$ Multiply both sides by $4x(x + 3)$.

$$4(x + 3) \quad + \quad 4x \quad = \quad x(x + 3)$$
$$4x + 12 + 4x = x^2 + 3x$$
$$x^2 - 5x - 12 = 0$$

$$x = \frac{5 \pm \sqrt{73}}{2}$$ Why should we discard the negative answer?

$$x = \frac{5 + \sqrt{73}}{2} \approx 6.77 \text{ hours}$$ Larger pipe

$$x + 3 = 9.77 \text{ hours}$$ Smaller pipe

Check

$$\frac{1}{6.77} + \frac{1}{9.77} \overset{?}{=} \frac{1}{4}$$

$$0.250\ 065 \overset{\checkmark}{\approx} 0.25$$

[*Note:* We do not expect the check to be exact, since we rounded the answers to two decimal places. An exact check would be produced by using $x = (5 + \sqrt{73})/2$. The latter is left to the reader.]

PROBLEM 40   Two pipes can fill a tank in 3 hours when used together. Alone, one can fill the tank 2 hours faster than the other. How long will it take each pipe to fill the tank alone? Compute the answers to two decimal places.

Answers to Matched Problems   **38.** 11 and 12   **39.** 15 and 20 miles/hour
40. 5.16 and 7.16 hours

Exercise 1-9 ■

APPLICATIONS

These problems are not grouped from easy (A) to difficult or theoretical (C). They are grouped according to type. However, the most difficult problems are double-starred (★★), moderately difficult problems are single-starred (★), and the easier problems are not marked.

Numbers

1. Find two numbers such that their sum is 21 and their product is 104.

2. Find all numbers with the property that when the number is added to itself the sum is the same as when the number is multiplied by itself.

3. Find two consecutive positive even integers whose product is 168.

★**4.** The sum of a number and its reciprocal is $\frac{10}{3}$. Find the number.

Geometry

5. If the length and width of a 4 × 2-inch rectangle are each increased by the same amount, the area of the new rectangle will be twice the old. What are the dimensions to two decimal places of the new rectangle?

6. Find the base and height of a triangle with an area of 2 square feet if its base is 3 feet longer than its height. ($A = \frac{1}{2}bh$.)

★**7.** Approximately how far is the horizon of the earth from a balloon 4 miles high? Assume the radius of the earth is 4,000 miles. Estimate the answer to the nearest mile. [*Hint:* See the figure.]

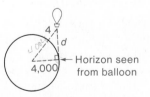

4

d

4,000

Horizon seen
from balloon

★★**8.** A flag has a white cross of uniform width on a red background. If the cross extends from edge to edge on a 4 × 3-feet flag, find its width if it takes up exactly half the total area of the flag.

Business and Economics

9. If P dollars are invested at $r\%$ compounded annually, at the end of 2 years the amount will be $A = P(1 + r)^2$. At what interest rate will $1,000 increase to $1,440 in 2 years? [*Note:* $A = \$1,440$ and $P = \$1,000$.]

$\star$**10.** In a certain city the demand equation for stereo tapes is $q_d = 3,000/p$, where q_d is the quantity of tapes demanded on a given day if the selling price is $p per tape. (Notice that, as the price goes up, the number of tapes people are willing to buy goes down, and vice versa.) On the other hand, the supply equation is $q_s = 1,000p - 500$, where q_s is the quantity of tapes a supplier is willing to supply at $p per tape. (Notice that, as the price goes up, the number of tapes a supplier is willing to sell goes up, and vice versa.) At what price will supply equal demand; that is, at what price will $q_d = q_s$? In economic theory the price at which supply equals demand is called the *equilibrium point*, the point at which the price ceases to change.

Rate-Time   **11.** Two boats travel at right angles to each other after leaving the same dock at the same time. One hour later they are 13 miles apart. If one travels 7 miles/hour faster than the other, what is the rate of each?

12. A speedboat takes 1 hour longer to go 24 miles up a river than to return. If the boat cruises at 10 miles/hour in still water, what is the rate of the current?

$\star$**13.** One pipe can fill a tank in 5 hours less than another; together they fill the tank in 5 hours. How long would it take each alone to fill the tank? Compute the answer to two decimal places.

$\star$**14.** A new automatic machine can do a job in 1 hour less than an older machine. Together they can do the same job in 1.2 hours. How long would it take each to do the job alone?

Physics and Engineering   **15.** The pressure p in pounds per square foot of wind blowing at v miles/hour is $p = 0.003v^2$. If a pressure gauge on a bridge registers a wind pressure of 14.7 pounds per square foot, what is the velocity of the wind?

16. If a projectile is shot vertically into the air (from the ground) with an initial velocity of 176 feet/second, its distance y above the ground t seconds after it is shot (neglecting air resistance) is given by $y = 176t - 16t^2$.
   (A)   Find the times when y is 0, and interpret the results physically.
   (B)   Find the times when the projectile is 16 feet off the ground. Compute the answers to two decimal places.

$\star$**17.** For a car traveling at a speed of v miles/hour, the least number of feet d under the best possible conditions necessary to stop it (including reaction time) is given by the empirical formula $d = 0.044v^2 + 1.1v$. Estimate the speed of a car that requires 165 feet to stop after the danger is realized.

★★18. A barrel 2 feet in diameter and 4 feet in height has a 1-inch diameter drainpipe in the bottom. It can be shown that the height h of the surface of the water above the bottom of the barrel at time t minutes after the drain has been opened is given by the formula $h = [\sqrt{h_0} - (5t/12)]^2$, where h_0 is the water level above the drain at time $t = 0$. If the barrel is full and the drain opened, how long will it take to empty half the contents? [*Hint:* The problem is very easily solved if the right side of the equation is not squared.]

Section 1-10 Equations Reducible to Quadratic Form

■ Equations Involving Radicals
■ Equations Involving Rational Exponents

■ **Equations Involving Radicals**

In solving an equation involving a radical such as

$$\sqrt{x - 1} = 2x - 3$$

it appears that we can remove the radical by squaring each side and then proceed to solve the resulting quadratic equation. Thus,

$$\left(\sqrt{x - 1}\right)^2 = (2x - 3)^2$$
$$x - 1 = 4x^2 - 12x + 9$$
$$4x^2 - 13x + 10 = 0$$
$$(4x - 5)(x - 2) = 0$$
$$x = \tfrac{5}{4}, 2$$

Checking, we find that 2 is a solution, but $\tfrac{5}{4}$ is not. These results are a special case of Theorem 6.

THEOREM 6

If both sides of an equation are squared, then the solution set of the original equation is a subset of the solution set of the new equation.

Equation	Solution set
$x = 3$	$\{3\}$
$x^2 = 9$	$\{-3, 3\}$

This theorem provides us with a method of solving some equations involving radicals. It is important to remember that any new equation obtained by raising both members of an equation to the same power may have solutions (called **extraneous solutions**) that are not solutions of the original equation. On the other hand, any solution of the original equation must be among those of the new equation. Thus, every solution of the new equation must be checked in the original equation to eliminate so-called extraneous solutions.

EXAMPLE 41　Solve.

(A) $x + \sqrt{x - 4} = 4$　　(B) $\sqrt{2x + 3} - \sqrt{x - 2} = 2$

Solution　(A)　$x + \sqrt{x - 4} = 4$

$\sqrt{x - 4} = 4 - x$　　Isolate radical on one side.

$x - 4 = 16 - 8x + x^2$　Square both members.

$x^2 - 9x + 20 = 0$

$(x - 5)(x - 4) = 0$

$x = 5, 4$

Checking shows that 4 is good and 5 is extraneous; thus,

$$\{x \mid x + \sqrt{x - 4} = 4\} = \{4\}$$

(B) $\sqrt{2x + 3} - \sqrt{x - 2} = 2$

$\sqrt{2x + 3} = \sqrt{x - 2} + 2$

$2x + 3 = x - 2 + 4\sqrt{x - 2} + 4$

$x + 1 = 4\sqrt{x - 2}$

$x^2 + 2x + 1 = 16(x - 2)$

$x^2 - 14x + 33 = 0$

$(x - 11)(x - 3) = 0$

$x = 3, 11$

The work is made a little easier by having one radical on each side before squaring. After squaring, isolate the remaining radical on one side and square again.

Both solutions check; hence,

$$\{x \mid \sqrt{2x + 3} - \sqrt{x - 2} = 2\} = \{3, 11\}$$

PROBLEM 41　Solve.

(A)　$x - 5 = \sqrt{x - 3}$　　(B)　$\sqrt{2x + 5} + \sqrt{x + 2} = 5$

■　Equations Involving Rational Exponents

If asked to solve the equation

$$x^{2/3} - x^{1/3} - 6 = 0$$

you might at first have trouble. But if you recognize that the equation is quadratic in $x^{1/3}$, you can solve for $x^{1/3}$ first and then solve for x. It may be convenient to make the substitution $u = x^{1/3}$, and then solve the equation

$$u^2 - u - 6 = 0$$

$$(u - 3)(u + 2) = 0$$

$$u = 3, \, -2$$

Replacing u with $x^{1/3}$, we obtain

$$x^{1/3} = 3 \qquad\qquad x^{1/3} = -2$$

$$x = 27 \qquad\qquad x = -8 \quad \text{Not } x = 3^{1/3} \text{and } x = (-8)^{1/3}$$

In general, if an equation that is not quadratic can be transformed to the form

$$au^2 + bu + c = 0$$

where u is an expression in some other variable, then the equation is said to be in *quadratic form*. Once recognized as a quadratic form, an equation can often be solved using quadratic methods.

EXAMPLE 42   Solve as far as possible using techniques we have developed up to this point. Some equations may have additional complex solutions that you will not be able to find without further study in the theory of equations.

(A) $x^{10} + 6x^5 - 16 = 0$ (B) $4y^{-4} - 37y^{-2} + 9 = 0$

Solution   (A) $x^{10} + 6x^5 - 16 = 0$

Let $u = x^5$ and solve:

$$u^2 + 6u - 16 = 0$$

$$(u + 8)(u - 2) = 0$$

$$u = -8, \, 2$$

Thus,

$$x^5 = -8 \qquad\qquad \text{or} \qquad x^5 = 2$$

$$x = \sqrt[5]{-8} = -\sqrt[5]{8} \qquad\qquad x = \sqrt[5]{2} \quad \text{Not } x = (-8)^5 \text{ and } x = 2^5$$

(B) $4y^{-4} - 37y^{-2} + 9 = 0$

Let $u = y^{-2}$, then

$$4u^2 - 37u + 9 = 0$$

$$(4u - 1)(u - 9) = 0$$

$$u = \tfrac{1}{4}, \, 9$$

$$y^{-2} = \tfrac{1}{4} \qquad y^{-2} = 9$$

$$\frac{1}{y^2} = \tfrac{1}{4} \qquad \frac{1}{y^2} = 9$$

$$y^2 = 4 \qquad y^2 = \tfrac{1}{9}$$

$$y = \pm 2 \qquad y = \pm \tfrac{1}{3}$$

PROBLEM 42   Solve as far as possible using techniques we have developed up to this point.

(A) $x^{2/3} - x^{1/3} - 12 = 0$ (B) $x^4 - 5x^2 + 4 = 0$
(C) $2x^{-2} - 5x^{-1} - 12 = 0$

Answers to Matched Problems   **41.** (A) 7   (B) 2

42. (A) 64, -27 (B) $\pm 1, \pm 2$ (C) $\tfrac{1}{4}, -\tfrac{2}{3}$

Exercise 1-10 ■

Find all solutions possible by the techniques that have been developed so far.

A   **1.** $\sqrt[3]{x + 5} = 3$ **2.** $\sqrt[4]{x - 3} = 2$

   3. $\sqrt{5n + 9} = n - 1$ **4.** $m - 13 = \sqrt{m + 7}$

   5. $\sqrt{x + 5} + 7 = 0$ **6.** $3 + \sqrt{2x - 1} = 0$

   7. $\sqrt{3x + 4} = 2 + \sqrt{x}$ **8.** $\sqrt{3w - 2} - \sqrt{w} = 2$

   9. $y^4 - 2y^2 - 8 = 0$ **10.** $x^4 - 7x^2 - 18 = 0$

   11. $x^{10} + 3x^5 - 10 = 0$ **12.** $x^{10} - 7x^5 - 8 = 0$

   13. $2x^{2/3} + 3x^{1/3} - 2 = 0$ **14.** $x^{2/3} - 3x^{1/3} - 10 = 0$

   15. $(m^2 - m)^2 - 4(m^2 - m) = 12$ **16.** $(x^2 + 2x)^2 - (x^2 + 2x) = 6$

B   **17.** $\sqrt{u - 2} = 2 + \sqrt{2u + 3}$ **18.** $\sqrt{3t + 4} + \sqrt{t} = -3$

   19. $\sqrt{3y - 2} = 3 - \sqrt{3y + 1}$ **20.** $\sqrt{2x - 1} - \sqrt{x - 4} = 2$

   21. $\sqrt{7x - 2} - \sqrt{x + 1} = \sqrt{3}$ **22.** $\sqrt{3x + 6} - \sqrt{x + 4} = \sqrt{2}$

   23. $3n^{-2} - 11n^{-1} - 20 = 0$ **24.** $6x^{-2} - 5x^{-1} - 6 = 0$

   25. $9y^{-4} - 10y^{-2} + 1 = 0$ **26.** $4x^{-4} - 17x^{-2} + 4 = 0$

   27. $y^{1/2} - 3y^{1/4} + 2 = 0$ **28.** $4x^{-1} - 9x^{-1/2} + 2 = 0$

   29. $(m - 5)^4 + 36 = 13(m - 5)^2$ **30.** $(x - 3)^4 + 3(x - 3)^2 = 4$

C   **31.** $\sqrt{5 - 2x} - \sqrt{x + 6} = \sqrt{x + 3}$

   32. $\sqrt{2x + 3} - \sqrt{x - 2} = \sqrt{x + 1}$

Solve Problems 33–36 two ways: by squaring and by substitution.

   33. $m - 7\sqrt{m} + 12 = 0$ **34.** $y - 6 + \sqrt{y} = 0$

   35. $t - 11\sqrt{t} + 18 = 0$ **36.** $x = 15 - 2\sqrt{x}$

Section 1-11   Chapter Review

IMPORTANT TERMS
AND SYMBOLS

1-1 Sets. Set; set notation: listing method, rule method; element; member; empty set; null set; subset, union, disjoint, intersection, universal set, complement, Venn diagram, $\in$, $\notin$, $\varnothing$, $A \subset B$, $A = B$, $A \cup B$, $A \cap B$, U, A'

1-2 Linear Equations. Equal to; not equal to; algebraic equation; basic properties of equality: reflexive property, symmetric property, transitive property, substitution principle; linear equation; replacement set; solution; root; solution set; solving an equation; equivalent equations; further properties of equality: addition property, subtraction property, multiplication property, division property; $=$; $\neq$

1-3 Applications. A strategy for solving word problems, number and geometric problems, rate-time problems, mixture problems

1-4 Real Number Line; Linear Inequalities. The real number system: real numbers, rational numbers, irrational numbers, integers, natural numbers; the real number line: coordinate, origin, positive real numbers, negative real numbers; inequality relations: less than, greater than, less than or equal to, greater than or equal to, double inequality, inequality notation, interval notation; inequality statements; solution set; solving an inequality; equivalent inequalities; inequality properties: addition property, subtraction property, multiplication property, division property; $a < b$; $b > a$; $a \leq b$; $b \geq a$

1-5 Absolute Value in Equations and Inequalities. Absolute value, distance between two points, absolute value in equations and inequalities, $|x|$, $|x - a|$, $d(A, B)$

1-6 Nonlinear Inequalities. Polynomial inequalities, rational inequalities, sign-analysis technique of solving

1-7 Complex Numbers. Complex number, pure imaginary number, imaginary unit, conjugate, equality, addition, subtraction, multiplication, division, complex numbers and radicals, $a + bi$, $i = \sqrt{-1}$, $i^2 = -1$, $a - bi$ is the conjugate of $a + bi$, $\sqrt{-a} = i\sqrt{a} \ (a > 0)$

1-8 Quadratic Equations. Quadratic equation, solution by factoring, solution by square root, solution by completing the square, solution by quadratic formula, $ax^2 + bx + c = 0 \ (a \neq 0)$, $x = (-b \pm \sqrt{b^2 - 4ac})/2a$

1-9 Applications. Applications leading to quadratic equations

1-10 Equations Reducible to Quadratic Form. Equations involving radicals, equations involving rational exponents

Exercise 1-11   Chapter Review

Work through all the problems in this chapter review and check answers in the back of the book. (Answers to all problems are there, and following each answer is a number in italics indicating the section in which that

*type of problem is discussed.) Where weaknesses show up, review
appropriate sections in the text. When you are satisfied that you know
the material, take the practice test following this review.*

A   *Problems 1 and 2 refer to the following sets:*

$$A = \{1, 2, 3, 4, 5, 6\} \qquad B = \{1, 3, 5\}$$
$$C = \{3, 4, 5\} \qquad D = \{2, 4, 6\} \qquad E = \{5, 1, 3\}$$

1. Find each of the following.
   (A) $B \cup C$ (B) $B \cap C$ (C) $B \cap D$
   (D) $A \cap C$ (E) B' relative to A

2. Indicate true (T) or false (F).
   (A) $B \subset C$ (B) $D \subset A$ (C) $3 \in A$
   (D) $3 \notin D$ (E) $B = E$ (F) $E \subset B$

3. $\{x \mid 6x^2 = 11x + 10\} = \{\text{List elements}\}$

4. Given the amount formula for simple interest $S = P + I$ and the
simple interest formula $I = Prt$, what equality property permits us
to write $S = P + Prt$?

Solve.

5. $0.05x + 0.25(30 - x) = 3.3$ **6.** $\dfrac{5x}{3} - \dfrac{4 + x}{2} = \dfrac{x - 2}{4} + 1$

Solve and graph Problems 7–11.

7. $3(2 - x) - 2 \leq 2x - 1$ **8.** $|y + 9| < 5$

9. $|3 - 2x| \leq 5$ **10.** $x^2 + x < 20$

11. $x^2 \geq 4x + 21$

12. Perform the indicated operations and write the answers in the
form $a + bi$.
   (A) $(-3 + 2i) + (6 - 8i)$ (B) $(3 - 3i)(2 + 3i)$

   (C) $\dfrac{13 - i}{5 - 3i}$

Solve Problems 13–18.

13. $2x^2 - 7 = 0$ **14.** $2x^2 = 4x$

15. $2x^2 = 7x - 3$ **16.** $m^2 + m + 1 = 0$

17. $y^2 = \frac{3}{2}(y + 1)$ **18.** $\sqrt{5x - 6} - x = 0$

B   **19.** Let H be the set of all numbers x such that $10x + 11 = 6/x$.
   (A) Denote H by the rule method.
   (B) Denote H by the listing method.

20. Let $A = \{x \mid -1 \le x < 2, x \text{ an integer}\}$

$B = \{x \mid x < 4, x \text{ a natural number}\}$

(A) Find $A \cup B$. (B) Find $A \cap B$. (C) Is $-3 \in B$?
(D) Is $\varnothing \subset A$? (E) Is $B \subset A$? (F) Is $A \subset B$?

21. Which of the following sets is the empty set: $\varnothing, \{0\}, \{\varnothing\}$?

Solve.

22. $\dfrac{7}{2 - x} = \dfrac{10 - 4x}{x^2 + 3x - 10}$

23. $\dfrac{u - 3}{2u - 2} = \dfrac{1}{6} - \dfrac{1 - u}{3u - 3}$

Solve and graph Problems 24–27.

24. $\dfrac{x + 3}{8} \le 5 - \dfrac{2 - x}{3}$

25. $|3x - 8| > 2$

26. $\dfrac{1}{x} < 2$

27. $\dfrac{3}{x - 4} \le \dfrac{2}{x - 3}$

28. If the coordinates of A and B on a real number line are -8 and -2, respectively, find:
(A) $d(A, B)$ (B) $d(B, A)$

29. Perform the indicated operations and write the final answers in the form $a + bi$.
(A) $(3 + i)^2 - 2(3 + i) + 3$ (B) i^{27}

30. Convert to $a + bi$ forms, perform the indicated operations, and write the final answers in $a + bi$ form.
(A) $(2 - \sqrt{-4}) - (3 - \sqrt{-9})$ (B) $\dfrac{2 - \sqrt{-1}}{3 + \sqrt{-4}}$

(C) $\dfrac{4 + \sqrt{-25}}{\sqrt{-4}}$

Find all solutions possible using techniques we have developed so far.

31. $\left(u + \dfrac{5}{2}\right)^2 = \dfrac{5}{4}$

32. $1 + \dfrac{3}{u^2} = \dfrac{2}{u}$

33. $\dfrac{x}{x^2 - x - 6} - \dfrac{2}{x - 3} = 3$

34. $2x^{2/3} - 5x^{1/3} - 12 = 0$

35. $m^4 + 5m^2 - 36 = 0$

36. $\sqrt{y - 2} - \sqrt{5y + 1} = -3$

Solve Problems 37 and 38 for the indicated variable in terms of the other variables.

37. $P = M - Mdt$ for M (mathematics of finance)

38. $P = EI - RI^2$ for I (electrical engineering)

C   **39.**   Is $A \cup B$ or $A \cap B$ defined by $\{x \mid x \in A \text{ and } x \in B\}$?

40.   Indicate true (T) or false (F).
(A)   If $A \cap B = A$, then $A \subset B$.
(B)   If $A \cup B = A$, then $A \subset B$.
(C)   If $A \subset B$, then $A \cup B = B$.

41.   Evaluate: $(a + bi)\left(\dfrac{a}{a^2 + b^2} - \dfrac{b}{a^2 + b^2}\, i\right),\quad a, b \neq 0$

Solve Problems 42–44.

42. $2x > \dfrac{x^2}{5} + 5$ **43.** $\dfrac{x^2}{4} + 4 \geq 2x$ **44.** $\left| x - \dfrac{8}{x} \right| \geq 2$

45.   Solve by substitution and also by squaring: $x - 8\sqrt{x} + 15 = 0$

APPLICATIONS   **46.**   *Chemistry.*   A chemist has 1,200 milliliters of a 60% acid solution. How much should be drained off and replaced with pure acid to obtain the same amount of a 75% acid solution?

47.   *Business.*   From a survey of 1,000 residences, it was found that 850 had dead-bolt locks, 350 had alarm systems, and 275 had both.
(A)   How many residences had either dead bolts or an alarm system?
(B)   How many residences had neither dead bolts nor an alarm system?

48.   *Numbers.*   Find a number such that when its reciprocal is subtracted from the number the difference is $\frac{16}{15}$.

49.   *Cost analysis.*   Cost equations for manufacturing companies are often quadratic in nature. (At very high or very low outputs, the costs are more per unit because of inefficiency of plant operation at these extremes.) If the cost equation for manufacturing transistor radios is $C = x^2 - 10x + 31$, where C is the cost of manufacturing x units per week (both in thousands), find (A) the output for a $15,000 weekly cost and (B) the output for a $6,000 weekly cost.

50.   *Break-even analysis.*   The manufacturing company in Problem 50 sells its transistor radios for $3 each. Thus, its revenue equation is $R = 3x$, where R is revenue and x is the number of units sold per week (both in thousands). Find the break-even points for the company—that is, the output at which revenue equals cost.

Practice Test   Chapter 1

Take this practice test as if it were a graded test. Allow yourself up to 50 minutes. Work the problems without looking back in the chapter. Correct your work using the answers (keyed to appropriate sections) in the back of the book.

Problems 1 and 2 refer to the following sets:

$$A = \{-3, -2, -1, 0, 1, 2, 3\} \qquad B = \{-2, 0, 2\}$$
$$C = \{1, 2, 3\} \qquad D = \{-3, -1, 1, 3\} \qquad E = \{-1, 1, -3, 3\}$$

1. One of the following is false. Indicate which one.
    (A) $B \cup C = \{-2, 0, 1, 2, 3\}$ (B) $B \cap C = \{0, 2\}$
    (C) $B \cap D = \emptyset$ (D) D' relative to $A = B$

2. One of the following is false. Indicate which one.
    (A) $B \subset A$ (B) $\emptyset \subset D$
    (C) $\{1, 2, 3\} \in A$ (D) $E \subset D$

3. Evaluate $\dfrac{x^2 + 3}{x}$ for $x = 1 - 2i$. Write the final answer in the form $a + bi$.

Solve and graph. Write each solution using inequality notation and interval notation.

4. $0.05x + 0.25(20 - x) \geq 1.4$ 5. $|3 - 2x| < 5$

6. $|x - 3| \geq 4$ 7. $\dfrac{(x - 2)^2}{4 - x} \leq 1$

Solve.

8. $\dfrac{3}{x^2 + x} + \dfrac{1}{x} + \dfrac{3}{x + 1} = 0$ 9. $1 + \dfrac{7}{x^2} = \dfrac{4}{x}$

10. $\dfrac{1}{x^2 - 4} + \dfrac{3x}{2 - x} = -2$ 11. $x^{2/5} - x^{1/5} - 2 = 0$

12. $4x^{-4} - 7x^{-2} - 2 = 0$ 13. $\sqrt{3x + 1} - \sqrt{x + 4} = 1$

14. An excursion boat takes 1 hour longer to go 24 miles up a river than to return. If the boat's speed in still water is 10 miles/hour, what is the rate of the current?

15. A chemical storeroom has an 80% alcohol solution and a 30% alcohol solution. How many milliliters of each should be used to obtain 50 milliliters of a 60% alcohol solution?

Graphs and Functions

A natural design of mathematical interest. Can you guess the source? See the back of the book.

Chapter 2 ▪ Graphs and Functions

Section 2-1 Rectangular Coordinate System; Graphing

- ▪ Cartesian Coordinate System
- ▪ Graphing: Point by Point
- ▪ Symmetry
- ▪ Distance between Two Points
- ▪ Circles

▪ Cartesian Coordinate System

Just as we formed a **real number line** by establishing a one-to-one correspondence between the points on a line and the elements in the set of real numbers, we can form a **real plane** by establishing a one-to-one correspondence between the points in a plane and elements in the set of all ordered pairs of real numbers. This can be done by means of a Cartesian coordinate system.*

Recall that to form a **Cartesian (rectangular) coordinate system**, we select two real number lines, one vertical and one horizontal, and let them cross through their origins (0's) as indicated in Figure 1. Up and to

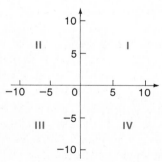

FIGURE 1 Cartesian coordinate system

* Named after René Descartes (1596–1650), the French philosopher-mathematician who is generally recognized as the founder of analytic geometry—the wedding of algebra and plane geometry.

the right are the usual choices for the positive directions. These two number lines are called the **vertical axis** and the **horizontal axis**, or (together) the **coordinate axes**. The coordinate axes divide the plane into four parts called **quadrants**. The quadrants are numbered counterclockwise from I to IV.

Pick a point P in the plane at random (Fig. 2). Pass horizontal and vertical lines through the point. The vertical line will intersect the horizontal axis at a point with coordinate a, and the horizontal line will intersect the vertical axis at a point with coordinate b. These two numbers written as the ordered pair*

(a, b)

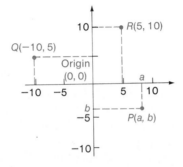

FIGURE 2

form the coordinates of the point P. In Figure 2, the coordinates of point Q are $(-10, 5)$ and of point R are $(5, 10)$. The first coordinate a of the coordinates of point P is also called the **abscissa** of P; the second coordinate b of the coordinates of point P is called the **ordinate** of P. The abscissa of Q in Figure 2 is -10, and the ordinate of Q is 5. The point with coordinates $(0,0)$ is called the **origin**.

We know that the coordinates (a, b) exist for each point in the plane, since every point on each axis has a real number associated with it. Hence, by the procedure just described, each point located in the plane can be labeled with a unique pair of real numbers. Conversely, by reversing the process, each pair of real numbers can be associated with a unique point in the plane. Thus, we have established *a one-to-one correspondence between the points in a plane and the elements in the*

---

* An **ordered pair** of real numbers is a pair of numbers in which the order is specified. We have now used (a, b) as the coordinates of a point and as an interval on a real number line. The context in which (a, b) is used will determine its meaning.

set of all ordered pairs of real numbers. This result is often referred to as the **fundamental theorem of analytic geometry**.

■ Graphing: Point by Point

Because of the fundamental theorem of analytic geometry, we are now in a position to look at algebraic forms geometrically and to look at geometric forms algebraically. We start by considering an equation in two variables, say

$$y = x^2 - 4 \tag{1}$$

A **solution** to equation (1) is an ordered pair of numbers (a, b) such that

$$b = a^2 - 4$$

The **solution set** of equation (1) is the set of all of its solutions. More formally,

Solution set of equation $(1) = \{(x, y) | y = x^2 - 4\}$

that is, the set of all ordered pairs (x, y) such that $y = x^2 - 4$.

To find the solutions of equation (1), we simply replace one of the variables with a number and solve for the other variable. For example, if $x = 2$, then $y = 2^2 - 4 = 0$, and the ordered pair $(2, 0)$ is a solution. Continuing in the same way, assigning different values for x and solving for y (or vice versa), we can obtain as many solutions in the solution set as we please. Now, each solution in the solution set (since it is an ordered pair of real numbers) is the coordinates of a point in a Cartesian coordinate system. The set of all points in a Cartesian coordinate system that have coordinates from the solution set form the **graph** of the given equation. Thus, **to graph an equation in two variables** is to graph its solution set.

Returning to equation (1), we find that its solution set has infinitely many elements and its graph will extend off any paper we might choose, no matter how large. Thus, **to sketch a graph of an equation**, we include enough points from its solution set so that what remains is apparent.

EXAMPLE 1   Sketch a graph of $y = x^2 - 4$.

Solution   We make up a table of ordered pairs (solutions) of numbers that satisfy the given equation:

x	-4	-3	-2	-1	0	1	2	3	4
y	12	5	0	-3	-4	-3	0	5	12

If, after plotting these, there remain certain regions of ambiguity in the graph, we plot enough additional solutions to resolve this ambiguity. These solutions are plotted and joined with a smooth curve.

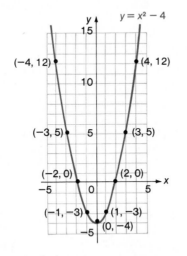

The resulting figure is called a **parabola**. Notice that if we fold the paper along the y axis, the right side will match the left side. We say that the graph is **symmetric with respect to the y axis** and call the y axis the **axis of the parabola**. (More will be said about parabolas later in the text.)

PROBLEM 1   Sketch a graph of $y^2 = x$.

The procedure used to sketch the graph of $y = x^2 - 4$ in Example 1 is called **point-by-point plotting**. As equations get more involved, this basic approach to graphing is substantially aided by the use of hand calculators. In addition to calculators, there are mathematical aids to graphing that can speed up the process significantly. A number of mathematical aids will be discussed in this book, and additional powerful aids to graphing are developed in courses on calculus.

■ Symmetry

We noticed that the graph of $y = x^2 - 4$ in Example 1 is *symmetric with respect to the y axis*; that is, the two parts of the graph coincide if the paper is folded along the y axis. Similarly, we say that a graph is *symmetric with respect to the x axis* if the parts above and below the x axis coincide when the paper is folded along the x axis. In general, we define symmetry with respect to the y axis, x axis, and origin as follows:

Symmetry

A graph is **symmetric with respect to**:

1. **The y axis** if $(-a, b)$ is on the graph whenever (a, b) is on the graph.
2. **The x axis** if $(a, -b)$ is on the graph whenever (a, b) is on the graph.
3. **The origin** if $(-a, -b)$ is on the graph whenever (a, b) is on the graph.

Figure 3 illustrates these three types of symmetry.

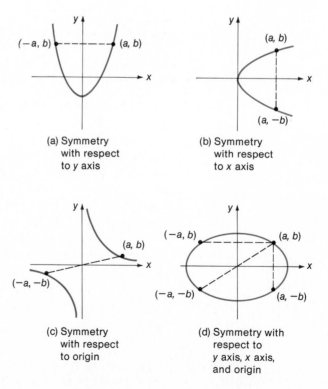

(a) Symmetry
with respect
to y axis

(b) Symmetry
with respect
to x axis

(c) Symmetry
with respect
to origin

(d) Symmetry with
respect to
y axis, x axis,
and origin

FIGURE 3   Symmetry

Given an equation, if we could determine the symmetry properties of its graph ahead of time, we could save a lot of time and energy in sketching the graph. For example, if we knew that the graph of $y = x^2 - 4$ in Example 1 were symmetric with respect to the y axis, we would have to carefully sketch only the right side of the graph; then reflect the result across the y axis to obtain the whole sketch—the point-by-point plotting would be cut in half!

The tests for symmetry (based on the preceding discussion) are given in Table 1. These tests are easily applied and are very helpful aids to graphing.

TABLE 1　TESTS FOR SYMMETRY

Symmetry with Respect to the	Equation Remains Unchanged if
y axis	x is replaced with $-x$
x axis	y is replaced with $-y$
Origin	x and y are replaced with $-x$ and $-y$

Using the tests on $y = x^2 - 4$ from Example 1, we replace x with $-x$ and observe that the equation does not change:

$$y = (-x)^2 - 4$$
$$= x^2 - 4$$

Thus, according to the test, the graph is symmetric with respect to the y axis. Note that if we replace y with $-y$, the equation will change:

$$-y = x^2 - 4$$

or

$$y = 4 - x^2$$

EXAMPLE 2　Test for symmetry and graph.

(A) $y = x^3$　　(B) $x^2 + 4y^2 = 36$

Solution　(A) Substitute $-x$ for x and $-y$ for y in $y = x^3$, and observe that the equation remains unchanged:

$$-y = (-x)^3$$
$$-y = -x^3$$
$$y = x^3$$

Thus, the graph is symmetric with respect to the origin. Note that positive values of x produce positive values for y, and negative values of x produce negative values for y; hence, the graph occurs in the first and third quadrants. We make a careful sketch in the first quadrant; then reflect these points through the origin to obtain the complete sketch.

x	0	1	2
y	0	1	8

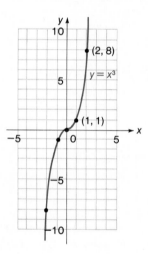

At a glance, the graph shows us how y varies as x varies. A graph is a visual aid and should be constructed to impart the maximum amount of information using the least amount of effort on the part of the observer. Label coordinate axes and indicate scales on both axes.

(B) Since both x and y occur to only even powers in $x^2 + 4y^2 = 36$, the equation will remain unchanged if x is replaced with $-x$ or if y is replaced with $-y$. Consequently, the graph is symmetric with respect to the y axis, x axis, and origin. We need to make a careful sketch in only the first quadrant, reflect this graph across the y axis, and then reflect everything across the x axis. To find first-quadrant solutions, we solve the equation for either y in terms of x or x in terms of y. We choose the latter because the result is simpler to work with.

$$x^2 + 4y^2 = 36$$
$$x^2 = 36 - 4y^2$$
$$x = \pm\sqrt{36 - 4y^2}$$

To obtain the first-quadrant portion of the graph, we sketch $x = \sqrt{36 - 4y^2}$ for $0 \le y \le 3$. Note that y cannot be larger than 3. (Why?)

x	6	$\sqrt{32} \approx 5.7$	$\sqrt{20} \approx 4.5$	0
y	0	1	2	3

Choose values for y and solve for x.

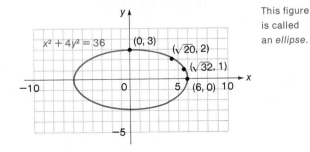

This figure is called an *ellipse*.

PROBLEM 2 Test for symmetry and graph.

(A) $y = x$ (B) $9x^2 + y^2 = 36$

■ **Distance between Two Points**

Analytic geometry is concerned with two basic problems:

1. Given an equation, find its graph.
2. Given a figure (line, circle, parabola, ellipse, etc.) in a coordinate system, find its equation.

So far we have concentrated on the first problem. We now introduce a basic tool that is used extensively in solving the second problem. This basic tool is the *distance-between-two-points formula*, which is easily derived using the Pythagorean theorem. Let $P_1(x_1, y_1)$ and $P_2(x_2, y_2)$ be two points in a rectangular coordinate system (the scale on each axis is assumed to be the same). Then referring to Figure 4, we see that

$$[d(P_1, P_2)]^2 = |x_2 - x_1|^2 + |y_2 - y_1|^2$$
$$= (x_2 - x_1)^2 + (y_2 - y_1)^2 \quad \text{Since } |N|^2 = N.$$

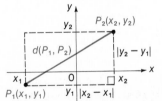

FIGURE 4

Thus:

Distance between $P_1(x_1, y_1)$ and $P_2(x_2, y_2)$

$$d(P_1, P_2) = \sqrt{(x_2 - x_1)^2 + (y_2 - y_1)^2}$$

EXAMPLE 3   Find the distance between $(-3, 5)$ and $(-2, -8)$.*

Solution   It does not matter which point we designate P_1 and P_2 because of the squaring in the formula. Let $(x_1, y_1) = (-3, 5)$ and $(x_2, y_2) = (-2, -8)$. Then

$$d = \sqrt{[(-2) - (-3)]^2 + [(-8) - (5)]^2} = \sqrt{170}$$

PROBLEM 3   Find the distance between $(6, -3)$ and $(-7, -5)$.

■   Circles

The distance-between-two-points formula would still be helpful if its only use were to find actual distances between points, such as in Example 3. However, its more important use is in finding equations of figures in a rectangular coordinate system. We will use it to derive the general equation of a circle. We start with a coordinate-free definition of a circle.

Definition of a Circle

A **circle** is the set of all points in a plane equidistant from a fixed point. The fixed distance is called the **radius**, and the fixed point is called the **center**.

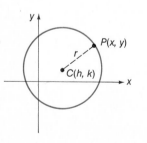

FIGURE 5

Let us find the equation of a circle with radius r $(r > 0)$ and center at (h, k) in a rectangular coordinate system (Fig. 5). The point $P(x, y)$ is on the circle if and only if $d(P, C) = r$; that is, if and only if

$$\sqrt{(x - h)^2 + (y - k)^2} = r \qquad r > 0$$

---

* We often speak of the point (a, b) when we are referring to the point with coordinates (a, b). This shorthand, though not accurate, causes little trouble, and we will continue the practice.

or, equivalently,

$$(x - h)^2 + (y - k)^2 = r^2$$

Standard Equations of a Circle

1. Circle with radius r ($r > 0$) and center at (h, k):

$$(x - h)^2 + (y - k)^2 = r^2$$

2. Circle with radius r ($r > 0$) and center at $(0, 0)$:

$$x^2 + y^2 = r^2$$

EXAMPLE 4 | Find the equation of a circle with radius 4 and center at (A) $(-3, 6)$ and (B) $(0, 0)$. Graph each equation.

Solution | (A) $(h, k) = (-3, 6)$ and $r = 4$

$$(x - h)^2 + (y - k)^2 = r^2$$
$$[x - (-3)]^2 + (y - 6)^2 = 4^2$$
$$(x + 3)^2 + (y - 6)^2 = 16$$

To graph the equation, locate the center $C(-3, 6)$ and draw a circle of radius 4.

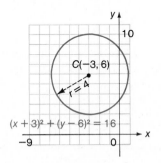

(B) $(h, k) = (0, 0)$ and $r = 4$

$$x^2 + y^2 = r^2$$
$$x^2 + y^2 = 4^2$$
$$x^2 + y^2 = 16$$

To graph the equation, locate the center at the origin and draw a circle of radius 4.

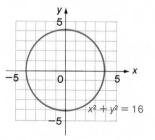

$x^2 + y^2 = 16$

PROBLEM 4   Find the equation of a circle with radius 3 and center at (A) (3, -2) and (B) (0, 0). Graph each equation.

EXAMPLE 5   Find the center and radius of the circle with equation $x^2 + y^2 + 6x - 4y = 23$.

Solution   We try to transform the equation into the form $(x - h)^2 + (y - k)^2 = r^2$ by completing the square relative to x and relative to y. From this standard form we can determine the center and radius.

$$x^2 + y^2 + 6x - 4y = 23$$
$$(x^2 + 6x \quad\ ) + (y^2 - 4y \quad\ ) = 23 \qquad \text{Complete the squares}$$
$$(x^2 + 6x + 9) + (y^2 - 4y + 4) = 23 + 9 + 4$$
$$(x + 3)^2 + (y - 2)^2 = 36$$
$$[x - (-3)]^2 + (y - 2)^2 = 6^2$$

Center: $C(h, k) = C(-3, 2)$
Radius: $r = \sqrt{36} = 6$

PROBLEM 5   Find the center and radius of the circle with equation $x^2 + y^2 - 8x + 10y = -25$.

Answers to Matched Problems   **1.**

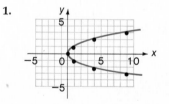

2.   (A)   Symmetric with respect to the origin

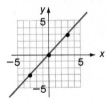

(B)   Symmetric with respect to the x axis, y axis, and origin

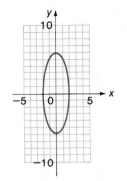

3. $d = \sqrt{173}$

4.   (A) $(x - 3)^2 + (y + 2)^2 = 9$ (B) $x^2 + y^2 = 9$

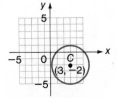   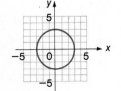

5. $(x - 4)^2 + (y + 5)^2 = 16$, radius: 4, center: $(4, -5)$

Exercise 2-1 ■   A   *Determine symmetry with respect to the x axis, y axis, or origin, if any exists, and graph.*

1. $y = 2x - 4$ **2.** $y = \frac{1}{2}x + 1$

3. $y = |x|$ **4.** $y = -|x|$

5. $|y| = x$ **6.** $|y| = -x$

7. $|x| = |y|$ **8.** $y = -x$

Find the distance between the indicated points. Leave the answer in radical form.

9. $(-6, -4)$, $(3, 4)$

10. $(-5, 4)$, $(6, -1)$

11. $(6, 6)$, $(4, -2)$

12. $(5, -3)$, $(-1, 4)$

Write the equation of a circle with the indicated center and radius.

13. $C(0, 0)$, $r = 7$

14. $C(0, 0)$, $r = 5$

15. $C(2, 3)$, $r = 6$

16. $C(5, 6)$, $r = 2$

17. $C(-4, 1)$, $r = \sqrt{7}$

18. $C(-5, 6)$, $r = \sqrt{11}$

19. $C(-3, -4)$, $r = \sqrt{2}$

20. $C(4, -1)$, $r = \sqrt{5}$

B *Determine symmetry with respect to the x axis, y axis, or origin, if any exists, and graph.*

21. $y^2 = x + 2$

22. $y^2 = x - 2$

23. $y = x^2 + 1$

24. $y + 2 = x^2$

25. $9x^2 + y^2 = 9$

26. $4x^2 + y^2 = 4$

27. $x^2 + 4y^2 = 4$

28. $x^2 + 9y^2 = 9$

29. $y^3 = x$

30. $y = x^4$

In Problems 31 and 32 determine whether the given points are vertices of a right triangle. (Recall, a triangle is a right triangle if and only if the square of the longest side is equal to the sum of the squares of the shorter sides.)

31. $(-3, 2)$, $(1, -2)$, $(8, 5)$

32. $(-4, -1)$, $(0, 7)$, $(6, -6)$

33. Find x such that $(x, 8)$ is thirteen units from $(2, -4)$.

34. Find y such that $(-2, y)$ is five units from $(-6, 6)$.

Find the center and radius of the circle with the given equation:

35. $(x - 3)^2 + (y - 5)^2 = 49$

36. $(x - 6)^2 + (y - 1)^2 = 64$

37. $(x + 4)^2 + (y - 2)^2 = 7$

38. $(x - 5)^2 + (y + 7)^2 = 15$

39. $x^2 + y^2 - 6x - 4y = 36$

40. $x^2 + y^2 - 2x - 10y = 55$

41. $x^2 + y^2 + 8x - 6y + 8 = 0$

42. $x^2 + y^2 + 4x + 10y + 15 = 0$

C *In Problems 43–48 determine symmetry with respect to the x axis, y axis, or origin, if any exists, and graph.*

43. $y^3 = |x|$

44. $|y| = x^3$

45. $xy = 1$

46. $xy = -1$

47. $y = 6x - x^2$

48. $y = x^2 - 6x$

49. Find the equation of the perpendicular bisector of the line segment joining $(-6, -2)$ and $(4, 4)$.

50. Show that

$$\left(\frac{x_1 + x_2}{2}, \frac{y_1 + y_2}{2}\right)$$

is the **midpoint formula** for the line segment joining (x_1, y_1) and (x_2, y_2) by using the distance-between-two-points formula.

Find the equation of a circle that has a diameter with the indicated end points. [Hint: See Problem 50.]

51. $(7, -3), \quad (1, 7)$ **52.** $(-3, 2), \quad (7, -4)$

CALCULATOR PROBLEMS

Find the perimeter to two decimal places of the triangle with the indicated vertices.

53. $(-3, 1), \quad (1, -2), \quad (4, 3)$ **54.** $(-2, 4), \quad (3, 1), \quad (-3, -2)$

Graph, using symmetry properties where appropriate and a hand calculator.

55. $y = 0.6x^2 - 4.5$ **56.** $x = 0.8y^2 - 3.5$

57. $y = \sqrt{17 - x^2}$ **58.** $y = \sqrt{100 - 4x^2}$

59. $y = x^{2/3}$ **60.** $y^{2/3} = x$

Section 2-2   Relations and Functions

- Introduction
- Relations and Functions
- Relations Specified by Equations
- Function Notation
- A Brief History of Function

Introduction

Relations among various sets of objects abound in one's daily activities. For example,

To each person there corresponds an age.

To each item in a drugstore there corresponds a price.

To each automobile there corresponds a license number.

To each circle there corresponds an area.

To each number there corresponds its cube.

To each nonzero real number there corresponds two square roots.

One of the most important aspects of science is establishing relations among various types of phenomena. Once a relation is known, predictions can be made. A chemist can use a gas law to predict the pressure of an enclosed gas given its temperature; an engineer can use a formula to predict the deflections of a beam subject to different loads; an economist would like to be able to predict interest rates given the rate of change of the money supply; and so on.

Establishing and working with relations are so fundamental to both pure and applied science that people have found it necessary to describe them in the precise language of mathematics. Special relations called *functions* represent one of the most important concepts in all of mathematics. Effort made to understand and use this concept correctly right from the beginning will be rewarded many times.

■ Relations and Functions

What do all the examples of relations given here have in common? Each deals with the matching of elements from a first set, called the **domain** of the relation, with elements in a second set, called the **range** of the relation.

Consider the accompanying table showing three relations involving the cube, square, and square root. (The choice of small domains enables us to introduce two important concepts in a relatively simple setting. Shortly, we will consider relations with infinite domains.) The first two relations are examples of functions. The third is not a function. These two very important terms, *relation and function*, are defined in the box on page 163.

RELATION 1		RELATION 2		RELATION 3	
Domain (Number)	Range (Cube)	Domain (Number)	Range (Square)	Domain (Number)	Range (Square Root)
0 ⟶ 0		−2	4	0 ⟶ 0	
1 ⟶ 1		−1	1	1	1, −1
2 ⟶ 8		0	0	4	2, −2
		1		9	3, −3
		2			

Definition of a Relation and of a Function: Rule Form

A **relation** is a rule (process or method) that produces a correspondence between a first set of elements called the **domain** and a second set of elements called the **range** such that to each element in the domain there corresponds *one or more* elements in the range.

A **function** is a relation with the added restriction that to each domain element there corresponds *one and only one* range element.

(All functions are relations, but some relations are not functions.)

   In the cube, square, and square root examples, we see that all three are relations according to the definition.* Relations 1 and 2 are also functions, since to each domain value there corresponds exactly one range value (for example, the square of -2 is 4 and no other number). On the other hand, relation 3 is not a function, since to at least one domain value there corresponds more than one range value (for example, to the domain value 9 there corresponds -3 and 3, both square roots of 9).
   Since in a relation (or function) elements in the range are paired with elements in the domain by some rule or process, this correspondence (pairing) can be illustrated using ordered pairs of elements where the first component represents a domain element and the second component a corresponding range element. Thus, we can write relations 1–3 as

Relation 1 $= \{(0, 0), (1, 1), (2, 8)\}$

Relation 2 $= \{(-2, 4), (-1, 1), (0, 0), (1, 1), (2, 4)\}$

Relation 3 $= \{(0, 0), (1, 1), (1, -1), (4, 2), (4, -2), (9, 3), (9, -3)\}$

   This suggests an alternative but equivalent way of defining relations and functions that provides additional insight into these concepts.

Definition of a Relation and of a Function: Set Form

A **relation** is any set of ordered pairs of elements.

A **function** is a relation with the added restriction that no two distinct ordered pairs can have the same first component.

The set of first components in a relation (or function) is called the **domain** of the relation, and the set of second components is called the **range**.

---

* We have used the word *relation* earlier as a word from our ordinary language. After the formal definition, the word *relation* becomes part of our technical mathematical vocabulary. From now on when we use the word *relation* in a mathematical context, it will have the meaning as specified.

According to this definition, we see (as before) that relation 3 is not a function, since there are two distinct ordered pairs [(1, 1) and (1, −1), for example] that have the same first component (more than one range element is associated with a given domain element).

The rule form of the definition of a relation and a function suggests a formula or a "machine" operating on domain values to produce range values—a dynamic process. On the other hand, the set definition of these concepts is closely related to graphs in a Cartesian coordinate system—a static form. Each approach has its advantages in certain situations.

One of the main objectives of this section is to expose you to the more common ways relations and functions are specified (including special notation) and to provide you with experience in determining whether a given relation is or is not a function.

As a consequence of the definitions, we find that a relation (or function) can be specified in many different ways: by an equation, by a table, by a set of ordered pairs of elements, and by a graph, to name a few of the more common ways (Table 2). All that matters is that we are given a set of elements called the domain and a rule (method or process) for obtaining corresponding range values for each domain value.

TABLE 2   COMMON WAYS OF SPECIFYING RELATIONS AND FUNCTIONS

Method	Illustration	Example
Equation	$y = x^2 - x, \quad x \in R^*$	$x = -1$ corresponds to $y = 2$
Table	$\begin{array}{c\|c\|c\|c} m & 1 & 2 & 3 \\ \hline n & 1 & 8 & 27 \end{array}$	$m = 2$ corresponds to $n = 8$
Sets of ordered pairs of elements	(a) $\{(1, 1), (2, 8), (3, 27)\}$ (b) $\{(x, y) \mid y = x^3, x \in R\}$	3 corresponds to 27 $x = -2$ corresponds to $y = -8$
Graph	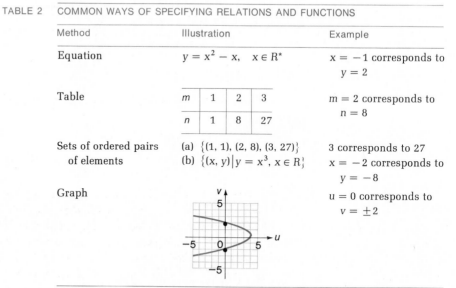	$u = 0$ corresponds to $v = \pm 2$

* Recall that R is the set of real numbers.

Which relation in Table 2 is not a function? The relation specified by the graph is not a function, since it is possible for a domain value to correspond to more than one range value. (What does $u = -5$ correspond to?)

It is very easy to determine whether a relation is a function if one has its graph.

Vertical Line Test for a Function

A relation is a function if each vertical line in the coordinate system passes through *at most* one point on the graph of the relation. (If a vertical line passes through two or more points on the graph of a relation, then the relation is not a function.)

■ Relations Specified by Equations

Most of the domains and ranges included in this text will be sets of numbers, and the rules associating range values with domain values will be equations in two variables.

Consider the equation

$$y = x^2 - x \qquad x \in R$$

For each **input** x we obtain one **output** y. For example,

If $x = 3$, then $y = 3^2 - 3 = 6$.

If $x = -\frac{1}{2}$, then $y = (-\frac{1}{2})^2 - (-\frac{1}{2}) = \frac{1}{4} + \frac{1}{2} = \frac{3}{4}$.

The input values are domain values and the output values are range values. The equation (a rule) assigns each domain value x a range value y. The variable x is called an *independent variable* (since values are "independently" assigned to x from the domain), and y is called a *dependent variable* (since y's value "depends" on the value assigned to x). In general, any variable used as a placeholder for domain values is called an **independent variable**; any variable that is used as a placeholder for range values is called a **dependent variable**.

Unless stated to the contrary, we shall adhere to the following convention regarding domains and ranges for relations and functions specified by equations.

Agreement on Domains and Ranges

If a relation or function is specified by an equation and the domain is not indicated, then we shall assume that the domain is the set of all real number replacements of the independent variable (inputs) that produce real values for the dependent variable (outputs). The range is the set of all outputs corresponding to input values.

Most equations in two variables specify relations, but when does an equation specify a function?

Equations and Functions

If, in an equation in two variables, there corresponds exactly one value of the dependent variable (output) for each value of the independent variable (input), then the equation specifies a function. If there is more than one output for at least one input, then the equation does not specify a function.

EXAMPLE 6   (A) Is the relation specified by the equation $y^2 = x + 1$ a function, assuming x is the independent variable?
(B) What is the domain of the relation?

Solution   (A) The relation is not a function, since, for example, if $x = 3$, then $y = \pm 2$.
(B) The domain of the relation (since it is not explicitly given) is the set of all real x that produce real y. Solving for y in terms of x, we obtain

$$y = \pm\sqrt{x + 1}$$

For y to be real, $x + 1$ must be greater than or equal to 0; that is,

$$x + 1 \geq 0$$
$$x \geq -1$$

Thus,

Domain: $x \geq -1$ or $[-1, \infty)$

PROBLEM 6   (A) Is the relation specified by the equation $x^2 + y^2 = 25$ a function, assuming x is the independent variable?
(B) What is the domain of the relation?

■ **Function Notation**

We have just seen that a function involves two sets of elements, a domain and a range, and a rule of correspondence that enables one to assign each element in the domain to exactly one element in the range. We use different letters to denote names for numbers; in essentially the same way, we will now use different letters to denote names for functions. For example, f and g may be used to name the two functions

$$f: \quad y = 2x + 1$$
$$g: \quad y = x^2 + 2x - 3$$

If x represents an element in the domain of a function f, then we will often use the symbol

$$f(x)$$

in place of y to designate the number in the range of the function f to which x is paired (Fig. 6).

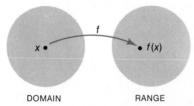

The function f "maps" the domain value x into the range value $f(x)$.

FIGURE 6   DOMAIN   RANGE

It is important not to think of $f(x)$ as the product of f and x. The symbol $f(x)$ is read "f of x," or "the value of f at x." The variable x is an independent variable; both y and $f(x)$ are dependent variables.

This function notation is extremely important, and its correct use should be mastered as early as possible. For example, in place of the more formal representation of the functions f and g, we can now write

$$f(x) = 2x + 1 \quad \text{and} \quad g(x) = x^2 + 2x - 3$$

The function symbols $f(x)$ and $g(x)$ have certain advantages over the variable y in certain situations. For example, if we write $f(3)$ and $g(5)$, then each symbol indicates in a concise way that these are range values of particular functions associated with particular domain values. Let us find $f(3)$ and $g(5)$.

To find $f(3)$, we replace x by 3 wherever x occurs in

$$f(x) = 2x + 1$$

and evaluate the right side:

$$f(3) = 2 \cdot 3 + 1$$
$$= 6 + 1$$
$$= 7$$

Thus,

$f(3) = 7$ The function *f* assigns the range value 7 to the domain value 3; the
ordered pair (3, 7) belongs to *f*.

To find g(5), we replace x by 5 wherever x occurs in

$$g(x) = x^2 + 2x - 3$$

and evaluate the right side:

$$g(5) = 5^2 + 2 \cdot 5 - 3$$
$$= 25 + 10 - 3$$
$$= 32$$

Thus,

$g(5) = 32$ The function *g* assigns the range value 32 to the domain value 5; the
ordered pair (5, 32) belongs to *g*.

It is very important to understand and remember the definition of $f(x)$:

The Function Symbol $f(x)$

For any element x in the domain of the function f, the function symbol

$$f(x)$$

represents the element in the range of f corresponding to x in the
domain of f. [If x is an input value, then $f(x)$ is an output value, or
symbolically, $f:$ $x \rightarrow f(x)$.] The ordered pair $(x, f(x))$ belongs to the
function f.

Figure 7, illustrating a "function machine," may give you additional
insight into the nature of functions and the function symbol $f(x)$. We
can think of a function machine as a device that produces exactly one
output (range) value for each input (domain) value based on a set of
instructions such as those found in an equation, graph, or table. (If more
than one output value is produced for an input value, then the machine
would be a "relation machine" and not a "function machine.")

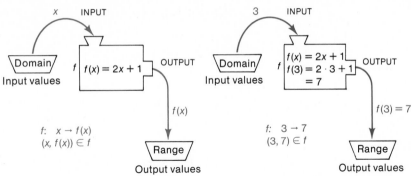

FIGURE 7   "Function Machine"—
exactly one output for each input

For the function $f(x) = 2x + 1$, the machine takes each domain value (input), multiplies it by 2, then adds 1 to the result to produce the range value (output). Different rules inside the machine result in different functions.

EXAMPLE 7   Let $f(x) = |x| - 1$, $g(x) = 1 - x^2$, $I(x) = x$, and $F(x) = 5$. Then

(A) $f(-2) = |-2| - 1 = 2 - 1 = 1$

(B) $g(-3) = 1 - (-3)^2 = 1 - 9 = -8$

(C) $F(7) = 5$ F is a constant function with domain the set of real numbers R and range {5}.

(D) $\dfrac{2f(-1) - 3I(3)}{g(-1)} = \dfrac{2\big[|-1| - 1\big] - 3[3]}{1 - (-1)^2} = \dfrac{2(0) - 9}{0} = \dfrac{-9}{0}$;   not defined

(E) $f(a) + g(a) = \big[|a| - 1\big] + \big[1 - a^2\big] = |a| - a^2$

(F) $\dfrac{f(a + h) - f(a)}{h} = \dfrac{\big[|a + h| - 1\big] - \big[|a| - 1\big]}{h} = \dfrac{|a + h| - |a|}{h}$

PROBLEM 7   For the functions f, g, I, and F in Example 7, find:

(A) $f(-6)$ (B) $g(-4)$ (C) $\dfrac{f(0) - g(0)}{2g(0)}$

(D) $4F(-2) + 2I(0)$ (E) $g(m) + F(m)$ (F) $\dfrac{g(a + h) - g(a)}{h}$

EXAMPLE 8   (A) Find the domain and range for the function

$$f = \{(-2, 3), (-1, 3), (0, 2), (1, 2)\}$$

Solution   Domain $= \{-2, -1, 0, 1\}$ Set of first components

Range $= \{2, 3\}$ Set of second components

(B) Find the domain for the function g where

$$g(x) = \frac{2x}{x^2 - 4}$$

Solution   Since the domain is not explicitly stated, it is understood to be the set of all real x that produce real g(x). The expression $2x/(x^2 - 4)$ represents a real number for all x except for

$$x^2 - 4 = 0$$
$$x^2 = 4$$
$$x = \pm 2$$

Domain: All real numbers except ± 2.

(C) Find the domain for the function H where

$$H(u) = \sqrt{25 - u^2}$$

Solution   The domain is the set of all real u such that $\sqrt{25 - u^2}$ is a real number—that is, such that $25 - u^2 \geq 0$. Solving the inequality $25 - u^2 = (5 - u)(5 + u) \geq 0$ by methods discussed in Section 1-6, we find

Domain: $-5 \leq u \leq 5$ or $[-5, 5]$

PROBLEM 8   (A) Find the domain and range for the function

$$h = \{(x, y) | y = x^2, \quad x \in \{-2, 0, 2\}\}$$

(B) Find the domain of the function G where

$$G(t) = \frac{2t + 1}{t^2 - t - 6}$$

(C) Find the domain of the function f where

$$f(x) = \sqrt{\frac{x - 2}{x + 3}}$$

Remark: It is useful to summarize the different ways a function can be specified. All the following specify the same function f:

$$f(x) = \sqrt{25 - x^2}$$
$$f: \quad y = \sqrt{25 - x^2}$$
$$f: \quad x^2 + y^2 = 25 \qquad y \geq 0$$
$$f: \quad x \rightarrow \sqrt{25 - x^2} \qquad \text{Read ``x is mapped into } \sqrt{25 - x^2}.\text{''}$$
$$f = \{(x, y) | y = \sqrt{25 - x^2}\}$$

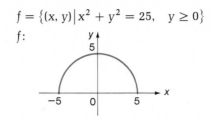

$$f = \{(x, y) \mid x^2 + y^2 = 25, \quad y \geq 0\}$$

EXAMPLE 9   A rectangular feeding pen for cattle is to be made with 100 meters of fencing.

(A) If x represents the width of the pen, express its area $A(x)$ in terms of x.

(B) What is the domain of the function A (determined by the physical restrictions)?

Solution   (A) Draw a figure and label the sides.

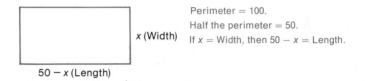

$$A(x) = (\text{Width})(\text{Length}) = x(50 - x) \quad \text{Area depends on width } x.$$

(B) To have a pen, x must be positive, but x must also be less than 50 (or the length will not exist). Thus,

Domain: $0 < x < 50$ Inequality notation

$(0, 50)$ Interval notation

PROBLEM 9   Work Example 9 with the added assumption that a large barn is to be used as one side of the pen.

■   A Brief History of Function

In reviewing the history of function, we are made aware of the tendency of mathematicians to extend and generalize a concept. The word *function* appears to have been first used by Leibniz in 1694 to stand for any quantity associated with a curve. By 1718, Johann Bernoulli considered a function any expression made up of constants and a variable. Later in the same century, Euler came to regard a function as any equation made up of constants and variables. Euler made extensive use of the extremely important notation $f(x)$, although its origin is generally attributed to Clairaut (1734).

The form of the definition of function that has been used until well into this century (many texts still contain this definition) was formulated by Dirichlet (1805–1859). He stated that, if two variables x and y are so related that for each value of x there corresponds exactly one value of y, then y is said to be a (single-valued) function of x. He called x, the variable to which values are assigned at will, the independent variable, and y, the variable whose values depend on the values assigned to x, the dependent variable. He called the values assumed by x the domain of the function, and the corresponding values assumed by y the range of the function.

Now, since set concepts permeate almost all mathematics, we have the more general definitions of function presented in this section in terms of sets of ordered pairs of elements. The function concept is one of the most important concepts in mathematics, and as such it plays a central and natural role as a guide for the selection and development of material in many mathematics courses (look at the section titles following this section and the chapter titles following this chapter).

Answers to Matched Problems

6. (A) No   (B) Domain: $-5 \le x \le 5$ or $[-5, 5]$

7. (A) 5   (B) -15 (C) -1 (D) 20   (E) $6 - m^2$
(F) $-2a - h$

8. (A) Domain $= \{-2, 0, 2\}$; range $= \{0, 4\}$
(B) Domain: All real numbers except -2 and 3
(C) Domain $x < -3$ or $x \ge 2$ Inequality notation
$(-\infty, -3) \cup [2, \infty)$ Interval notation

9. (A) $A(x) = x(100 - 2x)$
(B) Domain: $0 < x < 50$ Inequality notation
$(0, 50)$ Interval notation

Exercise 2-2 ■

A *Indicate whether each relation in Problems 1–12 is or is not a function.*

1. Domain   Range
-1 1
0   2
1   3

2. Domain   Range
2 $\longrightarrow$ 1
4 $\longrightarrow$ 3
6 $\longrightarrow$ 5

3. Domain   Range
1 $\longrightarrow$ 3
3 $\longrightarrow$ 5
     7
5 $\longrightarrow$ 9

4. Domain   Range
-1 $\longrightarrow$ 0
-2 $\longrightarrow$ 5
-3 $\longrightarrow$ 8

5. Domain   Range
-1
0
1   3
2

6. Domain   Range
2 $\longrightarrow$ 8
3
4
5 $\longrightarrow$ 9

7.
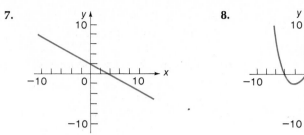

8.
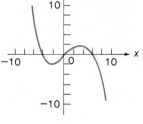

9.
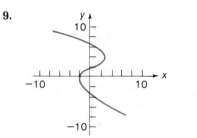

10.
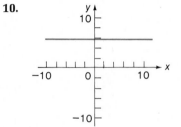

11.
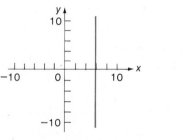

12.
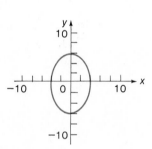

Problems 13–24 refer to

$$f(x) = 3x - 5 \qquad g(t) = 4 - t \qquad F(m) = 3m^2 \qquad G(u) = u - u^2$$

Evaluate as indicated.

13. $f(-1)$ **14.** $g(6)$

15. $G(-2)$ **16.** $F(-3)$

17. $F(-1) + f(3)$ **18.** $G(2) - g(-3)$

19. $2F(-2) - G(-1)$ **20.** $3G(-2) + 2F(-1)$

21. $\dfrac{f(0) \cdot g(-2)}{F(-3)}$ **22.** $\dfrac{g(4) \cdot f(2)}{G(1)}$

23. $\dfrac{f(2 + h) - f(2)}{h}$ **24.** $\dfrac{g(3 + h) - g(3)}{h}$

B   *Each equation specifies a relation. Which specify a function given that x is an independent variable?*

25. $y = 3 - x$ **26.** $y = 2x + 3$

27. $y = 2x^2 - 3x + 5$ **28.** $y = (2 - x)^3$

29. $y^2 - x = 2$ **30.** $y - x^2 = 2$

31. $y = |x - 2|$ *function*   **32.** $|y| = x + 2$

33. $x^2 + y^2 = 81$ *no function*   **34.** $16x^2 + y^2 = 16$

35. $y^3 = x$ *function*   **36.** $y^5 = x$

37. $y = \dfrac{x + 3}{x - 2}$ *function*   **38.** $y = \dfrac{x - 4}{2x^2 + 7x - 4}$

39. $y = \sqrt{x + 1}$ *function*   **40.** $y^2 = x + 1$

41. $y = \sqrt{\dfrac{x + 3}{x - 2}}$ *function*   **42.** $y = \sqrt{x^2 + x - 12}$

Given each relation in Problems 43–50, state its domain and range and indicate which are functions. The variable x is independent.

43. $f = \{(2, 4), (4, 2), (2, 0), (4, -2)\}$

44. $g = \{(-1, 3), (0, 1), (1, 3)\}$

45. $G = \{(-4, 1), (0, 1), (4, 1)\}$

46. $H = \{(-6, 3), (-4, 5), (-6, 0)\}$

47. $P = \{(x, y) \,|\, y^2 = x, \quad x \in \{0, 1, 4\}\}$

48. $Q = \{(x, y) \,|\, |y| = x, \quad x \in \{0, 1, 2\}\}$

49. $R = \{(x, y) \,|\, y = x^2 - x, \quad x \in \{-2, 2\}\}$

50. $h = \{(x, y) \,|\, x^2 + y^2 = 25, \quad x \in \{-5, 5\}\}$

For Problems 51–62, determine the domain of the relation in the indicated problem.

51. Problem 25	**52.** Problem 26	**53.** Problem 27
54. Problem 28	**55.** Problem 32	**56.** Problem 31
57. Problem 33	**58.** Problem 34	**59.** Problem 37
60. Problem 38	**61.** Problem 41	**62.** Problem 42

63. If $g(x) = 2 - x^2$, find: $\dfrac{g(3 + h) - g(3)}{h}$

64. If $f(x) = x^2$, find: $\dfrac{f(2 + h) - f(2)}{h}$

65. If $Q(t) = t^2 - 2t + 1$, find: $\dfrac{Q(1 + h) - Q(1)}{h}$

66. If $P(m) = 2m^2 + 3$, find: $\dfrac{P(2 + h) - P(2)}{h}$

C 67. If $f(x) = x^2 - 1$, find: $\dfrac{f(a + h) - f(a)}{h}$

68. If $g(x) = x^2 + x - 1$, find: $\dfrac{g(a + h) - g(a)}{h}$

69. If $f(x) = x^2 - 1$, find: $\dfrac{f(x + h) - f(x)}{h}$

70. If $g(x) = x^2 + x - 1$, find: $\dfrac{g(x + h) - g(x)}{h}$

71. If $h(x) = x^3$, find: $\dfrac{h(x + h) - h(x)}{h}$

72. If $g(x) = x^3 + x$, find: $\dfrac{g(x + h) - g(x)}{h}$

APPLICATIONS *Each relationship in Problems 73–78 can be described by a function. Write an equation that specifies the function.*

73. *Cost function.* The cost per day $C(x)$ for renting a car at $10 per day plus 12¢ per mile for x miles. (The cost depends on the number of miles driven.)

74. *Cost function.* The cost per day $C(x)$ of manufacturing x pairs of skis if fixed costs are $375 per day and variable costs are $68 per pair of skis manufactured. (The cost per day depends on the number of skis manufactured per day.)

75. *Area function.* The area $A(r)$ of a circle is π times the square of the radius r. (The area depends on the radius.)

76. *Earth science.* The pressure $P(d)$ in the ocean in pounds per square inch is found by dividing the depth d by 33, adding 1 to the quotient, and multiplying the final result by 15. (The pressure below sea level depends on the depth.)

77. *Manufacturing.* A candy box is to be made out of a piece of cardboard 8 by 12 inches. Equal-sized squares, x inches on a side, will be cut from each corner, and then the ends and sides will be folded up. Find a formula for the volume of the box $V(x)$ in terms of x. From practical considerations, what is the domain of the function V?

78. *Construction.* A rancher has 20 miles of fencing to fence a rectangular piece of grazing land along a straight river. If no fence is required along the river and the sides perpendicular to the river are x miles long, find a formula for the area $A(x)$ of the rectangle in terms of x. From practical considerations, what is the domain of the function A?

79. *Physics—rate.* The distance in feet that an object falls in a vacuum is given by $s(t) = 16t^2$, where t is time in seconds. Find:

(A) $s(0)$, $s(1)$, $s(2)$, $s(3)$

(B) $\dfrac{s(2 + h) - s(2)}{h}$

(C) What happens in (B) when h tends to 0? Interpret physically.

Section 2-3   Functions: Graphs and Properties

- Graphs of Functions
- Function Properties
- More Aids to Graphing

Each function that has a real number domain and range has a graph—the graph of the ordered pairs of real numbers that constitute the function. In this section we will identify several basic functions by name and sketch their graphs. Special function properties will be introduced through graphs and definitions, and additional aids to graphing will be presented.

- ## Graphs of Functions

When functions are graphed, domain values are usually associated with the horizontal axis and range values with the vertical axis. Thus, if we graph

$$y = f(x)$$

then x would be the independent variable and the abscissa of a point on the graph of the function f; y and $f(x)$ would be dependent variables and either the ordinate of a point on the graph of f (Fig. 8).

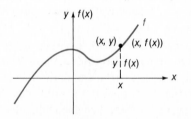

FIGURE 8

The **graph of a function f** is the same as the graph of the equation $y = f(x)$. Figure 9 illustrates the graphs of several basic functions with which you have had some experience in equation form.

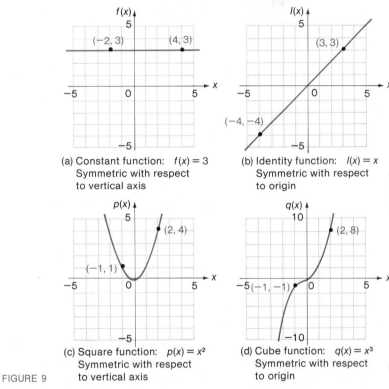

(a) Constant function: $f(x) = 3$
Symmetric with respect
to vertical axis

(b) Identity function: $I(x) = x$
Symmetric with respect
to origin

(c) Square function: $p(x) = x^2$
Symmetric with respect
FIGURE 9   to vertical axis

(d) Cube function: $q(x) = x^3$
Symmetric with respect
to origin

■ Function Properties

A function whose graph is symmetric with respect to the vertical axis is called an **even function**. A function whose graph is symmetric with respect to the origin is called an **odd function**. (Is it possible for a function with independent variable x to be symmetric with respect to the x axis?) The constant and square functions in Figure 9 are even functions, and the identity and cube functions are odd functions. Of course, if the graph of a function is not symmetric with respect to either the vertical axis or the origin, then it is neither even nor odd. As a consequence of the tests for symmetry discussed in Section 2-1, we have the following tests for even and odd functions:

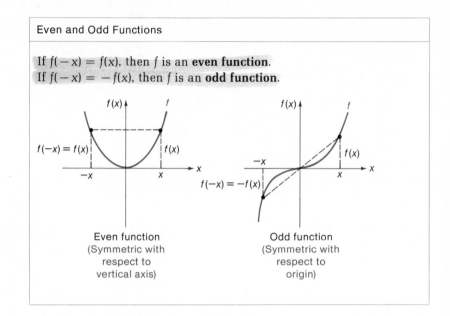

Even and Odd Functions

If $f(-x) = f(x)$, then f is an **even function**.
If $f(-x) = -f(x)$, then f is an **odd function**.

Even function
(Symmetric with
respect to
vertical axis)

Odd function
(Symmetric with
respect to
origin)

EXAMPLE 10   Without graphing, determine whether the functions f, g, and h are even, odd, or neither.

Solution   (A) $f(x) = |x|$ (B) $g(x) = x^3 + 1$ (C) $h(x) = \sqrt[3]{x}$

(A) $f(-x) = |-x| = |x| = f(x)$;   therefore, f is even.

(B) $g(x) = x^3 + 1$

$g(-x) = (-x)^3 + 1 = -x^3 + 1 \quad g(-x) \neq g(x)$

$-g(x) = -(x^3 + 1) = -x^3 - 1 \quad g(-x) \neq -g(x)$

Therefore, g is neither even nor odd.

(C) $h(-x) = \sqrt[3]{-x} = -\sqrt[3]{x} = -h(x)$;   therefore, h is odd.

PROBLEM 10   Without graphing, determine whether the functions F, G, and H are even, odd, or neither.

(A) $F(x) = x^3 + x$ (B) $G(x) = x^2 + 1$ (C) $H(x) = 2x + 4$

Why are we interested in knowing whether a function is even or odd? If we want to graph a function specified by an equation, then the even–odd test in the box provides a useful aid for graphing: If the function is even, then its graph is symmetric with respect to the vertical axis; if it is odd, then its graph is symmetric with respect to the origin.

In addition, certain problems and developments in calculus and more advanced mathematics are simplified if one recognizes the presence of either an even or odd function.

We now take a look at increasing–decreasing properties of functions. Intuitively, a function is increasing over an interval I in its domain if its graph rises as the independent variable increases over I; a function is decreasing over I if its graph falls as the independent variable increases over I (Fig. 10).

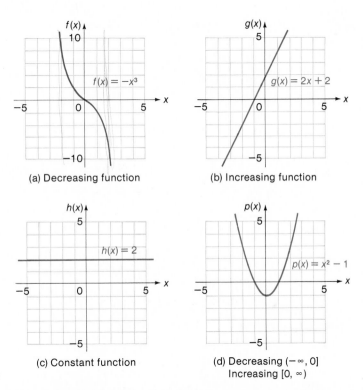

(a) Decreasing function

(b) Increasing function

(c) Constant function

(d) Decreasing $(-\infty, 0]$
Increasing $[0, \infty)$

FIGURE 10

More formally, we define increasing, decreasing, and constant functions as follows:

Increasing, Decreasing, and Constant Functions

Let I be an interval in the domain of a function f, then

1. **f is increasing** on I if $f(b) > f(a)$ whenever $b > a$ in I.
2. **f is decreasing** on I if $f(b) < f(a)$ whenever $b > a$ in I.
3. **f is constant** on I if $f(a) = f(b)$ for all a and b in I.

Another important property of functions is the **continuity property**. Our discussion of continuity must remain informal at this point and will rely heavily on geometric considerations. (A precise presentation of continuity can be found in calculus texts.) Nevertheless, it is useful to have an intuitive idea of the concept of continuity before a formal presentation is made. Let us introduce the idea through an interesting function called the *greater integer function*.

The **greatest integer** of a real number x, denoted by $[\![x]\!]$, is the integer n such that $n \le x < n + 1$ (that is, $[\![x]\!]$ is the largest integer less than or equal to x). For example,

$$[\![3.45]\!] = 3$$
$$[\![7]\!] = 7$$
$$[\![0]\!] = 0$$
$$[\![-2.13]\!] = -3 \quad \text{Not } -2$$
$$[\![-8]\!] = -8$$

The **greatest integer function** f (also called a **step function**) is determined by the equation $f(x) = [\![x]\!]$. The domain of f is the set of all real numbers and the range of f is the set of integers. A sketch of the graph of f is shown in Figure 11.

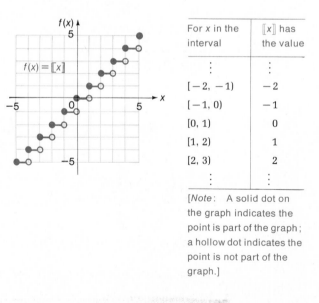

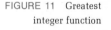

For x in the interval	$[\![x]\!]$ has the value
⋮	⋮
$[-2, -1)$	-2
$[-1, 0)$	-1
$[0, 1)$	0
$[1, 2)$	1
$[2, 3)$	2
⋮	⋮

[*Note*: A solid dot on the graph indicates the point is part of the graph; a hollow dot indicates the point is not part of the graph.]

FIGURE 11 Greatest integer function

We notice in Figure 11 that at each integer value for x there is a break in the graph, and between integer values for x there is no break. If the graph of a function is not broken (disconnected) at a point, then the function is said to be **continuous** at that point. A function whose

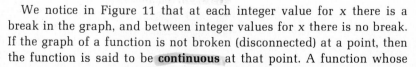

graph is broken (disconnected) at a certain point is said to be **discontinuous** at that point. A function is **continuous over an interval** if its graph is continuous (not broken) at each value on the interval. Thus, we see that the greatest integer function is discontinuous at each integer but is continuous in each interval that does not contain an integer.

EXAMPLE 11   Let a function f be defined as follows:

$$f(x) = \begin{cases} 0 & \text{for } x < 0 \\ -x + 2 & \text{for } 0 \le x < 2 \\ 2 & \text{for } x \ge 2 \end{cases}$$

Graph f and indicate points of discontinuity.

Solution   Note that f is defined by different formulas for different parts of its domain. This is a perfectly acceptable way to define a function and is used in many applications. (All that matters is that we have a way of determining a range value for each domain value.) The graph of f is given here. We see from the graph that the function f is discontinuous at $x = 0$ and $x = 2$.

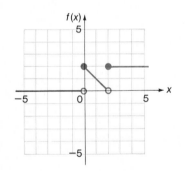

PROBLEM 11   Let a function g be defined as follows:

$$g(x) = \begin{cases} 1 & \text{for } x < -1 \\ x^2 & \text{for } -1 \le x < 2 \\ 0 & \text{for } x \ge 2 \end{cases}$$

Graph g and indicate points of discontinuity.

Remarks: In calculus it can be shown that:

1. Polynomial functions [functions defined by equations of the form $f(x) = (\text{Polynomial in } x)$] are continuous for all real numbers; that is, polynomial functions have no breaks in their graphs. Thus, f where $f(x) = 2x^3 - x^2 + 3x - 1$ is continuous for all real numbers. [*Note:* Domain is the set of all real numbers.]

2. Rational functions [functions defined by equations of the form $f(x) =$ (Polynomial in x)/(Polynomial in x)] are continuous for all real numbers except values of x that make a denominator 0. Thus, f where

$$f(x) = \frac{x - 1}{x^2 - x - 6} = \frac{x - 1}{(x - 3)(x + 2)}$$

is continuous for all real numbers, except $x = -2$ and $x = 3$; that is, f is discontinuous (there is a break in the graph) at $x = -2$ and at $x = 3$. [*Note:* Domain is the set of all real numbers except -2 and 3.]

We will have more to say about these remarks and their consequences in later sections.

■ More Aids to Graphing

There are situations in which it is possible to use known graphs of basic functions to graph related functions. Involved here are vertical and horizontal shifting, reflecting with respect to an axis, and expanding and contracting. These graphing aids will be illustrated through examples.

EXAMPLE 12   Graph $f(x) = x^2$, $F(x) = x^2 + 3$, and $G(x) = x^2 - 2$.

Solution   We already know that the graph of $y = x^2$ is a parabola opening upward; it passes through the origin and has as its axis the y axis [see figure (a)]. To graph $y = x^2 + 3$, we add 3 to each ordinate value for the graph of $y = x^2$ [see figure (b)]. To graph $y = x^2 - 2$, we subtract 2 from each ordinate value [see figure (c)]. The net result is that the graph of $y = x^2 + 3$ is just the graph of $y = x^2$ shifted upward three units, and the graph of $y = x^2 - 2$ is just the graph of $y = x^2$ shifted downward two units.

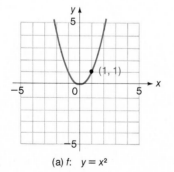

(a) f: $y = x^2$

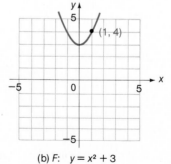

(b) F: $y = x^2 + 3$

The graph of $y = x^2 + 3$ is the same as the graph of $y = x^2$ shifted up three units.

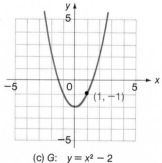

(c) G: $y = x^2 - 2$

The graph of $y = x^2 - 2$ is the same as the graph of $y = x^2$ shifted down two units.

Following the same line of reasoning as in Example 12, the results can be generalized as given in Table 3.

TABLE 3   VERTICAL SHIFTING (TRANSLATION)

To Graph	Shift the Graph of $y = f(x)$
$y = f(x) + k, \quad k > 0$	Up k units
$y = f(x) + k, \quad k < 0$	Down $|k|$ units

PROBLEM 12 Graph $f(x) = |x| + k$ for $k = 0$, $k = 2$, and $k = -3$.

We now turn to horizontal shifting.

EXAMPLE 13 Graph $f(x) = x^2$, $P(x) = (x + 2)^2$, and $Q(x) = (x - 3)^2$.

Solution   Observe the following:

$$f(x) = x^2 \qquad P(x) = (x + 2)^2 \qquad\qquad Q(x) = (x - 3)^2$$
$$f(a) = a^2 \qquad P(a - 2) = (a - 2 + 2)^2 \qquad Q(a + 3) = (a + 3 - 3)^2$$
$$(a, a^2) \in f \qquad\qquad = a^2 \qquad\qquad\qquad = a^2$$
$$(a - 2, a^2) \in P \qquad\qquad (a + 3, a^2) \in Q$$

Thus, the point with abscissa a on the graph of $y = f(x) = x^2$ has the same ordinate value, a^2, as the point with abscissa $a - 2$ on the graph of $y = P(x) = (x + 2)^2$ and the point with abscissa $a + 3$ on the graph of $y = Q(x) = (x - 3)^2$. We conclude that the graph of $y = (x + 2)^2$ is the same as the graph of $y = x^2$ shifted to the left two units [see figure (b)]. And the graph of $y = (x - 3)^2$ is the same as the graph of $y = x^2$ shifted to the right three units [see figure (c)].

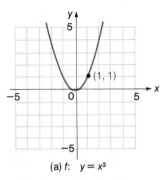

(a) f: $y = x^2$

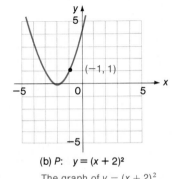

(b) P: $y = (x + 2)^2$

The graph of $y = (x + 2)^2$
is the same as the graph
of $y = x^2$ shifted to the left
two units.

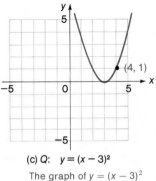

(c) Q: $y = (x - 3)^2$

The graph of $y = (x - 3)^2$
is the same as the graph
of $y = x^2$ shifted to the
right three units.

Following the same line of reasoning as in Example 13, the results can be generalized as given in Table 4.

TABLE 4

HORIZONTAL SHIFTING (TRANSLATION)

To Graph	Shift the Graph of $y = f(x)$
$y = f(x + h), \quad h > 0$	To the left h units
$y = f(x + h), \quad h < 0$	To the right $|h|$ units

Remark: The horizontal shift is just the opposite of what many people expect. A positive h is associated with a shift to the left; a negative h is associated with a shift to the right. On the other hand, relative to vertical shifts, a positive k [in $y = f(x) + k$] is associated with a shift upward, and a negative k is associated with a shift downward.

PROBLEM 13   Graph $f(x) = |x + h|$ for $h = 0$, $h = 2$, and $h = -3$.

We conclude this section by considering reflections, expansions, and contractions of graphs.

EXAMPLE 14   Graph $f(x) = x^2$, $R(x) = -x^2$, $S(x) = 2x^2$, and $T(x) = \frac{1}{2}x^2$.

Solution   If we take the negative of each ordinate value on the graph of $y = x^2$

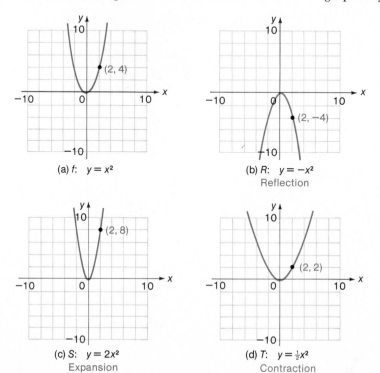

(a) f: $y = x^2$

(b) R: $y = -x^2$
Reflection

(c) S: $y = 2x^2$
Expansion

(d) T: $y = \frac{1}{2}x^2$
Contraction

[figure (a)], we will obtain the graph of $y = -x^2$ [figure (b)]. If we double each ordinate value on the graph of $y = x^2$, we will obtain the graph of $y = 2x^2$ [figure (c)]. And if we cut each ordinate value in half on the graph of $y = x^2$, we will obtain the graph of $y = \frac{1}{2}x^2$ [figure (d)].

Following the same line of reasoning as in Example 14, the results can be generalized as given in Table 5.

TABLE 5   REFLECTION, EXPANSION, CONTRACTION

Graph of	Relationship to the Graph of $y = f(x)$
$y = -f(x)$	Reflection of the graph of $y = f(x)$ across the x axis
$y = Cf(x), \quad C > 1$	All ordinate values are expanded by a factor of C
$y = Cf(x), \quad 0 < C < 1$	All ordinate values are contracted by a factor of C

PROBLEM 14   Graph $f(x) = C|x|$ for $C = 1$, $C = -1$, $C = 2$, and $C = \frac{1}{2}$.

Answers to Matched Problems   **10.** (A) Odd   (B) Even   (C) Neither

11. Discontinuous at $x = 2$

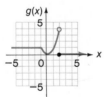

12.

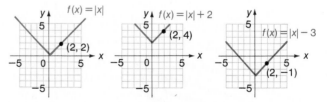

13.

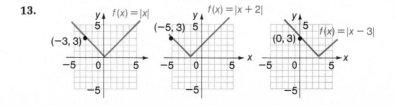

14.

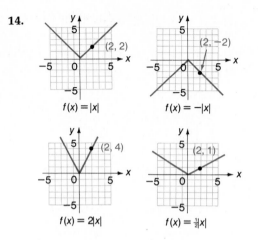

$f(x) = |x|$ $f(x) = -|x|$

$f(x) = 2|x|$ $f(x) = \frac{1}{2}|x|$

Exercise 2-3 ■ A   *Problems 1–12 refer to functions f, g, p, and q given by the following graphs. (Assume the graphs continue as indicated beyond the parts shown.)*

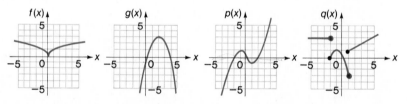

1. Which functions are symmetric with respect to the vertical axis?

2. Which functions are symmetric with respect to the origin?

3. Which functions are even?

4. Which functions are odd?

5. Indicate intervals over which f is:
    (A) Constant   (B) Increasing   (C) Decreasing

6. Indicate intervals over which g is:
    (A) Constant   (B) Increasing   (C) Decreasing

7. Indicate intervals over which q is:
    (A) Constant   (B) Increasing   (C) Decreasing

8. Indicate intervals over which p is:
    (A) Constant   (B) Increasing   (C) Decreasing

9. Which functions are continuous on the interval $(-3, 3)$?

10. Which functions are continuous on the interval $(2, 4)$?

11. Which functions have points of discontinuity? Name them.

12. Is function f discontinuous at $x = 0$?

B *Without graphing, indicate whether each function is even, odd, or neither.*

13. $g(x) = x^3 + x$

14. $f(x) = x^5 - x$

15. $m(x) = x^4 + 3x^2$

16. $h(x) = x^4 - x^2$

17. $F(x) = x^5 + 1$

18. $f(x) = x^5 - 3$

19. $G(x) = x^4 + 2$

20. $P(x) = x^4 - 4$

21. $q(x) = x^2 + x - 3$

22. $n(x) = 2x - 3$

whether neither (handwritten)

Graph as indicated.

23. $f(x) = \sqrt{x} + k$ for $k = 0, k = 2, k = -3$

24. $g(x) = -x^2 + k$ for $k = 0, k = 4, k = -1$

25. $F(x) = \sqrt{x + h}$ for $h = 0, h = 4, h = -1$

26. $G(x) = -(x + h)^2$ for $h = 0, h = 1, h = -2$

27. $p(x) = C\sqrt{x}$ for $C = 1, C = -1, C = 2, C = \frac{1}{2}$

28. $Q(x) = Cx$ for $C = 1, C = -1, C = 2, C = \frac{1}{2}$

Indicate how the graph of each function is related to the graph of $y = x^2$, $y = |x|$, or $y = \sqrt{x}$. Graph each function.

29. $g(x) = -(x + 2)^2$

30. $h(x) = -(x - 3)^2$

31. $G(x) = -|x + 2|$

32. $H(x) = -|x - 3|$

33. $P(x) = -\sqrt{x - 1}$

34. $Q(x) = -\sqrt{x + 2}$

Graph each function using the aids to graphing discussed in this section. Indicate if a function is either even or odd. Indicate any points of discontinuity.

35. $g(x) = -4$

36. $f(x) = -2$

37. $S(x) = -2|x + 2|$

38. $f(x) = -\frac{1}{2}(x - 2)^2$

39. $g(x) = \frac{1}{2}(x + 1)^2 - 2$

40. $h(x) = -2|x - 2| + 2$

41. $y = \dfrac{1}{|x|}$

42. $y = \dfrac{1}{x}$

43. $f(x) = \dfrac{|x|}{x}$

44. $g(x) = x + \dfrac{|x|}{x}$

45. $f(x) = \begin{cases} 1 & \text{if } x < 0 \\ x + 1 & \text{if } 0 \le x < 2 \\ 2 & \text{if } x \ge 2 \end{cases}$

46. $g(x) = \begin{cases} -x & \text{if } x < 0 \\ 2 & \text{if } 0 \le x < 2 \\ x - 2 & \text{if } x \ge 2 \end{cases}$

47. $p(x) = \begin{cases} 0 & \text{if } x < 0 \\ 4 - x^2 & \text{if } 0 \le x < 2 \\ 0 & \text{if } x \ge 2 \end{cases}$

48. $T(x) = \begin{cases} 0 & \text{if } x < 0 \\ x^2 & \text{if } 0 \le x < 2 \\ 0 & \text{if } x \ge 2 \end{cases}$

49. $p(x) = -[\![x]\!]$ **50.** $q(x) = [\![x]\!] + 2$

51. $m(x) = -[\![x - 1]\!]$ **52.** $g(x) = [\![x + 3]\!]$

C **53.** $m(x) = x^2 - 2|x|$ **54.** $n(x) = 2|x| - x^2$

55. $f(x) = x^3 - 3x$ **56.** $T(x) = 3x - x^3$

57. $r(x) = x - [\![x]\!]$ **58.** $S(x) = [\![x]\!] - x$

Section 2-4   Linear Relations and Functions

- Linear Functions
- Slope of a Line
- Equations of Lines—Standard Forms
- Parallel and Perpendicular Lines
- Concluding Remarks

We now turn to special relations and functions called *linear relations and functions*. Both are used extensively in mathematical developments and applications.

- Linear Functions

We start the discussion by defining a linear function:

Linear Function

A function f is a **linear function** if

$$f(x) = ax + b \qquad a \neq 0$$

where a and b are real numbers.

The algebraic expression $ax + b$, $a \neq 0$, is a *first-degree polynomial*; hence, a linear function is called a **first-degree polynomial function**. (In subsequent sections and chapters we will discuss second- and higher-degree polynomial functions.)

The word *linear* is used in naming the function, since its graph is a straight line. To see this, we will show that any three points having coordinates that satisfy

$$f(x) = ax + b, \qquad a \neq 0 \tag{1}$$

are **collinear** (that is, they lie on the same line). Pick three arbitrary but distinct values for x, and label them so that $x_1 < x_2 < x_3$. Then, using equation (1), we obtain

$$P_1(x_1, \ f(x_1)) = P_1(x_1, \ ax_1 + b)$$
$$P_2(x_2, \ f(x_2)) = P_2(x_2, \ ax_2 + b)$$
$$P_3(x_3, \ f(x_3)) = P_3(x_3, \ ax_3 + b)$$

These three points have coordinates that satisfy equation (1). We now show that P_1, P_2, and P_3 are collinear by showing that

$$d(P_1, P_2) + d(P_2, P_3) = d(P_1, P_3)$$

To accomplish this, we compute each of the distances and add the first two to obtain the third.*

$$\begin{aligned} d(P_1, P_2) &= \sqrt{(x_2 - x_1)^2 + [(ax_2 + b) - (ax_1 + b)]^2} \\ &= \sqrt{(x_2 - x_1)^2 + a^2(x_2 - x_1)^2} \\ &= \sqrt{(x_2 - x_1)^2(1 + a^2)} = |x_2 - x_1|\sqrt{1 + a^2} \\ &= (x_2 - x_1)\sqrt{1 + a^2} \quad \text{Since } x_2 > x_1 \end{aligned}$$

Similarly,

$$d(P_2, P_3) = (x_3 - x_2)\sqrt{1 + a^2}$$
$$d(P_1, P_3) = (x_3 - x_1)\sqrt{1 + a^2}$$

Thus,

$$\begin{aligned} d(P_1, P_2) + d(P_2, P_3) &= (x_2 - x_1)\sqrt{1 + a^2} + (x_3 - x_2)\sqrt{1 + a^2} \\ &= (x_2 - x_1 + x_3 - x_2)\sqrt{1 + a^2} \\ &= (x_3 - x_1)\sqrt{1 + a^2} = d(P_1, P_3) \end{aligned}$$

We conclude that P_1, P_2, and P_3 are collinear. We have just proved Theorem 1 (a result you no doubt guessed to be true from the graphing that was done earlier).

THEOREM 1

> The graph of a linear function f,
>
> $$f(x) = ax + b \qquad a \neq 0$$
>
> is a straight line.

---

* Recall from plane geometry that the shortest distance between two points is a straight line joining the two points.

Now that we know this theorem as a fact, graphing linear functions is very easy. Since two points determine a line, we have only to find two solutions to $y = ax + b$, plot them, and then draw a line through the two points using a straightedge. (Sometimes it is useful to find a third solution to $y = ax + b$ as a check point.)

EXAMPLE 15   Graph the linear function f given by

$$f(x) = -\tfrac{2}{3}x + 4$$

Solution

x	$f(x)$
0	4
3	2
6	0

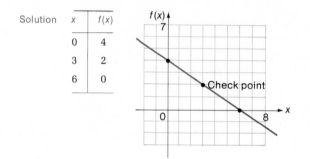

PROBLEM 15   Graph $g(x) = \tfrac{3}{2}x - 6$.

■   Slope of a Line

We next turn to a measure of the "steepness" of a line called *slope*. (We first note that a **vertical line** is a line parallel to the y axis, and a **horizontal line** is a line parallel to the x axis.)

Slope of a Line

If $P_1(x_1, y_1)$ and $P_2(x_2, y_2)$ are two distinct points on a nonvertical line ($x_1 \neq x_2$), then the **slope** of the line is given by

$$m = \frac{y_2 - y_1}{x_2 - x_1} \qquad x_1 \neq x_2$$

Interpreted geometrically, the slope is the ratio of the change in y (**rise**) to the change in x (**run**) as we move from P_1 to P_2. If the points are labeled so that $x_2 > x_1$, then m is positive if $y_2 > y_1$, 0 if $y_2 = y_1$, and negative if $y_2 < y_1$. On a vertical line $x_1 = x_2$; thus, its slope is not defined. All four cases are illustrated in Figure 12.

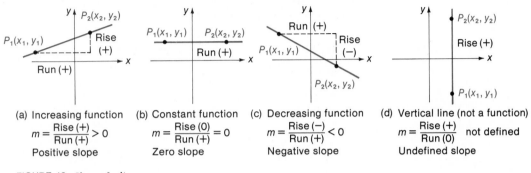

(a) Increasing function
$m = \dfrac{\text{Rise }(+)}{\text{Run }(+)} > 0$
Positive slope

(b) Constant function
$m = \dfrac{\text{Rise }(0)}{\text{Run }(+)} = 0$
Zero slope

(c) Decreasing function
$m = \dfrac{\text{Rise }(-)}{\text{Run }(+)} < 0$
Negative slope

(d) Vertical line (not a function)
$m = \dfrac{\text{Rise }(+)}{\text{Run }(0)}$ not defined
Undefined slope

FIGURE 12   Slope of a line

In using the formula to find the slope of a line through two points, it does not matter which point is labeled P_1 or P_2, since

$$\frac{y_2 - y_1}{x_2 - x_1} = \frac{y_1 - y_2}{x_1 - x_2}$$

In addition, it is important to note that the definition of slope does not depend on the two points chosen on the line as long as they are distinct. This follows from the fact that the ratios of corresponding sides of similar triangles are equal. (The details of a proof are left to the reader.)

EXAMPLE 16   Sketch a line through each pair of points and find the slope of each line.

(A) $(-3, -4)$, $(3, 2)$ (B) $(-2, 3)$, $(1, -3)$
(C) $(-4, 2)$, $(3, 2)$ (D) $(2, 4)$, $(2, -3)$

Solution   (A)

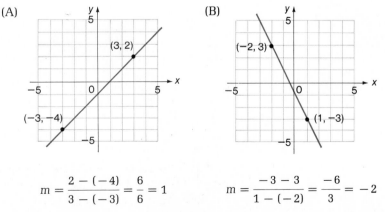

(B)

$$m = \frac{2 - (-4)}{3 - (-3)} = \frac{6}{6} = 1$$

$$m = \frac{-3 - 3}{1 - (-2)} = \frac{-6}{3} = -2$$

(C) 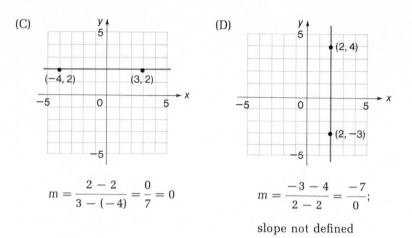 (D)

$$m = \frac{2-2}{3-(-4)} = \frac{0}{7} = 0 \qquad\qquad m = \frac{-3-4}{2-2} = \frac{-7}{0};$$

slope not defined

PROBLEM 16   Find the slope of the line through each pair of points. Do not graph.

(A) $(-3, -3), (2, -3)$ (B) $(-2, -1), (1, 2)$
(C) $(0, 4), (2, -4)$ (D) $(-3, 2), (-3, -1)$

■ Equations of Lines—Standard Forms

There are several standard equations of lines that we will now identify. (Each has a particular advantage for certain situations.) We will start with the simplest equations, those for horizontal and vertical lines. Consider the two equations

$$x + 0y = a \quad \text{or} \quad x = a \tag{2}$$
$$0x + \ y = b \quad \text{or} \quad y = b \tag{3}$$

In equation (2), y can be any number as long as $x = a$. Thus, the graph of $x = a$ is a vertical line crossing the x axis at $(a, 0)$. In equation (3), x can be any number as long as $y = b$. Thus, the graph of $y = b$ is a horizontal line crossing the y axis at $(0, b)$. We summarize these results as follows:

Vertical and Horizontal Lines

EQUATION
$x = a$ (short for $x + 0y = a$)
$y = b$ (short for $0x + y = b$)

GRAPH
Vertical line through $(a, 0)$
Horizontal line through $(b, 0)$

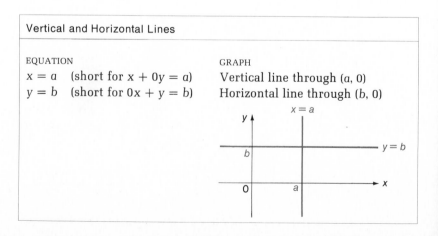

EXAMPLE 17   Find the equations of the horizontal and vertical lines that pass through $(-7, 12)$.

Solution   Horizontal line: $y = 12$

Vertical line: $x = -7$

PROBLEM 17   Find the equations of the horizontal and vertical lines that pass through $(5, -2)$.

Now suppose a nonvertical line passes through $P_1(x_1, y_1)$ and has slope m. What is its equation? Choose $P(x, y)$, $x \neq x_1$, as an arbitrary point in the coordinate system. The point P will be on the line if and only if the slope of the line through P_1 and P is m—that is, if and only if

$$\frac{y - y_1}{x - x_1} = m \tag{4}$$

or

$$y - y_1 = m(x - x_1) \tag{5}$$

Note that (x_1, y_1) also satisfies equation (5); hence, *any* point having coordinates that satisfy equation (5) is on the line that passes through (x_1, y_1) with slope m.

We have just obtained the **point–slope form** of the equation of a line.

Point–Slope Form

An equation of a line through $P_1(x_1, y_1)$ with slope m is

$$y - y_1 = m(x - x_1)$$

Remember that $P(x, y)$ is a variable point and $P_1(x_1, y_1)$ is fixed.

EXAMPLE 18   Find the equation of the line that passes through $(4, -1)$ and $(-8, 5)$. Write the resulting equation in the form $y = ax + b$.

Solution   First find the slope:

$$m = \frac{y_2 - y_1}{x_2 - x_1} = \frac{5 - (-1)}{-8 - 4} = \frac{6}{-12} = -\frac{1}{2}$$

Then use either point for (x_1, y_1).

$$(x_1, y_1) = (4, -1) \qquad \text{or} \qquad (x_1, y_1) = (-8, 5)$$

$$y - y_1 = m(x - x_1) \qquad\qquad\qquad y - y_1 = m(x - x_1)$$

$$y - (-1) = -\tfrac{1}{2}(x - 4) \qquad\quad y - 5 = -\tfrac{1}{2}[x - (-8)]$$

$$y + 1 = -\tfrac{1}{2}x + 2 \qquad\qquad\quad y - 5 = -\tfrac{1}{2}x - 4$$

$$y = -\tfrac{1}{2}x + 1 \qquad\qquad\qquad y = -\tfrac{1}{2}x + 1$$

Note that we obtained the same equation for both choices of (x_1, y_1).

PROBLEM 18 Find the equation of the line that passes through $(-2, 4)$ and $(1, -2)$. Write the answer in the form $y = mx + b$.

EXAMPLE 19 A sporting goods store sells a tennis racket that cost $60 for $82 and a pair of ski boots that cost $80 for $106.

    (A) If the markup policy of the store for items that cost more than $30 is assumed to be linear and is reflected in the pricing of these two items, write an equation that relates retail price R to cost C.

    (B) Use the equation to find the retail price for a pair of running shoes that cost $40.

Solution (A) If the retail price R is assumed linearly related to cost C, then we are looking for an equation whose graph passes through $(60, 82)$ and $(80, 106)$. We find the slope, and then use the point–slope form to find the equation.

$$m = \frac{106 - 82}{80 - 60} = \frac{24}{20} = 1.2$$

$$R - R_0 = m(C - C_0)$$

$$R - 82 = 1.2(C - 60)$$

$$R - 82 = 1.2C - 72$$

$$R = 1.2C + 10$$

    (B) $R = 1.2(40) + 10 = \$58$

PROBLEM 19 The management of a company that manufactures ball-point pens estimates costs for running the company to be $200 per day at zero output, and $700 per day at an output of 1,000 pens.

    (A) Assuming total cost per day C is linearly related to total output per day x, write an equation relating these two quantities.

    (B) What is the total cost per day for an output of 5,000 pens?

    We now observe that if the point–slope form is solved for y in terms of x, then we obtain another useful form:

$$y - y_1 = m(x - x_1)$$

$$y - y_1 = mx - mx_1$$

$$y = mx + (y_1 - mx_1)$$

which is of the form

$$y = mx + b$$

where $b = y_1 - mx_1$ is a constant. Geometrically, the coefficient of x represents the slope and b is the y coordinate of the point of crossing of the line with the y axis (generally called the **y intercept**). Thus, we have the slope–intercept form.

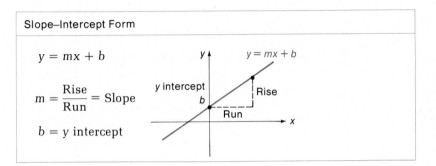

Slope–Intercept Form

$$y = mx + b$$

$$m = \frac{\text{Rise}}{\text{Run}} = \text{Slope}$$

$$b = y \text{ intercept}$$

Let us look at the equation

$$Ax + By = C \qquad A \text{ and } B \text{ not both } 0 \tag{6}$$

We now have enough information to conclude that the graph of this equation is always a line. If $B \neq 0$, then solving for y in terms of x, we obtain

$$y = -\frac{A}{B}x + \frac{C}{B}$$

which is the equation of a line with slope $-A/B$ and y intercept C/B. If $B = 0$ and $A \neq 0$, then equation (6) can be written in the form

$$x = \frac{C}{A}$$

which is the equation of a vertical line crossing the x axis at $(C/A, 0)$. Thus, we have sketched a partial proof of Theorem 2.

THEOREM 2

The graph of the equation

$$Ax + By = C \quad \text{Standard form}$$

where A, B, and C are constants (A and B not both 0) and x and y are variables, is a straight line. Any line in a rectangular coordinate system has an equation of this form.

A graph of an equation of the form $Ax + By = C$, where A, B, $C \neq 0$, can be quickly made by plotting the x and y intercepts of the graph. (Recall, the **x intercept** is the x coordinate of the point of intersection of the graph and the x axis, and the **y intercept** is the y coordinate of the point of intersection of the graph and the y axis.) To find the x intercept, set $y = 0$ and solve for x; to find the y intercept, set $x = 0$ and solve for y.

EXAMPLE 20   Graph $2x - 3y = 9$.

Solution   Find the x and y intercepts and draw a line through these two points with a straightedge.

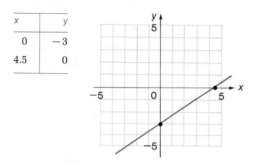

x	y
0	−3
4.5	0

PROBLEM 20   Graph $4x + 5y = 12$.

We conclude our identification of special equations of lines with an **intercept form**.

Intercept Form

The equation

$$\frac{x}{a} + \frac{y}{b} = 1$$

is the equation of a line with x intercept a and y intercept b.

To see this is the case, let $x = 0$ and solve for y; then let $y = 0$ and solve for x. The intercept form is particularly useful if it is desired to find the equation of a line with given intercepts.

EXAMPLE 21   Find the equation of a line with x intercept 4 and y intercept −2. Write the final answer in the form $Ax + By = C$, where A, B, and C are integers.

Solution $\dfrac{x}{4} + \dfrac{y}{-2} = 1$ Multiply both sides by 4.

$x - 2y = 4$

PROBLEM 21   Find the equation of a line with x intercept 3.5 and y intercept 7. Write the final answer in the form $Ax + By = C$, where $A, B,$ and C are integers.

■ **Parallel and Perpendicular Lines**

From geometric considerations, we know that two vertical lines are parallel to each other and that a horizontal line and a vertical line are perpendicular to each other. How can we tell when two nonvertical lines are parallel or perpendicular to each other? Theorem 3 (which we state without proof) provides a convenient test.

THEOREM 3   Given two (nonvertical) lines L_1 and L_2 with slopes m_1 and m_2, respectively, then

$L_1 \parallel L_2$ if and only if $m_1 = m_2$.

$L_1 \perp L_2$ if and only if $m_1 m_2 = -1$.

The symbols $\parallel$ and $\perp$ mean, respectively, "is parallel to" and "is perpendicular to." In the case of perpendicularity, the condition $m_1 m_2 = -1$ can also be written as

$$m_2 = -\frac{1}{m_1} \quad \text{or} \quad m_1 = -\frac{1}{m_2}$$

Thus, **two (nonvertical) lines are perpendicular if and only if their slopes are the negative reciprocals of each other.**

EXAMPLE 22   Given the line L: $3x - 2y = 5$ and the point $P(-3, 5)$, find an equation of a line through P that is (A) parallel to L, (B) perpendicular to L. Write the final answers in the form $y = mx + b$.

Solution   First find the slope of L by writing $3x - 2y = 5$ in the equivalent slope–intercept form $y = mx + b$:

$3x - 2y = 5$

$-2y = -3x + 5$

$y = \frac{3}{2}x - \frac{5}{2}$

Thus, the slope of L is $\frac{3}{2}$. The slope of a line parallel to L is the same, $\frac{3}{2}$, and that of a line perpendicular to L, $-\frac{2}{3}$. We can now find the equations of the two lines in parts (A) and (B) using the point–slope formula (which needs to be memorized).

(A) $y - y_1 = m(x - x_1)$ (B) $y - y_1 = m(x - x_1)$

$y - 5 = \frac{3}{2}(x + 3)$ $y - 5 = -\frac{2}{3}(x + 3)$

$y - 5 = \frac{3}{2}x + \frac{9}{2}$ $y - 5 = -\frac{2}{3}x - 2$

$y = \frac{3}{2}x + \frac{19}{2}$ $y = -\frac{2}{3}x + 3$

PROBLEM 22   Given the line L: $4x + 2y = 3$ and the point $P(2, -3)$, find an equation of a line through P that is (A) parallel to L, (B) perpendicular to L. Write the final answers in the form $y = mx + b$.

■   Concluding Remarks

Relative to lines in a rectangular coordinate system, linear equations in two variables, and linear functions, we have answered two of the fundamental questions in analytic geometry:

1. Given an equation, what is its graph?
2. Given a graph, what is its equation?

We now know how to graph linear equations on sight and how to write equations of lines given certain information about the lines.

Answers to Matched Problems

15.

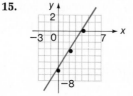

16. (A) $m = 0$ (B) $m = 1$ (C) $m = -4$
     (D) m not defined

17. Horizontal line: $y = -2$;
     vertical line: $x = 5$

18. $y = -2x$

19. (A) $C = 0.5x + 200$ (B)   \$2,700

20.

x	0	3
y	$\frac{12}{5}$	0

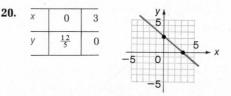

21. $2x + y = 7$

22. (A) $y = -2x + 1$ (B) $y = \frac{1}{2}x - 4$

Exercise 2-4 ■ **A** *Graph each equation and indicate the slope of the graph if it exists.*

1. $f(x) = 2x - 4$

2. $g(x) = -\frac{1}{2}x + 6$

3. $h(x) = -\frac{3}{5}x + 4$

4. $p(x) = -\frac{3}{2}x + 6$

5. $y = -\frac{3}{4}x$

6. $y = \frac{2}{3}x - 3$

7. $2x - 3y = 15$

8. $4x + 3y = 24$

9. $4x - 5y = -24$

10. $6x - 7y = -49$

11. $\dfrac{x}{2} - \dfrac{y}{3} = 1$

12. $\dfrac{x}{4} + \dfrac{y}{5} = 1$

13. $\dfrac{y}{8} - \dfrac{x}{4} = 1$

14. $\dfrac{y}{6} - \dfrac{x}{5} = 1$

15. $x = -3$

16. $y = -2$

17. $y = 3.5$

18. $x = 2.5$

B *Write an equation of the line that contains the indicated point(s) and/or has the indicated slope and/or has the indicated intercepts. Write the final equation in the form $y = mx + b$ or $x = c$.*

19. $(0, 4)$; $m = -3$

20. $(2, 0)$; $m = 2$

21. $(-5, 4)$; $m = -\frac{2}{5}$

22. $(3, -3)$; $m = -\frac{1}{3}$

23. $(5, 5)$; $m = 0$

24. $(-4, -2)$; $m = \frac{1}{2}$

25. $(1, 6)$, $(5, -2)$

26. $(-3, 4)$, $(6, 1)$

27. $(-4, 8)$, $(2, 0)$

28. $(2, -1)$, $(10, 5)$

29. $(-3, 4)$, $(5, 4)$

30. $(0, -2)$, $(4, -2)$

31. $(4, 6)$, $(4, -3)$

32. $(-3, 1)$, $(-3, -4)$

33. x intercept 6
y intercept 2

34. x intercept $3, 0$
y intercept 4

35. x intercept -4
y intercept 3

36. x intercept -4
y intercept -5

In Problems 37–48 write an equation of the line that contains the indicated point and meets the indicated condition(s). Write the final answer in the form $Ax + By = C$, $A > 0$.

37. $(-3, 4)$; parallel to $y = 3x - 5$

38. $(-4, 0)$; parallel to $y = -2x + 1$

39. $(2, -3)$; perpendicular to $y = -\frac{1}{3}x$

40. $(-2, -4)$;   perpendicular to $y = \frac{2}{3}x - 5$

41. $(2, 5)$;   parallel to y axis

42. $(7, 3)$;   parallel to x axis

43. $(3, -2)$;   vertical

44. $(-2, -3)$;   horizontal

45. $(5, 0)$;   parallel to $3x - 2y = 4$

46. $(3, 5)$;   parallel to $3x + 4y = 8$

47. $(0, -4)$;   perpendicular to $x + 3y = 9$

48. $(-2, 4)$;   perpendicular to $4x + 5y = 0$

49. Graph $f(x) = mx + 2$ for $m = 2$, $m = \frac{1}{2}$, $m = 0$, $m = -\frac{1}{2}$, and $m = -2$, all on the same coordinate system.

50. Graph $g(x) = -\frac{1}{2}x + b$ for $b = -3$, $b = 0$, and $b = 3$, all on the same coordinate system.

Problems 51–56 refer to the quadrilateral with vertices $A(0, 2)$, $B(4, -1)$, $C(1, -5)$, and $D(-3, 2)$.

51. Show that $AB \parallel DC$.   **52.** Show that $DA \parallel CB$.

53. Show that $AB \perp BC$.   **54.** Show that $AD \perp DC$.

C **55.** Find an equation of the perpendicular bisector of AD. [*Hint:* First find the midpoint of AD.]

56. Find an equation of the perpendicular bisector of AB.

57. If the graph of a linear function g has slope $-\frac{2}{3}$ and passes through $(-6, 4)$, find $g(x)$.

58. If the graph of a linear function f passes through $(-6, 2)$ and $(-4, -4)$, find $f(x)$.

59. If $f(-1) = 3$ and $f(3) = 5$ for a linear function f, find $f(x)$.

60. If $h(0) = 4$ and $h(-3) = -2$ for a linear function h, find $h(x)$.

APPLICATIONS **61.** *Physics—spring stretch.* It is known from physics (Hooke's law) that the relationship between the stretch s of a spring and the weight w causing the stretch is linear (a principle upon which all spring scales are constructed). A 10-pound weight stretches a spring 1 inch and with no weight the stretch of the spring is zero.

(A) Find a linear function f: $s = f(w) = mw + b$, that represents this relationship. [*Hint:* Both points $(10, 1)$ and $(0, 0)$ are on the graph of f.]

(B) Find $f(15)$ and $f(30)$—that is, the stretch of the spring for 15-pound and 30-pound weights.

(C) What is the slope of the graph of f? (The slope indicates the increase in stretch for each pound increase in weight.)

(D) Graph f for $0 \le w \le 40$.

62. *Business—depreciation.* An electronic computer was purchased by a company for \$20,000 and is assumed to have a salvage value of \$2,000 after 10 years (for tax purposes). Its value is depreciated linearly from \$20,000 to \$2,000.

(A) Find the linear function f: $V = f(t)$, that relates value V in dollars to time t in years.

(B) Find $f(4)$ and $f(8)$, the values of the computer after 4 and 8 years, respectively.

(C) Find the slope of the graph of f. (The slope indicates the decrease in value per year.)

(D) Graph f for $0 \le t \le 10$.

63. *Biology—nutrition.* A biologist needs to prepare a special diet for a group of experimental animals. Two food mixes, M and N, are available. If mix M contains 20% protein and mix N contains 10% protein, what combinations of each mix will provide exactly 20 grams of protein? Let x be the amount of M used and y the amount of N used. Then write a linear equation relating x, y, and 20. Graph this equation for $x \ge 0$ and $y \ge 0$.

64. *Psychology—motivation.* In an experiment on motivation (see J. S. Brown, *Journal of Comparative Physiology and Psychology*, 1948, **41**:450–465), J. S. Brown trained a group of rats to run down a narrow passage to receive food in a goal box. He then connected the rats, using a harness, to an overhead wire that was attached to a spring scale. A rat was placed at different distances d (in centimeters) from the goal box, and the pull p (in grams) of the rat toward the food was measured. Brown found that the relationship between the two variables was close to being linear and could be approximated by the linear function f: $p = f(d) = -\frac{1}{5}d + 70$, $30 \le d \le 175$.

(A) Find $f(30)$ and $f(175)$, the pull toward the goal box at 30 centimeters and 175 centimeters, respectively.

(B) What is the slope of the graph of f? (The slope indicates the change in pull per unit change in position away from the goal box.)

(C) Graph f.

Section 2-5   Graphing Polynomial Functions

■ Graphing by "Nested Factoring"
■ Quadratic Functions—Special Properties

A very important class of functions is polynomial functions. A function f defined by an equation of the form

$$f(x) = a_n x^n + a_{n-1} x^{n-1} + \cdots + a_1 x + a_0 \qquad a_n \neq 0$$

where the coefficients, a_i, are constants and n is a nonnegative integer, is called an **nth-degree polynomial function**. In the last section we discussed first-degree polynomial functions. The following equations define polynomial functions of various degrees:

$$f(x) = 2x - 3 \qquad\qquad g(x) = 2x^2 - 3x + 2$$
$$P(x) = x^3 - 2x^2 + x - 1 \qquad Q(x) = x^4 - 5$$

In Section 2-3 we discussed continuity properties of functions and noted that **polynomial functions are continuous everywhere**. Unless otherwise restricted, the domain of a polynomial function is the set of real numbers. (In the next chapter we will extend the domain to the set of complex numbers.)

In this section we will consider a particularly efficient way of graphing many polynomial functions; we will also consider some special properties of quadratic functions.

■ Graphing by "Nested Factoring"

Graphing polynomial functions by "nested factoring" is a device used to speed up the process of point-by-point plotting. It is particularly well suited for hand calculator use but is also effective for hand or mental calculations. An example will illustrate the process.

EXAMPLE 23   Graph: $P(x) = x^3 + 3x^2 - x - 3, \quad -4 \leq x \leq 2$

Solution   We first write $P(x)$ in a "nested factored" form as follows:

$$P(x) = x^3 + 3x^2 - x - 3 \qquad \text{Factor the first two terms, and repeat until you}$$
$$= (x + 3)x^2 - x - 3 \qquad \text{cannot go any further.}$$
$$= [(x + 3)x - 1]x - 3$$

This "nested factored" form is particularly convenient for evaluating $P(x)$ for various values of x by hand, and even more convenient for use with a hand calculator when x is not a small whole number. When using a hand calculator, store the chosen value of x and recall it as necessary as you proceed from left to right. In this case, all the calculations involving integers from -4 to 2 can be done mentally. If decimal values between the integers are desired (for increased graph clarity), then a hand calculator will be a considerable help.

Proceed (mentally from the "inside out") using

$$P(x) = [(x + 3)x - 1]x - 3$$

$P(-4) = [((-4) + 3)(-4) - 1](-4) - 3 = -15$

$P(-3) = [((-3) + 3)(-3) - 1](-3) - 3 = 0$

and so on

CALCULATOR COMPUTATION

A: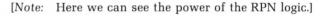

P:

[*Note:* Here we can see the power of the RPN logic.]

We construct a table of ordered pairs of numbers belonging to the function P. We then plot these points and join them with a smooth curve. (It is important to plot enough points so that it is clear what happens between the points when the points are joined by a smooth curve.)

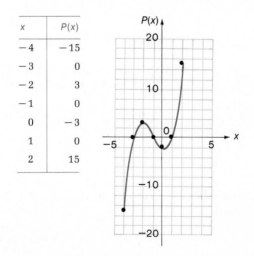

x	$P(x)$
-4	-15
-3	0
-2	3
-1	0
0	-3
1	0
2	15

PROBLEM 23   Graph $P(x) = x^3 - 4x^2 - 4x + 16$, $-3 \leq x \leq 5$,   using the nested factoring method.

Two nested factorings are shown here for polynomials with terms missing.

$$P(x) = x^3 - 2x^2 - 5 \qquad Q(x) = 2x^3 - 4x + 3$$
$$= (x - 2)x^2 - 5 \qquad\quad = (2x^2 - 4)x + 3$$
$$= [(x - 2)x]x - 5$$

■ Quadratic Functions—Special Properties

Let us now consider a special second-degree polynomial function, called a **quadratic function**, defined by an equation of the form

$$Q(x) = ax^2 + bx + c \qquad a \neq 0 \tag{1}$$

In earlier sections we graphed special cases of equation (1) and found, in particular, that the graph of $f(x) = x^2$ is as indicated in Figure 13. The graph is a parabola.

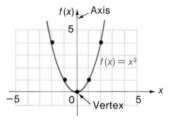

FIGURE 13   A parabola

It can be shown that the graph of any quadratic function is a parabola. In fact, the graph of $Q(x) = ax^2 + bx + c$, where $a \neq 0$, is the same as the graph of $f(x) = x^2$ modified by shifting it left, right, up, or down, and/or reflecting, and/or by expansion or contraction, depending on the values of the constants a, b, and c. The line of symmetry is called the **axis** of the parabola, and the point of intersection of the parabola and its axis is called the **vertex**.

To get a quick sketch of equation (1) we could proceed by the nested factoring method just discussed, and this is exactly what we would do in many cases. However, we can uncover some interesting and useful properties of quadratic functions by transforming equation (1) into the form

$$Q(x) = a(x + h)^2 + k \qquad a, h, k \text{ constants} \tag{2}$$

by completing a square. We look at the process through an example. Starting with

$$f(x) = 2x^2 - 8x + 5$$

we transform it into form (2) by completing a square as follows:

$$
\begin{aligned}
f(x) &= 2x^2 - 8x + 5 \\
&= 2(x^2 - 4x) + 5 \\
&= 2(x^2 - 4x + ?) + 5
\end{aligned}
$$

Factor the coefficient of x out of the first two terms.

Complete the square within the parentheses.

$$= 2(x^2 - 4x + 4) + 5 - 8$$

We added 4 to complete the square inside the parentheses. But because of the 2 outside of the parentheses, we have actually added 8, so we must subtract 8.

$$= 2(x - 2)^2 - 3$$

The transformation is complete. The graph is the same as the graph of $y = 2x^2$ shifted to the right two units and down three units.

Thus

$$f(x) = \underbrace{2(x - 2)^2}_{\text{Never negative (Why?)}} - 3$$

When $x = 2$, the first term on the right is 0, and we add 0 to -3 to obtain $f(2) = -3$. For *any* other value of x, we will add a positive number to -3, thus making $f(x)$ larger. Therefore,

$$f(2) = -3$$

is the minimum value of $f(x)$ for *all* x—a very important result! The vertical line $x = 2$ is the axis of the parabola and $(2, -3)$ are the coordinates of its vertex.

We plot the vertex and the axis and a couple of points on either side of the axis to complete the graph (Fig. 14).

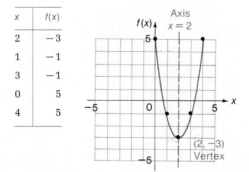

x	$f(x)$
2	-3
1	-1
3	-1
0	5
4	5

FIGURE 14

Note the important results we have obtained with this approach. We have found:

The axis of the parabola

Its vertex

The minimum value of $f(x)$

The graph of $y = f(x)$

If we had started with the general quadratic function defined by $f(x) = ax^2 + bx + c$, $a \neq 0$, following the same process as in the preceding example, we would have obtained the following general result.

Properties of $f(x) = ax^2 + bx + c, \quad a \neq 0$

1. Axis (of symmetry): $x = -\dfrac{b}{2a}$

2. Vertex: $\left(-\dfrac{b}{2a},\quad f\left(-\dfrac{b}{2a}\right)\right)$

3. Maximum or minimum value of $f(x)$:

$$f\left(-\dfrac{b}{2a}\right)\quad\begin{cases}\text{Minimum} & \text{if } a > 0 \\ \text{Maximum} & \text{if } a < 0\end{cases}$$

4. Graph:

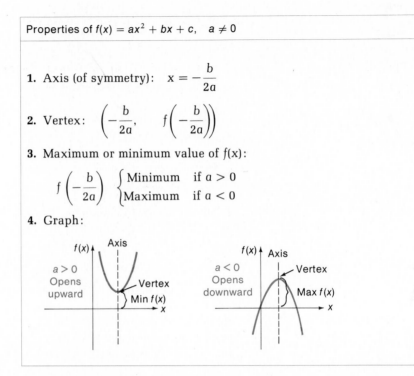

To graph a quadratic function using the method of completing the square, we can either actually complete the square as in the preceding example or use the properties in the box; some people can more readily remember a formula, others a process. We will use the properties in the box in the next example.

EXAMPLE 24   Graph, finding the axis, vertex, and maximum or minimum of $f(x)$:

Solution $f(x) = 12x - 2x^2$

$f(x) = -2x^2 + 12x$ Write in standard form, $f(x) = ax^2 + bx + c$, and note that $a = -2, b = 12$, and $c = 0$.

Axis of symmetry:

$$x = -\frac{b}{2a} = -\frac{12}{2(-2)} = 3$$

$x = 3$ Axis of symmetry

Vertex:

$$\left(-\frac{b}{2a},\quad f\left(-\frac{b}{2a}\right)\right) = (3, f(3)) = (3, 18)$$

Maximum value of $f(x)$ (since $a = -2 < 0$):

   Max $f(x) = f(3) = 18$

Graph of $y = f(x)$: To graph f, locate the axis and vertex, then locate a couple of points on either side of the axis.

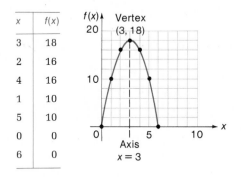

x	$f(x)$
3	18
2	16
4	16
1	10
5	10
0	0
6	0

PROBLEM 24   Graph, finding the axis, vertex, and maximum or minimum of $f(x)$.

   $f(x) = x^2 - 2x - 3$

Answers to Matched Problems   **23.**   **24.** Min $f(x) = f(1) = -4$

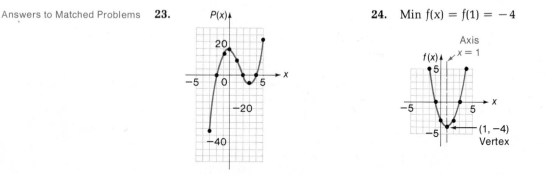

Exercise 2-5 ■ *Even though most of the problems in this exercise can be done by hand, a hand calculator will relieve much of the drudgery. The graphing problems involving nested factoring are particularly suited for calculator use.*

A *Graph each quadratic function using the nested factoring method.*

   1. $P(x) = x^2 - 4x - 5$ **2.** $P(x) = x^2 - 6x + 5$

   3. $P(x) = -x^2 + 6x$ **4.** $P(x) = -x^2 + 2x + 8$

Graph, finding the axis, vertex, and maximum or minimum of $f(x)$. [Hint: Look at pages 204 and 205 carefully.]

5. $f(x) = (x - 3)^2 + 2$ **6.** $f(x) = \frac{1}{2}(x + 2)^2 - 4$

7. $f(x) = -(x + 3)^2 - 2$ **8.** $f(x) = -(x - 2)^2 + 4$

B Graph by completing the square. Indicate the axis, vertex, and maximum or minimum value of $f(x)$.

9. $f(x) = x^2 + 6x + 11$ **10.** $f(x) = x^2 - 8x + 14$

11. $f(x) = -x^2 + 6x - 6$ **12.** $f(x) = -x^2 - 10x - 24$

Graph each polynomial function using the nested factoring method.

13. $P(x) = x^3 - 5x^2 + 2x + 8, \quad -2 \le x \le 5$

14. $P(x) = x^3 + 2x^2 - 5x - 6, \quad -4 \le x \le 3$

15. $P(x) = x^3 + 4x^2 - x - 4, \quad -5 \le x \le 2$

16. $P(x) = x^3 - 2x^2 - 5x + 6, \quad -3 \le x \le 4$

C Graph by completing the square. Indicate the axis, vertex, and maximum or minimum value of $f(x)$.

17. $f(x) = \frac{1}{2}x^2 + 2x$ **18.** $f(x) = 2x^2 - 12x + 14$

19. $f(x) = -2x^2 - 8x - 2$ **20.** $f(x) = -\frac{1}{2}x^2 + 4x - 4$

Graph each polynomial function using the nested factoring method.

21. $P(x) = x^4 - 2x^3 - 2x^2 + 8x - 8$

22. $P(x) = x^4 - 2x^2 + 16x - 15$

23. $P(x) = x^4 + 4x^3 - x^2 - 20x - 20$

24. $P(x) = x^4 - 4x^2 - 4x - 1$

APPLICATIONS **25.** *Construction.* A rectangular dog pen is to be made with 100 feet of fence wire.

(A) If x represents the width of the pen, express its area $A(x)$ in terms of x.

(B) Considering the physical limitations, what is the domain of the function A?

(C) Graph the function for this domain.

(D) Determine the dimensions of the rectangle that will make the area maximum.

26. *Construction.* Work Problem 25 with the added assumption that an existing property fence will be used for one side of the pen. (Let x equal the width—see the figure.)

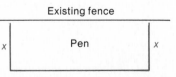

Existing fence

Pen

x x

27. *Packaging.* A candy box is to be made out of a rectangular piece of cardboard that measures 8 by 12 inches. Equal-sized squares (x by x inches) will be cut out of each corner, and then the ends and sides will be folded up to form a rectangular box.
    (A) Write the volume of the box V(x) in terms of x.
    (B) Considering the physical limitations, what is the domain of the function V?
    (C) Graph the function for this domain.
    (D) From the graph, estimate to the nearest half-inch the size square that must be cut from each corner to yield a box with the maximum volume. What is the maximum volume?

28. *Packaging.* A parcel delivery service will deliver only packages with length plus girth (distance around) not exceeding 108 inches. A packaging company wishes to design a box with a square base (x by x inches) that will have a maximum volume and will meet the delivery service's restrictions.
    (A) Write the volume of the box V(x) in terms of x.
    (B) Considering the physical limitations imposed by the delivery service, what is the domain of the function V?
    (C) Graph the function for this domain.
    (D) From the graph estimate to the nearest inch the dimensions of the box with the maximum volume. What is the maximum volume?

Section 2-6 Graphing Rational Functions

■ Rational Functions
■ Vertical and Horizontal Asymptotes
■ Graphing Rational Functions

■ **Rational Functions**

Just as rational numbers are defined in terms of quotients of integers, rational functions are defined in terms of quotients of polynomials. The following equations define rational functions:

$$f(x) = \frac{x-1}{x^2 - x - 6} \qquad g(x) = \frac{1}{x} \qquad h(x) = \frac{x^3 - 1}{x}$$

$$p(x) = 2x^2 - 3 \qquad q(x) = 3 \qquad r(x) = 0$$

In general, a function R is a **rational function** if

$$R(x) = \frac{P(x)}{Q(x)} \qquad Q(x) \neq 0$$

and P(x) and Q(x) are polynomials. The **domain of R** is the set of all real numbers x such that $Q(x) \neq 0$. In Section 2-3 we noted that rational functions are continuous (no breaks or holes in the graph) for all values of x except those for which the denominator Q(x) is 0. If $x = a$ and $Q(a) = 0$, then R is said to be **discontinuous at x = a**.

■ Vertical and Horizontal Asymptotes

Even though a rational function R may be discontinuous at $x = a$ (no graph for $x = a$), it is still useful to know what happens to the graph of R when x is close to a. Let us make this discussion concrete through a very simple rational function f defined by

$$f(x) = \frac{1}{x}$$

It is clear that the function f is discontinuous at $x = 0$. But what happens to $f(x)$ when x approaches 0 from either side of 0? A few table values will give us an idea of what happens to $f(x)$ when x gets close to 0 (Tables 6 and 7). As x approaches 0 from the right (denoted by $x \to 0^+$), x stays positive and $1/x$ gets larger and larger; that is, $1/x$ increases without bound (denoted by $1/x \to \infty$).* As x approaches 0 from the left (denoted by $x \to 0^-$), x stays negative and $1/x$ decreases without bound (denoted by $1/x \to -\infty$). Thus, the graph of $f(x) = 1/x$ approaches the y axis (but never touches it) as x gets closer to 0. The graph of $f(x) = 1/x$ for

TABLE 6
x APPROACHES 0
FROM THE RIGHT ($x \to 0^+$)

x	1	0.1	0.01	0.001	0.0001	0.000 01	0.000 001	. . .
$1/x$	1	10	100	1,000	10,000	100,000	1,000,000	. . .

TABLE 7
x APPROACHES 0
FROM THE LEFT ($x \to 0^-$)

x	-1	-0.0	-0.01	-0.001	-0.0001	$-0.000\ 01$	$-0.000\ 001$	. . .
$1/x$	-1	-10	-100	$-1,000$	$-10,000$	$-100,000$	$-1,000,000$	. . .

* The symbol ∞, called **infinity**, does not represent a real number. When we write $1/x \to \infty$, we mean that $1/x$ exceeds any given number N no matter how large N is chosen.

$-4 \leq x \leq 4$ is shown in Figure 15. Note that f is an odd function and the graph is symmetric with respect to the origin.

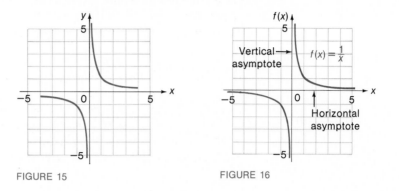

FIGURE 15

FIGURE 16

Now let us look at the behavior of $f(x) = 1/x$ as $|x|$ gets very large— that is, as $x \rightarrow \infty$ and as $x \rightarrow -\infty$. Consider Tables 8 and 9. As x increases without bound $(x \rightarrow \infty)$, $1/x$ stays positive and approaches 0 from above $(1/x \rightarrow 0^+)$. As x decreases without bound $(x \rightarrow -\infty)$, $1/x$ stays negative and approaches 0 from below $(1/x \rightarrow 0^-)$. The graph never touches the x or y axes; that is, there are no x or y intercepts. A sketch of the graph of $f(x) = 1/x$ is completed in Figure 16.

TABLE 8

$x \rightarrow \infty$

x	1	10	100	1,000	10,000	100,000	1,000,000	. . .
$1/x$	1	0.1	0.01	0.001	0.0001	0.000 01	0.000 001	. . .

TABLE 9

$x \rightarrow -\infty$

x	-1	-10	-100	$-1,000$	$-10,000$	$-100,000$	$-1,000,000$	. . .
$1/x$	-1	-0.1	-0.01	-0.001	-0.0001	$-0.000\ 01$	$-0.000\ 001$	. . .

The fixed lines that the graph approaches are called *asymptotes*. In this case the y axis is a *vertical asymptote* and the x axis is a *horizontal asymptote*. In general, a line $x = a$ is a **vertical asymptote** for the graph of $y = f(x)$ if $f(x)$ either increases or decreases without bound as $x \rightarrow a^+$ or $x \rightarrow a^-$. A line $y = b$ is a **horizontal asymptote** if $f(x)$ approaches b as $x \rightarrow -\infty$ or $x \rightarrow \infty$. Graphing certain kinds of functions is considerably aided by locating vertical and horizontal asymptotes first, if they exist. Using the same kind of reasoning as in the preceding example, we have the following general method of locating vertical asymptotes for rational functions.

Vertical Asymptotes

Let R be a rational function defined by

$$R(x) = \frac{P(x)}{Q(x)}$$

where $P(x)$ and $Q(x)$ are polynomials. If a is a real number such that $Q(a) = 0$ and $P(a) \neq 0$, then the line $x = a$ is a vertical asymptote of the graph $y = R(x)$.

To gain added insight in locating horizontal asymptotes for the graph of $R(x) = P(x)/Q(x)$, consider the following three examples.

1. Degree of numerator less than degree of denominator:

$$f(x) = \frac{2x - 1}{3x^2 - 2x}$$

Divide each term in numerator and denominator by x^2, the highest power of x that appears in the numerator and denominator.

$$= \frac{\dfrac{2x}{x^2} - \dfrac{1}{x^2}}{\dfrac{3x^2}{x^2} - \dfrac{2x}{x^2}}$$

Reduce internal fractions.

$$= \frac{\dfrac{2}{x} - \dfrac{1}{x^2}}{3 - \dfrac{2}{x}}$$

As $x \to \infty$ or $-\infty$, $2/x \to 0$ and $1/x^2 \to 0$; hence, $f(x) \to 0$. Thus, the line $y = 0$ is a horizontal asymptote.

2. Degree of numerator equal to degree of denominator:

$$g(x) = \frac{2x^2 - 1}{3x^2 - 2x}$$

Divide each term in numerator and denominator by x^2, the highest power of x that appears in the numerator and denominator.

$$= \frac{\dfrac{2x^2}{x^2} - \dfrac{1}{x^2}}{\dfrac{3x^2}{x^2} - \dfrac{2x}{x^2}}$$

Reduce internal fractions.

$$= \frac{2 - \dfrac{1}{x^2}}{3 - \dfrac{2}{x}}$$

As $x \to \infty$ or $-\infty$, $1/x^2 \to 0$ and $2/x \to 0$; hence, $g(x) \to \frac{2}{3}$. Thus, $y = \frac{2}{3}$ is a horizontal asymptote.

3. Degree of numerator greater than degree of denominator:

$$h(x) = \frac{2x^3 - 1}{3x^2 - 2x}$$

Divide each term in numerator and denominator by x^3, the highest power of x that appears in the numerator and denominator.

$$= \frac{\dfrac{2x^3}{x^3} - \dfrac{1}{x^3}}{\dfrac{3x^2}{x^3} - \dfrac{2x}{x^3}}$$

Reduce internal fractions.

$$= \frac{2 - \dfrac{1}{x^3}}{\dfrac{3}{x} - \dfrac{2}{x^2}}$$

As $x \to \infty$ or $-\infty$, the numerator approaches 2 and the denominator approaches 0. Thus, $|h(x)|$ increases without bound and there is no horizontal asymptote.

Reasoning in the same way as in these examples, one can establish the general method of locating horizontal asymptotes shown in the box.

Horizontal Asymptotes

Let R be a rational function defined by the quotient of two polynomials as follows:

$$R(x) = \frac{a_m x^m + \cdots + a_1 x + a_0}{b_n x^n + \cdots + b_1 x + b_0} \qquad a_m, b_n \neq 0$$

1. For $m < n$, the x axis $(y = 0)$ is a horizontal asymptote.
2. For $m = n$, the line $y = a_m/b_n$ is a horizontal asymptote.
3. For $m > n$, there are no horizontal asymptotes.

■ Graphing Rational Functions

We now use these new aids to graphing (along with other aids discussed earlier) to graph several rational functions. First, we outline a systematic approach to the problem of graphing rational functions:

To Graph a Rational Function

$$y = R(x) = \frac{P(x)}{Q(x)}$$

1. Determine symmetry with respect to the vertical axis and origin. (Is R even or odd?)
2. Find and plot x and y intercepts.
3. Find points of discontinuity.
4. Determine vertical and horizontal asymptotes and sketch them using broken lines.
5. Find where the graph of $y = R(x)$ is above and below the x axis. [Solve $R(x) > 0$ and $R(x) < 0$.]
6. Determine the behavior of the graph when it is close to its asymptotes. [What happens to $f(x)$ when $x \to \infty$ or $-\infty$? If the line $x = a$ is a vertical asymptote, what happens to $f(x)$ when $x \to a^+$ or a^-?]
7. Complete the sketch of the graph by plotting additional points as necessary and joining these points with a smooth curve. (Do not cross points of discontinuity.)

EXAMPLE 25   Graph: $y = f(x) = \dfrac{2x}{x - 3}$

Solution   **1.** Symmetry: Compute

$$f(-x) = \frac{-2x}{-x - 3}$$

f is neither even nor odd, since $f(-x) \neq f(x)$ and $f(-x) \neq -f(x)$.

2. Intercepts:

When $x = 0$, $y = \dfrac{2 \cdot 0}{0 - 3} = 0$ y intercept

When $y = 0$, $0 = \dfrac{2x}{x - 3}$ x intercept

$$x = 0$$

The graph crosses the coordinate axes only at the origin. Sketch the intercepts [figure (a)].

3. Points of discontinuity: f is discontinuous where $Q(x) = x - 3 = 0$—that is, at $x = 3$.

4. Asymptotes:

   Vertical: $x = 3$

   Horizontal: $y = \dfrac{2}{1} = 2$ Degrees of numerator and denominator are the same.

Sketch the asymptotes [figure (a)].

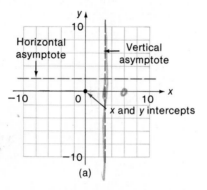

(a)

5. Where is the graph above and below the x axis? (Review Section 1-6.)

   Critical points: $x = 0, 3$

Sign of $2x$	$- \;-$	$+\;+\;+\;+$	$+\;+$
Sign of $(x-3)$	$-\;-$	$-\;-\;-\;-$	$+\;+$

   0   3

   $f(x) > 0$ for $x < 0$ or $x > 3$ Graph above x axis

   $f(x) < 0$ for $0 < x < 3$ Graph below x axis

6. Asymptotic behavior: Vertical asymptote ($x = 3$): With the help of steps 4 and 5, we see that

   $f(x) \to -\infty$ as $x \to 3^-$

   $f(x) \to \infty$ as $x \to 3^+$

Horizontal asymptote ($y = 2$): With the help of steps 4 and 5, we see that

   $f(x) \to 2$ as $x \to -\infty$ and as $x \to \infty$

7. Complete the sketch: By plotting a few additional points, we obtain the graph as in figure (b).

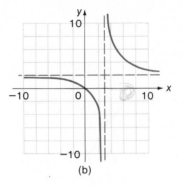

(b)

As one gains experience in graphing, many of the steps in Example 25 can be done mentally (or on scratch paper) and the process can be speeded up considerably.

PROBLEM 25   Proceed as in Example 25 and graph: $y = f(x) = \dfrac{3x}{x + 2}$

EXAMPLE 26   Graph: $y = f(x) = \dfrac{x - 1}{x^2 - 1}$

Solution   Factor the denominator:

$$y = f(x) = \frac{x - 1}{(x - 1)(x + 1)} = \frac{1}{x + 1} \qquad x \neq 1$$

1. Symmetry: f is neither even nor odd.

2. Intercepts:

When $x = 0$, $\quad y = \dfrac{1}{0 + 1} = 1 \quad$ y intercept

No x intercept

Graph the intercept [figure (a)].

3. Points of discontinuity: $x = -1$ and $x = 1$.

4. Asymptotes: Vertical: Find values of x such that the denominator $x^2 - 1 = 0$, but the numerator $x - 1 \neq 0$. This is true for $x = -1$. Thus, the line $x = -1$ is a vertical asymptote.
Horizontal: x axis (since the degree of the numerator is less than the degree of the denominator).

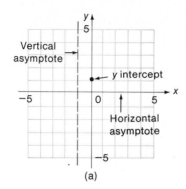

(a)

Sketch in asymptotes [figure (a)].

5. Where is graph above and below the x axis? By inspection,

$$f(x) = \frac{1}{x+1} > 0 \quad \text{when } x > -1 \quad \text{Graph above } x \text{ axis}$$

$$f(x) = \frac{1}{x+1} < 0 \quad \text{when } x < -1 \quad \text{Graph below } x \text{ axis}$$

6. Asymptotic behavior: Vertical asymptote ($x = 1$): With the help of steps 4 and 5, we see that

$$f(x) \to -\infty \quad \text{as } x \to 1^-$$
$$f(x) \to \infty \quad \text{as } x \to 1^+$$

Horizontal asymptote ($y = 0$): With the help of steps 4 and 5, we see that

$$f(x) \to 0 \quad \text{as } x \to -\infty \quad \text{and} \quad \text{as } x \to \infty$$

7. Complete the sketch: · By plotting a few additional points we obtain the graph in figure (b). Note what happens at the two points of discontinuity $x = -1$ and $x = 1$. At $x = -1$ we have a huge break in the graph; at $x = 1$ there is just a point missing, which, if replaced, would make the function continuous there.

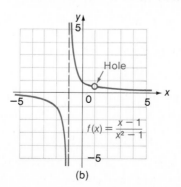

$$f(x) = \frac{x-1}{x^2-1}$$

(b)

PROBLEM 26   Graph: $y = f(x) = \dfrac{x + 1}{x^2 - 1}$

EXAMPLE 27   Graph: $y = g(x) = \dfrac{x^2 + 4}{x^2 - 4}$

Solution   Factor the denominator:

$$y = f(x) = \frac{x^2 + 4}{(x - 2)(x + 2)}$$

1. Symmetry: $f(-x) = f(x)$; hence, f is even and its graph is symmetric with respect to the y axis.

2. Intercepts:

$$\text{When } x = 0, \quad y = \frac{0 + 4}{0 - 4} = -1 \quad \text{y intercept}$$

No x intercept $x^2 + 4 \neq 0$ for all x

Graph the intercepts [figure (a)].

3. Points of discontinuity: $x = -2$ and $x = 2$.

4. Asymptotes:

Vertical: $x = -2$ and $x = 2$

Horizontal: $y = 1$ Degrees of numerator and
                                 denominator are the same.

Sketch the asymptotes on graph [figure (a)].

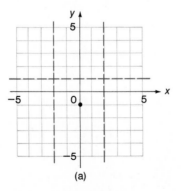

(a)

5. Where is the graph above and below the x axis?

$$f(x) = \frac{x^2 + 4}{(x - 2)(x + 2)}$$

Critical points: $x = -2$ and $x = 2$

Sign of $(x^2 + 4)$	+ +		+ + + +		+ +	Positive for all x
Sign of $(x - 2)$	− −		− − − −		+ +	
Sign of $(x + 2)$	− −		+ + + +		+ +	

$$f(x) > 0 \quad \text{for } x < -2 \text{ and } x > 2 \quad \text{Graph above } x \text{ axis}$$

$$f(x) < 0 \quad \text{for } -2 < x < 2 \quad\quad \text{Graph below } x \text{ axis}$$

6. Asymptotic behavior: Vertical asymptotes $(x = -2$ and $x = 2)$: With the help of steps 4 and 5, we see that

$$f(x) \to \infty \quad\quad \text{as } x \to -2^-$$
$$f(x) \to -\infty \quad \text{as } x \to -2^+$$
$$f(x) \to -\infty \quad \text{as } x \to 2^-$$
$$f(x) \to \infty \quad\quad \text{as } x \to 2^+$$

Horizontal asymptotes: With the help of steps 4 and 5, we see that

$$f(x) \to 1 \quad \text{as } x \to -\infty \text{ and as } x \to \infty$$

7. Complete the sketch: By plotting a few additional points we obtain the graph in figure (b).

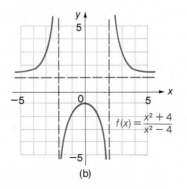

$$f(x) = \frac{x^2 + 4}{x^2 - 4}$$

(b)

PROBLEM 27 Graph: $f(x) = \dfrac{x^2}{x^2 - 1}$

Answers to Matched Problems

25.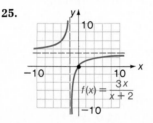

$$f(x) = \frac{3x}{x + 2}$$

26.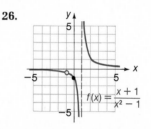

$$f(x) = \frac{x + 1}{x^2 - 1}$$

27.

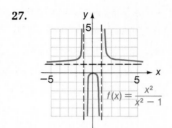

$$f(x) = \frac{x^2}{x^2 - 1}$$

Exercise 2-6 ■ Graph each function. Sketch in any horizontal or vertical asymptotes.

A **1.** $f(x) = \dfrac{1}{x - 4}$

2. $g(x) = \dfrac{1}{x + 3}$

3. $p(x) = \dfrac{-1}{x - 4}$

4. $m(x) = \dfrac{-1}{x + 3}$

5. $f(x) = \dfrac{x}{x + 1}$

6. $f(x) = \dfrac{3x}{x - 3}$

B **7.** $q(x) = \dfrac{2x - 1}{x}$

8. $r(x) = \dfrac{1 - x}{x}$

9. $h(x) = \dfrac{x}{2x - 2}$

10. $p(x) = \dfrac{3x}{4x + 4}$

11. $g(x) = \dfrac{1 - x^2}{x^2}$

12. $f(x) = \dfrac{x^2 + 1}{x^2}$

13. $f(x) = \dfrac{9}{x^2 - 9}$

14. $g(x) = \dfrac{6}{x^2 - x - 6}$

15. $f(x) = \dfrac{x}{x^2 - 1}$

16. $p(x) = \dfrac{x}{1 - x^2}$

17. $g(x) = \dfrac{2}{x^2 + 1}$

18. $f(x) = \dfrac{x}{x^2 + 1}$

19. $h(x) = \dfrac{2x^2}{x^2 + 1}$

20. $f(x) = \dfrac{-2x^4}{x^4 + 1}$

C **21.** $m(x) = \dfrac{x^2 - 1}{x^2 + x - 2}$

22. $g(x) = \dfrac{x^2 - x - 6}{x^2 - 4}$

Section 2-7   Composite and Inverse Functions

■ Composite Functions
■ Inverse Relations and Functions
■ One-to-One Correspondence and Inverses

In this section we will discuss two important methods of obtaining new relations and functions from known functions.

■ Composite Functions

Consider the function h given by the equation

$$h(x) = \sqrt{2x + 1}$$

Inside the radical is a first-degree polynomial that defines a linear function. So the function h is really a combination of a square root function and a linear function. We see this clearly as follows:

Let $\quad u = 2x + 1 = g(x)$
$\quad\quad y = \sqrt{u} = f(u)$

Then $\quad h(x) = f[g(x)]$

The function h is said to be the **composite** of the two simpler functions f and g. (Loosely speaking, we can think of h as a function of a function.) What can we say about the domain of h given the domains of f and g? In forming the composite

$$h(x) = f[g(x)]$$

x must be restricted so that x is in the domain of g and g(x) is in the domain of f. Since the domain of f, where $f(u) = \sqrt{u}$, is the set of non-negative real numbers,* we see that $g(x)$ must be nonnegative; that is,

$$g(x) \geq 0$$
$$2x + 1 \geq 0$$
$$x \geq -\tfrac{1}{2}$$

Thus, the domain of h is this restricted domain of g.

A special function symbol is often used to represent the **composite of two functions**, which we define in general terms in the box on page 222.

---

* Recall from Section 2-2 we said that if a function is specified by an equation and the domain is not indicated, then we shall assume that the domain is the set of all real replacements of the independent variable that produce real values for the dependent variable. The range is the set of all dependent variable values corresponding to independent variable values.

Composite Functions

Given functions f and g, then $f \circ g$ is called their **composite** and is defined by the equation

$$(f \circ g)(x) = f[g(x)]$$

The domain of $f \circ g$ is the set of all numbers x such that x is in the domain of g and $g(x)$ is in the domain of f.

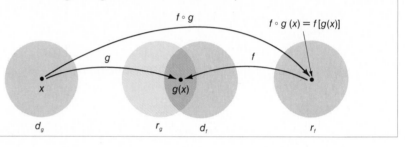

EXAMPLE 28   Find $(f \circ g)(x)$ and $(g \circ f)(x)$ and their domains for $f(x) = x^{10}$ and $g(x) = 3x^4 - 1$.

Solution
$$(f \circ g)(x) = f[g(x)] = [g(x)]^{10} = (3x^4 - 1)^{10}$$
$$(g \circ f)(x) = g[f(x)] = 3[f(x)]^4 - 1 = 3(x^{10})^4 - 1 = 3x^{40} - 1$$

The domain of f and the domain of g are the set of all real numbers R. Thus, for all real x, x is in the domain of g and $g(x)$ is in the domain of f, and x is in the domain of f and $f(x)$ is in the domain of g. Hence, the domain of $f \circ g$ and the domain of $g \circ f$ are R.

PROBLEM 28   Find $(f \circ g)(x)$ and $(g \circ f)(x)$ and their domains for $f(x) = 2x + 1$ and $g(x) = (x - 1)/2$.

■   Inverse Relations and Functions

We now turn to a second way of obtaining new functions from given functions. The process is used to obtain logarithmic functions from exponential functions and inverse trigonometric functions from trigonometric functions.

     Given a relation G, if we interchange the order of the components in each ordered pair belonging to G, we obtain a new relation G^{-1} called the **inverse of G**. [*Note:* G^{-1} is a relation–function symbol; it does not mean $1/G$.] For example, if

$$G = \{(2, 4), (-1, 3), (0, 4)\}$$

then by reversing the components in each ordered pair in G, we obtain

$$G^{-1} = \{(4, 2), (3, -1), (4, 0)\}$$

It follows from the definition (and is evident from the example) that the domains and ranges of G and G^{-1} are interchanged.

Inverse of G

If G is a relation, the inverse of G, denoted by G^{-1}, is given by

$$G^{-1} = \{(b, a) \mid (a, b) \in G\}$$

Domain of G^{-1} = Range of G

Range of G^{-1} = Domain of G

If a relation G is specified by an equation, say

$$G: \quad y = 2x - 1 \tag{1}$$

then how do we find G^{-1}? The answer is easy: We interchange the variables in equation (1). Thus,

$$G^{-1}: \quad x = 2y - 1 \tag{2}$$

or, solving for y,

$$G^{-1}: \quad y = \frac{x + 1}{2} \tag{3}$$

Any ordered pair of numbers that satisfies equation (1), when reversed in order, will satisfy equations (2) and (3). For example, (3, 5) satisfies equation (1) and (5, 3) satisfies equations (2) and (3), as can easily be checked.

If we sketch a graph of G, G^{-1}, and y = x on the same coordinate system (Fig. 17), we will observe something interesting. If we fold the

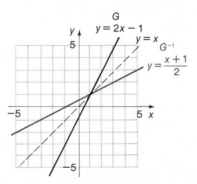

FIGURE 17

paper along the line $y = x$, then the graphs of G and G^{-1} match. Actually, we can graph G^{-1} by drawing G with wet ink and folding the paper along $y = x$ before the ink dries; G will then print G^{-1}. [To prove this in general, one has to show that the line $y = x$ is the perpendicular bisector of the line joining (a, b) and (b, a).] Knowing that the graphs of G and G^{-1} are symmetric relative to the line $y = x$ makes it easy to graph G^{-1} if G is known, and vice versa.

In Figure 17 observe that G and G^{-1} are both functions. This is not always the case, however. Inverses of some functions may not be functions. Consider the following example.

EXAMPLE 29   The relation f is given by $y = x^2$.

(A) Find f^{-1}.
(B) Graph f, f^{-1}, and $y = x$. Is either f or f^{-1} a function?
(C) Indicate the domain and range of f and f^{-1}.

Solution   (A) f^{-1}: $x = y^2$ or $y = \pm\sqrt{x}$

(B)

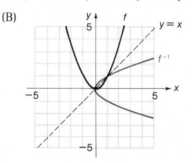

f is a function

f^{-1} is not a function

(C) Domain of f = R = Range of f^{-1}

Range of $f = [0, \infty)$ = Domain of f^{-1}

PROBLEM 29   The relation f is given by $y = |x| + 2$.

(A) Find f^{-1}.
(B) Graph f, f^{-1}, and $y = x$. Is either f or f^{-1} a function?
(C) Indicate the domain and range of f and f^{-1}.

■   One-to-One Correspondence and Inverses

If we are given a function f, how can we tell in advance whether its inverse f^{-1} will be a function? The answer is contained in the concept of one-to-one correspondence. A **one-to-one correspondence** exists between two sets if each element in the first set corresponds to exactly one element in the second set, and each element in the second set corresponds to exactly one element in the first set. Consider the two functions f and g and their inverses:

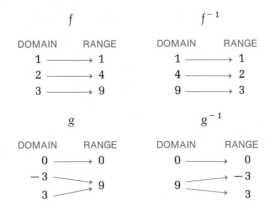

Function f has a one-to-one correspondence between domain and range values (notice that f^{-1} is also a function). Function g does not have a one-to-one correspondence between domain and range values (notice that g^{-1} is not a function).

THEOREM 4	Inverses

A function f has an inverse that is a function if and only if there exists a one-to-one correspondence between domain and range values of f. In this case,

$$(f \circ f^{-1})(y) = y \quad \text{and} \quad (f^{-1} \circ f)(x) = x$$

Theorem 4 is interpreted schematically in Figure 18. Figure 19 illustrates some functions that are one to one,* and Figure 20 illustrates

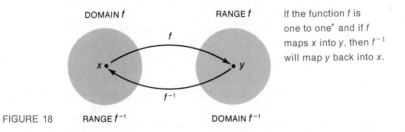

FIGURE 18

* When we refer to a function as being *one to one*, we mean there exists a one-to-one correspondence between its domain and range values.

some that are not. The following observation can be made (and proved in general): **All increasing functions and all decreasing functions are one to one and hence have inverses that are functions**.

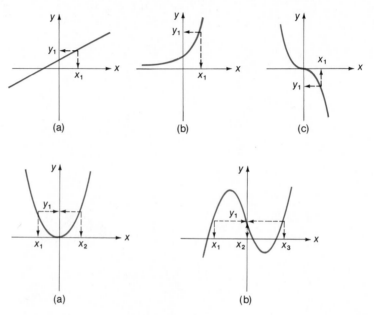

FIGURE 19   Functions that are one to one (each has an inverse that is a function).

FIGURE 20   Functions that are not one to one (neither has an inverse that is a function).

When a function is defined by an equation and the function is either increasing or decreasing (thus, it is one to one), we can often find its inverse in terms of an equation.

EXAMPLE 30   Given $f(x) = 3x + 2$, find:

(A) $f^{-1}(x)$ (B) $f^{-1}(5)$
(C) $(f^{-1} \circ f)(5)$ (D) $(f^{-1} \circ f)(x)$

Solution   Function f is linear with slope 3 and hence is increasing and has an inverse that is a function.

(A) f: $y = 3x + 2$ Replace $f(x)$ with y in $f(x) = 3x + 2$.

f^{-1}: $x = 3y + 2$ Interchange variables x and y to obtain f^{-1}.

$$y = \frac{x - 2}{3}$$ Solve for y in terms of x.

Thus,

$$f^{-1}(x) = \frac{x - 2}{3}$$ Replace y with $f^{-1}(x)$.

(B) $f^{-1}(5) = \dfrac{5 - 2}{3} = \dfrac{3}{3} = 1$

(C) $(f^{-1} \circ f)(5) = f^{-1}[f(5)]$ We are just verifying the results of Theorem 4; that is,
if f and f^{-1} are both functions and 5 is in the domain
$$= \dfrac{f(5) - 2}{3}$$ of $f^{-1} \circ f$, then $(f^{-1} \circ f)(5) = 5$, or, in general,
$(f^{-1} \circ f)(x) = x$ for all x in the domain of $f^{-1} \circ f$.

$$= \dfrac{17 - 2}{3}$$

$$= \tfrac{15}{3} = 5$$

(D) $(f^{-1} \circ f)(x) = f^{-1}[f(x)]$ See comment in part C.

$$= \dfrac{f(x) - 2}{3}$$

$$= \dfrac{(3x + 2) - 2}{3} = x$$

PROBLEM 30   Given $g(x) = \dfrac{x}{3} - 2$,   find:

(A) $g^{-1}(x)$ (B) $g^{-1}(-2)$
(C) $(g^{-1} \circ g)(3)$ (D) $(g^{-1} \circ g)(x)$

There are many cases where we start with a function that is not
one to one and restrict its domain so that the function is either increasing
or decreasing and hence becomes one to one. We do this so that we
can obtain an inverse that is a function. Suppose we start with $f(x) = x^2$.
Because f is not one to one, its inverse will not be a function [Fig. 21(a)].
There are many ways in which the domain of f can be restricted to
obtain either an increasing or a decreasing function. Figures 21(b) and
21(c) illustrate two such restrictions.

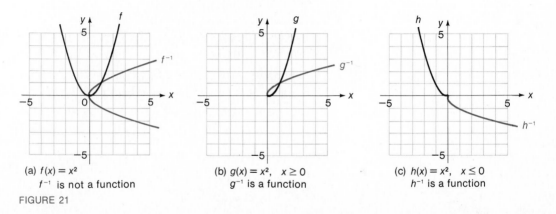

(a) $f(x) = x^2$ (b) $g(x) = x^2$, $x \geq 0$ (c) $h(x) = x^2$, $x \leq 0$
f^{-1} is not a function g^{-1} is a function h^{-1} is a function
FIGURE 21

EXAMPLE 31   Given $f(x) = x^2$, $x \geq 0$, find:

(A) The domain and range of f and f^{-1}
(B) $f^{-1}(x)$ (C) $f^{-1}(4)$ (D) $(f \circ f^{-1})(3)$ (E) $(f \circ f^{-1})(x)$

Solution   (A) Domain of $f = [0, \infty) = $ Range of f^{-1}
Range of $f = [0, \infty) = $ Domain of f^{-1}

(B) f: $y = x^2$ $x \geq 0$ Replace $f(x)$ with y in $f(x) = x^2$.
        f^{-1}: $x = y^2$ $y \geq 0$ Interchange variables x and y to obtain f^{-1}.
            $y = \sqrt{x}$ Solve for y in terms of x. [Note: $y \neq -\sqrt{x}$. Why?]
        $f^{-1}(x) = \sqrt{x}$ Replace y with $f^{-1}(x)$.

(C) $f^{-1}(4) = \sqrt{4} = 2$

(D) $(f \circ f^{-1})(3) = 3$ Direct use of Theorem 4.

(E) $(f \circ f^{-1})(x) = x$ Direct use of Theorem 4.

PROBLEM 31   Given $g(x) = x^2 - 1$, $x \geq 0$, find:

(A) The domain and range of g and g^{-1}
(B) $g^{-1}(x)$ (C) $g^{-1}(8)$ (D) $(g^{-1} \circ g)(5)$ (E) $(g^{-1} \circ g)(x)$

Answers to Matched Problems   **28.** $(f \circ g)(x) = x$, Domain $= R$; $(g \circ f)(x) = x$, Domain $= R$
29. (A) f^{-1}: $x = |y| + 2$ or $|y| = x - 2$ or $y = \pm(x - 2)$

(B)

f is a function
f^{-1} is not a function
(C)   Domain of $f = R = $ Range of f^{-1}
Range of $f = [2, \infty) = $ Domain of f^{-1}

30. (A) $3x + 6$ (B) 0   (C) 3   (D) x

31. (A) Domain of $g = [0, \infty) = $ Range of g^{-1}
Range of $g = [-1, \infty) = $ Domain of g^{-1}
(B) $\sqrt{x + 1}$ (C) 3   (D) 5   (E) x

Exercise 2-7 ■ A   Find $(f \circ g)(x)$ and $(g \circ f)(x)$ for functions f and g as indicated.

1. $f(x) = x^7$, $g(x) = x^2 - x + 1$
2. $f(x) = x^{12}$, $g(x) = 2x^3 - 5$

3. $f(x) = \sqrt{x}, \quad g(x) = 2x + 5$

4. $f(x) = \sqrt{x}, \quad g(x) = x - 4$

5. $f(x) = x^{1/2}, \quad g(x) = 1 - x^2$

6. $f(x) = x^{1/4}, \quad g(x) = 4 - x^2$

7. $f(x) = |x|, \quad g(x) = 3x - 2$

8. $f(x) = |x - 1|, \quad g(x) = \sqrt{x}$

9. $f(x) = x^{2/3}, \quad g(x) = x^3 - 4$

10. $f(x) = x^{3/5}, \quad g(x) = 3x^{10} + x^5$

11. $f(x) = |x|, \quad g(x) = -7$

12. $f(x) = \sqrt{x}, \quad g(x) = 4$

Graph each relation, its inverse, and $y = x$ on the same coordinate system. Identify all functions in addition to $y = x$. [Use the graph of the original relation to find the graph of the inverse. Remember, if $(a, b) \in f$, then $(b, a) \in f^{-1}$.]

13. $H = \{(-2, \frac{1}{4}), (0, 1), (1, 2), (2, 4)\}$

14. $G = \{(-4, 2), (-1, 2), (2, 2), (4, 2)\}$

15. $g: \quad y = 2x - 2$

16. $f: \quad y = -\frac{1}{2}x + 2$

17. $p: \quad y = x^2 + 1$

18. $q: \quad y = (x + 2)^2$

B *For the following functions f that are one to one, find:*
(A) The domain and range of f and f^{-1}
(B) $f^{-1}(x)$ (C) $f^{-1}(4)$ (D) $(f^{-1} \circ f)(2)$ (E) $(f^{-1} \circ f)(x)$

19. $f(x) = 2x - 2$

20. $f(x) = -\frac{1}{2}x + 2$

21. $f(x) = 3x - 5$

22. $f(x) = 2x - 4$

23. $f(x) = x^2 + 1$

24. $f(x) = (x + 2)^2$

25. $f(x) = x^2 + 1, \quad x \geq 0$

26. $f(x) = (x + 2)^2, \quad x \geq -2$

27. $f(x) = (x - 1)^2$

28. $f(x) = x^2 - 2$

29. $f(x) = (x - 1)^2, \quad x \geq 1$

30. $f(x) = x^2 - 2, \quad x \geq 0$

Find the domain of $f \circ g$ and of $g \circ f$, given functions f and g in:

31. Problem 1   32. Problem 2   33. Problem 3

34. Problem 4   35. Problem 5   36. Problem 6

37. Problem 7   38. Problem 8   39. Problem 9

40. Problem 10   41. Problem 11   42. Problem 12

C *For f as indicated in Problems 43–46, find:*
(A) $(f \circ f)(x)$ (B) $f^{-1}(x)$ (C) $(f \circ f^{-1})(x)$
(D) Any conclusions?

43. $f(x) = 4 - x$

44. $f(x) = \sqrt{2 - x}$

45. $f(x) = 1/x$ **46.** $f(x) = \sqrt{9 - x^2}, \quad x \geq 0$

47. For $f(x) = (x + 1)/(x - 2)$, find a function g so that $(g \circ f)(x) = x$.

48. For $f(x) = (3x + 1)/(x - 2)$, find a function g so that $(g \circ f)(x) = x$.

49. Given two functions F and G with common domain X, then $F = G$ if and only if $F(x) = G(x)$ for all x in X. Using this definition, show that $f^{-1} \circ f = I$, where I is the identity function that has the same domain as $f^{-1} \circ f$. This suggests why the notation f^{-1} is used for the inverse.

50. Does a constant function have an inverse that is a function? Explain.

APPLICATIONS **51.** *Geometry.* For a circle with circumference C, diameter D, and radius R, we have

$$C = f(D) = \pi D \quad \text{and} \quad D = g(R) = 2R$$

Find $(f \circ g)(R)$ and interpret.

52. *Cost analysis.* The cost C to produce x units of a given product per month is given by

$$C = f(x) = 19,200 + 160x$$

If the demand x each month at a selling price of \$p per unit is given by

$$x = g(p) = 200 - \frac{p}{4}$$

find $(f \circ g)(p)$ and interpret.

Section 2-8   Variation

- Direct Variation
- Inverse Variation
- Joint Variation
- Combined Variation

In reading scientific material, one is likely to come across statements such as "The pressure of an enclosed gas varies directly as the absolute temperature," or "The frequency of vibration of air in an organ pipe varies inversely as the length of the pipe," or even more complicated statements such as "The force of attraction between two bodies varies jointly as their masses and inversely as the square of the distance between the two bodies." These statements have precise mathematical meaning in that they represent particular types of functions. The purpose of this section is to investigate these special functions.

■ Direct Variation

The statement **y varies directly as x** means

$$y = kx \qquad k \neq 0$$

where k is a constant called the **constant of variation**. Similarly, the statement "y varies directly as the square of x" means

$$y = kx^2 \qquad k \neq 0$$

and so on. The first equation defines a linear function, and the second a quadratic function.

Direct variation is illustrated by the familiar formulas

$$C = \pi D \quad \text{and} \quad A = \pi r^2$$

where the first formula asserts that the circumference of a circle varies directly as the diameter, and the second that the area of a circle varies directly as the square of the radius. In both cases, π is the constant of variation.

EXAMPLE 32 Translate each statement into an appropriate equation, and find the constant of variation if $y = 16$ when $x = 4$.

(A) y varies directly as x.

Solution $y = kx$ Do not forget k.

To find the constant of variation k, substitute $x = 4$ and $y = 16$ and solve for k.

$$16 = k \cdot 4$$
$$k = \tfrac{16}{4} = 4$$

Thus, $k = 4$ and the equation of variation is

$$y = 4x$$

(B) y varies directly as the cube of x.

Solution $y = kx^3$ Do not forget k.

To find k, substitute $x = 4$ and $y = 16$.

$$16 = k \cdot 4^3$$
$$k = \tfrac{16}{64} = \tfrac{1}{4}$$

Thus, the equation of variation is

$$y = \tfrac{1}{4}x^3$$

232 2 Graphs and Functions

PROBLEM 32 If $y = 4$ when $x = 8$, find the equation of variation for each statement.

(A) y varies directly as x.
(B) y varies directly as the cube root of x.

■ Inverse Variation

The statement **y varies inversely as x** means

$$y = \frac{k}{x} \qquad k \neq 0$$

where k is a constant (the constant of variation). As in the case of direct variation, we also discuss y varying inversely as the square of x, and so on.

An illustration of inverse variation is given in the distance-rate-time formula $d = rt$ in the form $t = d/r$ for a fixed distance d. In driving a fixed distance, say $d = 400$ miles, time varies inversely as the rate; that is,

$$t = \frac{400}{r}$$

where 400 is the constant of variation—as the rate increases, the time decreases, and vice versa.

EXAMPLE 33 Translate each statement into an appropriate equation, and find the constant of variation if $y = 16$ when $x = 4$.

(A) y varies inversely as x.

Solution $y = \frac{k}{x}$ Do not forget k.

To find k, substitute $x = 4$ and $y = 16$.

$$16 = \frac{k}{4}$$

$$k = 64$$

Thus, the equation of variation is

$$y = \frac{64}{x}$$

(B) y varies inversely as the square root of x.

Solution

$$y = \frac{k}{\sqrt{x}}$$

To find k, substitute $x = 4$ and $y = 16$.

$$16 = \frac{k}{\sqrt{4}}$$

$$k = 32$$

Thus, the equation of variation is

$$y = \frac{32}{\sqrt{x}}$$

PROBLEM 33   If $y = 4$ when $x = 8$, find the equation of variation for each statement.

(A) y varies inversely as x.
(B) y varies inversely as the square of x.

■ Joint Variation

The statement **w varies jointly as x and y** means

$$w = kxy \qquad k \neq 0$$

where k is a constant (the constant of variation). Similarly, if

$$w = kxyz^2 \qquad k \neq 0$$

we would say that "w varies jointly as x, y, and the square of z," and so on. For example, the area of a rectangle varies jointly as its length and width (recall $A = lw$), and the volume of a right circular cylinder varies jointly as the square of its radius and its height (recall $V = \pi r^2 h$). What is the constant of variation in each case?

■ Combined Variation

The basic types of variation introduced above are often combined. For example, the statement "w varies jointly as x and y and inversely as the square of z" means

$$w = k\,\frac{xy}{z^2} \qquad k \neq 0 \qquad \text{We do not write:} \quad w = \frac{kxy}{kz^2}$$

Thus the statement, "The force of attraction F between two bodies varies jointly as their masses m_1 and m_2 and inversely as the square of the distance d between the two bodies," means

$$F = k\,\frac{m_1 m_2}{d^2} \qquad k \neq 0$$

If (assuming k is positive) either of the two masses is increased, the force of attraction increases; on the other hand, if the distance is increased, the force of attraction decreases.

EXAMPLE 34   The pressure P of an enclosed gas varies directly as the absolute temperature T and inversely as the volume V. If 500 cubic feet of gas yields a pressure of 10 pounds per square foot at a temperature of 300 K (absolute temperature*), what will be the pressure of the same gas if the volume is decreased to 300 cubic feet and the temperature increased to 360 K?

Solution   *Method 1.* Write the equation of variation $P = k(T/V)$,   and find k using the first set of values:

$$10 = k\left(\tfrac{300}{500}\right)$$
$$k = \tfrac{50}{3}$$

Hence, the equation of variation for this particular gas is $P = \tfrac{50}{3}(T/V)$. Now find the new pressure P using the second set of values:

$$P = \tfrac{50}{3}\left(\tfrac{360}{300}\right) = 20 \text{ pounds per square foot}$$

Method 2 (generally faster than Method 1). Write the equation of variation $P = k(T/V)$; then convert to the equivalent form:

$$\frac{PV}{T} = k$$

* A Kelvin (absolute) and a Celsius degree are the same size, but 0 on the Kelvin scale is $-273°$ on the Celsius scale. This is the point at which molecular action is supposed to stop and is called *absolute zero*.

If P_1, V_1, and T_1 are the first set of values for the gas, and P_2, V_2, and T_2 are the second set, then

$$\frac{P_1 V_1}{T_1} = k \quad \text{and} \quad \frac{P_2 V_2}{T_2} = k$$

Hence

$$\frac{P_1 V_1}{T_1} = \frac{P_2 V_2}{T_2}$$

Since all values are known except P_2, substitute and solve. Thus,

$$\frac{(10)(500)}{300} = \frac{P_2(300)}{360}$$

$$P_2 = 20 \text{ pounds per square foot}$$

PROBLEM 34   The length L of skid marks of a car's tires (when brakes are applied) varies directly as the square of the speed v of the car. If skid marks of 20 feet are produced at 30 miles/hour, how fast would the same car be going if it produced skid marks of 80 feet? Solve in two ways (see Example 34).

EXAMPLE 35   The frequency of pitch f of a given musical string varies directly as the square root of the tension T and inversely as the length L. What is the effect on the frequency if the tension is increased by a factor of 4 and the length is cut in half?

Solution   Write the equation of variation:

$$f = \frac{k\sqrt{T}}{L} \quad \text{or equivalently} \quad \frac{f_2 L_2}{\sqrt{T_2}} = \frac{f_1 L_1}{\sqrt{T_1}}$$

We are given that $T_2 = 4T_1$ and $L_2 = 0.5L_1$. Substituting in the second equation, we have

$$\frac{f_2 0.5L_1}{\sqrt{4T_1}} = \frac{f_1 L_1}{\sqrt{T_1}} \quad \text{Solve for } f_2.$$

$$\frac{f_2 0.5L_1}{2\sqrt{T_1}} = \frac{f_1 L_1}{\sqrt{T_1}}$$

$$f_2 = \frac{2\sqrt{T_1} f_1 L_1}{0.5L_1 \sqrt{T_1}} = 4f_1$$

Thus, the frequency of pitch is increased by a factor of 4.

PROBLEM 35   The weight w of an object on or above the surface of the earth varies
inversely as the square of the distance d between the object and the
center of the earth. If an object on the surface of the earth is moved
into space so as to double its distance from the earth's center, what
effect will this move have on its weight?

Answers to Matched Problems   **32.** (A) $y = \frac{1}{2}x$ (B) $y = 2\sqrt[3]{x}$

33. (A) $y = 32/x$ (B) $y = 256/x^2$

34. $v = 60$ miles/hour

35. It will be one-fourth as heavy.

Exercise 2-8 ■
set up eqquasion

A   *Translate each problem into an equation using k as the constant of
variation.*

1. F varies directly as the square of v.

2. u varies directly as v.

3. The pitch or frequency f of a guitar string of a given length varies
directly as the square root of the tension T of the string.

4. Geologists have found in studies of earth erosion that the erosive
force (sediment-carrying power) P of a swiftly flowing stream varies
directly as the sixth power of the velocity v of the water.

5. y varies inversely as the square root of x.

6. I varies inversely as t.

7. The biologist Reaumur suggested in 1735 that the length of time t
that it takes fruit to ripen during the growing season varies inversely as the sum T of the average daily temperatures during the
growing season.

8. In a study on urban concentration, F. Auerbach discovered an
interesting law. After arranging all the cities of a given country
according to their population size, starting with the largest, he
found that the population P of a city varied inversely as the number
n indicating its position in the ordering.

9. R varies jointly as S, T, and V.

10. g varies jointly as x and the square of y.

11. The volume of a cone V varies jointly as its height h and the square
of the radius r of its base.

12. The amount of heat put out by an electrical appliance (in calories)
varies jointly as time t, resistance R in the circuit, and the square
of the current I.

Solve using either of the two methods illustrated in Example 34.

13. u varies directly as the square root of v. If $u = 2$ when $v = 2$, find u when $v = 8$.

14. y varies directly as the square of x. If $y = 20$ when $x = 2$, find y when $x = 5$.

15. L varies inversely as the square root of M. If $L = 9$ when $M = 9$, find L when $M = 3$.

16. I varies inversely as the cube of t. If $I = 4$ when $t = 2$, find I when $t = 4$.

B *Translate each problem into an equation using k as the constant of variation.*

17. U varies jointly as a and b and inversely as the cube of c.

18. w varies directly as the square of x and inversely as the square root of y.

19. The maximum safe load L for a horizontal beam varies jointly as its width w and the square of its height h, and inversely as its length l.

20. Joseph Cavanaugh, a sociologist, found that the number of long-distance phone calls n between two cities in a given time period varied (approximately) jointly as the populations P_1 and P_2 of the two cities, and inversely as the distance d between the two cities.

Solve using either of the two methods illustrated in Example 34.

21. Q varies jointly as m and the square of n, and inversely as P. If $Q = -4$ when $m = 6$, $n = 2$, and $P = 12$, find Q when $m = 4$, $n = 3$, and $P = 6$.

22. w varies jointly as x, y, and z and inversely as the square of t. If $w = 2$ when $x = 2$, $y = 3$, $z = 6$, and $t = 3$, find w when $x = 3$, $y = 4$, $z = 2$, and $t = 2$.

23. The weight w of an object on or above the surface of the earth varies inversely as the square of the distance d between the object and the center of the earth. If a girl weighs 100 pounds on the surface of the earth, how much would she weigh (to the nearest pound) 400 miles above the earth's surface? (Assume the radius of the earth is 4,000 miles.)

24. A child was struck by a car in a crosswalk. The driver of the car had slammed on his brakes and left skid marks 160 feet long. He told the police he had been driving at 30 miles/hour. The police know that the length of skid marks L (when brakes are applied) varies directly as the square of the speed of the car v, and that at 30 miles/hour (under ideal conditions) skid marks would be 40 feet long. How fast was the driver actually going before he applied his brakes?

25. Ohm's law states that the current I in a wire varies directly as the electromotive force E and inversely as the resistance R. If $I = 22$ amperes when $E = 110$ volts and $R = 5$ ohms, find I if $E = 220$ volts and $R = 11$ ohms.

26. Anthropologists, in their study of race and human genetic groupings, often use an index called the *cephalic index*. The cephalic index C varies directly as the width w of the head and inversely as the length l of the head (both when viewed from the top). If an Indian in Baja California (Mexico) has measurements of $C = 75$, $w = 6$ inches, and $l = 8$ inches, what is C for an Indian in northern California with $w = 8.1$ inches and $l = 9$ inches?

C **27.** If the horsepower P required to drive a speedboat through water varies directly as the cube of the speed v of the boat, what change in horsepower is required to double the speed of the boat?

28. The intensity of illumination E on a surface varies inversely as the square of its distance d from a light source. What is the effect on the total illumination on a book if the distance between the light source and the book is doubled?

29. The frequency of vibration f of a musical string varies directly as the square root of the tension T and inversely as the length L of the string. If the tension of the string is increased by a factor of 4 and the length of the string is doubled, what is the effect on the frequency?

30. In an automobile accident the destructive force F of a car varies (approximately) jointly as the weight w of the car and the square of the speed v of the car. (This is why accidents at high speed are generally so serious.) What would be the effect on the destructive force of a car if its weight were doubled and its speed were doubled?

ADDITIONAL APPLICATIONS   *The following problems include significant applications from many different areas and are arranged according to subject area. The more difficult problems are marked with two stars (★★), the moderately difficult problems are marked with one star (★), and the easier problems are not marked.*

Astronomy   **31.** The square of the time t required for a planet to make one orbit around the sun varies directly as the cube of its mean (average) distance d from the sun. Write the equation of variation, using k as the constant of variation.

★**32.** The centripetal force F of a body moving in a circular path at constant speed varies inversely as the radius r of the path. What happens to F if r is doubled?

33. The length of time t a satellite takes to complete a circular orbit of the earth varies directly as the radius r of the orbit and inversely as the orbital velocity v of the satellite. If $t = 1.42$ hours when $r = 4,050$ miles and $v = 18,000$ miles/hour (Sputnik I), find t for $r = 4,300$ miles and $v = 18,500$ miles/hour.

Life Science **34.** The number N of gene mutations resulting from x-ray exposure varies directly as the size of the x-ray dose r. What is the effect on N if r is quadrupled?

35. In biology there is an approximate rule, called the *bioclimatic rule* for temperate climates, which states that the difference d in time for fruit to ripen (or insects to appear) varies directly as the change in altitude h. If $d = 4$ days when $h = 500$ feet, find d when $h = 2,500$ feet.

Physics—Engineering **36.** Over a fixed distance d, speed r varies inversely as time t. Police use this relationship to set up speed traps. (The graph of the resulting function is a hyperbola.) If in a given speed trap $r = 30$ miles/hour when $t = 6$ seconds, what would be the speed of a car if $t = 4$ seconds?

★**37.** The length L of skid marks of a car's tires (when the brakes are applied) varies directly as the square of the speed v of the car. How is the length of skid marks affected by doubling the speed?

38. The time t required for an elevator to lift a weight varies jointly as the weight w and the distance d through which it is lifted, and inversely as the power P of the motor. Write the equation of variation, using k as the constant of variation.

39. The total pressure P of the wind on a wall varies jointly as the area of the wall A and the square of the velocity of the wind v. If $P = 120$ pounds when $A = 100$ square feet and $v = 20$ miles/hour, find P if $A = 200$ square feet and $v = 30$ miles/hour.

★★**40.** The thrust T of a given type of propeller varies jointly as the fourth power of its diameter d and the square of the number of revolutions per minute n it is turning. What happens to the thrust if the diameter is doubled and the number of revolutions per minute is cut in half?

Psychology **41.** In early psychological studies on sensory perception (hearing, seeing, feeling, and so on), the question was asked: "Given a certain level of stimulation S, what is the minimum amount of added stimulation ΔS that can be detected?" A German physiologist, E. H. Weber (1795–1878) formulated, after many experiments, the famous law that now bears his name: "The amount of change ΔS

that will be just noticed varies directly as the magnitude S of the stimulus."

(A) Write the law as an equation of variation.

(B) If a person lifting weights can just notice a difference of 1 ounce at the 50-ounce level, what will be the least difference she will be able to notice at the 500-ounce level?

(C) Determine the just noticeable difference in illumination a person is able to perceive at 480 candlepower if he is just able to perceive a difference of 1 candlepower at the 60-candle-power level.

42. Psychologists in their study of intelligence often use an index called IQ. IQ varies directly as mental age MA and inversely as chronological age CA (up to the age of 15). If a 12-year-old boy with a mental age of 14.4 has an IQ of 120, what will be the IQ of an 11-year-old girl with a mental age of 15.4?

Music **43.** The frequency of vibration of air in an open organ pipe varies inversely as the length of the pipe. If the air column in an open 32-foot pipe vibrates 16 times per second (low C), then how fast would the air vibrate in a 16-foot pipe?

44. The frequency of pitch f of a musical string varies directly as the square root of the tension T and inversely as the length l and the diameter d. Write the equation of variation using k as the constant of variation. (It is interesting to note that if pitch depended on only length, then pianos would have to have strings varying from 3 inches to 38 feet.)

Photography **45.** The f-stop numbers N on a camera, known as focal ratios, vary directly as the focal length F of the lens and inversely as the diameter d of the diaphragm opening (effective lens opening). Write the equation of variation using k as the constant of variation.

★**46.** In taking pictures using flashbulbs, the lens opening (f-stop number) N varies inversely as the distance d from the object being photographed. What adjustment should you make on the f-stop number if the distance between the camera and the object is doubled?

Chemistry ★**47.** Atoms and molecules that make up the air constantly fly about like microscopic missiles. The velocity v of a particular particle at a fixed temperature varies inversely as the square root of its molecular weight w. If an oxygen molecule in air at room temperature has an average velocity of 0.3 mile/second, what will be the average velocity of a hydrogen molecule, given that the hydrogen molecule is one-sixteenth as heavy as the oxygen molecule?

48. The Maxwell–Boltzmann equation says that the average velocity v of a molecule varies directly as the square root of the absolute temperature T and inversely as the square root of its molecular weight w. Write the equation of variation using k as the constant of variation.

Business   **49.** The amount of work A completed varies jointly as the number of workers W used and the time t they spend. If ten workers can finish a job in eight days, how long will it take four workers to do the same job?

50. The simple interest I earned in a given time varies jointly as the principal p and the interest rate r. If $100 at 4% interest earns $8, how much will $150 at 3% interest earn in the same period?

Geometry **★51.** The volume of a sphere varies directly as the cube of its radius r. What happens to the volume if the radius is doubled?

★52. The surface area S of a sphere varies directly as the square of its radius r. What happens to the area if the radius is cut in half?

Section 2-9   Chapter Review

IMPORTANT TERMS
AND SYMBOLS
2-1 Rectangular Coordinate System; Graphing.   Cartesian coordinate system; coordinate axes; quadrants; coordinates of a point; abscissa; ordinate; point-by-point graphing; graph of an equation in two variables; symmetry with respect to the x axis, the y axis, and the origin; distance-between-two-points formula; circle; equation of a circle

2-2 Relations and Functions.   Relation; function; domain; range; relations specified by equations, tables, ordered pairs, and graphs; independent variable; dependent variable; function notation; $f(x)$; f: $y = f(x)$

2-3 Functions: Graphs and Properties.   Graph of a function f, constant function, identity function, square function, cube function, even function, odd function, increasing function, decreasing function, greatest integer function $[\![x]\!]$, discontinuity at a point, continuity at a point, continuity over an interval, vertical and horizontal shifting, reflecting with respect to an axis, expanding and contracting

2-4 Linear Relations and Functions.   Linear function, first-degree polynomial function, graph of a linear function, slope, vertical line, horizontal line, point–slope form, slope–intercept form, intercept form, parallel lines, perpendicular lines

2-5 Graphing Polynomial Functions.   Polynomial function, continuity, graphing by nested factoring, quadratic functions, graphs of quadratic functions, axis, vertex, maximum or minimum

2-6 Graphing Rational Functions.   Rational function, points of discontinuity, infinity, horizontal asymptote, vertical asymptote, graphing rational functions, ∞

2-7 Composite and Inverse Functions.   Composite function, inverse relation, one-to-one correspondence, f^{-1}

2-8 Variation.   Direct variation, inverse variation, joint variation, combined variation, constant of variation

Exercise 2-9   Chapter Review

Work through all the problems in this chapter review and check answers in the back of the book. (Answers to all problems are there, and following each answer is a number in italics indicating the section in which that type of problem is discussed.) Where weaknesses show up, review appropriate sections in the text. When you are satisfied that you know the material, take the practice test following this review.

A   **1.**   How are the graphs of the following related to the graph of $y = x^2$?
   (A) $y = -x^2$ (B) $y = x^2 - 3$ (C) $y = (x + 3)^2$

2.   Given the points $A(-2, 3)$ and $B(4, 0)$, find:
   (A)   The distance between A and B
   (B)   The slope of AB
   (C)   The slope of a line perpendicular to AB

3.   Write the equations of the vertical and horizontal lines passing through $(-3, 4)$. What is the slope of each?

Problems 4–8 refer to the following functions:

$$f(x) = 3x + 5 \qquad g(x) = 4 - x^2 \qquad h(x) = 5 \qquad m(x) = 2|x| - 1$$

Find the indicated quantities or expressions.

4. $f(2) + g(-2) + h(0)$

5. $\dfrac{m(-2) + 1}{g(2) + 4}$

6. $\dfrac{f(2 + h) - f(2)}{h}$

7. $\dfrac{g(a + h) - g(a)}{h}$

8. $(f \circ g)(x)$ and $(g \circ f)(x)$

9.   Given $f(x) = 4x - 1$, find:
   (A) $f^{-1}(x)$ (B) $f^{-1}(7)$ (C) $(f^{-1} \circ f)(x)$

10.   Graph $3x + 2y = 9$ and indicate its slope.

11.   Write an equation of a line with x intercept 6 and y intercept 4. Write the final answer in the form $Ax + By = C$, where A, B, and C are integers.

12. Graph $f(x) = 1/(x + 2)$. Indicate any vertical or horizontal asymptotes with broken lines.

13. y varies directly as x, and inversely as z.
   (A) Write the equation of variation.
   (B) If $y = 4$ when $x = 6$ and $z = 2$, find y when $x = 4$ and $z = 4$.

B **14.** Write an equation of a line through $A(-6, -4)$ and $B(4, 1)$. Write the final answer in the form $Ax + By = C$, where A, B, and C are integers.

15. Find the equation of a circle with radius 2 and center at $(3, -2)$.

16. If the slope of a line is negative, is the function represented by the graph increasing, decreasing, or constant?

17. Discuss the graph of $4x^2 + 9y^2 = 36$ relative to symmetry with respect to the x axis, y axis, and origin.

18. Given:

$$f(x) = \sqrt[3]{x} \qquad g(x) = \frac{x^2}{x^2 - 1} \qquad h(x) = x + 1$$

   (A) Which are odd?
   (B) Which are even?
   (C) Which are neither even nor odd?

19. Graph: $f(x) = -|x + 1| - 1$

20. Write an equation of the line (A) parallel to, or (B) perpendicular to the line $6x + 3y = 5$ and passing through the point $(-2, 1)$. Write final answers in the form $y = mx + b$.

21. Graph $f(x) = x^2 - 6x + 5$. Show the axis and vertex, and find the maximum or minimum value of $f(x)$.

22. Given: $f(x) = \dfrac{x - 1}{x + 2}$

   (A) Is f even or odd?
   (B) Find the x and y intercepts.
   (C) Find the points of discontinuity.
   (D) Find the equations of horizontal and vertical asymptotes.
   (E) Where is the graph above and below the x axis?
   (F) Sketch a graph of f (include asymptotes).

23. What are the coordinates of the center and the radius of the circle given by $x^2 + y^2 - 6x + 8y = 0$?

24. Write $P(x) = x^3 - 2x^2 - 5x + 6$ in a nested factored form and graph for $-3 \le x \le 4$.

25. Each of the following equations defines a function. Which have inverses that are functions?

     (A) $f(x) = x^3$ (B) $g(x) = (x - 2)^2$
     (C) $h(x) = 2x - 3$ (D) $F(x) = (x + 3)^2$, $x \geq -3$

26.   Given: f: $y = 2^x$, $x \in \{-2, -1, 0, 1, 2\}$
     (A)   Find the domain and range of f and f^{-1}.
     (B)   Graph f, f^{-1}, and $y = x$ on the same coordinate system.
     (C)   Which of f and f^{-1} is a function?

27.   Given $f(x) = x^2 - 1$, $x \geq 0$, find:
     (A)   The domain and range of f and f^{-1}
     (B) $f^{-1}(x)$ (C) $f^{-1}(3)$ (D) $(f^{-1} \circ f)(4)$
     (E) $(f^{-1} \circ f)(x)$

28.   The time t required for an elevator to lift a weight varies jointly as the weight w and the distance d through which it is lifted, and inversely as the power P of the motor. Write the equation of variation using k as the constant of variation.

C   **29.**   Find the equations for the horizontal and vertical asymptotes for the graphs of:

     (A) $f(x) = \dfrac{3x}{2x + 3}$ (B) $g(x) = \dfrac{5x}{x^2 - x - 6}$

30.   For what values of x are f and g discontinuous in Problem 29?

31.   For $f(x) = (x + 2)/(x - 3)$, find:
     (A) $f^{-1}(x)$ (B) $f^{-1}(3)$ (C) $(f^{-1} \circ f)(x)$

32.   Find the equation of the set of points equidistant from $(3, 3)$ and $(6, 0)$. What is the name of the geometric figure formed by this set?

33.   For what values of x is the graph of $f(x) = (x - 1)/(x + 2)$ below its horizontal asymptote? Above its horizontal asymptote?

34.   The total force F of a wind on a wall varies jointly as the area of the wall A and the square of the velocity of the wind. How is the total force on the wall affected if the area is cut in half and the velocity is doubled?

Practice Test   Chapter 2

Take this practice test as if it were a graded test. Allow yourself up to 50 minutes. Work the problems without looking back in the chapter. Correct your work using the answers (keyed to appropriate sections) in the back of the book.

   1.   (A)   Find an equation of the line through $P(-4, 3)$ and $Q(0, -3)$. Write the final answer in the form $Ax + By = C$, where A, B, and C are integers with $A > 0$.
        (B)   Find $d(P, Q)$.

2. Given the line L: $3x + 4y = 10$, find an equation of the line through $(4, -2)$ that is (A) parallel to L, or (B) perpendicular to L. Write the final answers in the form $y = mx + b$.

3. Find the center and radius of the circle given by $x^2 + y^2 + 4x - 6y = 3$.

Problems 4 and 5 refer to the following functions:

$$f(x) = 3x \qquad g(x) = 2|x| - 1 \qquad F(x) = 3 \quad 2x^2$$

4. (A) $\dfrac{f(2) + g(0)}{h(2)} = ?$ (B) $\dfrac{F(a + h) - F(a)}{h} = ?$

5. (A) List the even functions.
   (B) Which have graphs that are symmetric with respect to the origin?
   (C) Which are one to one and thus have inverses that are functions?

6. Find the maximum or minimum value of $f(x) = x^2 - 6x + 11$ without graphing. What are the coordinates of the vertex of the graph?

Problems 7–9 refer to the function f defined by: $\quad f(x) = \dfrac{x - 1}{2x + 2}$

7. (A) Is f even or odd?
   (B) Where is f discontinuous?
   (C) Find the x and y intercepts for the graph of f.
   (D) Find the equations of the horizontal and vertical asymptotes.

8. For what values of x is the graph of f above the x axis? Below the x axis?

9. Sketch a graph of f. Draw vertical and horizontal asymptotes with broken lines.

10. Given $f(x) = \sqrt{x} - 8$ and $g(x) = |x|$.
    (A) Find $(f \circ g)(x)$ and $(g \circ f)(x)$.
    (B) Find the domains for $f \circ g$ and $g \circ f$.

11. Given $f(x) = 3x + 7$, find:
    (A) $f^{-1}(x)$ (B) $f^{-1}(5)$ (C) $(f^{-1} \circ f)(x)$
    (D) Is f a decreasing or increasing function?

12. Given $f(x) = \sqrt{x - 1}$.
    (A) Find the domain and range for f and f^{-1}.
    (B) Graph f, f^{-1}, and $y = x$ on the same coordinate system and indicate which of f and f^{-1} is a function.
    (C) Find $f^{-1}(x)$, if it exists.

13. Suppose H varies directly as n and inversely as the square of m.
  (A) Write the equation of variation.
  (B) If $H = 4$ when $n = 2$ and $m = 3$, find H when $n = 4$ and $m = 2$.
  (C) What happens to H if both n and m are doubled?

14. How is the graph of $f(x) = -(x - 2)^2 - 1$ related to the graph of $g(x) = x^2$?

15. (A) Write $f(x) = x^3 - 3x^2 - x + 3$ in a nested factored form.
  (B) Using (A), graph f for $-2 \le x \le 4$.

Polynomial Functions and Theory of Equations ■ 3

A natural design of mathematical interest. Can you guess the source? See the back of the book.

Chapter 3 ▪ Polynomial Functions and Theory of Equations

Section 3-1   Introduction

We know how to solve first- and second-degree polynomial equations (linear and quadratic equations). For example, for the linear equations

$$ax + b = 0 \qquad a \neq 0$$

$$x = -\frac{b}{a}$$

and for the quadratic equation

$$ax^2 + bx + c = 0 \qquad a \neq 0$$

$$x = \frac{-b \pm \sqrt{b^2 - 4ac}}{2a}$$

What about third- and higher-degree polynomial equations? For example, how do we solve equations such as

$$2x^3 - 3x^2 + x - 5 = 0$$

and

$$x^7 - 6x^4 + 3x - 1 = 0$$

It turns out that there are direct methods (though complicated) for finding all solutions for any third- or fourth-degree polynomial equation. However, Evariste Galois (1811–1832), a Frenchman, proved at the age of 20 that for polynomial equations of degree greater than four there was no finite step-by-step process that would always yield all solutions.* This does not mean that we give up looking for solutions to higher-degree polynomials. In this chapter you will find that solutions always exist for all polynomial equations of degree greater than or equal to one.

---

* Galois' contribution, using the new concept of "group," was of the highest mathematical significance and originality. However, his contemporaries hardly read his papers, dismissing them as "almost unintelligible." At the age of 21, involved in political agitation, Galois met an untimely death in a duel. A short but fascinating account of Galois' tragic life can be found in E. T. Bell's *Men of Mathematics* (New York: Simon & Schuster, 1937), pp. 362–377.

We will develop methods for finding or approximating all real solutions of polynomials with real coefficients.

To aid this endeavor, we will find it helpful to switch our emphasis from polynomial equations to polynomial functions. We will then uncover some important properties of polynomial functions that will lead directly to solutions of certain polynomial equations. The following definitions make a useful connecting link between polynomial functions and polynomial equations.

For the nth-degree polynomial function P given by

$$P(x) = a_n x^n + a_{n-1} x^{n-1} + \cdots + a_1 x + a_0 \qquad a_n \neq 0$$

where the coefficients are real or complex, r is said to be a **zero of the function P**, or a **zero of the polynomial P(x)**, or a **solution or root of the equation P(x) = 0**, if

$$P(r) = 0$$

A zero of a polynomial may or may not be 0; a zero of a polynomial is *any* number (real or complex) that makes a polynomial 0. If we consider the graph of $y = P(x)$, then a real zero of $P(x)$ is simply an x intercept. Consider the polynomial

$$P(x) = x^2 - 4x + 3$$

The graph of P is shown in Figure 1.

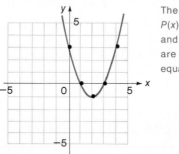

The x intercepts 1 and 3 are zeros of $P(x) = x^2 - 4x + 3$, since $P(1) = 0$ and $P(3) = 0$. The x intercepts 1 and 3 are also solutions or roots for the equation $x^2 - 4x + 3 = 0$.

FIGURE 1

In general:

Zeros and Roots

The x intercepts of the graph of $y = P(x)$ are real **zeros** of P and $P(x)$ and real solutions or **roots** for the equation $P(x) = 0$.

Section 3-2   Synthetic Division

- Algebraic Long Division
- Synthetic Division

We now digress for a moment to discuss algebraic long division. This apparent digression actually provides us with a very useful tool that we will use throughout most of this chapter.

■ Algebraic Long Division

We can find quotients of polynomials by a long-division process similar to that used in arithmetic. An example will illustrate the process.

EXAMPLE 1   Divide $5 + 4x^3 - 3x$ by $2x - 3$.

Solution

$$
\begin{array}{r}
2x^2 + 3x + 3 \\
2x - 3 \overline{)\,4x^3 + 0x^2 - 3x + 5} \\
\underline{4x^3 - 6x^2} \\
6x^2 - 3x \\
\underline{6x^2 - 9x} \\
6x + 5 \\
\underline{6x - 9} \\
14 = R
\end{array}
$$

Remainder

Arrange the dividend and the divisor in descending powers of the variable. Insert, with 0 coefficients, any missing terms of degree less than three. Divide the first term of the divisor into the first term of the dividend. Multiply the divisor by $2x^2$, line up like terms, subtract as in arithmetic, and bring down $-3x$. Repeat the process until the degree of the remainder is less than that of the divisor.

Thus,

$$
\frac{4x^3 - 3x + 5}{2x - 3} = 2x^2 + 3x + 3 + \frac{14}{2x - 3}
$$

Check

$$
(2x - 3)\left[(2x^2 + 3x + 3) + \frac{14}{2x - 3} \right] = (2x - 3)(2x^2 + 3x + 3) + 14
$$

$$
= 4x^3 - 3x + 5
$$

PROBLEM 1   Divide $6x^2 - 30 + 9x^3$ by $3x - 4$.

■ Synthetic Division

Being able to divide a polynomial $P(x)$ by a linear polynomial of the form $x - r$ quickly and accurately will be of great aid to us (as strange as it may seem now) in the search for zeros of higher-degree polynomials functions. This kind of division can be carried out efficiently by a method called **synthetic division**. The method is most easily understood through an example. Let us start by dividing $P(x) = 2x^4 + 3x^3 - x - 5$ by $x + 2$, using ordinary long division. The critical parts of the process are indicated in color.

$$
\begin{array}{r}
2x^3 - 1x^2 + 2x - 5 \qquad \text{Quotient} \\
\text{Divisor} \quad x + 2 \overline{\smash{)}\,2x^4 + 3x^3 + 0x^2 - 1x - 5} \qquad \text{Dividend} \\
2x^4 + 4x^3 \\
\overline{-1x^3 + 0x^2} \\
-1x^3 - 2x^2 \\
\overline{2x^2 - 1x} \\
2x^2 + 4x \\
\overline{-5x - 5} \\
-5x - 10 \\
\overline{5 \qquad \text{Remainder}}
\end{array}
$$

The numerals printed in color, which represent the essential part of the division process, are arranged more conveniently as:

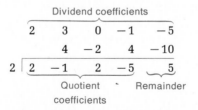

Mechanically, we see that the second and third rows of numerals are generated as follows. The first coefficient 2 of the dividend is brought down and multiplied by 2 from the divisor, and the product 4 is placed under the second dividend coefficient 3 and subtracted. The difference -1 is again multiplied by the 2 from the divisor, and the product is placed under the third coefficient from the dividend and subtracted. This process is repeated until the remainder is reached. The process can be made a little faster, and less prone to sign errors, by changing $+2$ from the divisor to -2 and adding instead of subtracting. Thus,

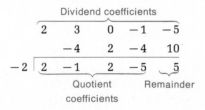

> **Key Steps in the Synthetic Division Process**
>
> 1. Arrange the coefficients of $P(x)$ in order of descending powers of x (write 0 as the coefficient for each missing power).
> 2. After writing the divisor in the form $x - r$, use r to generate the second and third rows of numbers as follows. Bring down the first coefficient of the dividend and multiply it by r; then add the product to the second coefficient of the dividend. Multiply this sum by r, and add the product to the third coefficient of the dividend. Repeat the process until a product is added to the constant term of $P(x)$. [*Note:* This process is well suited to hand calculator use. Store r; then proceed from left to right recalling r and using it as indicated.]
> 3. The last number in the third row of numbers is the remainder; the other numbers in the third row are the coefficients of the quotient, which is of degree 1 less than $P(x)$.

EXAMPLE 2   Use synthetic division to find the quotient and remainder resulting from dividing $P(x) = 4x^5 - 30x^3 - 50x - 2$ by $x + 3$. Write the answer in the form $Q(x) + R/(x - r)$, where R is a constant.

Solution $x + 3 = x - (-3)$; therefore, $r = -3$.

$$
\begin{array}{r|rrrrrr}
 & 4 & 0 & -30 & 0 & -50 & -2 \\
 & & -12 & 36 & -18 & 54 & -12 \\
\hline
-3 & 4 & -12 & 6 & -18 & 4 & -14
\end{array}
$$

The quotient is $4x^4 - 12x^3 + 6x^2 - 18x + 4$ with a remainder of -14. Thus,

$$\frac{P(x)}{x + 3} = 4x^4 - 12x^3 + 6x^2 - 18x + 4 + \frac{-14}{x + 3}$$

PROBLEM 2   Repeat Example 2 with $P(x) = 3x^4 - 11x^3 - 18x + 8$ and divisor $x - 4$.

Answers to Matched Problems   **1.** $3x^2 + 6x + 8 + \dfrac{2}{3x - 4}$

2. $\dfrac{P(x)}{x - 4} = 3x^3 + x^2 + 4x - 2 + \dfrac{0}{x - 4}$

$\qquad = 3x^3 + x^2 + 4x - 2$

Exercise 3-2 ■ A *Divide, using algebraic long division. Write the quotient and indicate the remainder.*

    1. $(4m^2 - 1) \div (2m - 1)$

    2. $(y^2 - 9) \div (y + 3)$

    3. $(6 - 6x + 8x^2) \div (2x + 1)$

    4. $(11x - 2 + 12x^2) \div (3x + 2)$

    5. $(x^3 - 1) \div (x - 1)$

    6. $(a^3 + 27) \div (a + 3)$

    7. $(3y - y^2 + 2y^3 - 1) \div (y + 2)$

    8. $(3 + x^3 - x) \div (x - 3)$

Use algebraic long division and synthetic division to write the quotient $P(x) \div (x - r)$ in the form $P(x)/(x - r) = Q(x) + R/(x - r)$ where R is a constant.

    9. $(x^2 + 3x - 7) \div (x - 2)$

    10. $(x^2 + 3x - 3) \div (x - 3)$

    11. $(4x^2 + 10x - 9) \div (x + 3)$

    12. $(2x^2 + 7x - 5) \div (x + 4)$

    13. $(2x^3 - 3x + 1) \div (x - 2)$

    14. $(x^3 + 2x^2 - 3x - 4) \div (x + 2)$

B *Divide, using synthetic division. Write the quotient and indicate the remainder. As coefficients get more involved, a hand calculator will be very helpful.*

    15. $(3x^4 - x - 4) \div (x + 1)$

    16. $(5x^4 - 2x^2 - 3) \div (x - 1)$

    17. $(x^5 + 1) \div (x + 1)$

    18. $(x^4 - 16) \div (x - 2)$

    19. $(2x^3 + 4x^2 - 9x - 11) \div (x + 3)$

    20. $(x^4 - 3x^3 - 5x^2 + 6x - 3) \div (x - 4)$

    21. $(2x^4 - 13x^3 + 14x^2 + 15) \div (x - 5)$

    22. $(x^5 + 10x^2 + 5x + 2) \div (x + 2)$

    23. $(4x^4 + 2x^3 - 6x^2 - 5x + 1) \div (x + \frac{1}{2})$

    24. $(2x^3 - 5x^2 + 6x + 3) \div (x - \frac{1}{2})$

    25. $(4x^3 + 4x^2 - 7x - 6) \div (x + \frac{3}{2})$

    26. $(3x^3 - x^2 + x + 2) \div (x + \frac{2}{3})$

    27. $(x^3 - 2x^2 + 3x - 1) \div (x - 0.3)$

    28. $(2x^3 + 3x^2 - 2x + 1) \div (x - 0.2)$

29. $(3x^3 + 2x - 4) \div (x - 0.2)$

30. $(4x^3 - 2x^2 - 1) \div (x - 0.3)$

C *Divide, using algebraic long division. Write the quotient and indicate the remainder.*

31. $(16x - 5x^3 - 8 + 6x^4 - 8x^2) \div (2x - 4 + 3x^2)$

32. $(8x^2 - 7 - 13x + 24x^4) \div (3x + 5 + 6x^2)$

Divide, using synthetic division. Write the quotient and indicate the remainder. A hand calculator will be very helpful in Problems 33–36.

33. $(2x^3 - 5x^2 - 8x + 6) \div (x - 3.3)$

34. $(2x^4 - x^2 + 3x - 2) \div (x - 0.6)$

35. $(2x^3 - 5x^2 - 8x + 6) \div (x + 1.4)$

36. $(x^3 + 2x - 4) \div (x - 1.2)$

37. $(x^3 - 3x^2 + x - 3) \div (x - i)$

38. $(x^3 - 2x^2 + x - 2) \div (x + i)$

39. (A) Divide $P(x) = a_2 x^2 + a_1 x + a_0$ by $x - r$, using synthetic division and the long-division process, and compare the coefficients of the quotient and the remainder produced by each method.

(B) Expand the expression representing the remainder. What do you observe?

40. Repeat Problem 39 for $P(x) = a_3 x^3 + a_2 x^2 + a_1 x + a_0$.

Section 3-3 Remainder and Factor Theorems

- Division Algorithm
- Remainder Theorem
- Graphing Polynomials
- Factor Theorem

- **Division Algorithm**

If we divide $P(x) = 2x^4 - 5x^3 - 4x^2 + 13$ by $x - 3$, we obtain

$$\frac{2x^4 - 5x^3 - 4x^2 + 13}{x - 3} = 2x^3 + x^2 - x - 3 + \frac{4}{x - 3} \qquad x \neq 3$$

If we multiply both members by $x - 3$, then

$$2x^4 - 5x^3 - 4x^2 + 13 = (x - 3)(2x^3 + x^2 - x - 3) + 4$$

This last equation is an identity in that the left side is equal to the right side for *all* replacements of x by real or complex numbers, including $x = 3$. This example suggests the important **division algorithm**, which we state as Theorem 1 without proof.

THEOREM 1

Division Algorithm

For each polynomial $P(x)$ of degree one or greater and each number r, there exists a unique polynomial $Q(x)$ of degree 1 less than $P(x)$ and a unique number R (which may be 0) such that

$P(x) = (x - r)Q(x) + R$

The polynomial $Q(x)$ is called the **quotient**, $x - r$ the **divisor**, and R the **remainder**.

■ Remainder Theorem

We now use the division algorithm in Theorem 1 to prove the important and useful remainder theorem.

The equation in Theorem 1

$P(x) = (x - r)Q(x) + R$

is an identity; that is, it is true for all real or complex replacements for x. In particular, if we let $x = r$, then we observe a very interesting and extremely useful relationship:

$$P(r) = (r - r)Q(r) + R$$
$$= 0 \cdot Q(r) + R$$
$$= 0 + R$$
$$= R$$

In words, the value of a polynomial $P(x)$ at $x = r$ is the same as the remainder R one obtains by dividing $P(x)$ by $x - r$. We have proved the well-known remainder theorem (Theorem 2).

THEOREM 2

Remainder Theorem

If R is the remainder after dividing the polynomial $P(x)$ by $x - r$, then

$P(r) = R$

EXAMPLE 3   If $P(x) = 4x^4 + 10x^3 + 19x + 5$, find $P(-3)$ by (A) using the remainder theorem and synthetic division, and (B) evaluating $P(-3)$ directly.

Solution   (A)

$$\begin{array}{r}
4 \quad\ \ 10 \quad\ \ 0 \quad\ \ 19 \quad\ \ 5 \\
-12 \quad\ 6 \quad -18 \quad -3 \\
\hline
-3\,|\ \ 4 \quad -2 \quad\ 6 \quad\ \ 1 \quad\ \ 2 = R = P(-3)
\end{array}$$

(B) $P(-3) = 4(-3)^4 + 10(-3)^3 + 19(-3) + 5$
$= 2$

PROBLEM 3   Repeat Example 3 for $P(x) = 3x^4 - 16x^2 - 3x + 7$ and $x = -2$.

■ Graphing Polynomials

The remainder theorem and synthetic division provide us with an efficient way of graphing polynomials. In terms of the mechanics, the process is equivalent to the "nested factoring" method of graphing discussed in Section 2-5 (see Problems 39–42 in Exercise 3-3). The following example illustrates the process.

EXAMPLE 4   Graph: $P(x) = x^3 + 3x^2 - x - 3$, $-4 \le x \le 2$

Solution   We evaluate $P(x)$ from $x = -4$ to $x = 2$, for selected values of x, using synthetic division and the remainder theorem. The process is speeded by forming a synthetic division table. The second row is left blank and the computation for succeeding rows is done either mentally or on a hand-held calculator—the hand calculator becomes increasingly useful as the coefficients become more numerous or complicated. The table also provides other important information, as will be seen in subsequent sections.

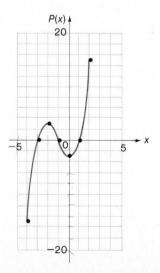

	1	3	-1	-3	
-4	1	-1	3	-15	$= P(-4)$
-3	1	0	-1	0	$= P(-3)$
-2	1	1	-3	3	$= P(-2)$
-1	1	2	-3	0	$= P(-1)$
0	1	3	-1	-3	$= P(0)$
1	1	4	3	0	$= P(1)$
2	1	5	9	15	$= P(2)$

PROBLEM 4　Graph $P(x) = x^3 - 4x^2 - 4x + 16$,　$-3 \le x \le 5$.　Find points using synthetic division and the remainder theorem.

■　Factor Theorem

The equation $P(x) = (x - r)Q(x) + R$ in Theorem 1 may, because of the remainder theorem, be written in a form where R is replaced by $P(r)$. Thus,

$$P(x) = (x - r)Q(x) + P(r)$$

It is easy to see that $x - r$ is a factor of $P(x)$ if and only if $P(r) = 0$—that is, if and only if r is a zero of the polynomial $P(x)$ [or a root or solution of the polynomial equation $P(x) = 0$]. This result is known as the **factor theorem** (Theorem 3).

THEOREM 3

Factor Theorem
If r is a zero of the polynomial $P(x)$, then $x - r$ is a factor of $P(x)$; conversely, if $x - r$ is a factor of $P(x)$, then r is a zero of $P(x)$.

If we can find a zero of a polynomial, then we can find one of its factors. On the other hand, if we can find a linear factor of a polynomial, we can find a zero of the polynomial.

EXAMPLE 5　(A)　Use the factor theorem to show that $x + 1$ is a factor of $P(x) = x^{25} + 1$.
(B)　What are the zeros of $P(x) = 3(x - 5)(x + 2)(x - 3)$?

Solution　(A)　$x + 1 = x - (-1)$; thus, $r = -1$.

$$P(r) = P(-1) = (-1)^{25} + 1 = -1 + 1 = 0$$

Hence, -1 is a zero of $P(x) = x^{25} + 1$. Thus, $x - (-1) = x + 1$ is a factor of $x^{25} + 1$.
(B)　5, -2, and 3 are zeros of $P(x)$, since $(x - 5)$, $(x + 2)$, and $(x - 3)$ are all factors of $P(x)$.

PROBLEM 5　(A)　Use the factor theorem to show that $x - 1$ is a factor of $P(x) = x^{54} - 1$.
(B)　What are the zeros of the polynomial

$$P(x) = 2(x + 3)(x + 7)(x - 8)(x + 1)?$$

Answers to Matched Problems　**3.**　$P(-2) = -3$ for both parts, as it should.

4.

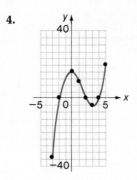

5. (A) $r = 1$ and $P(1) = 1^{54} - 1 = 1 - 1 = 0$; therefore, $x - r = x - 1$ is a factor of $P(x) = x^{54} - 1$.

  (B) $-3, -7, 8, -1$

Exercise 3-3 ∎ ▦

A hand calculator will prove helpful in some of the following problems.

A Use synthetic division and the remainder theorem in each of the following problems.

1. Find $P(-2)$, given $P(x) = 3x^2 - x - 10$.

2. Find $P(-3)$, given $P(x) = 4x^2 + 10x - 8$.

3. Find $P(2)$, given $P(x) = 2x^3 - 5x^2 + 7x - 7$.

4. Find $P(5)$, given $P(x) = 2x^3 - 12x^2 - x + 30$.

5. Find $P(-4)$, given $P(x) = x^4 - 10x^2 + 25x - 2$.

6. Find $P(-7)$, given $P(x) = x^4 + 5x^3 - 13x^2 - 30$.

Find the zeros for the following polynomials using the factor theorem.

7. $P(x) = (x - 3)(x + 5)$

8. $P(x) = (x + 2)(x - 7)$

9. $P(x) = 2(x + \frac{1}{2})(x - 8)(x + 2)$

10. $P(x) = 3(x - \frac{2}{3})(x - 5)(x + 7)$ $\frac{2}{3}, 5, -7$

Determine whether the second polynomial is a factor of the first polynomial without dividing or using synthetic division. [Hint: Evaluate directly and use the factor theorem.]

11. $x^{18} - 1$; $x - 1$

12. $x^{18} - 1$; $x + 1$

13. $3x^3 - 7x^2 - 8x + 2$; $x + 1$

14. $3x^4 - 2x^3 + 5x - 6$; $x - 1$

B Use synthetic division and the remainder theorem in each of the following problems.

15. Find $P(\frac{1}{2})$, given $P(x) = 4x^3 - 8x^2 + 5x - 4$.

16. Find $P(\frac{1}{3})$, given $P(x) = 6x^3 + 4x^2 - 5x - 4$.

17. Find $P(0.3)$ for $P(x) = x^3 - 2x + 1$.

18. Find $P(0.7)$ for $P(x) = 2x^3 + 3x^2 - 5x + 2$.

Graph each polynomial function using synthetic division and the remainder theorem.

19. $P(x) = x^3 - 5x^2 + 2x + 8$, $\quad -2 \le x \le 5$

20. $P(x) = x^3 + 2x^2 - 5x - 6$, $\quad -4 \le x \le 3$

21. $P(x) = x^3 + 4x^2 - x - 4$, $\quad -5 \le x \le 2$

22. $P(x) = x^3 - 2x^2 - 5x + 6$, $\quad -3 \le x \le 4$

Find three solutions for each equation.

23. $(x + 4)(x + 8)(x - 1) = 0$

24. $(x - 2)(x + 5)(x - 3) = 0$

25. $7(x - \frac{1}{8})(x + \frac{3}{5})(x + 4) = 0$

26. $4(x + \frac{3}{4})(x - 5)(x - \frac{2}{3}) = 0$

Use the quadratic formula and the factor theorem to factor each polynomial.

27. $P(x) = x^2 - 3x + 1$

28. $P(x) = x^2 - 4x - 2$

29. $P(x) = x^2 - 6x + 10$

30. $P(x) = x^2 - 4x + 5$

Determine whether the second polynomial is a factor of the first polynomial without dividing or using synthetic division.

31. $x^n - a^n$; $\quad x - a$

32. $x^n - a^n$, n even; $\quad x + a$

33. $4x^7 - 2x^6 + x^2 + 2x + 5$; $\quad x - 1$

34. $2x^5 - 5x^2 - x + 4$; $\quad x + 1$

C Graph each polynomial function in Problems 35–38 using synthetic division and the remainder theorem. A hand calculator may prove useful.

35. $P(x) = x^4 - 2x^3 - 2x^2 + 8x - 8$

36. $P(x) = x^4 - 2x^2 + 16x - 15$

37. $P(x) = x^4 + 4x^3 - x^2 - 20x - 20$

38. $P(x) = x^4 - 4x^2 - 4x - 1$

39. Recall from Section 2-5 that polynomials can also be conveniently evaluated using a nested factoring scheme. For example, to evaluate $P(x) = x^4 - 6x^3 + 19x^2 - 26x + 18$ for $x = 2$, we would first

write $P(x) = \{[(x - 6)x + 19]x - 26\}x + 18$. Find $P(2)$ using this "factored" form, and by synthetic division using the remainder theorem. How do the two methods compare step for step?

40. Repeat Problem 39 for the equation $P(x) = 3x^4 - 10x^2 + 5x - 2$ and $x = -2$.

41. (A) Write $P(x) = a_2x^2 + a_1x + a_0$ in the form $P(x) = (a_2x + a_1)x + a_0$ and find $P(r)$ using the latter.

     (B) Find $P(r)$ using synthetic division and the remainder theorem, and compare with part (A).

42. Repeat Problem 41 for $P(x) = a_3x^3 + a_2x^2 + a_1x + a_0$.

Section 3-4 Fundamental Theorem of Algebra

- Fundamental Theorem of Algebra
- *n* Zeros Theorem
- Complex Zeros
- Remarks

In our search for zeros of polynomial functions it would be useful to know at the outset how many zeros to expect for a given function. The following two theorems tell us exactly how many zeros exist for a polynomial function of a given degree. Even though the theorems do not tell us how to find the zeros, it is still very helpful to know that what we are looking for exists. These theorems were first proved in 1797 by Carl Friedrich Gauss, one of the greatest mathematicians of all time, at the age of 20.

- Fundamental Theorem of Algebra

Theorem 4, often referred to as the **fundamental theorem of algebra**, requires a proof that is beyond the scope of this book, so we state it without proof.

THEOREM 4

Fundamental Theorem of Algebra
Every polynomial $P(x)$ of degree $n \geq 1$, with real or complex coefficients, has at least one real or complex zero.

■ *n* Zeros Theorem

If $P(x) = a_n x^n + a_{n-1} x^{n-1} + \cdots + a_1 x + a_0$ is a polynomial of degree $n \geq 1$, then, according to this theorem, it has at least one zero, say r_1. According to the factor theorem, $x - r_1$ is a factor of $P(x)$. Thus,

$$P(x) = (x - r_1)Q_1(x)$$

where $Q_1(x)$ is a polynomial of degree $n - 1$. If $n - 1 = 0$, then $Q_1(x) = a_n$. If $n - 1 \geq 1$, then, by the fundamental theorem of algebra, $Q_1(x)$ has at least one zero, say r_2. And

$$Q_1(x) = (x - r_2)Q_2(x)$$

where $Q_2(x)$ is a polynomial of degree $n - 2$. Thus,

$$P(x) = (x - r_1)(x - r_2)Q_2(x)$$

If $n - 2 = 0$, then $Q_2(x) = a_n$. If $n - 2 \geq 1$, then $Q_2(x)$ has at least one zero, say r_3. And

$$Q_2(x) = (x - r_3)Q_3(x)$$

where $Q_3(x)$ is a polynomial of degree $n - 3$.

We continue in this way until $Q_k(x)$ is of degree zero—that is, until $k = n$. At this point, $Q_n(x) = a_n$ and we have

$$P(x) = (x - r_1)(x - r_2) \cdot \cdots \cdot (x - r_n)a_n$$

Thus, $r_1, r_2, \ldots, r_n$ are n zeros (not necessarily distinct) of $P(x)$. Is it possible for $P(x)$ to have more than these n zeros? Let us assume that r is a number different from the zeros above; then

$$P(r) = a_n(r - r_1)(r - r_2) \cdot \cdots \cdot (r - r_n) \neq 0$$

since r is not equal to any of the zeros. Hence, r is not a zero and we conclude that $r_1, r_2, \ldots, r_n$ are the only zeros of $P(x)$. We have just sketched a proof of Theorem 5.

THEOREM 5 | *n* Zeros Theorem

> Every polynomial $P(x)$ of degree $n \geq 1$, with real or complex coefficients, can be expressed as the product of n linear factors (hence, has exactly n zeros—not necessarily distinct).

If $P(x)$ is represented as the product of linear factors and $x - r$ occurs m times, then r is called a **zero of multiplicity m**. For example, if

$$P(x) = 4(x - 5)^3(x + 1)^2(x - i)(x + i)$$

then this seventh-degree polynomial has seven zeros, not all distinct. Five is a zero of multiplicity 3 (or a triple zero); -1 is a zero of multiplicity 2 (or a double zero). Thus, this seventh-degree polynomial has exactly seven zeros if we count 5 and -1 with their respective multiplicities.

EXAMPLE 6   If -2 is a double zero of $P(x) = x^4 - 7x^2 + 4x + 20$, write $P(x)$ as a product of first-degree factors.

Solution   Since -2 is a double zero of $P(x)$, we can write

$$P(x) = (x + 2)^2 Q(x)$$
$$= (x^2 + 4x + 4)Q(x)$$

and find $Q(x)$ by dividing $P(x)$ by $x^2 + 4x + 4$. Carrying out the division, we obtain

$$Q(x) = x^2 - 4x + 5$$

The zeros of $Q(x)$ are found, using the quadratic formula, to be $2 - i$ and $2 + i$. Thus, $P(x)$ written as a product of linear factors is

$$P(x) = (x + 2)^2 [x - (2 - i)][x - (2 + i)]$$

PROBLEM 6   If 3 is a double zero of $P(x) = x^4 - 12x^3 + 55x^2 - 114x + 90$, write $P(x)$ as a product of first-degree factors.

■ Complex Zeros

Something interesting happens if we restrict the coefficients of a polynomial to real numbers. Let us use the quadratic formula to find the zeros of the polynomial

$$P(x) = x^2 - 6x + 10$$

To find the zeros of $P(x)$, we solve $P(x) = 0$:

$$x^2 - 6x + 10 = 0$$

$$x = \frac{6 \pm \sqrt{36 - 40}}{2}$$

$$= \frac{6 \pm \sqrt{-4}}{2} = \frac{6 \pm 2i}{2} = 3 \pm 2i$$

The zeros of $P(x)$ are $3 - 2i$ and $3 + 2i$, conjugate complex numbers (see Section 1-7). Also observe that the complex zeros in Example 6 are the conjugate complex numbers $2 - i$ and $2 + i$.

In general, one can prove the following theorem:

THEOREM 6

Complex Zeros Theorem
Nonreal complex zeros of polynomials with real coefficients, if they exist, occur in conjugate pairs.

As a consequence of Theorems 5 and 6, we immediately know (think this through) that:

Real Zeros and Odd-Degree Polynomials
A polynomial of odd degree with real coefficients always has at least one real zero.

EXAMPLE 7   Let $P(x)$ be a third-degree polynomial with real coefficients. One of the following statements is false; indicate which one.

(A) $P(x)$ has at least one real zero.
(B) $P(x)$ has three zeros.
(C) $P(x)$ can have two real zeros and one complex zero.

Solution   Statement (C) is false, since complex zeros of polynomials with real co-efficients *must* occur in conjugate pairs. If $P(x)$ has two real zeros, then we know that the third zero must also be real.

PROBLEM 7   Let $P(x)$ be a polynomial of fourth degree with real coefficients. One of the following statements is false; indicate which one.

(A) $P(x)$ has four zeros.
(B) $P(x)$ has at least two real zeros.
(C) If we know $P(x)$ has three real zeros, then the fourth zero must be real.

■ Remarks

The fundamental theorem of algebra tells us that in the set of complex numbers not only $x^2 + 1 = 0$ has a solution, but every polynomial equation with real or complex coefficients has a solution.

This important and useful result does not come free. In extending the real numbers to a number system that provides solutions for all polynomial equations we have to give up something—namely, an ordering of the number system. The complex numbers cannot be ordered; that is, in general, we cannot say that one complex number is less than or greater than another.

Answers to Matched Problems

6. $P(x) = (x - 3)^2[x - (3 - i)][x - (3 + i)]$

7. (B) is false. According to the three theorems in this section, the possible combinations of real and complex zeros for $P(x)$ are as follows: (1) four complex, (2) two real and two complex, (3) four real. So $P(x)$ may not have any real zeros.

Exercise 3-4 ■ A

Write the zeros of each polynomial, and indicate the multiplicity of each if over one. What is the degree of the polynomial?

1. $P(x) = (x + 8)^3(x - 6)^2$

2. $P(x) = (x - 5)(x + 7)^2$

3. $P(x) = 3(x + 4)^3(x - 3)^2(x + 1)$

4. $P(x) = 5(x - 2)^3(x + 3)^2(x - 1)$

Find a polynomial $P(x)$ of lowest degree, with leading coefficient 1, that has the indicated set of zeros. (Leave the answer in a factored form.) Indicate the degree of the polynomial.

5. 3 (multiplicity 2) and -4

6. -2 (multiplicity 3) and 1 (multiplicity 2)

7. -7 (multiplicity 3), $\frac{2}{3}$, and -5

8. $\frac{1}{3}$ (multiplicity 2), 5, and -1

9. $(2 - 3i)$, $(2 + 3i)$, -4 (multiplicity 2)

10. $i\sqrt{3}$ (multiplicity 2), $-i\sqrt{3}$ (multiplicity 2), and 4 (multiplicity 3)

B

Given the indicated polynomials, what are the possible combinations of real and complex zeros?

11. $P(x) = 2x^3 - 3x^2 + x - 5$

12. $P(x) = 2x^4 - 2x^3 + x - 8$

13. $P(x) = 3x^6 - 5x^5 + 3x^2 - 4$

14. $P(x) = x^5 - 2x^3 + 5x^2 - 6$

Write $P(x)$ as a product of first-degree factors.

15. $P(x) = x^3 + 9x^2 + 24x + 16$; -1 is a zero

16. $P(x) = x^3 - 4x^2 - 3x + 18$;   3 is a double zero

17. $P(x) = x^4 - 1$;   1 and -1 are zeros

18. $P(x) = x^4 + 2x^2 + 1$; i is a double zero

19. $P(x) = 2x^3 - 17x^2 + 90x - 41$; $\frac{1}{2}$ is a zero

20. $P(x) = 3x^3 - 10x^2 + 31x + 26$; $-\frac{2}{3}$ is a zero

Given the indicated equations, what are the possible combinations of real and complex solutions?

21. $x^4 - 3x^3 + 5x - 6 = 0$

22. $2x^3 - 4x^2 - x + 3 = 0$

23. $x^6 - 3x^4 + x^3 - x - 7 = 0$

24. $4x^5 + x^4 - 5x^2 - x + 3 = 0$

Multiply.

25. $[x - (4 - 5i)][x - (4 + 5i)]$ **26.** $[x - (2 - 3i)][x - (2 + 3i)]$

27. $[x - (a + bi)][x - (a - bi)]$ **28.** $(x - bi)(x + bi)$

C In Problems 19–32 find two other zeros of $P(x)$, given the indicated zeros.

29. $P(x) = x^3 - 5x^2 + 4x + 10$; $3 - i$ is one zero $3+i$

30. $P(x) = x^3 + x^2 - 4x + 6$; $1 + i$ is one zero $1-i$

31. $P(x) = x^3 - 3x^2 + 25x - 75$; $-5i$ is one zero

32. $P(x) = x^3 + 2x^2 + 16x + 32$; $4i$ is one zero $-4i$

33. The solutions to the equation $x^3 - 1 = 0$ are all the cube roots of 1.
(A) How many cube roots of 1 are there?
(B) 1 is obviously a cube root of 1; find all others.

34. The solutions to the equation $x^3 - 8 = 0$ are all the cube roots of 8.
(A) How many cube roots of 8 are there?
(B) 2 is obviously a cube root of 8; find all others.

35. If P is a polynomial function of degree n, with n odd, then what is the maximum number of times the graph of $y = P(x)$ can cross the x axis? What is the minimum number of times?

36. Answer the questions in Problem 35 for n even.

37. Given $P(x) = x^2 + 2ix - 5$ with $2 - i$ a zero, show that $2 + i$ is not a zero of $P(x)$. Does this contradict Theorem 6? Explain.

38. If $P(x)$ and $G(x)$ are two polynomials of degree n, and if $P(x) = G(x)$ for more than n values oi x, then how are $P(x)$ and $Q(x)$ related?

Section 3-5 Isolating Real Zeros

- Descartes' Rule of Signs
- Bounding Real Zeros
- Sign Changes in $P(x)$

For the rest of this chapter we will focus on the problem of finding real zeros of polynomials with real coefficients. Three theorems will help us greatly in this regard. The first theorem gives us useful information about the possible number of real zeros of a given polynomial; the second theorem tells us how to determine a finite interval that contains all the real zeros, if they exist; and the third theorem will help us isolate particular zeros further within this interval.

- ### Descartes' Rule of Signs

When the terms of a polynomial with real coefficients are arranged in order of descending powers, we say that a **variation in sign** occurs if two successive terms have opposite signs. Missing terms (terms with 0 coefficients) are ignored. For a given polynomial $P(x)$ we are going to be interested in the total number of variations in sign in both $P(x)$ and $P(-x)$.

EXAMPLE 8   If $P(x) = 3x^4 - 2x^3 + 3x - 5$, how many variations in sign are in $P(x)$ and in $P(-x)$?

Solution $P(x) = 3x^4 - 2x^3 + 3x - 5$ Three variations in sign

$$P(-x) = 3x^4 + 2x^3 - 3x - 5 \quad \text{One variation in sign}$$

PROBLEM 8   If $P(x) = 2x^5 - x^4 - x^3 + x + 5$, how many variations in sign are in $P(x)$ and in $P(-x)$?

The number of variations in sign for $P(x)$ and for $P(-x)$ gives us useful information about the number of real zeros of a polynomial with real coefficients. In 1636, René Descartes, a French philosopher-mathematician, gave the first proof of a simplified version of a theorem that now bears his name. We state Theorem 7 without proof, since a proof is beyond the scope of this book.

THEOREM 7 | Descartes' Rule of Signs

Given a polynomial $P(x)$ with real coefficients:

1. *Positive real zeros.* The number of positive real zeros of $P(x)$ is never greater than the number of variations in sign in $P(x)$ and, if less, then always by an even number.
2. *Negative real zeros.* The number of negative real zeros of $P(x)$ is never greater than the number of variations in sign in $P(-x)$ and, if less, then always by an even number.

EXAMPLE 9   What can you say about the number of positive and negative real zeros of:

(A) $P(x) = 3x^4 - 2x^3 + 3x - 5$
(B) $Q(x) = 2x^6 + x^4 - x + 3$

Solution   (A) $P(x) = 3x^4 - 2x^3 + 3x - 5$ Three variations in sign
                  $P(x) = 3x^4 + 2x^3 - 3x - 5$ One variation in sign

Positive real zeros: three or one

Negative real zeros: one

(B) $Q(x) = 2x^6 + x^4 - x + 3$ Two variations in sign
          $Q(-x) = 2x^6 + x^4 + x + 3$ No variations in sign

Positive real zeros: two or zero

Negative real zeros: none

PROBLEM 9   What can you say about the number of positive and negative real zeros of:

(A) $P(x) = 4x^5 + 2x^4 - x^3 + x - 5$ (B) $Q(x) = x^3 + 3x^2 + 5$

■ Bounding Real Zeros

Any number that is greater than or equal to the largest zero of a polynomial is called an **upper bound of the zeros** of the polynomial; any number that is less than or equal to the smallest zero of a polynomial is called a **lower bound of the zeros** of the polynomial. Theorem 8 enables us to determine upper and lower bounds of all real zeros of a polynomial with real coefficients.

THEOREM 8

Upper and Lower Bounds of Real Zeros

Given a polynomial $P(x)$ with real coefficients, degree $n \geq 1$, and the coefficient of the nth-degree term positive. Let $P(x)$ be divided by $x - r$ using synthetic division.

1. *Upper bound.* If $r > 0$ and all numbers in the quotient row of the synthetic division are nonnegative, then r is an upper bound of the zeros of $P(x)$.
2. *Lower bound.* If $r < 0$ and all numbers in the quotient row of the synthetic division alternate in sign, then r is a lower bound of the zeros of $P(x)$.

[*Note:* In this lower-bound test, if 0 appears in one or more places in the quotient row, the sign in front of it can be considered either positive or negative.]

We sketch a proof of part 1 of Theorem 8. The proof of part 2 is similar, only a little more difficult.

If all the numbers in the quotient row of the synthetic division are nonnegative after dividing $P(x)$ by $x - r$, then

$$P(x) = (x - r)Q(x) + R$$

where the coefficients of $Q(x)$ are nonnegative and R is nonnegative. If $x > r > 0$, then $x - r > 0$ and $Q(x) > 0$; hence,

$$P(x) = (x - r)Q(x) + R > 0$$

Thus, $P(x)$ cannot be 0 for any x greater than r, and r is an upper bound for the real zeros of $P(x)$.

EXAMPLE 10   Find the smallest positive integer and the largest negative integer that, by Theorem 8, are upper and lower bounds, respectively, for the real zeros of

$$P(x) = x^3 - 3x^2 - 18x + 4$$

Solution   An easy way to locate these upper and lower bounds, particularly if the coefficients of $P(x)$ are not too large, is to test $r = 1, 2, 3, \ldots$ until the quotient row turns nonnegative; then test $r = -1, -2, -3, \ldots$ until the quotient row alternates in sign. The resulting table will provide side benefits as we will see later.

		1	-3	-18	4
	1	1	-2	-20	-16
	2	1	-1	-20	-36
	3	1	0	-18	-50
	4	1	1	-14	-52
	5	1	2	-8	-36
UB	6	1	3	0	4
	-1	1	-4	-14	18
	-2	1	-5	-8	20
	-3	1	-6	0	4
LB	-4	1	-7	10	-36

← $\begin{cases}\text{This quotient row is nonnegative; hence,}\\ \text{6 is an upper bound (UB).}\end{cases}$ (at row 6)

← $\begin{cases}\text{This quotient row alternates in sign; hence,}\\ -4 \text{ is a lower bound (LB).}\end{cases}$ (at row -4)

Because of Theorem 8, we now know that all real zeros of $P(x) = x^3 - 3x^2 - 18x + 4$ (or all real solutions of $x^3 - 3x^2 - 18x + 4 = 0$) lie between -4 and 6.

PROBLEM 10   Repeat Example 10 for $P(x) = x^3 - 4x^2 - 5x + 8$.

■ **Sign Changes in $P(x)$**

Observing sign changes in a polynomial $P(x)$ with real coefficients as x is replaced with different real numbers leads to the further isolation of real zeros of $P(x)$. Recall that a polynomial function P (with real coefficients) is continuous everywhere; that is, the graph of $y = P(x)$ has no holes or breaks (Fig. 2). This property of polynomial functions is the basis of Theorem 9.

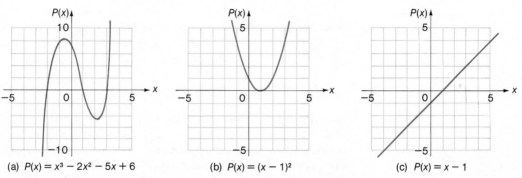

(a) $P(x) = x^3 - 2x^2 - 5x + 6$ (b) $P(x) = (x - 1)^2$ (c) $P(x) = x - 1$

FIGURE 2   The x intercepts of the graph of $y = P(x)$ are the real zeros of $P(x)$

THEOREM 9	Location Theorem
If $P(x)$ is a polynomial with real coefficients, and if $P(a)$ and $P(b)$ are of opposite sign, then there is at least one real zero between a and b.

Notice in Figure 2(a) that $P(-3) < 0$ and $P(4) > 0$ and there are three real zeros between -3 and 4. Also, in Figure 2(c) $P(-3) < 0$ and $P(4) > 0$ and there is one zero between -3 and 4. Since the graph of a polynomial function P with real coefficients is continuous, and if $P(a)$ and $P(b)$ are of opposite sign, the graph of $y = P(x)$ must cross the x axis at least once for x between a and b.

The converse of Theorem 9 is false; that is, if $P(x)$ has at least one real zero, then $P(x)$ may or may not change sign as x crosses a zero. Compare Figure 2(b) and (c). Both functions have zeros at $x = 1$, but the first is never negative, whereas the second is negative for $x < 1$ and positive for $x > 1$.

EXAMPLE 11   Show that there is at least one real zero of

$$P(x) = x^4 - 2x^3 - 6x^2 + 6x + 9$$

between 1 and 2.

Solution   Show that $P(1)$ and $P(2)$ have opposite signs.

	1	-2	-6	6	9
1	1	-1	-7	-1	$8 = P(1)$
2	1	0	-6	-6	$-3 = P(2)$

Since $P(1)$ and $P(2)$ have opposite signs, there is at least one real zero between 1 and 2 (Theorem 9).

PROBLEM 11   Show that there is at least one real zero of

$$P(x) = 2x^4 - 3x^3 - 3x - 4$$

between 2 and 3.

Answers to Matched Problems
8. Two in $P(x)$ and three in $P(-x)$

9. (A) Positive real zeros: three or one
             Negative real zeros: two or none
        (B) Positive real zeros: none
             Negative real zeros: one

10. Lower bound: -2; upper bound: 5

11. $P(2) = -2$ and $P(3) = 68$, and the conclusion follows from Theorem 9.

Exercise 3-5 ■ A *Using Descartes' rule of signs, what can you say about the number of positive and negative zeros of each of the following polynomials?*

1. $P(x) = 2x^2 + x - 4$ **2.** $Q(x) = 3x^2 - x - 5$

3. $M(x) = 7x^2 + 2x + 4$ **4.** $N(x) = -3x^2 - 2x - 1$

5. $Q(x) = 2x^3 - 4x^2 + x - 3$ **6.** $P(x) = x^3 + 7x^2 - x + 2$

Find the smallest positive integer and the largest negative integer that, by Theorem 8, are upper and lower bounds, respectively, for the real zeros of each of the following polynomials.

7. $P(x) = x^2 - 2x + 3$ **8.** $Q(x) = x^2 - 3x - 2$

9. $M(x) = x^3 - 3x + 5$ **10.** $R(x) = x^3 - 2x^2 + 3$

11. $M(x) = x^4 - x^2 + 3x + 2$ **12.** $N(x) = x^4 - 2x^3 + 4x - 3$

Show, using Theorem 9, that for each polynomial there is at least one real zero between the given values of a and b.

Location Theorem

13. $P(x) = x^2 - 3x - 2;\quad a = 3,\quad b = 4$

14. $Q(x) = x^2 - 3x - 2;\quad a = -1,\quad b = 0$

15. $P(x) = x^3 - 3x + 5;\quad a = -3,\quad b = -2$

16. $P(x) = x^3 - 2x^2 - 4;\quad a = 2,\quad b = 3$

17. $Q(x) = x^3 - 3x^2 - 3x + 9;\quad a = 1,\quad b = 2$

18. $G(x) = x^3 - 3x^2 - 3x + 9;\quad a = -2,\quad b = -1$

B *For each polynomial P(x),*

  (A) *Discuss the possible number of real zeros using Descartes' rule of signs.*

  (B) *Find the smallest and largest integers that are, respectively, upper and lower bounds of the zeros of P(x) according to Theorem 8.*

  (C) *Discuss the location of real zeros within the lower and upper bound interval by applying Theorem 9 to integer values of x within this interval.*

19. $P(x) = x^3 - x^2 - 6x + 6$ **20.** $P(x) = x^3 - 3x^2 - 2x + 6$

21. $P(x) = x^3 - 2x - 6$ **22.** $P(x) = x^3 - 3x^2 - 5$

23. $P(x) = x^4 + 4x^3 - 2x^2 - 12x - 3$

24. $P(x) = x^4 - 4x^3 + 8x - 4$

25. $P(x) = x^5 - 3x^3 + 2x - 5$

26. $P(x) = 2x^5 - 5x^4 - 2x + 5$

27. Prove that $P(x) = x^4 + 3x^2 - x - 5$ has two complex and two real zeros, without finding the zeros.

28. Prove that $P(x) = x^3 + 3x^2 + 5$ has one negative real zero and two complex zeros, without finding the zeros.

29. Prove that the graph of $P(x) = x^5 + 3x^3 + x$ crosses the x axis only once without graphing $y = P(x)$.

30. Prove that the graph of $P(x) = x^4 + 3x^2 + 7$ does not cross the x axis at all. Do not graph $y = P(x)$.

Section 3-6   Finding Rational Zeros

■ Rational Zero Theorem
■ Strategy for Finding Rational Zeros

■ **Rational Zero Theorem**

We start our investigation with a quadratic function whose zeros can be found easily by factoring. From this example, we will point out a relationship that generally holds for all polynomials with integer coefficients.

$$P(x) = 6x^2 - 13x - 5 = (2x - 5)(3x + 1)$$

Zeros of $P(x)$: $\dfrac{5}{2}$, $-\dfrac{1}{3}$ or $\dfrac{-1}{3}$

Notice that the numerator of each zero (5 and -1) is a factor of -5, the constant term in $P(x)$. The denominator of each zero (2 and 3) is a factor of 6, the coefficient of the highest-degree term in $P(x)$. These observations are generalized in Theorem 10.

THEOREM 10 | **Rational Zero Theorem**

If the rational number b/c, in lowest terms, is a zero of the polynomial

$$P(x) = a_n x^n + a_{n-1} x^{n-1} + \cdots + a_1 x + a_0$$

with integer coefficients, then b must be a factor of a_0 [the constant term in $P(x)$] and c must be a factor of a_n [the coefficient of the highest-degree term in $P(x)$].

The proof of Theorem 10 is not difficult, and is instructive, so we sketch it here. Since b/c is a zero of $P(x)$,

$$a_n\left(\frac{b}{c}\right)^n + a_{n-1}\left(\frac{b}{c}\right)^{n-1} + \cdots + a_1\left(\frac{b}{c}\right) + a_0 = 0 \tag{1}$$

If we multiply both members of equation (1) by c^n, we obtain

$$a_n b^n + a_{n-1} b^{n-1} c + \cdots + a_1 bc^{n-1} + a_0 c^n = 0 \qquad (2)$$

which can be written in the form

$$a_n b^n = c(-a_{n-1} b^{n-1} - \cdots - a_0 c^{n-1}) \qquad (3)$$

Thus c is a factor of $a_n b^n$, since the expression in parentheses is an integer. (Why?) And since b and c are **relatively prime** (that is, have no common factors other than ± 1), b^n and c must be relatively prime; hence c must be a factor of a_n. [Divide both sides of equation (3) by c to see why.]

Now, if we solve equation (2) for $a_0 c^n$ and factor b out of the right side, we have

$$a_0 c^n = b(-a_n b^{n-1} - \cdots - a_1 c^{n-1})$$

we see that b is a factor of $a_0 c^n$ and hence a factor of a_0, since b and c are relatively prime.

We emphasize that **Theorem 10 does not say that a polynomial with integers as coefficients has rational zeros; it simply states that if it does, then they must meet the conditions stated in the theorem. In short, it enables us to list a set of rational numbers that must include all rational zeros if they exist.**

EXAMPLE 12　List all possible rational zeros for

$$P(x) = 2x^4 - 3x^3 + x - 9$$

Solution　If b/c (in lowest terms) is a rational zero of $P(x)$, then b must be a factor of -9 and c must be a factor of 2.

Possible values of b (factors of -9):　$\pm 1, \pm 3, \pm 9$ 　　(4)

Possible values of c (factors of 2):　$\pm 1, \pm 2$ 　　(5)

Thus, writing all possible fractions b/c where b is from (4) and c is from (5), we have

Possible rational zeros for $P(x)$:　$\pm 1, \pm 3, \pm 9, \pm \frac{1}{2}, \pm \frac{3}{2}, \pm \frac{9}{2}$ 　　(6)

$$\left[\text{Note:} \quad \frac{\pm 1}{\pm 1} = \pm 1, \quad \frac{\pm 3}{\pm 1} = \pm 3, \quad \text{etc.} \right]$$

Thus, if $P(x)$ has rational zeros, they must be in list (6).

PROBLEM 12　List all possible rational zeros for

$$P(x) = x^3 + 2x^2 - 5x - 6$$

■ **Strategy for Finding Rational Zeros**

If a polynomial is of first or second degree, then we can always find *all* of its zeros using methods discussed in Chapter 1. With all the tools we have developed in this chapter, we are now ready to state a strategy that will efficiently lead to all rational zeros of polynomials of degree $n \geq 3$ with integer coefficients, if they exist. Of course, we could just test each of the possible rational zeros that result from the rational zero theorem. However, we can make the process more efficient by using many of the other properties of polynomials discussed earlier in this chapter. In addition, some of these same properties and procedures will help us locate irrational zeros in the next section.

Strategy for Finding Rational Zeros

Assume $P(x)$ is a polynomial of degree $n \geq 3$ with integer coefficients.

Step 1. List the possible rational zeros of $P(x)$ using Theorem 10.

Step 2. List the possible number of positive and negative real zeros using Descartes' rule of signs.

Step 3. Test the possible rational zeros from the list in step 1 being guided by the results of step 2 and the following steps a–e:

[*Note:* If a rational zero r is found in any of the steps a through e below, write

$$P(x) = (x - r)Q(x)$$

and immediately proceed to find the rational zeros for $Q(x)$, the reduced polynomial relative to $P(x)$. If $Q(x)$ is of degree $n \geq 3$, return to step 1 using $Q(x)$ in place of $P(x)$. If $Q(x)$ is quadratic, find *all* of its zeros using standard methods for solving quadratic equations.]

  a. Use the results of Descartes' rule of signs (step 2) in conjunction with steps b through e to further eliminate possible rational zeros from the list in step 1.
  b. Form a synthetic division table by testing the possible integer zeros from the list in step 1.
  c. Watch for sign changes in $P(x)$ and test any fractions from the list in step 1 that are between integers that produce sign changes in $P(x)$.
  d. Watch for test values that are lower or upper bounds for the real zeros of $P(x)$ — add more integer values if necessary to locate lower and upper bounds. Eliminate any rational numbers from the list in step 1 that are below the lower bound or above the upper bound.
  e. Test the remaining possible rational zeros from step 1. If none of the possible rational zeros from step 1 are zeros, we conclude that $P(x)$ has no rational zeros. The zeros must then be irrational or nonreal complex.

Let us see how the strategy works in several concrete examples.

EXAMPLE 13   Find all rational zeros of $P(x) = 2x^3 - x^2 - 8x + 4$.

Solution   *Step 1.*   List the possible rational zeros.

$$\pm 1, \quad \pm 2, \quad \pm 4, \quad \pm \tfrac{1}{2}$$

Step 2.   List the possible number of positive and negative real zeros.

$$P(x) = 2x^3 - x^2 - 8x + 4 \qquad \text{Two variations in sign}$$
$$P(-x) = -2x^3 - x^2 + 8x + 4 \qquad \text{One variation in sign}$$

Positive real zeros: two or none

Negative real zeros: one

Step 3.   Test the possible rational zeros listed in step 1.

	2	−1	−8	4	
1	2	1	−7	−3	
2	2	3	−2	0	2 is a real zero

We have found a real zero, so we write:

$$P(x) = (x - r)Q(x)$$
$$= (x - 2)(2x^2 + 3x - 2)$$

The zeros of $Q(x)$, since it is quadratic, can be found by solving $Q(x) = 0$:

$$2x^2 + 3x - 2 = 0$$
$$(2x - 1)(x + 2) = 0$$
$$2x - 1 = 0 \quad \text{or} \quad x + 2 = 0$$
$$x = \tfrac{1}{2} \qquad\qquad x = -2$$

Thus, the rational zeros of $P(x)$ are ± 2 and $\tfrac{1}{2}$.

PROBLEM 13   Find all rational zeros for $P(x) = 3x^3 + 10x^2 + x - 6$.

EXAMPLE 14   Find *all* zeros for $P(x) = 2x^3 - 5x^2 - 8x + 6$.

Solution   *Step 1.*   List the possible rational zeros.

$$\pm 1, \quad \pm 2, \quad \pm 3, \quad \pm 6, \quad \pm \tfrac{1}{2}, \quad \pm \tfrac{3}{2}$$

Step 2.   List the possible number of positive and negative real zeros.

$$P(x) = 2x^3 - 5x^2 - 8x + 6 \qquad \text{Two variations in sign}$$
$$P(-x) = -2x^3 - 5x^2 + 8x + 6 \qquad \text{One variation in sign}$$

Positive real zeros: two or none

Negative zeros: one

Step 3.   Test the possible rational zeros listed in step 1.

[*Note:* We may include other numbers as needed to isolate zeros further.]

	2	−5	−8	6	
1	2	−3	−11	−5	
2	2	−1	−10	−14	
3	2	1	−5	−9	There is an irrational zero between 3 and 4
UB 4	2	3	4	22	(Why?); also 4 is an upper bound.
−1	2	−7	−1	7	Real zero between −1 and 1
0	2	−5	−8	6	Real zero between 0 and 1
$\frac{1}{2}$	2	−4	−10	1	Irrational zero between $\frac{1}{2}$ and 1 (Why?)
LB −2	2	−9	10	−14	Real zero between −1 and −2; also, −2 is a lower bound
$-\frac{3}{2}$	2	−8	4	0	$-\frac{3}{2}$ is a rational zero

We now know that $P(x)$ has three real zeros: $-\frac{3}{2}$ is the negative zero, and the table tells us that there are two positive irrational zeros, one between $\frac{1}{2}$ and 1 and the other between 3 and 4. We write

$$P(x) = (x - r)Q(x) = (x + \tfrac{3}{2})(2x^2 - 8x + 4)$$

and find the zeros of the reduced polynomial $Q(x)$ by solving $Q(x) = 0$:

$$2x^2 - 8x + 4 = 0$$
$$x^2 - 4x + 2 = 0$$

$$x = \frac{4 \pm \sqrt{16 - 4(1)(2)}}{2}$$

$$= \frac{4 \pm 2\sqrt{2}}{2} = 2 \pm \sqrt{2}$$

The zeros of $P(x)$ are $-\frac{3}{2}$ and $2 \pm \sqrt{2}$ or, to two decimal places:

−1.5,   0.59,   3.41   Notice that 0.59 is between $\frac{1}{2}$ and 1, and 3.41 is between 3 and 4, as was predicted from the synthetic division table.

PROBLEM 14   Find *all* zeros for $P(x) = 2x^3 - 7x^2 + 6x + 5$.

EXAMPLE 15   Find all rational zeros for $P(x) = x^4 - 7x^3 + 17x^2 - 17x + 6$.

Solution   *Step 1.*   List the possible rational zeros.

$$\pm 1, \quad \pm 2, \quad \pm 3, \quad \pm 6$$

Step 2.   List the possible number of positive and negative real zeros.

$$P(x) = x^4 - 7x^3 + 17x^2 - 17x + 6 \quad \text{Four variations in sign}$$
$$P(-x) = x^4 + 7x^3 + 17x^2 + 17x + 6 \quad \text{No variation in sign}$$

Positive real zeros: four, two, or none

Negative real zeros: none

We can eliminate all the negative numbers from the list in step 1, since there are no negative real zeros.

Step 3.   Test the possible rational zeros listed in step 1.

$$\begin{array}{r|rrrrr} & 1 & -7 & 17 & -17 & 6 \\ \hline 1 & 1 & -6 & 11 & -6 & 0 \end{array} \quad \text{1 is a zero}$$

Write

$$\begin{aligned} P(x) &= (x - r)Q(x) \\ &= (x - 1)(x^3 - 6x^2 + 11x - 6) \end{aligned}$$

and return to step 1 for the reduced polynomial

$$Q(x) = x^3 - 6x^2 + 11x - 6$$

Step 1.   List the possible rational zeros.

1,   2,   3,   6   The negatives of these were eliminated in step 2 above.

We now go directly to step 3, since step 2 does not add a lot of additional information.

Step 3.   Test the possible rational zeros listed above.

$$\begin{array}{r|rrrr} & 1 & -6 & 11 & -6 \\ \hline 1 & 1 & -5 & 6 & 0 \end{array} \quad \text{1 is a zero}$$

Write

$$\begin{aligned} Q(x) &= (x - r)Q_1(x) \\ &= (x - 1)(x^2 - 5x + 6) \end{aligned}$$

and find the zeros of the reduced polynomial $Q_1(x)$ by solving $Q_1(x) = 0$:

$$x^2 - 5x + 6 = 0$$
$$(x - 2)(x - 3) = 0$$
$$x = 2 \quad \text{or} \quad 3$$

The zeros of $P(x)$ are 1 (multiplicity 2), 2, and 3.

PROBLEM 15   Find all rational zeros for $P(x) = x^4 + 8x^3 + 23x^2 + 28x + 12.$

With a little practice and ingenuity (educated guessing) you will be able to reduce the number of steps and effort required to find rational zeros, if they exist. To develop this efficiency, you must work problems yourself. The more you work, the easier and faster the process will become.

Answers to Matched Problems   **12.** $\pm 1, \quad \pm 2, \quad \pm 3, \quad \pm 6$ **13.** $-3, -1, \frac{2}{3}$

14. $-\frac{1}{2}, 2 \pm i$ **15.** $-3, -2$ (multiplicity 2), -1

Exercise 3-6 ■

A   *For each polynomial:*
(A) List all possible rational zeros (Theorem 10).
(B) Find all rational zeros. If there are no rational zeros, say so.

1. $P(x) = x^3 - 2x^2 - 5x + 6$

2. $P(x) = x^3 + 3x^2 - 6x - 8$

3. $P(x) = 3x^3 - 11x^2 + 8x + 4$

4. $P(x) = 2x^3 + x^2 - 4x - 3$

5. $P(x) = 12x^3 - 16x^2 - 5x + 3$

6. $P(x) = 2x^3 - 9x^2 + 14x - 5$

7. $P(x) = 3x^3 + 7x^2 - 10x - 4$

8. $P(x) = 2x^3 - 5x^2 - 2x + 15$

B   **9.** $P(x) = x^3 - 3x^2 + 6$

10. $P(x) = x^3 - 3x + 1$

11. $P(x) = x^4 - 2x^3 - 2x^2 + 8x - 8$

12. $P(x) = 2x^4 + 5x^3 - 7x^2 - 6x + 4$

13. $P(x) = 3x^4 - 8x^3 - 6x^2 + 17x + 6$

14. $P(x) = 12x^4 - 8x^3 - 37x^2 + 7x + 6$

Find all roots (rational, irrational, and complex) for each polynomial equation.

15. $2x^3 - 5x^2 + 1 = 0$

16. $2x^3 - 10x^2 + 12x - 4 = 0$

17. $x^4 + 4x^3 - x^2 - 20x - 20 = 0$

18. $x^4 - 4x^2 - 4x - 1 = 0$

19. $2x^5 - 3x^4 - 2x + 3 = 0$

20. $x^4 - 2x^2 - 16x - 15 = 0$

C   *Write each polynomial as a product of linear factors.*

21. $P(x) = 6x^3 + 13x^2 - 4$
22. $P(x) = 6x^3 - 17x^2 - 4x + 3$
23. $P(x) = x^3 + 2x^2 - 9x - 4$
24. $P(x) = x^3 - 8x^2 + 17x - 4$

Show that each of the following real numbers is not rational by writing an appropriate polynomial and making use of Theorem 10.

25. $\sqrt{6}$ **26.** $\sqrt{12}$ **27.** $\sqrt[3]{5}$ **28.** $\sqrt[5]{8}$

Solve each inequality using the factoring method discussed in Section 1-6. [Hint: Find the zeros of the corresponding polynomial function; then use the factor theorem.]

29. $x^2 \le 4x - 1$ **30.** $x^2 > 2x + 1$
31. $2x^3 + 6 \ge 13x - x^2$ **32.** $5x^3 - 3x^2 < 10x - 6$

Section 3-7   Approximating Irrational Zeros

How do we find irrational zeros of polynomials of degree greater than two? If the polynomial has no rational zeros but has irrational zeros, then the rational zero theorem discussed in the last section will be of little help. Since general methods for finding zeros of third- and fourth-degree polynomials are long and involved, and no general method exists for finding zeros of polynomials of degree higher than four, we introduce the *method of successive approximation*, which will enable us to approximate irrational zeros to any decimal accuracy desired.

Practically speaking, this method is used only to approximate irrational zeros to a couple of decimal places, since more efficient methods are available, particularly after you have had some calculus. The method of successive approximation does have the advantage, however, of being easily understood and easily remembered, and of providing a foundation for understanding the more refined methods that you are likely to encounter later. The tedious aspect of its application is reduced sharply by use of a hand calculator, even an inexpensive one.

We outline a general strategy for finding all real zeros (rational zeros exactly and irrational zeros approximately).

Strategy for Finding All Real Zeros of a Polynomial $P(x)$

Step 1. Find all rational zeros by the methods of Section 3-6 and set them aside.

Step 2. Write the final reduced polynomial $Q(x)$ resulting from step 1.

Step 3. Use Descartes' rule of signs to determine the possible number of positive and negative real zeros left after step 1.

Step 4. Isolate real zeros further. Form a synthetic division table using $Q(x)$ to:
  a. Locate lower and upper bounds for irrational zeros.
  b. Isolate irrational zeros, if possible, between successive integers by observing sign changes in $Q(x)$.
  c. A graph of $y = Q(x)$ between lower and upper bounds may be useful.

Step 5. Approximate irrational zeros (located approximately in step 4b) to desired accuracy using the *method of successive approximation* (described in Example 16).

EXAMPLE 16 Find all real zeros for $P(x) = 2x^4 + x^3 + 4x^2 - 6x - 4$. (Approximate irrational zeros to two decimal places.)

Solution *Step 1.* Find rational zeros, if any.

Using methods of the preceding section, we find $-\frac{1}{2}$ to be the only rational zero:

$$-\tfrac{1}{2} \,\big|\; \begin{array}{ccccc} 2 & 1 & 4 & -6 & -4 \\ \hline 2 & 0 & 4 & -8 & 0 \end{array}$$

We write

$P(x) = (x + \tfrac{1}{2})(2x^3 + 4x - 8)$ Factor 2 out of the second factor.

$\quad = 2(x + \tfrac{1}{2})(x^3 + 2x - 4)$

Step 2. Write the final reduced polynomial from step 1.

$Q(x) = x^3 + 2x - 4$

Step 3. Determine the possible positive and negative real zeros for $Q(x)$.

$Q(x) = x^3 + 2x - 4$ One variation in sign

$Q(-x) = -x^3 - 2x - 4$ No variation in sign

Positive real zeros: one

Negative real zeros: none

Now we know for certain that there is one positive irrational zero.

Step 4. Isolate real zeros further. We form a synthetic division table to try to isolate the irrational zero between two integers by observing sign changes in $P(x)$. We start with $x = 0$, since the irrational zero is positive (thus, 0 is a lower bound).

From the table, we see that the irrational zero must be between 1 and 2. To obtain a clearer picture of where the zero lies in this interval, we sketch a graph of $y = Q(x)$ for the interval [0, 2]—see figure (a). (This graphing step is optional in practice.)

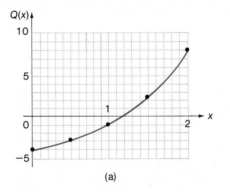

(a)

Step 5. Approximate the irrational zero.

We approximate this irrational zero to two decimal places using the *method of successive approximation.*

The Method of Successive Approximation

The first step is to divide the unit interval containing the zero into tenths. To determine the first decimal of the zero, we locate the interval in this subdivision within which $P(x)$ changes sign. We repeat this process, dividing this subinterval into ten parts, to locate the second decimal place, and so on. The process can be repeated as long as desired (barring fatigue) to produce a decimal approximation of an irrational zero to any accuracy desired. To obtain an accuracy of two decimal places, we go to the third and round back to the second. In general, we go to one more place than the accuracy desired, and then round back one place.

The irrational zero of $Q(x)$ is between 1 and 2. We divide the interval [1, 2] into tenths, locate $(1, P(1))$ and $(2, P(2))$, and join these two points with a straight line to determine approximately where the zero of $Q(x)$ lies [figure (b)]. [Generally we draw only the straight line; figure (b) contains the approximating straight line and the actual graph of $y = Q(x)$ for comparison.]

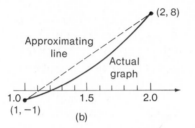

(b)

From figure (b), it appears that the zero is closer to 1 than to 2, so we start from that end. Using synthetic division and a hand calculator, we find $P(1.0)$, $P(1.1)$, $P(1.2)$, and so on, until a sign change occurs:

$$
\begin{array}{r|rrrr}
 & 1 & 0 & 2 & -4 \\
\hline
1.0 & 1 & 1 & 3 & -1 \\
1.1 & 1 & 1.1 & 3.21 & -0.47 \\
1.2 & 1 & 1.2 & 3.44 & 0.13 \\
\end{array}
$$

Zero $\begin{cases} 1.1 \\ 1.2 \end{cases}$ $\left. \begin{array}{c} -0.47 \\ 0.13 \end{array} \right\}$ Sign change

The zero is between 1.1 and 1.2. We now divide the interval from 1.1 to 1.2 into tenths, locate $(1.1, P(1.1))$ and $(1.2, P(1.2))$ approximately, and join these two points with a straight line to determine approximately where the graph of $y = Q(x)$ crosses the x axis [figure (c)].

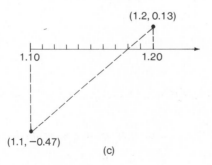

(c)

Test values close to the point of intersection to determine the sign change in $Q(x)$.

	1	0	2	-4
1.16	1	1.16	3.346	-0.119

$$\text{Zero}\begin{cases}1.17 \\ 1.18\end{cases}$$

	1	0	2	-4
1.17	1	1.17	3.369	-0.058 ⎫ Sign change
1.18	1	1.18	3.392	0.003 ⎭

We now have the zero between 1.17 and 1.18. Continuing, we divide this interval into tenths and proceed as above [figure (d)].

	1	0	2	-4
1.178	1	1.178	3.3877	-0.0093

$$\text{Zero}\begin{cases}1.179 \\ 1.180\end{cases}$$

	1	0	2	-4
1.178	1	1.178	3.3877	-0.0093
1.179	1	1.179	3.3900	-0.0031 ⎫ Sign change
1.180	1	1.180	3.3924	0.0030 ⎭

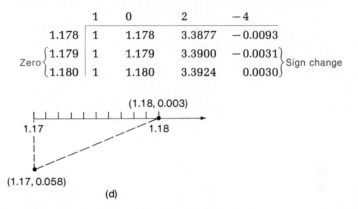

(d)

It is now clear that the irrational zero to two decimal places is 1.18, so the real zeros for the original polynomial

$$P(x) = 2x^4 + x^3 + 4x^2 - 6x - 4$$

are -0.5 and 1.18.

[*Note:* It is useful to observe that the method of successive approximations can be used to find real zeros for functions other than polynomial functions.]

PROBLEM 16   Find all real zeros for $P(x) = x^3 + 2x - 7$. Approximate irrational zeros to one decimal place.

Answers to Matched Problems   **16.** 1.6

Exercise 3-7 ▪   *A hand calculator will prove useful in most of the problems in this exercise.*

A   *Find the irrational zero to one decimal place in the indicated interval.*

1. $P(x) = x^3 - 5x^2 + 3$; [4, 5]
2. $P(x) = x^3 - 5x + 3$; [0, 1]
3. $P(x) = x^3 + x - 1$; [0, 1]
4. $P(x) = x^3 - x^2 - x - 1$; [1, 2]

B *Find all real zeros of each polynomial. Approximate irrational zeros to one-decimal-place accuracy.*

5. $P(x) = x^4 - 6x - 7$

6. $P(x) = x^4 - x^3 + 10x^2 - 28x + 18$

7. $P(x) = 2x^5 - 5x^4 - 7x^3 + 4x^2 + 21x + 9$

8. $P(x) = 2x^4 + 3x^3 + 6x^2 - x - 15$

C *In Problems 9 and 10, isolate each irrational root between successive integers; then approximate the largest root to two decimal places.*

9. $x^3 - 5x^2 + 3 = 0$ **10.** $x^3 + x^2 - 6x - 2 = 0$

11. Show that even though $P(2)$ and $P(3)$ have the same sign, $P(x) = x^4 - 4x^3 + x^2 + 6x + 2$ has at least two real zeros between 2 and 3.

12. Approximate the largest zero of $P(x)$ in Problem 11 in the interval $[2, 3]$ to one decimal place.

APPLICATIONS **13.** *Construction.* An open metal container is to be made from a rectangular piece of sheet metal, 11×9 inches, by cutting out squares of the same size from each corner and bending up the sides. If the volume of the container is to be 72 cubic inches, how large a square should be cut from each corner?

14. *Construction.* A rectangular box has dimensions of $1 \times 1 \times 2$ feet. If each dimension is increased by the same amount, how much should this amount be to triple the volume of the box? Approximate the answer to one decimal place.

15. *Physics—engineering.* In physics it can be shown that a solid buoy in the form of a sphere, with radius r and specific gravity s, $0 < s < 1$, will sink in water to a depth of x as given by the equation

$$x^3 - 3rx^2 + 4r^3s = 0$$

How far will a plastic buoy of radius 1 foot and specific gravity $s = 0.1$ sink? Give the answer accurate to one decimal place.

Section 3-8 Partial Fraction Decomposition

■ Preliminaries
■ Partial Fraction Decomposition

■ Preliminaries

You have now had some experience in combining two or more algebraic fractions into single fractions by addition or subtraction. For example, problems such as

$$\frac{2}{x + 5} + \frac{3}{x - 4} = \frac{5x + 7}{x^2 + x - 20}$$

should be routine. There are several places in more advanced courses, particularly in calculus and differential equations, where it is a great advantage to be able to reverse the process—that is, to be able to express the quotient of two polynomials as the sum of two or more simpler quotients called **partial fractions**. This process of decomposing a quotient into partial fractions, like many reverse processes, is more difficult than the original.

We confine our attention to quotients of the form $P(x)/D(x)$, where $P(x)$ and $D(x)$ are real polynomials. In addition, we will assume that the degree of $P(x)$ is less than the degree of $D(x)$. If the degree of $P(x)$ is greater than or equal to that of $D(x)$, we have only to divide $P(x)$ by $D(x)$ to obtain

$$\frac{P(x)}{D(x)} = Q(x) + \frac{R(x)}{D(x)}$$

where the degree of $R(x)$ is less than that of $D(x)$. For example,

$$\frac{x^4 - 3x^3 + 2x^2 - 5x + 1}{x^2 - 2x + 1} = x^2 - x - 1 + \frac{-6x + 2}{x^2 - 2x + 1}$$

If the degree of $P(x)$ is less than that of $D(x)$, then $P(x)/D(x)$ is called a **proper fraction**. Our task now is to figure out a systematic way to decompose proper fractions into the sum of two or more partial fractions. The following three theorems take care of the problem completely. The first and third theorems are stated without proof.

THEOREM 11

> Two polynomials are equal to each other if and only if the coefficients of like-degree terms are equal.

For example, if

$$5x - 3 = (A + 2B)x + B$$

then

$$B = -3$$
$$A + 2B = 5$$
$$A + 2(-3) = 5$$
$$A = 11$$

THEOREM 12	For a polynomial with real coefficients, there always exists a complete factoring involving only prime linear and/or quadratic factors relative to the set of real numbers.

That Theorem 12 is true can be seen as follows: From earlier theorems in this chapter, we know that an nth-degree polynomial $P(x)$ has n zeros and n linear factors. If the coefficients of $P(x)$ are real, complex zeros occur in conjugate pairs. Thus, if we multiply the factors corresponding to each pair of conjugate complex zeros (when they exist), we will obtain quadratic factors with real coefficients, as can be readily seen as follows (where a and b are real numbers):

Let $a \pm bi$ be two conjugate complex zeros of $P(x)$, then $[x - (a + bi)]$ and $[x - (a - bi)]$ are two linear factors of $P(x)$. Multiplying these two factors, we have

$$[x - (a + bi)][x - (a - bi)] = x^2 - 2ax + (a^2 + b^2)$$

The quadratic is a polynomial with real coefficients.

■ Partial Fraction Decomposition

We are now ready to state Theorem 13, which forms the basis for partial fraction decompositions.

THEOREM 13	Partial Fraction Decomposition

Any reduced proper fraction $P(x)/D(x)$ can be decomposed into the sum of partial fractions as follows:

1. If $D(x)$ has a nonrepeating linear factor of the form $ax + b$, then the partial fraction decomposition of $P(x)/D(x)$ contains a term of the form

$$\frac{A}{ax + b} \qquad A \text{ a constant}$$

2. If $D(x)$ has a k-repeating linear factor of the form $(ax + b)^k$, then the partial fraction decomposition of $P(x)/D(x)$ contains terms of the form

$$\frac{A_1}{ax + b} + \frac{A_2}{(ax + b)^2} + \cdots + \frac{A_k}{(ax + b)^k} \qquad A_1, A_2, \ldots, A_k \text{ constants}$$

---

3. If $D(x)$ has a nonrepeating quadratic factor of the form $ax^2 + bx + c$, the partial fraction decomposition of $P(x)/D(x)$ contains a term of the form

$$\frac{Ax + B}{ax^2 + bx + c} \qquad A \text{ and } B \text{ constants}$$

4. If $D(x)$ has a k-repeating quadratic factor of the form $(ax^2 + bx + c)^k$, then the partial fraction decomposition of $P(x)/D(x)$ contains terms of the form

$$\frac{A_1 x + B_1}{ax^2 + bx + c} + \frac{A_2 x + B_2}{(ax^2 + bx + c)^2} + \cdots + \frac{A_k x + B_k}{(ax^2 + bx + c)^k}$$

$$A_1, \ldots, A_k, \quad B_1, \ldots, B_k \text{ constants}$$

Let us see how the theorem is used to obtain partial fraction decompositions in several examples.

EXAMPLE 17   Decompose $\dfrac{5x + 7}{x^2 + 2x - 3}$ into partial fractions.

Solution   We first try to factor the denominator. If it is irreducible in the real numbers, then we will not be able to go further. In this example the denominator factors, so we apply step 1 from Theorem 13:

$$\frac{5x + 7}{(x - 1)(x + 3)} = \frac{A}{x - 1} + \frac{B}{x + 3} \tag{1}$$

To find the constants A and B we combine the right member of equation (1) to form a single fraction:

$$\frac{A(x + 3) + B(x - 1)}{(x - 1)(x + 3)}$$

and equate the numerator to $5x + 7$. Thus,

$$5x + 7 = A(x + 3) + B(x - 1) \tag{2}$$

We could multiply the right member and find A and B by using Theorem 11, but in this case it is easier to take advantage of the fact that equation (2) is an identity; that is, it must hold for all values of x. In particular, we note that if we let $x = 1$, then the second term of the right member drops out and we can solve for A.

$$5 \cdot 1 + 7 = A(1 + 3) + B(1 - 1)$$

$$12 = 4A$$

$$A = 3$$

Similarly, if we let $x = -3$, the first term will drop out and we find

$$-8 = -4B$$
$$B = 2$$

Hence,

$$\frac{5x + 7}{x^2 + 2x - 3} = \frac{3}{x - 1} + \frac{2}{x + 3}$$

as can easily be checked.

PROBLEM 17   Decompose $\dfrac{7x + 6}{x^2 + x - 6}$ into partial fractions.

EXAMPLE 18   Decompose $\dfrac{6x^2 - 14x - 27}{(x + 2)(x - 3)^2}$ into partial fractions.

Solution   Using steps 1 and 2 from Theorem 13, we write

$$\frac{6x^2 - 14x - 27}{(x + 2)(x - 3)^2} = \frac{A}{x + 2} + \frac{B}{x - 3} + \frac{C}{(x - 3)^2}$$

$$= \frac{A(x - 3)^2 + B(x + 2)(x - 3) + C(x + 2)}{(x + 2)(x - 3)^2}$$

Thus for all x,

$$6x^2 - 14x - 27 = A(x - 3)^2 + B(x + 2)(x - 3) + C(x + 2)$$

If $x = 3$, then

$$-15 = 5C$$
$$C = -3$$

If $x = -2$, then

$$25 = 25A$$
$$A = 1$$

If $x = 0$, then

$$-27 = 9 - 6B - 6$$
$$B = 5$$

Thus,

$$\frac{6x^2 - 14x - 27}{(x + 2)(x - 3)^2} = \frac{1}{x + 2} + \frac{5}{x - 3} - \frac{3}{(x - 3)^2}$$

PROBLEM 18   Decompose $\dfrac{x^2 + 11x + 15}{(x - 1)(x + 2)^2}$ into partial fractions.

EXAMPLE 19   Decompose $\dfrac{5x^2 - 8x + 5}{(x - 2)(x^2 - x + 1)}$ into partial fractions.

Solution   First we see that the quadratic in the denominator is irreducible in the real numbers and then use steps 1 and 3 from Theorem 13 to write

$$\frac{5x^2 - 8x + 5}{(x - 2)(x^2 - x + 1)} = \frac{A}{x - 2} + \frac{Bx + C}{x^2 - x + 1}$$

$$= \frac{A(x^2 - x + 1) + (Bx + C)(x - 2)}{(x - 2)(x^2 - x + 1)}$$

Thus, for all x,

$$5x^2 - 8x + 5 = A(x^2 - x + 1) + (Bx + C)(x - 2)$$

If $x = 2$, then

$$9 = 3A$$
$$A = 3$$

If $x = 0$, then

$$5 = 3 - 2C$$
$$C = -1$$

If $x = 1$, then

$$2 = 3 + (B - 1)(-1)$$
$$B = 2$$

Hence,

$$\frac{5x^2 - 8x + 5}{(x - 2)(x^2 - x + 1)} = \frac{3}{x - 2} + \frac{2x - 1}{x^2 - x + 1}$$

PROBLEM 19   Decompose $\dfrac{7x^2 - 11x + 6}{(x - 1)(2x^2 - 3x + 2)}$ into partial fractions.

EXAMPLE 20   Decompose $\dfrac{x^3 - 4x^2 + 9x - 5}{(x^2 - 2x + 3)^2}$ into partial fractions.

Solution   Since $x^2 - 2x + 3$ is irreducible in the real numbers, we proceed to use step 4 from Theorem 13 to write

$$\frac{x^3 - 4x^2 + 9x - 5}{(x^2 - 2x + 3)^2} = \frac{Ax + B}{x^2 - 2x + 3} + \frac{Cx + D}{(x^2 - 2x + 3)^2}$$

$$= \frac{(Ax + B)(x^2 - 2x + 3) + Cx + D}{(x^2 - 2x + 3)^2}$$

Thus, for all x,

$$x^3 - 4x^2 + 9x - 5 = (Ax + B)(x^2 - 2x + 3) + Cx + D$$

Multiplying out and rearranging the right member, we obtain

$$x^3 - 4x^2 + 9x - 5 = Ax^3 + (B - 2A)x^2 + (3A - 2B + C)x + (3B + D)$$

Now we use Theorem 11 to equate coefficients of like-powered terms.

$$A = 1$$
$$B - 2A = -4$$
$$3A - 2B + C = 9$$
$$3B + D = -5$$

From these equations we easily find that $A = 1$, $B = -2$, $C = 2$, and $D = 1$. And now we can write

$$\frac{x^3 - 4x^2 + 9x - 5}{(x^2 - 2x + 3)^2} = \frac{x - 2}{x^2 - 2x + 3} + \frac{2x + 1}{(x^2 - 2x + 3)^2}$$

PROBLEM 20   Decompose $\dfrac{3x^3 - 6x^2 + 7x - 2}{(x^2 - 2x + 2)^2}$ into partial fractions.

It should be clear that one of the key problems in decomposing quotients of polynomials into partial fractions is factoring the denominator into linear and quadratic factors with real coefficients. The material in the earlier parts of this chapter can be put to effective use in this regard.

Answers to Matched Problems

17. $\dfrac{4}{x - 2} + \dfrac{3}{x + 3}$

18. $\dfrac{3}{x - 1} - \dfrac{2}{x + 2} + \dfrac{1}{(x + 2)^2}$

19. $\dfrac{2}{x - 1} + \dfrac{3x - 2}{2x^2 - 3x + 2}$

20. $\dfrac{3x}{x^2 - 2x + 2} + \dfrac{x - 2}{(x^2 - 2x + 2)^2}$

Exercise 3-8 ■ A

Find constants A, B, C, and D so that the right member is equal to the left.

1. $\dfrac{7x - 14}{(x - 4)(x + 3)} = \dfrac{A}{x - 4} + \dfrac{B}{x + 3}$

2. $\dfrac{9x + 21}{(x + 5)(x - 3)} = \dfrac{A}{x + 5} + \dfrac{B}{x - 3}$

3. $\dfrac{17x - 1}{(2x - 3)(3x - 1)} = \dfrac{A}{2x - 3} + \dfrac{B}{3x - 1}$

4. $\dfrac{x - 11}{(3x + 2)(2x - 1)} = \dfrac{A}{3x + 2} + \dfrac{B}{2x - 1}$

5. $\dfrac{3x^2 + 7x + 1}{x(x + 1)^2} = \dfrac{A}{x} + \dfrac{B}{x + 1} + \dfrac{C}{(x + 1)^2}$

6. $\dfrac{x^2 - 6x + 11}{(x + 1)(x - 2)^2} = \dfrac{A}{x + 1} + \dfrac{B}{x - 2} + \dfrac{C}{(x - 2)^2}$

7. $\dfrac{3x^2 + x}{(x - 2)(x^2 + 3)} = \dfrac{A}{x - 2} + \dfrac{Bx + C}{x^2 + 3}$

8. $\dfrac{5x^2 - 9x + 19}{(x - 4)(x^2 + 5)} = \dfrac{A}{x - 4} + \dfrac{Bx + C}{x^2 + 5}$

9. $\dfrac{2x^2 + 4x - 1}{(x^2 + x + 1)^2} = \dfrac{Ax + B}{x^2 + x + 1} + \dfrac{Cx + D}{(x^2 + x + 1)^2}$

10. $\dfrac{3x^3 - 3x^2 + 10x - 4}{(x^2 - x + 3)^2} = \dfrac{Ax + B}{x^2 - x + 3} + \dfrac{Cx + D}{(x^2 - x + 3)^2}$

B *Decompose into partial fractions.*

11. $\dfrac{-x + 22}{x^2 - 2x - 8}$

12. $\dfrac{-x - 21}{x^2 + 2x - 15}$

13. $\dfrac{3x - 13}{6x^2 - x - 12}$

14. $\dfrac{11x - 11}{6x^2 + 7x - 3}$

15. $\dfrac{x^2 - 12x + 18}{x^3 - 6x^2 + 9x}$

16. $\dfrac{5x^2 - 36x + 48}{x(x - 4)^2}$

17. $\dfrac{5x^2 + 3x + 6}{x^3 + 2x^2 + 3x}$

18. $\dfrac{6x^2 - 15x + 16}{x^3 - 3x^2 + 4x}$

19. $\dfrac{2x^3 + 7x + 5}{x^4 + 4x^2 + 4}$

20. $\dfrac{-5x^2 + 7x - 18}{x^4 + 6x^2 + 9}$

21. $\dfrac{x^3 - 7x^2 + 17x - 17}{x^2 - 5x + 6}$

22. $\dfrac{x^3 + x^2 - 13x + 11}{x^2 + 2x - 15}$

C 23. $\dfrac{4x^2 + 5x - 9}{x^3 - 6x - 9}$

24. $\dfrac{4x^2 - 8x + 1}{x^3 - x + 6}$

25. $\dfrac{x^2 + 16x + 18}{x^3 + 2x^2 - 15x - 36}$

26. $\dfrac{5x^2 - 18x + 1}{x^3 - x^2 - 8x + 12}$

27. $\dfrac{-x^2 + x - 7}{x^4 - 5x^3 + 9x^2 - 8x + 4}$

28. $\dfrac{-2x^3 + 12x^2 - 20x - 10}{x^4 - 7x^3 + 17x^2 - 21x + 18}$

29. $\dfrac{4x^5 + 12x^4 - x^3 + 7x^2 - 4x + 2}{4x^4 + 4x^3 - 5x^2 + 5x - 2}$

30. $\dfrac{6x^5 - 11x^4 + x^3 - 10x^2 - 2x - 2}{6x^4 - 7x^3 + x^2 + x - 1}$

Section 3-9   Chapter Review

IMPORTANT TERMS
AND SYMBOLS 3-1 Introduction.   Zero of a function P, zero of a polynomial $P(x)$, solution or root of the equation $P(x) = 0$

3-2 Synthetic Division.   Algebraic long division, synthetic division.

3-3 Remainder and Factor Theorems.   Division algorithm, remainder theorem, graphing polynomials, factor theorem

3-4 Fundamental Theorem of Algebra.   Fundamental theorem of algebra, n zeros theorem, complex zeros

3-5 Isolating Real Zeros.   Variation in sign, Descartes' rule of signs, upper and lower bounds of real zeros, sign changes in $P(x)$

3-6 Finding Rational Zeros.   Rational zero theorem, strategy for finding rational zeros

3-7 Approximating Irrational Zeros.   Strategy for finding all real zeros of a polynomial $P(x)$ with real coefficients, method of successive approximation

3-8 Partial Fraction Decomposition.   Partial fractions, partial fraction decomposition

Exercise 3-9   Chapter Review

Work through all the problems in this chapter review and check answers in the back of the book. (Answers to all problems are there, and following each answer is a number in italics indicating the section in which that type of problem is discussed.) Where weaknesses show up, review appropriate sections in the text. When you are satisfied that you know the material, take the practice test following this review.

A   **1.**   Use synthetic division to divide $P(x) = 2x^3 + 3x^2 - 1$ by $D(x) = x + 2$, and write the answer in the form $P(x) = D(x)Q(x) + R$.

2.   If $P(x) = x^5 - 4x^4 + 9x^2 - 8$, find $P(3)$ using the remainder theorem and synthetic division.

3.   What are the zeros of $P(x) = 3(x - 2)(x + 4)(x + 1)$?

4.   If $P(x) = x^2 - 2x + 2$ and $P(1 + i) = 0$, find another zero of $P(x)$.

5.   Using Descartes' rule of signs, what can you say about the number of positive and negative zeros of
(A) $P(x) = x^3 - x^2 + x + 3$ (B) $P(x) = x^5 + x^3 + 4$

6.   According to the upper and lower bound theorem in this chapter, which of the following are upper or lower bounds of zeros of $P(x) = x^3 - 4x^2 + 2$: $-2, -1, 3, 4$?

7. How do you know that $P(x) = 2x^3 - 3x^2 + x - 5$ has at least one real zero between 1 and 2?

8. Write the possible rational zeros for $P(x) = x^3 - 4x^2 + x + 6$.

9. Find all rational zeros for $P(x) = x^3 - 4x^2 + x + 6$.

10. Decompose $(7x - 11)/(x - 3)(x + 2)$ into partial fractions.

B 11. Use synthetic division to divide $P(x) = 3x^3 + 4x^2 - 7x - 3$ by $x - \frac{2}{3}$, and write the answer in the form $P(x) = D(x)Q(x) + R$.

12. If $P(x) = 4x^3 - 8x^2 - 3x - 3$, find $P(-\frac{1}{2})$ using the remainder theorem and synthetic division.

13. Use the quadratic formula and the factor theorem to factor $P(x) = x^2 - 2x - 1$.

14. Is $x + 1$ a factor of $P(x) = x^{25} + 1$? Explain without dividing or using synthetic division.

15. For $P(x) = 2x^4 + 3x^3 - 14x^2 + 2x + 4$,
  (A) Using Descartes' rule of signs, discuss the possible number of real zeros.
  (B) Find the smallest and largest integers that are, respectively, upper and lower bounds of zeros of $P(x)$ according to Theorem 8.
  (C) Discuss the location of real zeros within the lower and upper bound interval.

16. Determine all rational zeros of $P(x) = 2x^3 - 3x^2 - 18x - 8$.

17. Factor the polynomial in Problem 16 into linear factors.

18. Find all rational zeros of $P(x) = x^3 - 3x^2 + 5$.

19. Find all zeros (rational, irrational, and complex) for $P(x) = 2x^3 - 3x^2 + 3x - 1$.

20. Factor the polynomial in Problem 19 into linear factors.

21. Find the real zero of $P(x) = x^4 - x^2 - 2$ between 1 and 2 to one-decimal-place accuracy.

22. Decompose $\dfrac{-x^2 + 3x + 4}{x(x - 2)^2}$ into partial fractions.

23. Decompose $\dfrac{8x^2 - 10x + 9}{2x^3 - 3x^2 + 3x}$ into partial fractions.

C 24. Use synthetic division to divide $P(x) = x^3 + 3x + 2$ by $[x - (1 + i)]$, and write the answer in the form $P(x) = D(x)Q(x) + R$.

25. Find a polynomial of lowest degree with leading coefficient 1 that has zeros $-\frac{1}{2}$ (multiplicity 2), -3, and 1 (multiplicity 3). (Leave the answer in factored form.) What is the degree of the polynomial?

26. Repeat Problem 25 for a polynomial $P(x)$ with zeros -5, $2 - 3i$, and $2 + 3i$.

27. Find all real roots of $2x^4 - x^3 - 12x^2 - 14x + 10 = 0$ (irrational roots to one-decimal-place accuracy).

28. Decompose $\dfrac{5x^2 + 2x + 9}{x^4 - 3x^3 + x^2 - 3x}$ into partial fractions.

Practice Test   Chapter 3

Take this practice test as if it were a graded test. Allow yourself up to 50 minutes. Work the problems without looking back in the chapter. Correct your work using the answers (keyed to appropriate sections) in the back of the book.

1. If $P(x) = 8x^4 - 14x^3 - 13x^2 - 4x + 7$, find $Q(x)$ and R such that $P(x) = (x - \frac{1}{4})Q(x) + R$. What is $P(\frac{1}{4})$?

2. Is $x + 1$ a factor of $P(x) = 9x^{26} - 11x^{17} + 8x^{11} - 5x^4 - 7$? Explain.

Problems 3–6 refer to $P(x) = 2x^3 - 7x^2 + 2x + 6$.

3. (A) List all possible rational zeros of $P(x)$.
   (B) Discuss the possible number of positive and negative real zeros of $P(x)$ using Descartes' rule of signs.

4. (A) Find all intervals of the form $[a, b]$, where a and b are successive integers, that contain at least one real zero of $P(x)$.
   (B) Find the smallest and largest integers that are, respectively, lower and upper bounds of zeros for $P(x)$.

5. Find all zeros of $P(x)$.

6. Factor $P(x)$ as a product of first-degree factors.

7. How do we know that $x^3 + 3x - 5 = 0$ has exactly one real root and that root is positive?

8. Find the real root of the equation in Problem 7 to one decimal place.

9. Decompose $\dfrac{x^2 - 2x + 10}{(x + 2)(x - 1)^2}$ into partial fractions.

10. Decompose $\dfrac{3x^2 + 2x + 4}{(x + 1)(x^2 + 4)}$ into partial fractions.

Exponential and Logarithmic Functions

A natural design of mathematical interest. Can you guess the source? See the back of the book.

Chapter 4 ∎ Exponential and Logarithmic Functions

Most of the functions we have considered have been **algebraic functions**—that is, functions defined by means of the basic algebraic operations on variables and constants. In this chapter we will define and investigate the properties of two new and important classes of functions: exponential and logarithmic functions.

Section 4-1 Exponential Functions

- Exponential Functions
- Graphing an Exponential Function
- Typical Exponential Graphs
- Base e
- Basic Exponential Properties

∎ Exponential Functions

In this and the next section we will consider two new kinds of functions that use variable exponents in their definitions. To start, note that

$$f(x) = 2^x \quad \text{and} \quad g(x) = x^2$$

are not the same function. The function g is a quadratic function, which we have already discussed; the function f is a new function called an *exponential function*. An **exponential function** is a function defined by an equation of the form:

Exponential Function
$f(x) = b^x \qquad b > 0, \quad b \neq 1$

where b is a constant, called the **base**, and the exponent x is a variable. The replacement set for the exponent, the **domain of f**, is the set of real numbers R. The **range of f** is the set of positive real numbers. We require b to be positive to avoid complex numbers such as $(-2)^{1/2}$.

■ Graphing an Exponential Function

Many students, if asked to graph an exponential function such as $f(x) = 2^x$, would not hesitate at all. They would likely make up a table by assigning integers to x, plot the resulting points, and then join these points with a smooth curve (Fig. 1). The only catch is that 2^x has not

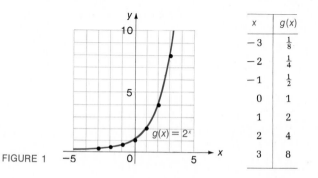

x	$g(x)$
-3	$\frac{1}{8}$
-2	$\frac{1}{4}$
-1	$\frac{1}{2}$
0	1
1	2
2	4
3	8

FIGURE 1

been defined at this point for all real numbers. We know what 2^5, 2^{-3}, $2^{2/3}$, $2^{-3/5}$, $2^{1.4}$, and $2^{-3.15}$ all mean (that is, 2^p, where p is a rational number), but what does

$$2^{\sqrt{2}}$$

mean? The question is not easy to answer at this time. In fact, a precise definition of $2^{\sqrt{2}}$ must wait for more advanced courses, where we can show that

$$b^x$$

names a real number for b a positive real number and x any real number, and that the graph of $g(x) = 2^x$ is as indicated in Figure 1. We can also show that for x irrational, b^x can be approximated as closely as we like by using rational number approximations for x. Since $\sqrt{2} = 1.414213\ldots$, for example, the sequence

$$2^{1.4},\ 2^{1.41},\ 2^{1.414},\ \ldots$$

approximates $2^{\sqrt{2}}$, and as we move to the right the approximation improves.

■ Typical Exponential Graphs

It is useful to compare the graphs of $y = 2^x$ and $y = (\frac{1}{2})^x = 2^{-x}$ by plotting both on the same coordinate system [Fig. 2(a)]. The graph of

$f(x) = b^x \qquad b > 1$ [Fig. 2(b)]

will look very much like the graph of $y = 2^x$, and the graph of

$f(x) = b^x \qquad 0 < b < 1$ [Fig. 2(b)]

will look very much like the graph of $y = (\frac{1}{2})^x$. Note in both cases that the x axis is a horizontal asymptote and the graphs will never touch it.

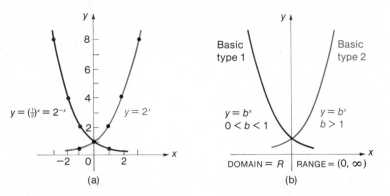

FIGURE 2   (a)   (b)

[*Note:* An exponential function is either increasing or decreasing, and hence is one to one and has an inverse that is a function. This fact will be important to us in the next section when we define a logarithmic function as an inverse of an exponential function.]

EXAMPLE 1   Graph $y = \frac{1}{2} 4^x$ for $-3 \le x \le 3$.

Solution

x	y
−3	0.01
−2	0.03
−1	0.13
0	0.50
1	2.00
2	8.00
3	32.00

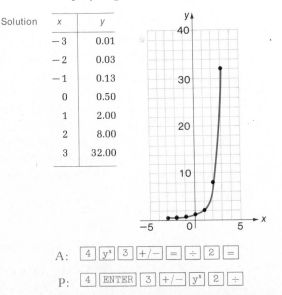

A: [4] [yˣ] [3] [+/−] [=] [÷] [2] [=]

P: [4] [ENTER] [3] [+/−] [yˣ] [2] [÷]

PROBLEM 1 Graph $y = \frac{1}{2}4^{-x}$ for $-3 \leq x \leq 3$.

Exponential functions are often referred to as *growth functions* because of their widespread use in describing different kinds of growth phenomena. These functions are used to describe population growth of people, animals, and bacteria; radioactive decay (negative growth); growth of a new chemical substance in a chemical reaction; increase or decline in the temperature of a substance being heated or cooled; growth of money at compound interest; light absorption (negative growth) as it passes through air, water, or glass; decline of atmospheric pressure as altitude is increased; and growth of learning a skill such as swimming or typing relative to practice.

■ Base *e*

For introductory purposes, the bases 2 and $\frac{1}{2}$ were convenient choices; however, a certain irrational number, denoted by *e*, is by far the most frequently used exponential base for both theoretical and practical purposes. In fact,

$$f(x) = e^x$$

is often referred to as *the* exponential function because of its widespread use. The reasons for the preference for *e* as a base are made clear in more advanced courses. And at that time, it is shown that *e* is approximated by $(1 + 1/n)^n$ to any decimal accuracy desired by making *n* (an integer) sufficiently large. The irrational number *e* to eight decimal places is

$$e \approx 2.71\text{8 } 281 \text{ } 83$$

Similarly, e^x can be approximated by using $(1 + 1/n)^{nx}$ for sufficiently large n. Because of the importance of e^x and e^{-x} tables for their evaluation are readily available. In fact, many hand calculators can evaluate these functions directly. A short table (Table I) for e^x and e^{-x} can be found in the back of this book for those not using a calculator.

The important constant *e* along with two other important constants $\sqrt{2}$ and π are shown on the number line in Figure 3.

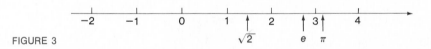

FIGURE 3

EXAMPLE 2 Graph $y = 10e^{-0.5x}$, $-3 \leq x \leq 3$, using a hand calculator or Table I.

Solution

x	y
−3	44.82
−2	27.18
−1	16.49
0	10.00
1	6.07
2	3.68
3	2.23

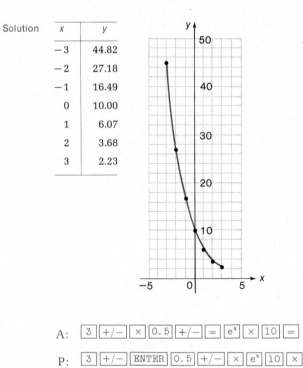

A: [3] [+/−] [×] [0.5] [+/−] [=] [e^x] [×] [10] [=]

P: [3] [+/−] [ENTER] [0.5] [+/−] [×] [e^x] [10] [×]

PROBLEM 2   Graph $y = 10e^{0.5x}$, $-3 \leq x \leq 3$, using a hand calculator or Table I.

EXAMPLE 3   If $P are invested at $100r\%$ compounded continuously, then the amount A in the account at the end of t years is given by (from mathematics of finance):

$$A = Pe^{rt}$$

If $100 is invested at 12% compounded continuously, graph the amount in the account relative to time for a period of 10 years.

Solution   We wish to graph

$$A = 100e^{0.12t} \qquad 0 \leq t \leq 10$$

We make up a table of values using a calculator or Table I, graph the points from the table, and join the points with a smooth curve. The table and graph are shown at the top of the next page.

PROBLEM 3   Repeat Example 3 with $5,000 being invested at 20% compounded continuously.

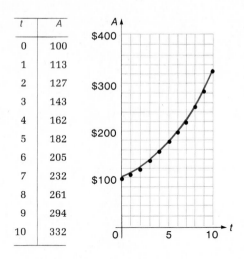

t	A
0	100
1	113
2	127
3	143
4	162
5	182
6	205
7	232
8	261
9	294
10	332

■ **Basic Exponential Properties**

Earlier (see Section 0-3) we discussed five laws of exponents for rational exponents. It can be shown that these same laws hold for irrational exponents. Thus, we now assume that all five laws of exponents hold for any real exponents as long as the bases involved are positive.

As a consequence of exponential functions being either increasing or decreasing and thus one to one, we have:

$$b^m = b^n \quad \text{if and only if} \quad m = n, b > 0, b \neq 1$$

Thus, if $2^{15} = 2^{3x}$, then $3x = 15$ and $x = 5$.

Answers to Matched Problems

1. $y = \frac{1}{2}4^{-x}$

x	y
-3	32.00
-2	8.00
-1	2.00
0	0.50
1	0.13
2	0.03
3	0.01

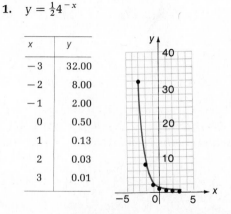

2. $y = 10e^{0.5x}$

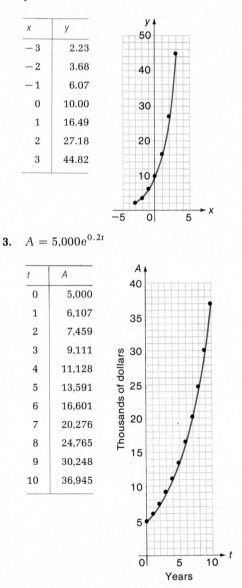

x	y
−3	2.23
−2	3.68
−1	6.07
0	10.00
1	16.49
2	27.18
3	44.82

3. $A = 5,000e^{0.2t}$

t	A
0	5,000
1	6,107
2	7,459
3	9,111
4	11,128
5	13,591
6	16,601
7	20,276
8	24,765
9	30,248
10	36,945

Exercise 4-1 ■ A *Graph each exponential function for* $-3 \leq x \leq 3$. *Plot points using integers for x, and join the points with a smooth curve.*

1. $y = 3^x$ **2.** $y = 2^x$

3. $y = (\frac{1}{3})^x = 3^{-x}$ **4.** $y = (\frac{1}{2})^x = 2^{-x}$

5. $y = 5 \cdot 3^x$ *[Note:* $4 \cdot 3^x \neq 12^x$]

6. $y = 5 \cdot 2^x$

B   **7.** $y = 2^{x+3}$ **8.** $y = 3^{x+1}$

       9. $y = 7(\frac{1}{2})^{2x} = 7 \cdot 2^{-2x}$ **10.** $y = 11 \cdot 2^{-2x}$

Graph Problems 11–14 for $-3 \le x \le 3$. Use a calculator or Table I.

11. $y = e^x$ **12.** $y = e^{-x}$

13. $y = 10e^{-0.12x}$ **14.** $y = 100e^{0.25x}$

C   **15.** Graph $y = 10 \cdot 2^{-x^2}$ for $-2 \le x \le 2$.

      16. Graph $y = e^{-x^2}$ for $x = -1.5, -1.0, -0.5, 0, 0.5, 1.0, 1.5$, and join these points with a smooth curve.

      17. Graph $y = y_0 2^x$, where y_0 is the value of y when $x = 0$. (Express the vertical scale in terms of y_0.)

      18. Graph $y = y_0 e^{-0.22x}$, where y_0 is the value of y when $x = 0$. (Express the vertical scale in terms of y_0.)

      19. Graph $y = 2^x$ and $x = 2^y$ on the same coordinate system.

      20. Graph $f(x) = 10^x$ and $y = f^{-1}(x)$ on the same coordinate system.

APPLICATIONS   **21.** If we start with 2¢ and double the amount each day, at the end of n days we will have 2^n¢. Graph $f(n) = 2^n$ for $1 \le n \le 10$. (Pick the scale on the vertical axis so that the graph will not go off the paper.)

      22. *Compound interest.* If a certain amount of money P, called the principal, is invested at $100r\%$ interest compounded annually, the amount of money A after t years is given by

$$A = P(1 + r)^t$$

Graph this equation for $P = \$10$, $r = 0.10$, and $0 \le t \le 10$.

      23. *Earth science.* The atmospheric pressure P, in pounds per square inch, can be calculated approximately using the formula

$$P = 14.7e^{-0.21x}$$

where x is altitude relative to sea level in miles. Graph the equation for $-1 \le x \le 5$.

      24. *Bacterial growth.* If bacteria in a certain culture double every hour, write a formula that gives the number of bacteria N in the culture after n hours, assuming the culture has N_0 bacteria to start with.

      25. *Radioactive decay.* Radioactive strontium-90 has a half-life of 28 years; that is, in 28 years one-half of any amount of strontium-90 will change to another substance because of radioactive decay. If we place a bar containing 100 milligrams of strontium-90 in a nuclear reactor, the amount of strontium-90 that will be left after

t years is given by $A = 100(\frac{1}{2})^{t/28}$. Graph this exponential function for $t = 0$, 28, 2(28), 3(28), 4(28), 5(28), and 6(28), and join these points with a smooth curve.

26. *Radioactive decay.* Radioactive argon-39 has a half-life of 4 minutes; that is, in 4 minutes one-half of any amount of argon-39 will change to another substance because of radioactive decay. If we start with A_0 milligrams of argon-39, the amount left after t minutes is given by $A = A_0(\frac{1}{2})^{t/4}$. Graph this exponential function for $A_0 = 100$ and $t = 0$, 4, 8, 12, 16, and 20, and join these points with a smooth curve.

27. *Sociology—small-group analysis.* Sociologists Stephan and Mischler found that, when the members of a discussion group of ten were ranked according to the number of times each participated, the number of times $N(i)$ the ith-ranked person participated was given approximately by the exponential function

$$N(i) = N_1 e^{-0.11(i-1)} \qquad 1 \le i \le 10$$

where N_1 is the number of times the top-ranked person participated in the discussion. Graph the exponential function using $N_1 = 100$.

Section 4-2   Logarithmic Functions

■ Logarithmic Functions
■ From Logarithmic to Exponential and Vice Versa
■ Finding x, b, or y in $y = \log_b x$
■ Logarithmic–Exponential Identities

■ Logarithmic Functions

We now define a new class of functions, called **logarithmic functions**, as inverses of exponential functions. (Since exponential functions are one to one, their inverses are functions.) Here you will see why we placed special emphasis on the general concept of inverse functions in Section 2-7. If you know quite a bit about a function, then (knowing about inverses in general) you will automatically know quite a bit about its inverse. For example, the graph of f^{-1} is the graph of f reflected across the line $y = x$, and the domain and range of f^{-1} are, respectively, the range and domain of f.

If we start with the exponential function

$$f: \quad y = 2^x$$

and interchange the variables x and y, we obtain the inverse of f:

$$f^{-1}: \quad x = 2^y$$

The graphs of f and f^{-1} (along with $y = x$) are shown in Figure 4. This new function is given the name **logarithmic function with base 2**, and is symbolized as follows (since we cannot "algebraically" solve $x = 2^y$ for y):

$$y = \log_2 x$$

Thus,

$$y = \log_2 x \quad \text{is equivalent to} \quad x = 2^y$$

that is, $\log_2 x$ is the power to which 2 must be raised to obtain x. (Symbolically, $x = 2^y = 2^{\log_2 x}$.)

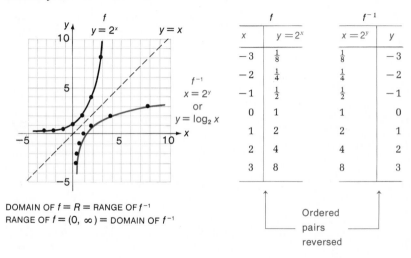

f		f^{-1}	
x	$y = 2^x$	$x = 2^y$	y
-3	$\frac{1}{8}$	$\frac{1}{8}$	-3
-2	$\frac{1}{4}$	$\frac{1}{4}$	-2
-1	$\frac{1}{2}$	$\frac{1}{2}$	-1
0	1	1	0
1	2	2	1
2	4	4	2
3	8	8	3

Ordered pairs reversed

FIGURE 4   DOMAIN OF $f = R =$ RANGE OF f^{-1}
RANGE OF $f = (0, \infty) =$ DOMAIN OF f^{-1}

In general, we define the **logarithmic function with base b** to be the inverse of the exponential function with base b ($b > 0$, $b \neq 1$).

Definition of Logarithmic Function

For $b > 0$ and $b \neq 1$,

$$y = \log_b x \quad \text{is equivalent to} \quad x = b^y$$

(The log to the base b of x is the power to which b must be raised to obtain x.)

$$y = \log_{10} x \quad \text{is equivalent to} \quad x = 10^y$$
$$y = \log_e x \quad \text{is equivalent to} \quad x = e^y$$

It is very important to remember that $y = \log_b x$ and $x = b^y$ define the same function, and as such can be used interchangeably.

Since the domain of an exponential function includes all real numbers and its range is the set of positive real numbers, the **domain** of a logarithmic function is the set of all positive real numbers and its **range** is the set of all real numbers. Thus, $\log_{10} 3$ is defined, but $\log_{10} 0$ and $\log_{10} (-5)$ are not defined (3 is a logarithmic domain value, but 0 and -5 are not). Typical logarithmic curves are shown in Figure 5.

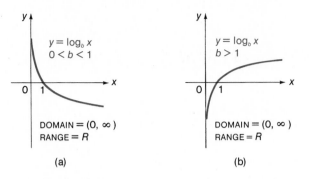

FIGURE 5 Typical logarithmic
graphs

(a)   (b)

■ **From Logarithmic to Exponential and Vice Versa**

We now look into the matter of converting logarithmic forms to equivalent exponential forms, and vice versa.

EXAMPLE 4   From logarithmic form to exponential form:

(A) $\log_2 8 = 3$ is equivalent to $8 = 2^3$
(B) $\log_{25} 5 = \frac{1}{2}$ is equivalent to $5 = 25^{1/2}$
(C) $\log_2 \frac{1}{4} = -2$ is equivalent to $\frac{1}{4} = 2^{-2}$

PROBLEM 4   Change to equivalent exponential form.

(A) $\log_3 27 = 3$ (B) $\log_{36} 6 = \frac{1}{2}$ (C) $\log_3 (\frac{1}{9}) = -2$

EXAMPLE 5   From exponential form to logarithmic form:

(A) $49 = 7^2$ is equivalent to $\log_7 49 = 2$
(B) $3 = \sqrt{9}$ is equivalent to $\log_9 3 = \frac{1}{2}$
(C) $\frac{1}{5} = 5^{-1}$ is equivalent to $\log_5 (\frac{1}{5}) = -1$

PROBLEM 5   Change to equivalent logarithmic form.

(A) $64 = 4^3$ (B) $2 = \sqrt[3]{8}$ (C) $\frac{1}{16} = 4^{-2}$

■ Finding x, b, or y in $y = \log_b x$

To gain a little deeper understanding of logarithmic functions and their relationship to the exponential functions, we will look at a few problems where one is to find x, b, or y in $y = \log_b x$, given the other two values. All values were chosen so that the problems can be solved without tables or a calculator.

EXAMPLE 6   Find x, b, or y as indicated.

(A) Find y: $y = \log_4 8$.

Solution   Write $y = \log_4 8$ in equivalent exponential form:

$$8 = 4^y \qquad \text{Write each number to the same base 2.}$$
$$2^3 = 2^{2y} \qquad \text{Recall } b^m = b^n \text{ if and only if } m = n.$$
$$2y = 3$$
$$y = \tfrac{3}{2}$$

Thus, $\tfrac{3}{2} = \log_4 8$.

(B) Find x: $\log_3 x = -2$.

Solution   Write $\log_3 x = -2$ in equivalent exponential form:

$$x = 3^{-2}$$

$$x = \frac{1}{3^2} = \frac{1}{9}$$

Thus, $\log_3 \left(\tfrac{1}{9}\right) = -2$.

(C) Find b: $\log_b 1{,}000 = 3$.

Solution   Write $\log_b 1{,}000 = 3$ in equivalent exponential form:

$$1{,}000 = b^3 \qquad \text{Write 1,000 as a third power.}$$
$$10^3 = b^3$$
$$b = 10$$

Thus, $\log_{10} 1{,}000 = 3$.

PROBLEM 6   Find x, b, or y as indicated.

(A) $y = \log_9 27$ (B) $\log_2 x = -3$ (C) $\log_b 100 = 2$

■ Logarithmic–Exponential Identities

Recall from Section 2-7 that if f and f^{-1} are both functions (that is, if f is one to one), then

$$f^{-1}[f(x)] = x \quad \text{and} \quad f[f^{-1}(x)] = x$$

Applying these general properties to $f(x) = b^x$ and $f^{-1}(x) = \log_b x$, we see that

$$f^{-1}[f(x)] = x \qquad f[f^{-1}(x)] = x$$
$$\log_b [f(x)] = x \qquad b^{f^{-1}(x)} = x$$
$$\log_b b^x = x \qquad b^{\log_b x} = x$$

Thus, we have the useful logarithmic–exponential identities:

Logarithmic–Exponential Identities

For $b > 0$, $b \neq 1$,

1. $\log_b b^x = x$
2. $b^{\log_b x} = x \qquad x > 0$

EXAMPLE 7
(A) $\log_{10} 10^5 = 5$ (B) $\log_{10} 0.01 = \log_{10} 10^{-2} = -2$
(C) $\log_e e^{2x+1} = 2x + 1$ (D) $\log_4 1 = \log_4 4^0 = 0$
(E) $10^{\log_{10} 7} = 7$ (F) $e^{\log_e x^2} = x^2$

PROBLEM 7
Find each of the following:

(A) $\log_{10} 10^{-5}$ (B) $\log_5 25$ (C) $\log_{10} 1$
(D) $\log_e e^{m+n}$ (E) $10^{\log_{10} 4}$ (F) $e^{\log_e (x^4+1)}$

Answers to Matched Problems

4. (A) $27 = 3^3$ (B) $6 = 36^{1/2}$ (C) $\frac{1}{9} = 3^{-2}$
5. (A) $\log_4 64 = 3$ (B) $\log_8 2 = \frac{1}{3}$ (C) $\log_4 (\frac{1}{16}) = -2$
6. (A) $y = \frac{3}{2}$ (B) $x = \frac{1}{8}$ (C) $b = 10$
7. (A) -5 (B) 2 (C) 0 (D) $m + n$ (E) 4
   (F) $x^4 + 1$

Exercise 4-2 ■ A

Rewrite in equivalent exponential form.

1. $\log_3 9 = 2$ 2. $\log_2 4 = 2$ 3. $\log_3 81 = 4$
4. $\log_5 125 = 3$ 5. $\log_{10} 1{,}000 = 3$ 6. $\log_{10} 100 = 2$
7. $\log_e 1 = 0$ 8. $\log_8 1 = 0$

Rewrite in equivalent logarithmic form.

9. $64 = 8^2$ 10. $25 = 5^2$ 11. $10{,}000 = 10^4$
12. $1{,}000 = 10^3$ 13. $u = v^x$ 14. $a = b^c$
15. $9 = 27^{2/3}$ 16. $8 = 4^{3/2}$

Find each of the following.

17. $\log_{10} 10^5$
18. $\log_5 5^3$
19. $\log_2 2^{-4}$
20. $\log_{10} 10^{-7}$
21. $\log_6 36$
22. $\log_3 9$
23. $\log_{10} 1,000$
24. $\log_{10} 0.001$

Find x, y, or b as indicated.

25. $\log_2 x = 2$
26. $\log_3 x = 2$
27. $\log_4 16 = y$
28. $\log_8 64 = y$
29. $\log_b 16 = 2$
30. $\log_b 10^{-3} = -3$

B *Rewrite in equivalent exponential form.*

31. $\log_{10} 0.001 = -3$
32. $\log_{10} 0.01 = -2$
33. $\log_{81} 3 = \frac{1}{4}$
34. $\log_4 2 = \frac{1}{2}$
35. $\log_{1/2} 16 = -4$
36. $\log_{1/3} 27 = -3$
37. $\log_a N = e$
38. $\log_k u = v$

Rewrite in equivalent logarithmic form.

39. $0.01 = 10^{-2}$
40. $0.001 = 10^{-3}$
41. $1 = e^0$
42. $1 = (\frac{1}{2})^0$
43. $\frac{1}{8} = 2^{-3}$
44. $\frac{1}{8} = (\frac{1}{2})^3$
45. $\frac{1}{3} = 81^{-1/4}$
46. $\frac{1}{2} = 32^{-1/5}$
47. $7 = \sqrt{49}$
48. $11 = \sqrt{121}$

Find each of the following.

49. $\log_b b^u$
50. $\log_b b^{uv}$
51. $\log_e e^{1/2}$
52. $\log_e e^{-3}$
53. $\log_2 \sqrt{8}$
54. $\log_5 \sqrt[3]{5}$
55. $\log_{23} 1$
56. $\log_{17} 1$
57. $\log_4 8$
58. $\log_4 (\frac{1}{4})$

Find x, y, or b as indicated.

59. $\log_4 x = \frac{1}{2}$
60. $\log_{25} x = \frac{1}{2}$
61. $\log_{1/3} 9 = y$
62. $\log_{49} (\frac{1}{7}) = y$
63. $\log_b 1,000 = \frac{3}{2}$
64. $\log_b 4 = \frac{2}{3}$
65. $\log_b 1 = 0$
66. $\log_b b = 1$

C 67. For $f = \{(x, y) | y = 1^x\}$, discuss the domain and range for f and f^{-1}. Are both relations functions?

68. Why is 1 not a suitable logarithmic base? [*Hint:* Try to find $\log_1 5$.]

69. (A) For $f = \{(x, y) | y = 10^x\}$, graph f and f^{-1} using the same coordinate axes.

(B)   Discuss the domain and range of f and f^{-1}.

(C)   What other name could you use for the inverse of f?

70.   Prove that $\log_b (1/x) = -\log_b x$.

Find the inverse of:

71. $f(x) = 5^{3x-1} + 4$

72. $g(x) = 3^{2x-3} - 2$

73. $g(x) = 3 \log_b (5x - 2)$

74. $f(x) = 2 + \log_b (2x - 3)$

Section 4-3   Properties of Logarithmic Functions

- Basic Logarithmic Properties
- Use of the Logarithmic Properties

Basic Logarithmic Properties

Logarithmic functions have several very useful properties that follow directly from the fact that they are inverses of exponential functions. These properties will enable us to convert multiplication problems into addition problems, division problems into subtraction problems, and power and root problems into multiplication problems. In addition, we will be able to solve exponential equations such as $2 = 10^x$.

THEOREM 1

Properties of Logarithmic Functions

If b, M, and N are positive real numbers, $b \neq 1$, and p is a real number, then

1. $\log_b b^u = u$

2. $\log_b MN = \log_b M + \log_b N$

3. $\log_b \dfrac{M}{N} = \log_b M - \log_b N$

4. $\log_b M^p = p \log_b M$

5. $\log_b 1 = 0$

The first property in Theorem 1 follows directly from the definition of a logarithmic function. The proof of the second property is based on the laws of exponents. To bring exponents into the proof, we let

$$u = \log_b M \quad \text{and} \quad v = \log_b N$$

and convert these to the equivalent exponential forms

$$M = b^u \quad \text{and} \quad N = b^v$$

Now, see if you can provide the reasons for each of the following steps:

$$\log_b MN = \log_b b^u b^v = \log_b b^{u+v} = u + v = \log_b M + \log_b N$$

The other properties are established in a similar manner.

■ Use of the Logarithmic Properties

We now see how logarithmic properties can be used to convert multiplication problems into addition problems, division problems into subtraction problems, and power and root problems into multiplication problems.

EXAMPLE 8

(A) $\log_{10} 10^5 = 5$ $\log_b b^u = u$

(B) $\log_b 3x = \log_b 3 + \log_b x$ $\log_b MN = \log_b M + \log_b N$

(C) $\log_b \dfrac{x}{5} = \log_b x - \log_b 5$ $\log_b \dfrac{M}{N} = \log_b M - \log_b N$

(D) $\log_b x^7 = 7 \log_b x$ $\log_b M^p = p \log_b M$

(E) $\log_b \dfrac{mn}{pq} = \log_b mn - \log_b pq$ $\log_b \dfrac{M}{N} = \log_b M - \log_b N$

$\qquad\qquad = \log_b m + \log_b n$ $\log_b MN = \log_b M + \log_b N$
$\qquad\qquad\quad - (\log_b p + \log_b q)$
$\qquad\qquad = \log_b m + \log_b n$
$\qquad\qquad\quad - \log_b p - \log_b q$

(F) $\log_b (mn)^{2/3} = \frac{2}{3} \log_b mn$ $\log_b M^p = p \log_b M$
$\qquad\qquad = \frac{2}{3}(\log_b m + \log_b n)$ $\log_b MN = \log_b M + \log_b N$

(G) $\log_b \dfrac{x^8}{y^{1/5}} = \log_b x^8 - \log_b y^{1/5}$ $\log_b \dfrac{M}{N} = \log_b M - \log_b N$
$\qquad\qquad = 8 \log_b x - \frac{1}{5} \log_b y$ $\log_b M^p = p \log_b M$

PROBLEM 8   Write in terms of simpler logarithmic forms, as in Example 8.

(A) $\log_b \left(\dfrac{r}{uv} \right)$ (B) $\log_b \left(\dfrac{m}{n} \right)^{3/5}$ (C) $\log_b \left(\dfrac{u^{1/3}}{v^5} \right)$

EXAMPLE 9   If $\log_e 3 = 1.10$ and $\log_e 7 = 1.95$, find

(A) $\log_e \left(\frac{7}{3} \right)$

Solution $\log_e (\frac{7}{3}) = \log_e 7 - \log_e 3 = 1.95 - 1.10 = 0.85$

(B) $\log_e \sqrt[3]{21}$

Solution $\log_e \sqrt[3]{21} = \log_e (21)^{1/3} = \frac{1}{3} \log_e (3 \cdot 7) = \frac{1}{3}(\log_e 3 + \log_e 7)$
$$= \frac{1}{3}(1.10 + 1.95) = 1.02$$

PROBLEM 9   If $\log_e 5 = 1.609$ and $\log_e 8 = 2.079$, find:

(A) $\log_e \left(\dfrac{5^{10}}{8} \right)$ (B) $\log_e \sqrt[4]{\dfrac{8}{5}}$

Finally, we note that since logarithmic functions are one to one,

$$\log_b m = \log_b n \qquad \text{if and only if} \qquad m = n$$

Thus, if $\log_{10} x = \log_{10} 32.15$, then $x = 32.15$.

The following example and problem, though somewhat artificial, will give you additional practice in using the properties in Theorem 1.

EXAMPLE 10   Find x so that $\log_b x = \frac{2}{3} \log_b 27 + 2 \log_b 2 - \log_b 3$ without using a calculator or table.

Solution $\log_b x = \frac{2}{3} \log_b 27 + 2 \log_b 2 - \log_b 3$ Express right side in terms of a single log.

$= \log_b 27^{2/3} + \log_b 2^2 - \log_b 3$ Property 4

$= \log_b 9 + \log_b 4 - \log_b 3$ $27^{2/3} = 9, \; 2^2 = 4$

$= \log_b \dfrac{9 \cdot 4}{3} = \log_b 12$ Properties 2 and 3

Thus,

$\log_b x = \log_b 12$

Hence,

$x = 12$

PROBLEM 10   Find x so that $\log_b x = \frac{2}{3} \log_b 8 + \frac{1}{2} \log_b 9 - \log_b 6$ without using a calculator or table.

Answers to Matched Problems   **8.** (A) $\log_b r - \log_b u - \log_b v$ (B) $\frac{3}{5}(\log_b m - \log_b n)$
(C) $\frac{1}{3} \log_b u - 5 \log_b v$

9. (A) 14.01 (to four significant digits)   **10.** $x = 2$
(B) 0.1175 (to four significant digits)

Exercise 4-3 ■ A *Write in terms of simpler logarithmic forms (going as far as you can with logarithmic properties—see Example 8).*

1. $\log_b uv$ 2. $\log_b rt$ 3. $\log_b (A/B)$

4. $\log_b (p/q)$ 5. $\log_b u^5$ 6. $\log_b w^{25}$

7. $\log_b N^{3/5}$ 8. $\log_b u^{-2/3}$ 9. $\log_b \sqrt{Q}$

10. $\log_b \sqrt[5]{M}$ 11. $\log_b uvw$ 12. $\log_b (u/vw)$

Write each expression in terms of a single logarithm with a coefficient of 1.

Example: $\log_b u^2 - \log_b v = \log_b (u^2/v)$.

13. $\log_b A + \log_b B$ 14. $\log_b P + \log_b Q + \log_b R$

15. $\log_b X - \log_b Y$ 16. $\log_b x^2 - \log_b y^3$

17. $\log_b w + \log_b x - \log_b y$ 18. $\log_b w - \log_b x - \log_b y$

If $\log_b 2 = 0.69$, $\log_b 3 = 1.10$, and $\log_b 5 = 1.61$, find the logarithm to the base b of each of the following numbers.

19. $\log_b 30$ 20. $\log_b 6$ 21. $\log_b \left(\frac{2}{5}\right)$

22. $\log_b \left(\frac{5}{3}\right)$ 23. $\log_b 27$ 24. $\log_b 16$

B *Write in terms of simpler logarithmic forms (going as far as you can with logarithmic properties—see Example 8).*

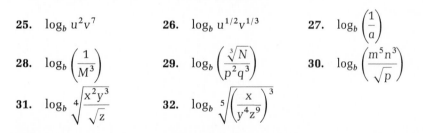

25. $\log_b u^2 v^7$ 26. $\log_b u^{1/2} v^{1/3}$ 27. $\log_b \left(\dfrac{1}{a}\right)$

28. $\log_b \left(\dfrac{1}{M^3}\right)$ 29. $\log_b \left(\dfrac{\sqrt[3]{N}}{p^2 q^3}\right)$ 30. $\log_b \left(\dfrac{m^5 n^3}{\sqrt{p}}\right)$

31. $\log_b \sqrt[4]{\dfrac{x^2 y^3}{\sqrt{z}}}$ 32. $\log_b \sqrt[5]{\left(\dfrac{x}{y^4 z^9}\right)^3}$

Write each expression in terms of a single logarithm with a coefficient of 1.

33. $2\log_b x - \log_b y$ 34. $\log_b m - \frac{1}{2}\log_b n$

35. $3\log_b x + 2\log_b y - 4\log_b z$ 36. $\frac{1}{3}\log_b w - 3\log_b x - 5\log_b y$

37. $\frac{1}{5}(2\log_b x + 3\log_b y)$ 38. $\frac{1}{3}(\log_b x - \log_b y)$

If $\log_b 2 = 0.69$, $\log_b 3 = 1.10$, and $\log_b 5 = 1.61$, find the logarithm to the base b of the following numbers.

39. $\log_b 7.5$ 40. $\log_b 1.5$ 41. $\log_b \sqrt[3]{2}$

42. $\log_b \sqrt{3}$ 43. $\log_b \sqrt{0.9}$ 44. $\log_b \sqrt[3]{\frac{3}{2}}$

C   **45.**   Find x so that $\frac{3}{2} \log_b 4 - \frac{2}{3} \log_b 8 + 2 \log_b 2 = \log_b x$.

    46.   Find x so that $3 \log_b 2 + \frac{1}{2} \log_b 25 - \log_b 20 = \log_b x$.

    47.   Write $\log_b y - \log_b c + kt = 0$ in exponential form free of logarithms.

    48.   Write $\log_e x - \log_e 100 = -0.08t$ in exponential form free of logarithms.

    49.   Prove that $\log_b (M/N) = \log_b M - \log_b N$ under the hypotheses of Theorem 1.

    50.   Prove that $\log_b M^p = p \log_b M$ under the hypotheses of Theorem 1.

    51.   Prove that $\log_b MN = \log_b M + \log_b N$ by starting with $M = b^{\log_b M}$ and $N = b^{\log_b N}$.

    52.   Prove that $\log_b (M/N) = \log_b M - \log_b N$ by starting with $M = b^{\log_b M}$ and $N = b^{\log_b N}$.

Section 4-4   Logarithms to Various Bases

- Common and Natural Logarithms—Calculator Evaluation
- Common Logarithms—Table Evaluation (Optional)
- Natural Logarithms—Table Evaluation (Optional)
- Change-of-Base Formula

John Napier (1550–1617) is credited with the invention of logarithms. They evolved out of an interest in reducing the computational strain in astronomy research. This new computational tool was immediately accepted by the scientific world. Now, with the availability of inexpensive hand calculators, logarithms have lost most of their importance as a computational device. However, the logarithmic concept has been greatly generalized since its conception, and logarithmic functions are used widely in both theoretical and applied sciences. For example, even with a very good scientific hand calculator, we still need logarithmic functions to solve the simple-looking exponential equation from population growth studies and the mathematics of finance:

$$2 = 1.08^x$$

Of all possible logarithmic bases, the base e and the base 10 are used almost exclusively. Before we can use logarithms in certain practical problems, we need to be able to approximate the logarithm of any number to either base 10 or base e. And conversely, if we are given the logarithm of a number to base 10 or base e, we need to be able to approximate the number. Historically, tables such as Table II and Table III at the back of the book were used for this purpose, but now with inexpensive scientific hand calculators readily available, most people will choose a calculator, since it is faster and far more accurate than any table you might use.

■ Common and Natural Logarithms—Calculator Evaluation

Common logarithms (also called **Briggsian logarithms**) are logarithms with base 10. **Natural logarithms** (also called **Napierian logarithms**) are logarithms with base e. Most scientific calculators have a button labeled "log" (or "LOG") and a button labeled "ln" (or "LN"). The former represents a common (base 10) logarithm and the latter a natural (base e) logarithm. In fact, "log" and "ln" are both used extensively in mathematical literature, and whenever you see either used in this book without a base indicated, they will be interpreted as follows:

Logarithmic Notation
$\log x = \log_{10} x$ $\ln x = \log_e x$

To find the common or natural logarithm using a scientific calculator is very easy: You simply enter a number from the domain of the function and push the log or ln button.

EXAMPLE 11   Use a scientific calculator to find each to six decimal places.

(A) log 3,184   (B) ln 0.000 349   (C) log (−3.24)

Solution

ENTER	PRESS	DISPLAY
(A) 3,184	$\boxed{\texttt{log}}$	3.502973
(B) 0.000 349	$\boxed{\texttt{ln}}$	−7.960439
(C) −3.24	$\boxed{\texttt{log}}$	Error

Why is an error indicated in (C)? Because −3.24 is not in the domain of the log function.

PROBLEM 11   Use a scientific calculator to find each to six decimal places.

(A) log 0.013 529   (B) ln 28.693 28   (C) ln (−0.438)

EXAMPLE 12   Use a scientific calculator to evaluate each to three decimal places.

(A) $n = \dfrac{\log 2}{\log 1.1}$ (B) $n = \dfrac{\ln 3}{\ln 1.08}$

Solution   (A) First note that $(\log 2)/(\log 1.1) \neq \log 2 - \log 1.1$. Recall (see Section 4-3) that $\log_b (M/N) = \log_b M - \log_b N$, which is, of course, not the same as $(\log_b M)/(\log_b N)$.

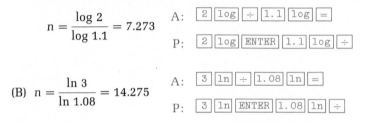

$$n = \dfrac{\log 2}{\log 1.1} = 7.273 \qquad \text{A:} \quad \boxed{2}\,\boxed{\log}\,\boxed{\div}\,\boxed{1.1}\,\boxed{\log}\,\boxed{=}$$

$$\text{P:} \quad \boxed{2}\,\boxed{\log}\,\boxed{\text{ENTER}}\,\boxed{1.1}\,\boxed{\log}\,\boxed{\div}$$

(B) $n = \dfrac{\ln 3}{\ln 1.08} = 14.275 \qquad \text{A:} \quad \boxed{3}\,\boxed{\ln}\,\boxed{\div}\,\boxed{1.08}\,\boxed{\ln}\,\boxed{=}$

$$\text{P:} \quad \boxed{3}\,\boxed{\ln}\,\boxed{\text{ENTER}}\,\boxed{1.08}\,\boxed{\ln}\,\boxed{\div}$$

PROBLEM 12   Use a scientific calculator to evaluate each to two decimal places.

(A) $n = \dfrac{\ln 2}{\ln 1.1}$ (B) $n = \dfrac{\log 3}{\log 1.08}$

We now turn to the second problem: Given the logarithm of a number, find the number. We make direct use of the logarithmic–exponential relationships that were discussed in Section 4-2.

Logarithmic–Exponential Relationships
$\log x = y$ is equivalent to $x = 10^y$
$\ln x = y$ is equivalent to $x = e^y$

EXAMPLE 13   Find x to three significant digits, given the indicated logarithms.

(A) $\log x = -9.315$ (B) $\ln x = 2.386$

Solution   (A) $\log x = -9.315$ Change to equivalent exponential form.

$x = 10^{-9.315}$ $\boxed{9.315}\,\boxed{+/-}\,\boxed{10^x}$

$x = 4.84 \times 10^{-10}$ Notice the answer is displayed in scientific notation in the calculator.

(B) $\ln x = 2.386$ Change to equivalent exponential form.

$x = e^{2.386}$ $\boxed{2.386}\,\boxed{e^x}$

$x = 10.9$

PROBLEM 13   Find x to four significant figures, given the indicated logarithms.

(A) $\ln x = -5.062$ (B) $\log x = 12.0821$

■ Common Logarithms—Table Evaluation (Optional)

We now show how Table II in the back of this book can be used to approximate common logarithms. Recalling that any decimal fraction can be written in scientific notation (see Section 0-2), we see that

$$
\begin{aligned}
\log_{10} 33{,}800 &= \log_{10} (3.38 \times 10^4) \\
&= \log_{10} 3.38 + \log_{10} 10^4 \\
&= \log_{10} 3.38 + 4
\end{aligned}
$$

and that

$$
\begin{aligned}
\log_{10} 0.003\ 51 &= \log_{10} (3.51 \times 10^{-3}) \\
&= \log_{10} 3.51 + \log_{10} 10^{-3} \\
&= \log_{10} 3.51 - 3
\end{aligned}
$$

In general:

If a number N is written in scientific notation

   $N = r \times 10^k$ $1 \le r < 10,$ k an integer

then

   $$
   \begin{aligned}
   \log N &= \log (r \times 10^k) \\
   &= \log r + \log 10^k \\
   &= \log r + k
   \end{aligned}
   $$

   Mantissa   Characteristic

Thus, if common logarithms of r, $1 \le r < 10$, are given in a table, we will be able to approximate the common logarithm of any positive decimal fraction to the accuracy of the table.

Using methods of advanced mathematics, a table of common logarithms of numbers from 1 to 10 can be computed to any decimal accuracy desired. Table II in the back of this book is such a table to four-decimal-place accuracy. It is useful to remember that if x is between 1 and 10, then log x is between 0 and 1 (Fig. 6).

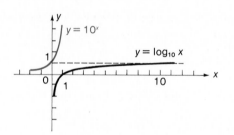

FIGURE 6

To illustrate the use of Table II, a small portion of it is reproduced in Table 1. To find log 3.47, for example, we first locate 3.4 under the x heading; then we move across to the column headed 7, where we find .5403. Thus, log 3.47 = 0.5403.*

TABLE 1

x	0	1	2	3	4	5	6	7	8	9
3.2	.5051	.5065	.5079	.5092	.5105	.5119	.5132	.5145	.5159	.5172
3.3	.5185	.5198	.5211	.5224	.5237	.5250	.5263	.5276	.5289	.5302
3.4	.5315	.5328	.5340	.5353	.5366	.5378	.5391	.5403	.5416	.5428
3.5	.5441	.5453	.5465	.5478	.5490	.5502	.5514	.5527	.5539	.5551

Now let us finish finding the common logarithms of 33,800 and 0.003 51.

EXAMPLE 14   (A) $\log 33{,}800 = \log (3.38 \times 10^4)$

$$= \log 3.38 + \log 10^4 \quad \text{Use Table 1.}$$
$$= 0.5289 + 4$$

$$= 4.5289$$

(B) $\log 0.003\ 51 = \log (3.51 \times 10^{-3})$

$$= \log 3.51 + \log 10^{-3} \quad \text{Use Table 1.}$$
$$= 0.5453 - 3$$

$$= -2.4547$$

PROBLEM 14   Use Table 1 to find:

(A) log 328,000   (B) log 0.000 342

---

* Throughout the rest of this chapter we will use = in place of $\approx$ in many places, realizing that values are only approximately equal. Occasionally, we will use $\approx$ when a special emphasis is desired.

Now let us reverse the problem; that is, given the log of a number, find the number. To find the number, we first write the log of the number in the form

$$m + c$$

where m (the mantissa) is a nonnegative number between 0 and 1, and c (the characteristic) is an integer; then reverse the process illustrated in Example 14.

EXAMPLE 15 (A) If log x = 2.5224, find x.

Solution
$$\log x = 2.5224$$
$$= 0.5224 + 2$$

Write 2.5224 in the form $m + c$, $0 \leq m < 1$ and c an integer. Look for 0.5224 in the body of Table 1. Thus, we see that $0.5224 = \log 3.33$.

$$= \log 3.33 + \log 10^2$$
$$= \log (3.33 \times 10^2)$$

Thus,

$$x = 3.33 \times 10^2 \quad \text{or} \quad 333$$

(B) If log x = 0.5172 − 4, find x.

Solution
$$\log x = 0.5172 - 4$$

$$= \log 3.29 + 10^{-4}$$
$$= \log (3.29 \times 10^{-4})$$

Use Table 1.

Thus,

$$x = 3.29 \times 10^{-4} \quad \text{or} \quad 0.000\ 329$$

(C) If log x = −4.4685, find x.

Solution
$$\log x = -4.4685$$
$$= 0.5315 - 5$$

Convert -4.4685 to $m + c$ form with $0 \leq m < 1$ by adding and subtracting 5:

$$\begin{array}{r} 5.0000 - 5 \\ -4.4685 \\ \hline 0.5315 - 5 \end{array}$$

$$= \log 3.40 + \log 10^{-5}$$
$$= \log (3.40 \times 10^{-5})$$

Thus,

$$x = 3.40 \times 10^{-5} \quad \text{or} \quad 0.000\ 0340$$

PROBLEM 15 Use Table 1 to find x if

(A) log x = 5.5378   (B) log x = 0.5289 − 3   (C) log x = −2.4921

What if a number has more significant digits than are included in a table? We then use the nearest table value, or, if more accuracy is desired, we use a calculator.

EXAMPLE 16   (A) $\log 32{,}683 \approx \log (3.27 \times 10^4)$ Round 3.2683 to nearest table value—
                                                                          that is, to 3.26.
$$= \log 3.27 + \log 10^4$$

$$= 0.5145 + 4$$

$$= 4.5145$$

(B) To find x if $\log x = 0.5241 - 3$, we observe (in Table 1) that 0.5241
     is between 0.5237 and 0.5250, but is closer to 0.5237. Thus, we write

$$\log x = 0.5241 - 3$$ Since 0.5241 is not in the body of the
$$\approx 0.5237 - 3$$ table, select the value in the table that
                                 is closest; that is, select 0.5237.
$$= \log 3.34 + \log 10^{-3}$$

$$= \log (3.34 \times 10^{-3})$$

Thus,

$$x = 3.34 \times 10^{-3} \quad \text{or} \quad 0.003\ 34$$

PROBLEM 16   Find:
(A) $\log 0.034\ 319$ (B) x if $\log x = 6.5473$
(C) x if $\log x = -4.4942$

- Natural Logarithms—Table Evaluation (Optional)

Approximating natural logarithms using Table III in the back of this
book proceeds in much the same way as finding common logarithms
using Table II, except the arithmetic is a little more complicated. A
couple examples will illustrate the process.

EXAMPLE 17   (A) $\ln 52{,}400 = \ln (5.24 \times 10^4)$

$$= \ln 5.24 + 4 \ln 10$$

[ln 5.24 is read out of the main table (Table III), and 4 ln 10 is obtained
from the list at the top of the table.]

$$= 1.6563 + 9.2103$$

$$= 10.8666$$

(B) $\ln 0.002\ 78 = \ln (2.78 \times 10^{-3})$

$$= \ln 2.78 - 3 \ln 10$$

$$= 1.0225 - 6.9078$$

$$= -5.8853$$

PROBLEM 17   Use Table III to find:
(A) $\ln 0.000\ 683$ (B) $\ln 328{,}000$ (C) $\ln 23{,}582$

■ **Change-of-Base Formula**

If we have a means (through either a calculator or table) of finding logarithms of numbers to one base, then by means of the change-of-base formula we can find the logarithm of a number to any other base.

Change-of-Base Formula

$$\log_b N = \frac{\log_a N}{\log_a b}$$

Can you supply the reasons for each step in the following derivation of this formula?

$$y = \log_b N$$

$$N = b^y$$

$$\log_a N = \log_a b^y$$

$$\log_a N = y \log_a b$$

$$y = \frac{\log_a N}{\log_a b}$$

$$\log_b N = \frac{\log_a N}{\log_a b} \qquad \text{Since } y = \log_b N$$

EXAMPLE 18 Find $\log_5 14$ using common logarithms.

Solution $\log_5 14 = \dfrac{\log_{10} 14}{\log_{10} 5} = 1.640$ A: $\boxed{14}\,\boxed{\log}\,\boxed{\div}\,\boxed{5}\,\boxed{\log}\,\boxed{=}$

P: $\boxed{14}\,\boxed{\log}\,\boxed{\text{ENTER}}\,\boxed{5}\,\boxed{\log}\,\boxed{\div}$

PROBLEM 18 Find $\log_7 729$.

Answers to Matched Problems

11. (A) $-1.868\ 734$ (B) $3.356\ 663$ (C) Not possible

12. (A) 7.27 (B) 14.27

13. (A) $0.006\ 333$ (B) 1.21×10^{12}

14. (A) 5.5159 (B) $0.5340 - 4 = -3.4660$

15. (A) 3.45×10^5 (B) 3.38×10^{-3} (C) 3.22×10^{-3}

16. (A) -1.4647 (B) 3.53×10^6 (C) 3.20×10^{-5}

17. (A) -7.2890 (B) 12.7008
        (C) Approx. $\ln 23{,}600 = 10.0690$

18. 3.3874

Exercise 4-4 ■

Hand calculators are used extensively in this exercise set.

A *Use a calculator to find each to four decimal places.*

1. log 82,734   **2.** log 843,250   **3.** log 0.001 439

4. log 0.035 604   **5.** ln 43.046   **6.** ln 2,843,100

7. ln 0.081 043   **8.** ln 0.000 0324

Use Table II to find each. (Optional)

9. log 7.29   **10.** log 6.37   **11.** log 2,040

12. log 327   **13.** log 0.0413   **14.** log 0.000 927

B *Use a calculator to find x to four significant digits, given:*

15. $\log x = 5.3027$ **16.** $\log x = 1.9168$

17. $\log x = -3.1773$ **18.** $\log x = -2.0411$

19. $\ln x = 3.8655$ **20.** $\ln x = 5.0884$

21. $\ln x = -0.3916$ **22.** $\ln x = -4.1083$

Use the nearest values in Table II to approximate each of the following. (Optional)

23. log 304,918   **24.** log 82,734

25. log 0.004 769   **26.** log 0.061 94

Find x using the nearest values in Table II. (Optional)

27. $\log x = 7.437 \ 15$ **28.** $\log x = 9.113 \ 64$

29. $\log x = -4.8013$ **30.** $\log x = -3.4128$

Evaluate each of the following to three decimal places using a calculator.

31. $n = \dfrac{\log 2}{\log 1.15}$ **32.** $n = \dfrac{\log 2}{\log 1.12}$ **33.** $n = \dfrac{\ln 3}{\ln 1.15}$

34. $n = \dfrac{\ln 4}{\ln 1.2}$ **35.** $x = \dfrac{\ln 0.5}{-0.21}$ **36.** $t = \dfrac{\log 200}{2 \log 2}$

Use the change-of-base formula and a calculator with either log or ln to find each to four decimal places.

37. $\log_5 372$ **38.** $\log_4 23$ **39.** $\log_8 0.0352$

40. $\log_2 0.005 \ 439$ **41.** $\log_3 0.1483$ **42.** $\log_{12} 435.62$

Find each of the following using the nearest values in Table III. (Optional)

43. ln 2.35 **44.** ln 7.02 **45.** ln 603,517

46. ln 5,233 **47.** ln 0.003 1687 **48.** ln 0.071 33

C *Find x to five significant digits using a calculator.*

49. $x = \log (5.3147 \times 10^{12})$ **50.** $x = \log (2.0991 \times 10^{17})$

51. $x = \ln (6.7917 \times 10^{-12})$ **52.** $x = \ln (4.0304 \times 10^{-8})$

53. $\log x = 32.068\ 523$ **54.** $\log x = -12.731\ 64$

55. $\ln x = -14.667\ 13$ **56.** $\ln x = 18.891\ 143$

Section 4-5 Exponential and Logarithmic Equations

- Exponential Equations
- Logarithmic Equations

Equations involving exponential and logarithmic functions, such as

$$2^{3x-2} = 5 \quad \text{and} \quad \log (x + 3) + \log x = 1$$

are called **exponential** and **logarithmic equations**, respectively. Logarithmic properties play a central role in their solution.

■ Exponential Equations

The following examples illustrate the use of logarithmic properties in solving exponential equations.

EXAMPLE 19 Solve $2^{3x-2} = 5$ for x to four decimal places.

Solution

$$2^{3x-2} = 5$$

How can we get x out of the exponent? Use logs! If two positive quantities are equal, their logs are equal.

$$\log 2^{3x-2} = \log 5$$
$$(3x - 2)\log 2 = \log 5$$

Use $\log_b N^p = p \log_b N$ to get $3x - 2$ out of the exponent position.

$$3x - 2 = \frac{\log 5}{\log 2}$$

Remember: $\log 5/\log 2 \neq \log 5 - \log 2$

$$x = \frac{1}{3}\left(2 + \frac{\log 5}{\log 2}\right)$$

A: ⁵ 🔲log🔲 ÷ 🔲2🔲 🔲log🔲 = + 🔲2🔲 = ÷ 🔲3🔲 =

or

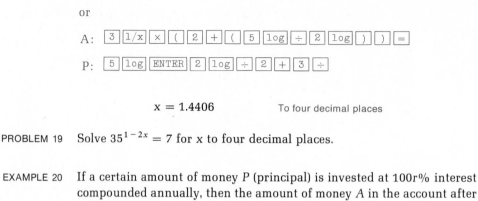

$$x = 1.4406 \qquad \text{To four decimal places}$$

PROBLEM 19   Solve $35^{1-2x} = 7$ for x to four decimal places.

EXAMPLE 20   If a certain amount of money P (principal) is invested at $100r\%$ interest compounded annually, then the amount of money A in the account after n years, assuming no withdrawals, is given by

$$A = P(1 + r)^n$$

How long will it take the money to double if it is invested at 6% compounded annually?

Solution   To find the doubling time, we replace A in $A = P(1.06)^n$ with 2P and solve for n.

$$2P = P(1.06)^n \qquad \text{Divide both sides by } P.$$

$$2 = 1.06^n \qquad \text{Take the common or natural log of both sides.}$$

$$\log 2 = \log 1.06^n$$

$$\log 2 = n \log 1.06 \qquad \text{Note how log properties are used to get } n \text{ out of the exponent position.}$$

$$n = \frac{\log 2}{\log 1.06}$$

$$= 12 \text{ years} \qquad \text{To the nearest year}$$

PROBLEM 20   Repeat Example 20 changing the interest rate from 6% compounded annually to 9% compounded annually.

EXAMPLE 21   The atmospheric pressure P (in pounds per square inch) at x miles above sea level is given approximately by

$$P = 14.7e^{-0.21x}$$

At what height will the atmospheric pressure be half of the sea-level pressure? Compute the answer to two significant digits.

Solution   Sea-level pressure is the pressure at $x = 0$. Thus,

$$P = 14.7e^0 = 14.7$$

One-half of sea-level pressure is $14.7/2 = 7.35$. Now our problem is to find x so that $P = 7.35$; that is, we solve $7.35 = 14.7e^{-0.21x}$ for x.

$$7.35 = 14.7e^{-0.21x} \qquad \text{Divide both sides by 14.7 to simplify.}$$

$$0.5 = e^{-0.21x} \qquad \text{Take the natural log of both sides.}$$

$$\ln 0.5 = \ln e^{-0.21x} \qquad \text{Why use natural logs? Compare with common log to}$$

$$\ln 0.5 = -0.21x \qquad \text{see why.}$$

$$x = \frac{\ln 0.5}{-0.21} \qquad \text{Use a hand calculator (or Table II).}$$

$$= 3.3 \text{ miles} \qquad \text{To two significant digits}$$

PROBLEM 21 Using the formula in Example 21, find the altitude in miles to two significant digits so that the atmospheric pressure will be one-eighth that at sea level.

The graph of

$$y = \frac{e^x + e^{-x}}{2} \qquad\qquad (1)$$

is a curve called a *catenary* (Fig. 7). A uniform cable suspended between two fixed points is a physical example of such a curve.

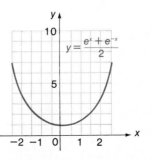

FIGURE 7 Catenary

EXAMPLE 22 Given equation (1), find x for y = 2.5. Compute the answer to four decimal places.

Solution
$$y = \frac{e^x + e^{-x}}{2}$$

$$2.5 = \frac{e^x + e^{-x}}{2}$$

$$5 = e^x + e^{-x} \qquad \text{Multiply both sides by } e^x.$$

$$5e^x = e^{2x} + 1$$

$$e^{2x} - 5e^x + 1 = 0 \qquad \text{This is a quadratic in } e^x.$$

Let $u = e^x$, then

$$u^2 - 5u + 1 = 0$$

$$u = \frac{5 \pm \sqrt{25 - 4(1)(1)}}{2}$$

$$= \frac{5 \pm \sqrt{21}}{2} \qquad \text{Replace } u \text{ with } e^x \text{ and solve for } x.$$

$$e^x = \frac{5 \pm \sqrt{21}}{2} \qquad \text{Take the natural log of both sides (both values on the right are positive).}$$

$$\ln e^x = \ln \frac{5 \pm \sqrt{21}}{2}$$

$$x = \ln \frac{5 \pm \sqrt{21}}{2}$$

$$= -1.5668, 1.5668$$

PROBLEM 22   Given $y = (e^x - e^{-x})/2$, find x for $y = 1.5$. Compute the answer to three decimal places.

■   Logarithmic Equations

The next two examples illustrate approaches to solving some types of logarithmic equations.

EXAMPLE 23   Solve $\log (x + 3) + \log x = 1$ and check.

Solution

$$\log (x + 3) + \log x = 1 \qquad \text{Combine left side using } \log M + \log N = \log MN.$$
$$\log [x(x + 3)] = 1 \qquad \text{Change to equivalent exponential form.}$$
$$x(x + 3) = 10^1 \qquad \text{Write in } ax^2 + bx + c = 0 \text{ form.}$$
$$x^2 + 3x - 10 = 0 \qquad \text{Solve.}$$
$$(x + 5)(x - 2) = 0$$
$$x = -5, 2$$

Check $x = -5$:

$\log (-5 + 3) + \log (-5)$ is not defined (Why?)

$x = 2$:

$$\log (2 + 3) + \log 2 = \log 5 + \log 2$$
$$= \log (5 \cdot 2) = \log 10 = 1$$

Remember, answers should be checked in the original equation to see whether any should be discarded.

PROBLEM 23   Solve $\log (x - 15) = 2 - \log x$ and check.

EXAMPLE 24   Solve $(\ln x)^2 = \ln x^2$.

Solution

$$(\ln x)^2 = \ln x^2$$

$$(\ln x)^2 = 2 \ln x$$

$$(\ln x)^2 - 2 \ln x = 0$$

$$(\ln x)(\ln x - 2) = 0$$

This is a quadratic equation in ln x. Move all nonzero terms to the left and factor.

$$\ln x = 0 \quad \text{or} \quad \ln x - 2 = 0$$

$$x = e^0 \qquad\qquad \ln x = 2$$

$$x = 1 \qquad\qquad\quad x = e^2$$

PROBLEM 24   Solve $\log x^2 = (\log x)^2$.

Answers to Matched Problems

19. 0.2263

20. More than double in 9 years, but not quite double in 8 years

21. 9.9 miles   **22.** 1.195

23. 20   **24.** 1, 100

Exercise 4-5 ▪

🖩 *A scientific hand calculator will prove useful in many of the problems in this exercise.*

A *Solve to three significant digits.*

1. $10^{-x} = 0.0347$ **2.** $10^x = 14.3$ **3.** $10^{3x+1} = 92$

4. $10^{5x-2} = 348$ **5.** $e^x = 3.65$ **6.** $e^{-x} = 0.0142$

7. $e^{2x-1} = 405$ **8.** $e^{3x+5} = 23.8$ **9.** $5^x = 18$

10. $3^x = 4$ **11.** $2^{-x} = 0.238$ **12.** $3^{-x} = 0.074$

Solve exactly.

13. $\log 5 + \log x = 2$ **14.** $\log x - \log 8 = 1$

15. $\log x + \log (x - 3) = 1$ **16.** $\log (x - 9) + \log 100x = 3$

B *Solve to three significant digits.*

17. $2 = 1.05^x$ **18.** $3 = 1.06^x$

19. $e^{-1.4x} = 13$ **20.** $e^{0.32x} = 632$

21. $123 = 500e^{-0.12x}$ **22.** $438 = 200e^{0.25x}$

Solve exactly.

23. $\log x - \log 5 = \log 2 - \log (x - 3)$

24. $\log (6x + 5) - \log 3 = \log 2 - \log x$

25. $(\ln x)^3 = \ln x^4$ **26.** $(\log x)^3 = \log x^4$

27. $\ln (\ln x) = 1$ **28.** $\log (\log x) = 1$

29. $x^{\log x} = 100x$ **30.** $3^{\log x} = 3x$

C   *In Problems 31–36 solve for the indicated letter in terms of all others using common or natural logs, whichever produces the simplest results.*

31. $I = I_0 e^{-kx}$ for x (x-ray absorption)

32. $A = P(1 + i)^n$ for n (compound interest)

33. $N = 10 \log \left(\dfrac{I}{I_0} \right)$ for I (sound intensity—decibels)

34. $t = \dfrac{-1}{k} (\ln A - \ln A_0)$ for A (radioactive decay)

35. $I = \dfrac{E}{R} (1 - e^{-Rt/L})$ for t (electric circuits)

36. $S = R \dfrac{(1 + i)^n - 1}{i}$ for n (future value of an annuity)

37. Find the fallacy.

$$3 > 2$$
$$(\log \tfrac{1}{2})3 > (\log \tfrac{1}{2})2$$
$$3 \log \tfrac{1}{2} > 2 \log \tfrac{1}{2}$$
$$\log (\tfrac{1}{2})^3 > \log (\tfrac{1}{2})^2$$
$$(\tfrac{1}{2})^3 > (\tfrac{1}{2})^2$$
$$\tfrac{1}{8} > \tfrac{1}{4}$$

38. Find the fallacy.

$$-2 < -1$$
$$\ln e^{-2} < \ln e^{-1}$$
$$2 \ln e^{-1} < \ln e^{-1}$$
$$2 < 1$$

APPLICATIONS   **39.** *Compound interest.* How long will it take a sum of money to double if it is invested at 15% interest compounded annually (see Example 20)?

40. *Compound interest.* How long will it take money to quadruple if it is invested at 20% interest compounded annually (see Example 20)?

41. *Bacterial growth.* A single cholera bacterium divides every $\tfrac{1}{2}$ hour to produce two complete cholera bacteria. If we start with a colony of 5,000 bacteria, then after t hours we will have

$$A = 5,000 \cdot 2^{2t}$$

bacteria. How long will it take for A to equal 1,000,000?

42. *Astronomy.*   An optical instrument is required to observe stars beyond the sixth magnitude, the limit of ordinary vision. However, even optical instruments have their limitations. The limiting magnitude L of any optical telescope with lens diameter D in inches is given by

$$L = 8.8 + 5.1 \log D$$

(A) Find the limiting magnitude for a homemade 6-inch reflecting telescope.
(B) Find the diameter of a lens that would have a limiting magnitude of 20.6.

43. *World population.*   A mathematical model for world population growth over short periods of time is given by

$$P = P_0 e^{rt}$$

where

P_0 = Population at $t = 0$
r = Rate compounded continuously
t = Time in years
P = Population at time t

How long will it take the earth's population to double if it continues to grow at its current rate of 2% per year (compounded continuously)? [*Hint:*   Given $r = 0.02$, find t so that $P = 2P_0$.]

44. *World population.*   If the world population is now 4 billion people and if it continues to grow at 2% per year (compounded continuously), how long will it be before there is only 1 square yard of land per person? Use the formula in Problem 43 and the fact that there is 1.7×10^{14} square yards of land on earth.

45. *Nuclear reactors—strontium-90.*   Radioactive strontium-90 is used in nuclear reactors and decays according to

$$A = Pe^{-0.0248t}$$

where P is the amount present at $t = 0$, and A is the amount remaining after t years. Find the half-life of strontium-90; that is, find t so that $A = 0.5P$.

46. *Archaeology—carbon-14 dating.*   Cosmic-ray bombardment of the atmosphere produces neutrons, which in turn react with nitrogen to produce radioactive carbon-14. Radioactive carbon-14 enters all

living tissues through carbon dioxide, which is first absorbed by plants. As long as a plant or animal is alive, carbon-14 is maintained in a constant amount in its tissues. Once dead, however, it ceases taking in carbon and, to the slow beat of time, the carbon-14 diminishes by radioactive decay according to the equation

$$A = A_0 e^{-0.000\ 124t}$$

where t is time in years. Estimate the age of a skull uncovered in an archaeological site if 10% of the original amount of carbon-14 is still present. [Hint: Find t such that $A = 0.1A_0$.]

47.   *Sound intensity—decibels.*   Because of the extraordinary range of sensitivity of the human ear (a range of over 1,000 million million to 1), it is helpful to use a logarithmic scale to measure sound intensity over this range rather than an absolute scale. The unit of measure is called the *decibel*, after the inventor of the telephone, Alexander Graham Bell. If we let N be the number of decibels, I the power of the sound in question in watts per cubic centimeter, and I_0 the power of sound just below the threshold of hearing (approximately 10^{-16} watt/square centimeter),

$$I = I_0 10^{N/10}$$

show that this formula can be written in the form

$$N = 10 \log \left(\frac{I}{I_0} \right)$$

48.   *Sound intensity—decibels.*   Use the formula in Problem 47 (with $I_0 = 10^{-16}$ watt/square centimeter) to find the decibel ratings of the following sounds:
(A)   Whisper (10^{-13} watt/square centimeter)
(B)   Normal conversation (3.16×10^{-10} watt/square centimeter)
(C)   Heavy traffic (10^{-8} watt/square centimeter)
(D)   Jet plane with afterburner (10^{-1} watt/square centimeter)

49.   *Earth science.*   For relatively clear bodies of freshwater or salt water, light intensity is reduced according to the exponential function

$$I = I_0 e^{-kd}$$

where I is the intensity at d feet below the surface, and I_0 is the intensity at the surface; k is called the coefficient of extinction. Two of the clearest bodies of water in the world are the freshwater Crystal Lake in Wisconsin ($k = 0.0485$) and the saltwater Sargasso Sea off the West Indies ($k = 0.009\ 42$). Find the depths (to the nearest foot) in these two bodies of water at which the light is reduced to 1% of that at the surface.

50. *Psychology—learning.* In learning a particular task, such as typing or swimming, one progresses faster at the beginning and then levels off. If you plot the level of performance against time, you will obtain a curve of the type shown in the figure. This is called a *learning curve* and can be very closely approximated by an exponential equation of the form $y = a(1 - e^{-cx})$, where a and c are positive constants. Curves of this type have applications in psychology, education, and industry. Suppose a particular person's history of learning to type is given by the exponential equation $N = 80(1 - e^{-0.08n})$, where N is the number of words per minute typed after n weeks of instruction. Approximately how many weeks did it take the person to learn to type sixty words per minute?

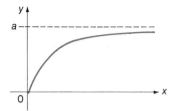

Section 4-6   Chapter Review

IMPORTANT TERMS AND SYMBOLS

4-1 Exponential Functions.   Exponential function, base, domain, range, graphs, base e, properties, b^x ($b > 0$, $b \neq 1$)

4-2 Logarithmic Functions.   Logarithmic function, base, domain, range, graphs, logarithmic–exponential identities, $\log_b x$ ($b > 0$, $b \neq 1$)

4-3 Properties of Logarithmic Functions.   Basic properties, use of properties

4-4 Logarithms to Various Bases.   Common logarithm, natural logarithm, change-of-base formula

4-5 Exponential and Logarithmic Equations.   Exponential equations, logarithmic equations

Exercise 4-6   Chapter Review

Work through all the problems in this chapter review and check answers in the back of the book. (Answers to all problems are there, and following each answer is a number in italics indicating the section in which that type of problem is discussed.) Where weaknesses show up, review appropriate sections in the text. When you are satisfied that you know the material, take the practice test following this review.

▦ *You will find a scientific hand calculator useful in many of the problems in this exercise.*

A   1. Write $m = 10^n$ in logarithmic form with base 10.

2. Write $\log x = y$ in exponential form.

Solve for x exactly. Do not use a calculator or table.

3. $\log_2 x = 3$ 4. $\log_x 25 = 2$ 5. $\log_3 27 = x$

Solve for x to three significant digits.

6. $10^x = 17.5$ 7. $e^x = 143,000$

In Problems 8 and 9, solve for x exactly. Do not use a calculator or table.

8. $\log x - 2 \log 3 = 2$ 9. $\log x + \log (x - 3) = 1$

B   10. Write $\ln y = x$ in exponential form.

11. Write $x = e^y$ in logarithmic form with base e.

Solve for x exactly. Do not use a calculator or table.

12. $\log_{1/4} 16 = x$ 13. $\log_x 9 = -2$ 14. $\log_{16} x = \frac{3}{2}$

15. $\log_x e^5 = 5$ 16. $10^{\log_{10} x} = 33$ 17. $\ln x = 0$

Solve for x to three significant digits.

18. $25 = 5(2)^x$ 19. $4,000 = 2,500 e^{0.12x}$

20. $0.01 = e^{-0.05x}$

In Problems 21–24 solve for x exactly. Do not use a table or calculator.

21. $\log 3x^2 - \log 9x = 2$

22. $\log x - \log 3 = \log 4 - \log (x + 4)$

23. $(\log x)^3 = \log x^9$

24. $\ln (\log x) = 1$

25. Calculate $\log_5 23$ to three significant digits.

C   26. Write $\ln y = -5t + \ln c$ in an exponential form free of logarithms; then solve for y in terms of the other letters.

27. For $f = \{(x, y) | y = \log_2 x\}$, graph f and f^{-1} using the same coordinate system. What are the domains and ranges for f and f^{-1}?

28. Explain why 1 cannot be used as a logarithmic base.

29. Prove that $\log_b (M/N) = \log_b M - \log_b N$.

APPLICATIONS   30. *Population growth.*   Many countries in the world have a population growth rate of 3% (or more) per year. At this rate how long, to the nearest year, will it take a population to double? Use the population growth model

$$P = P_0(1.03)^t$$

which assumes annual compounding. Compute the answer to three significant digits.

31. *Population growth.* Repeat Problem 30 using the continuous population growth model

$$P = P_0 e^{0.03t}$$

which assumes continuous compounding. Compute the answer to three significant digits.

32. *Carbon-14 dating.* How long, to three significant digits, will it take for the carbon-14 to diminish to 1% of the original amount after the death of a plant or animal?

$$A = A_0 e^{-0.000\ 124t} \qquad \text{where } t \text{ is time in years}$$

33. *x-ray absorption.* Solve $x = -(1/k) \ln (I/I_0)$ for I in terms of the other letters.

34. *Amortization—time payments.* Solve $r = P\{i/[1 - (1 + i)^{-n}]\}$ for n in terms of the other letters.

Practice Test   Chapter 4

Take this practice test as if it were a graded test. Allow yourself up to 50 minutes. Work the problems without looking back in the chapter. Correct your work using the answers (keyed to appropriate sections) in the back of the book.

1. Write $\ln y = -x^2$ in equivalent exponential form.

2. Write $\frac{1}{2} \log_b x + 3 \log_b y - 2 \log_b z$ in terms of a single logarithm with a coefficient of 1.

Solve for x to three significant digits.

3. $10^x = 42.6$ **4.** $5^{2x-3} = 7.08$

5. $125 = 500 e^{-0.000\ 124x}$

In Problems 6–10 solve for x exactly. Do not use a calculator or table.

6. $\log_{1/5} 125 = x$ **7.** $\log_{16} x = \frac{3}{4}$

8. $\log_x 100 = -2$ **9.** $\log (x + 4) = 1 - \log (x - 5)$

10. $\log (\ln x) = 1$

11. Write $f^{-1}(x)$ if $f(x) = e^x$. What are the domain and range of f and f^{-1}?

12. Find $\log_{12} 8$ to three significant digits.

13. *Continuous compound interest.* How long, to three significant digits, would it take money to quadruple if invested at 15% interest compounded continuously? ($A = Pe^{rt}$)

14. *Continuous compound interest.* Solve $A = Pe^{rt}$ for r in terms of the other letters. Use either common or natural logs, whichever produces the simplest result.

15. *Present value of an annuity.* Solve $P = R\{[1 - (1 + i)^{-n}]/i\}$ for n in terms of the other letters.

Systems of Equations and Inequalities ■ 5

A natural design of mathematical interest. Can you guess the source? See the back of the book.

Chapter 5 ■ Systems of Equations and Inequalities

In this chapter we will first review how systems of equations are solved using techniques learned in elementary algebra. These techniques are suitable for systems involving two or three variables, but they are not suitable for systems involving larger numbers of variables. After this review, we will introduce techniques that are more suitable for solving systems with larger numbers of variables. These new techniques form the basis for computer solutions of large-scale systems.

Section 5-1 Systems of Linear Equations—A Review

■ Systems in Two Variables
■ Systems in Three Variables
■ Application

■ Systems in Two Variables

To establish basic concepts, consider the following simple example: If two children have a combined weight of 80 kilograms and one weighs 20 kilograms more than the other, what is the weight of each?

$$\text{Let} \quad x = \text{Weight of heavier child}$$
$$y = \text{Weight of lighter child}$$
$$\text{Then} \quad x + y = 80$$
$$x - y = 20$$

We now have a system of two equations and two unknowns. To solve this system we find all ordered pairs of real numbers that satisfy both equations. In general, we are interested in solving linear systems of the type

$$ax + by = h$$
$$cx + dy = k$$

where a, b, c, d, h, and k are real constants. The **solution set** of this system is the set of all ordered pairs of numbers such that each ordered pair satisfies each equation in the system. We will consider three methods of solving such systems, each with certain advantages, depending on the situation.

SOLUTION BY GRAPHING To solve the weight problem by graphing, we graph both equations in the same coordinate system. Then the coordinates of any points that the graphs have in common must be solutions to the system, since they must satisfy both equations.

EXAMPLE 1 Solve the weight problem by graphing:

$$x + y = 80$$
$$x - y = 20$$

Solution

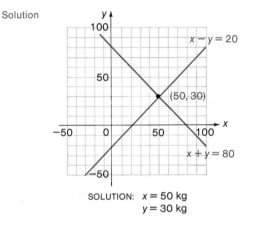

SOLUTION: $x = 50$ kg
$y = 30$ kg

PROBLEM 1 Solve by graphing:

$$x + y = 12$$
$$x - y = 4$$

It is clear that the preceding example (and problem) has exactly one solution, since the lines have exactly one point of intersection. In general, lines in a rectangular coordinate system are related to each other in one of the three ways illustrated in the next example.

EXAMPLE 2 Solve each of the following systems by graphing.

(A) $2x - 3y = 2$ (B) $4x + 6y = 12$ (C) $2x - 3y = -6$
    $x + 2y = 8$ $2x + 3y = -6$ $-x + \frac{3}{2}y = 3$

Solution   (A)

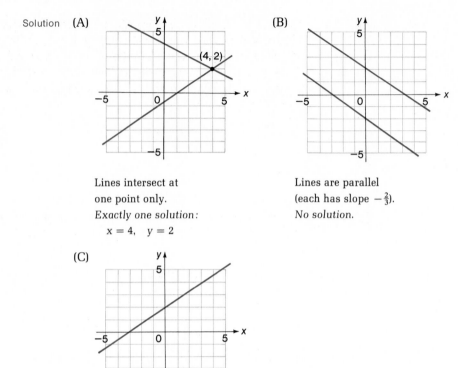

Lines intersect at
one point only.
Exactly one solution:
   $x = 4$, $y = 2$

Lines are parallel
(each has slope $-\frac{2}{3}$).
No solution.

(C)

Lines coincide.
Infinitely many
solutions.

PROBLEM 2   Solve each of the following systems by graphing.

    (A) $2x + 3y = 12$ (B) $x - 3y = -3$ (C) $2x - 3y = 12$
         $x - 3y = -3$ $-2x + 6y = 12$ $-x + \frac{3}{2}y = -6$

    By interpreting a system of two linear equations in two unknowns
geometrically, we gain useful information about what to expect in the
way of solutions to the system. Since two lines in a coordinate system
must intersect in exactly one point, be parallel, or coincide, we conclude
that the system has (1) exactly one solution, (2) no solution, or (3) infinitely many solutions. In addition, graphs frequently reveal relationships in problems that might otherwise be hidden. Generally, however,
graphic methods give us only rough approximations of solutions. The
methods of elimination by substitution and elimination by addition to
be considered next will yield solutions to any decimal accuracy desired—assuming solutions exist.

SOLUTION BY ELIMINATION USING SUBSTITUTION

Choose one of the two equations in a system and solve for one variable in terms of the other (make a choice that avoids fractions, if possible). Then substitute the result into the other equation and solve the resulting linear equation in one variable. Now substitute this result back into either of the original equations to find the second variable. An example should make the process clear.

EXAMPLE 3 Solve by elimination using substitution.

$$2x - 3y = 7$$
$$3x - y = 7$$

Solution Solve either equation for one variable in terms of the other; then substitute into the remaining equation. In this problem we can avoid fractions if we solve for y in terms of x in the second equation.

$$3x - y = 7 \qquad \text{Solve the second equation for } y \text{ in terms of } x.$$
$$-y = -3x + 7$$
$$y = \underbrace{3x - 7} \qquad \text{Substitute into the first equation to eliminate } y.$$

$$2x - 3y = 7 \qquad \text{First equation}$$
$$2x - 3(3x - 7) = 7$$
$$2x - 9x + 21 = 7$$
$$-7x = -14$$
$$x = 2$$

Now replace x with 2 in $y = 3x - 7$ to find y:

$$y = 3x - 7$$
$$= 3(2) - 7$$
$$y = -1$$

Thus, $(2, -1)$ is the unique solution to the original system.

Check
$$2x - 3y = 7 \qquad\qquad 3x - y = 7$$
$$2(2) - 3(-1) \overset{?}{=} 7 \qquad 3(2) - (-1) \overset{?}{=} 7$$
$$7 \overset{\checkmark}{=} 7 \qquad\qquad 7 \overset{\checkmark}{=} 7$$

PROBLEM 3 Solve by elimination using substitution.

$$3x - 4y = 18$$
$$2x + y = 1$$

SOLUTION BY ELIMINATION USING ADDITION

Now we turn to **elimination using addition**. This is probably the most important method of solution, since it is readily generalized to higher-order systems. The method involves the replacement of systems of equations with simpler *equivalent systems* (by performing appropriate operations) until we obtain a system with an obvious solution. **Equivalent systems** of equations are, as you would expect, systems that have exactly the same solution set. Theorem 1 lists operations that produce equivalent systems.

THEOREM 1

Producing Equivalent Systems
Equivalent systems of equations result if: **1.** Two equations are interchanged.   **2.** An equation is multiplied by a nonzero constant.   **3.** A constant multiple of another equation is added to a given equation.

EXAMPLE 4　Solve by elimination using addition.

$$3x - 2y = 8$$
$$2x + 5y = -1$$

Solution　We use Theorem 1 to eliminate one of the variables and thus obtain a system with an obvious solution.

$$3x - 2y = 8$$
$$2x + 5y = -1$$

If we multiply the top equation by 5, the bottom by 2, and then add, we can eliminate y.

$$15x - 10y = 40$$
$$\underline{4x + 10y = -2}$$
$$19x \qquad = 38$$
$$x = 2$$

Now substitute $x = 2$ back into either of the original equations, say the second equation, and solve for y ($x = 2$ paired with either of the two original equations produces an equivalent system).

$$2(2) + 5y = -1$$
$$5y = -5$$
$$y = -1$$

Check

$$3x - 2y = 8 \qquad\qquad 2x + 5y = -1$$
$$3(2) - 2(-1) \overset{?}{=} 8 \qquad 2(2) + 5(-1) \overset{?}{=} -1$$
$$8 \overset{\vee}{=} 8 \qquad\qquad -1 \overset{\vee}{=} -1$$

PROBLEM 4   Solve by elimination using addition.

$$6x + 3y = 3$$
$$5x + 4y = 7$$

Let us see what happens in the elimination process when a system either has no solution or has infinitely many solutions. Consider the following system:

$$2x + 6y = -3$$
$$x + 3y = 2$$

Multiplying the second equation by -2 and adding, we obtain

$$2x + 6y = -3$$
$$\underline{-2x - 6y = -4}$$
$$0 = -7$$

We have obtained a contradiction. An assumption that the original system has solutions must be false (otherwise, we have proved that $0 = -7$). Thus, the system has no solutions. The graphs of the equations are parallel. Systems with no solutions are said to be **inconsistent**. Systems of equations that have solutions are said to be **consistent**.

Now consider the system

$$x - \tfrac{1}{2}y = 4$$
$$-2x + y = -8$$

If we multiply the top equation by 2 and add the result to the bottom equation, we get

$$2x - y = 8$$
$$\underline{-2x + y = -8}$$
$$0 = 0$$

Obtaining $0 = 0$ by addition implies that the equations are equivalent. (Why?) Hence, the two equations have the same solution set, and the system has infinitely many solutions. If $x = k$, then $y = 2k - 8$; that is, $(k, 2k - 8)$ is a solution for any real number k. Such a system is said to be **dependent**. The variable k is called a **parameter**; replacing it with any real number produces a particular solution to the system.

■   Systems in Three Variables

Now that we know how to solve systems of linear equations in two variables, there is no reason to stop there. Systems of the form

$$a_1x + b_1y + c_1z = k_1$$
$$a_2x + b_2y + c_2z = k_2 \tag{1}$$
$$a_3x + b_3y + c_3z = k_3$$

as well as higher-order systems are encountered frequently. In fact, systems of equations are so important in solving real-world problems that there are whole courses devoted to this one topic. A triplet of numbers $x = x_0$, $y = y_0$, and $z = z_0$ [also written as an ordered triplet (x_0, y_0, z_0)] is a **solution** of system (1) if each equation is satisfied by this triplet. The set of all such ordered triplets of numbers is called the **solution set** of the system. Two systems are said to be **equivalent** if they have the same solution set. Linear equations in three variables represent planes in a three-dimensional space. Trying to visualize how three planes can intersect will give you insight as to what kind of solution sets are possible for system (1).

In this section we will use an extension of the method of elimination to solve systems in the form of (1). In the next section we will consider techniques for solving linear systems that are more compatible with a computer approach to solving such systems. In practice, most linear systems involving more than three variables are usually solved with the aid of a computer.

Steps in Solving Systems of Form (1)

1. Choose two equations from the system and eliminate one of the three variables using elimination by addition. The result is generally one equation in two unknowns.
2. Now eliminate the same variable from the unused equation and one of those used in step 1. We (generally) obtain another equation in two variables.
3. The two equations from steps 1 and 2 form a system of two equations and two unknowns. Solve as described in the first part of this section.
4. Substitute the solution from step 3 into any of the three original equations and solve for the third variable to complete the solution of the original system.

EXAMPLE 5   Solve.

$$3x - 2y + 4z = 6 \tag{2}$$
$$2x + 3y - 5z = -8 \tag{3}$$
$$5x - 4y + 3z = 7 \tag{4}$$

Solution   *Step 1.*   We look at the coefficients of the variables and choose to eliminate y from equations (2) and (4) because of the convenient coefficients

-2 and -4. Multiply equation (2) by -2 and add to equation (4):

$$-6x + 4y - 8z = -12 \quad -2[\text{Equation (2)}]$$
$$\underline{5x - 4y + 3z = \quad 7 \quad \text{Equation (4)}}$$
$$-x \qquad - 5z = \quad -5 \tag{5}$$

Step 2. Now we eliminate y (the same variable) from equations (2) and (3):

$$9x - 6y + 12z = \quad 18 \quad 3[\text{Equation (2)}]$$
$$\underline{4x + 6y - 10z = -16 \quad 2[\text{Equation (3)}]}$$
$$13x \qquad + \ 2z = \quad 2 \tag{6}$$

Step 3. From steps 1 and 2 we obtain the system

$$-x - 5z = -5 \quad \text{These equations, along with (2), (3), or (4), form} \tag{5}$$
$$13x + 2z = \quad 2 \quad \text{a system equivalent to the original system.} \tag{6}$$

We solve this system as in the first part of this section:

$$-13x - 65z = -65 \quad 13[\text{Equation (5)}]$$
$$\underline{13x + \ 2z = \quad 2 \quad \text{Equation (6)}}$$
$$-63z = -63$$
$$z = 1$$

Substitute $z = 1$ back into either equation (5) or (6) [we choose equation (5)] to find x:

$$-x - 5z = -5 \tag{5}$$
$$-x - 5(1) = -5$$
$$-x = 0$$
$$x = 0$$

Step 4. Substitute $x = 0$ and $z = 1$ back into any of the three original equations [we choose equation (2)] to find y:

$$3x - 2y + 4z = 6 \tag{2}$$
$$3(0) - 2y + 4(1) = 6$$
$$-2y + 4 = 6$$
$$-2y = 2$$
$$y = -1$$

Thus, the solution to the original system is $(0, -1, 1)$, or $x = 0$, $y = -1$, $z = 1$.

Check   To check the solution, we must check *each* equation in the original system:

$$3x - 2y + 4z = 6 \qquad\qquad 2x + 3y - 5z = -8$$
$$3(0) - 2(-1) + 4(1) \overset{?}{=} 6 \qquad 2(0) + 3(-1) - 5(1) \overset{?}{=} -8$$
$$6 \overset{\checkmark}{=} 6 \qquad\qquad -8 \overset{\checkmark}{=} -8$$

$$5x - 4y + 3z = 7$$
$$5(0) - 4(-1) + 3(1) \overset{?}{=} 7$$
$$7 \overset{\checkmark}{=} 7$$

PROBLEM 5   Solve.

$$3x - 4y + 2z = -9$$
$$2x + 5y - 3z = 5$$
$$4x - 2y - 4z = -12$$

In the process just described, if we encounter an equation that states a contradiction, such as $0 = -2$, then we must conclude that the system has no solution (that is, the system is inconsistent). If, on the other hand, one of the equations turns out to be $0 = 0$, either the system has infinitely many solutions or it has none. We must proceed further to determine which. Notice how this last result differs from the two-equation–two-unknown case. There, when we obtained $0 = 0$, we *knew* that there were infinitely many solutions. We will have more to say about this in the following sections.

■ Application

We now consider a real-world problem that leads to a system of equations.

EXAMPLE 6   *Production scheduling.* A small manufacturing plant makes three types of inflatable boats: one-person, two-person, and four-person models. Each boat requires the services of three departments, as listed in the table. The cutting, assembly, and packaging departments have available a maximum of 380, 330, and 120 work-hours per week, respectively. How many boats of each type must be produced each week for the plant to operate at full capacity?

	ONE-PERSON BOAT	TWO-PERSON BOAT	FOUR-PERSON BOAT
Cutting department	0.6 hr	1.0 hr	1.5 hr
Assembly department	0.6 hr	0.9 hr	1.2 hr
Packaging department	0.2 hr	0.3 hr	0.5 hr

Solution   Let x = Number of one-person boats produced per week

y = Number of two-person boats produced per week

z = Number of four-person boats produced per week

Then $0.6x + 1.0y + 1.5z = 380$ Cutting department

$0.6x + 0.9y + 1.2z = 330$ Assembly department

$0.2x + 0.3y + 0.5z = 120$ Packaging department

We can clear the system of decimals, if desired, by multiplying each side of each equation by 10. Thus,

$$6x + 10y + 15z = 3,800 \tag{7}$$

$$6x + 9y + 12z = 3,300 \tag{8}$$

$$2x + 3y + 5z = 1,200 \tag{9}$$

Let us start by eliminating x from equations (7) and (8):

$$\text{Add} \begin{cases} 6x + 10y + 15z = 3,800 & \text{Equation (7)} \\ -6x - 9y - 12z = -3,300 & -1[\text{Equation (8)}] \end{cases}$$
$$y + 3z = 500$$

Now we eliminate x from equations (7) and (9):

$$\text{Add} \begin{cases} 6x + 10y + 15z = 3,800 & \text{Equation (7)} \\ -6x - 9y - 15z = -3,600 & -3[\text{Equation (9)}] \end{cases}$$
$$y = 200$$

Substituting $y = 200$ into $y + 3z = 500$, we can solve for z:

$$200 + 3z = 500$$

$$3z = 300$$

$$z = 100$$

Now use equation (7), (8), or (9) to find x [we use (9)]:

$$2x + 3y + 5z = 1,200$$

$$2x + 3(200) + 5(100) = 1,200$$

$$2x = 100$$

$$x = 50$$

Thus, each week, the company should produce 50 one-person boats, 200 two-person boats, and 100 four-person boats to operate at full capacity. The check of the solution is left to the reader.

PROBLEM 6   Repeat Example 6 assuming the cutting, assembly, and packaging departments have available a maximum of 260, 234, and 82 work-hours per week, respectively.

1. $(8, 4)$, or $x = 8$ and $y = 4$

2. (A) $(3, 2)$, or $x = 3$ and $y = 2$ (B) No solution
   (C) Infinite number of solutions

3. $(2, -3)$, or $x = 2$ and $y = -3$

4. $(-1, 3)$, or $x = -1$ and $y = 3$

5. $(-1, 2, 1)$, or $x = -1$, $y = 2$, and $z = 1$

6. 100 one-person boats, 140 two-person boats, and 40 four-person boats

Exercise 5-1 ■ A *Solve by graphing.*

1. $3x - 2y = 12$
   $7x + 2y = 8$

2. $3x - y = 2$
   $x + 2y = 10$

3. $3u + 5v = 15$
   $6u + 10v = -30$

4. $m + 2n = 4$
   $2m + 4n = -8$

Solve by elimination using substitution.

5. $x - y = 4$
   $x + 3y = 12$

6. $2x - y = 3$
   $x + 2y = 14$

7. $3x - y = 7$
   $2x + 3y = 1$

8. $2x + y = 6$
   $x - y = -3$

Solve by elimination using addition.

9. $2x + 3y = 1$
   $3x - y = 7$

10. $2m - n = 10$
    $m - 2n = -4$

11. $4x + 3y = 26$
    $3x - 11y = -7$

12. $9x - 3y = 24$
    $11x + 2y = 1$

Solve by elimination using either substitution or addition.

13. $3x - 6y = -9$
    $-2x + 4y = 6$

14. $2x - 3y = -2$
    $-4x + 6y = 7$

15. $7m + 12n = -1$
    $5m - 3n = 7$

16. $3x + 8y = 4$
    $15x + 10y = -10$

17. $2x + 4y = -8$
    $x + 2y = 4$

18. $-6x + 10y = -30$
    $3x - 5y = 15$

19. $y = 0.08x$
    $y = 100 + 0.04x$

20. $y = 0.07x$
    $y = 80 + 0.05x$

B *Solve.*

21. $0.3u - 0.6v = 0.18$
    $0.5u + 0.2v = 0.54$

22. $0.2x - 0.5y = 0.07$
    $0.8x - 0.3y = 0.79$

23. $x - 3y + z = 4$
$-x + 4y - 4z = 1$
$2x - y + 5z = -3$

24. $2x + y - z = 5$
$x - 2y - 2z = 4$
$3x + 4y + 3z = 3$

25. $3u - 2v + 3w = 11$
$2u + 3v - 2w = -5$
$u + 4v - w = -5$

26. $2a + 4b + 3c = 6$
$a - 3b + 2c = -7$
$-a + 2b - c = 5$

C **27.** $3x - 2y - 4z = -8$
$4x + 3y - 5z = -5$
$6x - 5y + 2z = -17$

28. $2x - 3y + 3z = -15$
$3x + 2y - 5z = 19$
$5x - 4y - 2z = -2$

29. $-x + 2y - z = -4$
$2x + 5y - 4z = -16$
$x + y - z = -4$

30. $x - 8y + 2z = -1$
$x - 3y + z = 1$
$2x - 11y + 3z = 2$

APPLICATIONS   **31.**   *Puzzle.*   A friend of yours came out of the post office having spent $8.80 on 20¢ and 15¢ stamps. If she bought forty-seven stamps in all, how many of each type did she buy?

32.   *Puzzle.*   A parking meter contains only nickels and dimes worth $6.05. If there are eighty-nine coins in all, how many of each type are there?

33.   *Chemistry.*   A chemist has two concentrations of hydrochloric acid in stock: a 50% solution and an 80% solution. How much of each should she mix to obtain 100 milliliters of a 68% solution?

34.   *Chemistry.*   Repeat Problem 33 assuming the 50% stock solution is replaced with a 60% stock solution.

35.   *Business.*   A jeweler has two bars of gold alloy in stock, one 12 carat and the other 18 carat (24-carat gold is pure gold, 12-carat gold is 12/24 pure, 18-carat gold is 18/24 pure, and so on). How many grams of each alloy must be mixed to obtain 10 grams of 14-carat gold?

36.   *Business.*   Repeat Problem 35 assuming the jeweler has only 10-carat and pure gold in stock.

37.   *Nutrition.*   Animals in an experiment are to be kept on a strict diet. Each animal is to receive, among other things, 20 grams of protein and 6 grams of fat. The laboratory technician is able to purchase two food mixes with the compositions shown in the table. How many grams of each mix should be used to obtain the right diet for a single animal?

MIX	PROTEIN (%)	FAT (%)
A	10	6
B	20	2

38. *Nutrition.*   A biologist in a nutrition experiment wants to prepare a special diet for her experimental animals. She requires a food mixture that contains, among other things, 20 ounces of protein and 6 ounces of fat. Food mixes are available with the compositions shown in the table. How many ounces of each mix should be used to prepare the diet mix?

MIX	PROTEIN (%)	FAT (%)
A	20	2
B	10	6

39. *Earth science.*   An earthquake emits a primary wave and a secondary wave. Near the surface of the earth the primary wave travels at about 5 miles/second, and the secondary wave at about 3 miles/second. From the time lag between the two waves arriving at a given station, it is possible to estimate the distance to the quake. (The *epicenter* can be located by obtaining distance bearings at three or more stations.) Suppose a station measured a time difference of 16 seconds between the arrival of the two waves. How long did each wave travel, and how far was the earthquake from the station?

40. *Earth science.*   A ship using sound-sensing devices above and below water recorded a surface explosion 6 seconds sooner by its underwater device than its above-water device. Sound travels in air at about 1,100 feet/second and in seawater at about 5,000 feet/second.
(A)   How long did it take each sound wave to reach the ship?
(B)   How far was the explosion from the ship?

41. *Production scheduling.*   A garment industry manufactures three shirt styles. Each style of shirt requires the services of three departments, as listed in the table. The cutting, sewing, and packaging departments have available a maximum of 1,160, 1,560, and 480 work-hours per week, respectively. How many of each style shirt must be produced each week for the plant to operate at full capacity?

	STYLE A	STYLE B	STYLE C
Cutting department	0.2 hr	0.4 hr	0.3 hr
Sewing department	0.3 hr	0.5 hr	0.4 hr
Packaging department	0.1 hr	0.2 hr	0.1 hr

42. *Production scheduling.* Repeat Problem 41 with the cutting, sewing, and packaging departments having available a maximum of 1,180, 1,560, and 510 work-hours per week, respectively.

43. *Diet.* In an experiment involving mice, a zoologist finds she needs a food mix that contains, among other things, 23 grams of protein, 6.2 grams of fat, and 16 grams of moisture. She has on hand mixes with the compositions shown in the table. How many grams of each mix should she use to get the desired diet mix?

MIX	PROTEIN (%)	FAT (%)	MOISTURE (%)
A	20	2	15
B	10	6	10
C	15	5	5

44. *Diet.* Repeat Problem 43 assuming the diet mix is to contain 18.5 grams of protein, 4.9 grams of fat, and 13 grams of moisture.

Section 5-2 Systems and Augmented Matrices—An Introduction

- Introduction
- Augmented Matrices
- Solving Linear Systems Using Augmented Matrix Methods

Introduction

Most linear systems of any consequence involve large numbers of equations and unknowns. These systems are solved using computers, since hand methods would be impractical (try solving even a five-equation–five-unknown problem and you will understand why). However, even if you have a computer facility to help solve your problem, it is still important for you to know how to formulate the problem so that it can be solved by a computer. In addition, it is helpful to have at least a general idea of how computers solve these problems. And, finally, it is important for you to know how to interpret the results.

Even though the procedures and notation introduced in this and the next section are more involved than those used in the preceding section, it is important to keep in mind that our objective is not to find an efficient hand method for solving large-scale systems (there is none), but rather to find a process that generalizes readily for computer use. It turns out that you will receive an added bonus for your efforts, since several of the processes developed in this and the next section will be of additional use in Section 6-3.

■ Augmented Matrices

In solving systems of equations by elimination in the preceding section the coefficients of the variables and constant terms played a central role. The process can be made more efficient for generalization and computer work by the introduction of a mathematical form called a *matrix*. A **matrix** is a rectangular array of numbers written within brackets. Some examples are

$$\begin{bmatrix} 3 & 5 \\ 0 & -2 \end{bmatrix} \qquad \begin{bmatrix} 2 \\ -3 \\ 0 \end{bmatrix} \qquad [1 \quad -1 \quad 0 \quad 5]$$

$$\begin{bmatrix} -1 & 2 & -5 & 0 \\ 0 & 3 & 2 & 1 \end{bmatrix} \qquad \begin{bmatrix} 1 & 0 & 0 \\ 0 & 1 & 0 \\ 0 & 0 & 1 \end{bmatrix}$$

Each number in a matrix is called an **element** of the matrix.

Associated with each linear system of the form*

$$a_1x_1 + b_1x_2 = k_1$$
$$a_2x_1 + b_2x_2 = k_2 \tag{1}$$

where x_1 and x_2 are variables, is a matrix called the **augmented matrix** of the system:

$$\begin{array}{ll} & \text{Column 1 } (C_1) \\ & \text{Column 2 } (C_2) \\ & \text{Column 3 } (C_3) \end{array}$$

$$\begin{bmatrix} a_1 & b_1 & k_1 \\ a_2 & b_2 & k_2 \end{bmatrix} \begin{array}{l} \longleftarrow \text{Row 1 } (R_1) \\ \longleftarrow \text{Row 2 } (R_2) \end{array} \tag{2}$$

This matrix contains the essential parts of system (1). The vertical bar is included only to separate the coefficients of the variables from the constant terms. Our objective is to learn how to manipulate augmented matrices in such a way that a solution to system (1) will result, if a solution exists. The manipulative process is a direct outgrowth of the elimination process discussed in Section 5-1.

---

* We are gradually shifting notation for variables and constants to **subscript notation**. Subscript notation is more convenient for the generalization of these concepts, since in large systems one soon runs out of letters.

Recall that two linear systems are said to be **equivalent** if they have exactly the same solution set. How did we transform linear systems into equivalent linear systems? We used Theorem 1, which we restate here for convenient reference.

THEOREM 1

Producing Equivalent Systems
A system of linear equations is transformed into an equivalent system if: **1.** Two equations are interchanged. **2.** An equation is multiplied by a nonzero constant. **3.** A constant multiple of another equation is added to a given equation.

Paralleling the previous discussion, we say that two augmented matrices are **row-equivalent**, denoted by the symbol $\sim$ between the two matrices, if they are augmented matrices of equivalent systems of equations. (Think about this.) How do we transform augmented matrices into row-equivalent matrices? We use Theorem 2, which is a direct consequence of Theorem 1.

THEOREM 2

Producing Row-Equivalent Matrices
An augmented matrix is transformed into a row-equivalent matrix if: **1.** Two rows are interchanged $(R_i \leftrightarrow R_j)$. **2.** A row is multiplied by a nonzero constant $(kR_i \rightarrow R_i)$. **3.** A constant multiple of another row is added to a given row $(R_i + kR_j \rightarrow R_i)$. [*Note:* The arrow $\rightarrow$ means "replaces."]

■ Solving Linear Systems Using Augmented Matrix Methods

The use of Theorem 2 in solving systems in the form of (1) is best illustrated by examples.

EXAMPLE 7   Solve, using augmented matrix methods.

$$3x_1 + 4x_2 = 1$$
$$x_1 - 2x_2 = 7$$

(3)

Solution   We start by writing the augmented matrix corresponding to (3).

$$\left[\begin{array}{cc|c} 3 & 4 & 1 \\ 1 & -2 & 7 \end{array}\right] \tag{4}$$

Our objective is to use row operations from Theorem 2 to try to transform (4) into the form

$$\left[\begin{array}{cc|c} 1 & 0 & m \\ 0 & 1 & n \end{array}\right] \tag{5}$$

where m and n are real numbers. The solution to system (3) will then be obvious, since matrix (5) will be the augmented matrix of the following system:

$$x_1 \quad = m$$
$$\quad x_2 = n$$

We now proceed to use row operations to transform (4) into form (5).

Step 1.   To get a 1 in the upper left corner, we interchange rows 1 and 2 (Theorem 2-1).

$$\left[\begin{array}{cc|c} 3 & 4 & 1 \\ 1 & -2 & 7 \end{array}\right] \quad \begin{array}{c} R_1 \leftrightarrow R_2 \\ \sim \end{array} \quad \left[\begin{array}{cc|c} 1 & -2 & 7 \\ 3 & 4 & 1 \end{array}\right]$$
Now you see why we wanted Theorem 1-1.

Step 2.   To get a 0 in the lower left corner, we multiply R_1 by (-3) and add to R_2 (Theorem 2-3)—this changes R_2 but not R_1. Some people find it useful to write $(-3)R_1$ outside the matrix to help reduce errors in arithmetic, as shown.

$$\begin{array}{ccc} -3 & 6 & -21 \leftarrow\text{------} \end{array}$$
$$\left[\begin{array}{cc|c} 1 & -2 & 7 \\ 3 & 4 & 1 \end{array}\right] \quad \begin{array}{c} R_2 + (-3)R_1 \to R_2 \\ \sim \end{array} \quad \left[\begin{array}{cc|c} 1 & -2 & 7 \\ 0 & 10 & -20 \end{array}\right]$$

Step 3.   To get a 1 in the second row, second column, we multiply R_2 by $\frac{1}{10}$ (Theorem 2-2).

$$\left[\begin{array}{cc|c} 1 & -2 & 7 \\ 0 & 10 & -20 \end{array}\right] \quad \begin{array}{c} \frac{1}{10}R_2 \to R_2 \\ \sim \end{array} \quad \left[\begin{array}{cc|c} 1 & -2 & 7 \\ 0 & 1 & -2 \end{array}\right]$$

Step 4.   To get a 0 in the first row, second column, we multiply R_2 by 2 and add the result to R_1 (Theorem 2-3)—this changes R_1 but not R_2.

$$\begin{array}{ccc} 0 & 2 & -4 \leftarrow\text{------} \end{array}$$
$$\left[\begin{array}{cc|c} 1 & -2 & 7 \\ 0 & 1 & -2 \end{array}\right] \quad \begin{array}{c} R_1 + 2R_2 \to R_1 \\ \sim \end{array} \quad \left[\begin{array}{cc|c} 1 & 0 & 3 \\ 0 & 1 & -2 \end{array}\right]$$

We have accomplished our objective! The last matrix is the augmented matrix for the system

$$x_1 \quad = 3$$
$$x_2 = -2$$

(6)

Since system (6) is equivalent to system (3), our starting system, we have solved (3); that is, $x_1 = 3$ and $x_2 = -2$.

Check

$$3x_1 + 4x_2 = 1 \qquad x_1 - 2x_2 = 7$$

$$3(3) + 4(-2) \overset{?}{=} 1 \qquad 3 - 2(-2) \overset{?}{=} 7$$

$$9 - 8 \overset{\vee}{=} 1 \qquad 3 + 4 \overset{\vee}{=} 7$$

This process is written more compactly as follows:

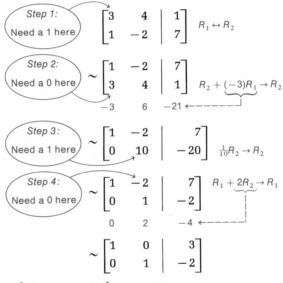

Therefore, $x_1 = 3$ and $x_2 = -2$.

PROBLEM 7   Solve, using augmented matrix methods.

$$2x_1 - \ \ x_2 = -7$$
$$x_1 + 2x_2 = 4$$

EXAMPLE 8   Solve, using augmented matrix methods:

$$2x_1 - 3x_2 = 7$$
$$3x_1 + 4x_2 = 2$$

Solution

Step 1:
Need a 1 here
$$\begin{bmatrix} 2 & -3 & | & 7 \\ 3 & 4 & | & 2 \end{bmatrix} \qquad \tfrac{1}{2}R_1 \to R_1$$

Step 2:
Need a 0 here
$$\sim \begin{bmatrix} 1 & -\tfrac{3}{2} & | & \tfrac{7}{2} \\ 3 & 4 & | & 2 \end{bmatrix} \qquad R_2 + (-3)R_1 \to R_2$$

$$-3 \qquad \tfrac{9}{2} \qquad -\tfrac{21}{2}$$

Step 3:
Need a 1 here
$$\sim \begin{bmatrix} 1 & -\tfrac{3}{2} & | & \tfrac{7}{2} \\ 0 & \tfrac{17}{2} & | & -\tfrac{17}{2} \end{bmatrix} \qquad \tfrac{2}{17}R_2 \to R_2$$

Step 4:
Need a 0 here
$$\sim \begin{bmatrix} 1 & -\tfrac{3}{2} & | & \tfrac{7}{2} \\ 0 & 1 & | & -1 \end{bmatrix} \qquad R_1 + \tfrac{3}{2}R_2 \to R_1$$

$$0 \qquad -\tfrac{3}{2} \qquad -\tfrac{3}{2}$$

$$\sim \begin{bmatrix} 1 & 0 & | & 2 \\ 0 & 1 & | & -1 \end{bmatrix}$$

Thus, $x_1 = 2$ and $x_2 = -1$.

PROBLEM 8 Solve, using augmented matrix methods.

$$5x_1 - 2x_2 = 12$$
$$2x_1 + 3x_2 = 1$$

EXAMPLE 9 Solve, using augmented matrix methods.

$$2x_1 - x_2 = 4$$
$$-6x_1 + 3x_2 = -12$$

Solution
$$\begin{bmatrix} 2 & -1 & | & 4 \\ -6 & 3 & | & -12 \end{bmatrix} \qquad \begin{array}{l} \tfrac{1}{2}R_1 \to R_1 \text{ (this produces a 1 in the upper left corner)} \\ \tfrac{1}{3}R_2 \to R_2 \text{ (this simplifies } R_2) \end{array}$$

$$\sim \begin{bmatrix} 1 & -\tfrac{1}{2} & | & 2 \\ -2 & 1 & | & -4 \end{bmatrix} \qquad \begin{array}{l} R_2 + 2R_1 \to R_2 \text{ (this produces a 0} \\ \text{in the lower left corner)} \end{array}$$

$$2 \qquad -1 \qquad 4$$

$$\sim \begin{bmatrix} 1 & -\tfrac{1}{2} & | & 2 \\ 0 & 0 & | & 0 \end{bmatrix}$$

The last matrix corresponds to the system

$$x_1 - \tfrac{1}{2}x_2 = 2$$
$$0x_1 + 0x_2 = 0$$

Thus, $x_1 = \tfrac{1}{2}x_2 + 2$. Hence, for any real number t,

$$x_2 = t \quad \text{and} \quad x_1 = \tfrac{1}{2}t + 2$$

that is, $(\frac{1}{2}t + 2, t)$ is a solution. For example, if $t = 6$, then $(5, 6)$ is a solution; if $t = -2$, then $(1, -2)$ is a solution; and so on. Geometrically, the graphs of the two original equations coincide and there are infinitely many solutions. In general, if we end up with a row of 0's in an augmented matrix for a two-equation–two-unknown system, the system is dependent and there are infinitely many solutions.

PROBLEM 9 Solve, using augmented matrix methods.

$$-2x_1 + 6x_2 = 6$$
$$3x_1 - 9x_2 = -9$$

EXAMPLE 10 Solve, using augmented matrix methods.

$$2x_1 + 6x_2 = -3$$
$$x_1 + 3x_2 = 2$$

Solution

$$\begin{bmatrix} 2 & 6 & \bigm| & -3 \\ 1 & 3 & \bigm| & 2 \end{bmatrix} \quad R_1 \leftrightarrow R_2$$

$$\sim \begin{bmatrix} 1 & 3 & \bigm| & 2 \\ 2 & 6 & \bigm| & -3 \end{bmatrix} \quad R_2 + (-2)R_1 \rightarrow R_2$$

$$\begin{matrix} -2 & -6 & & -4 \end{matrix}$$

$$\sim \begin{bmatrix} 1 & 3 & \bigm| & 2 \\ 0 & 0 & \bigm| & -7 \end{bmatrix} \quad R_2 \text{ implies the contradiction: } 0 = -7$$

The system is inconsistent and has no solution—otherwise, we have proved that $0 = -7$! Thus, if in a row of an augmented matrix we obtain all 0's to the left of the vertical bar and a nonzero number to the right of the bar, then the system is inconsistent and there are no solutions.

PROBLEM 10 Solve, using augmented matrix methods.

$$2x_1 - x_2 = 3$$
$$4x_1 - 2x_2 = -1$$

Summary

	FORM 2	FORM 3
FORM 1	INFINITELY MANY SOLUTIONS	NO SOLUTION
A UNIQUE SOLUTION	(DEPENDENT)	(INCONSISTENT)
$\begin{bmatrix} 1 & 0 & \bigm| & m \\ 0 & 1 & \bigm| & n \end{bmatrix}$	$\begin{bmatrix} 1 & m & \bigm| & n \\ 0 & 0 & \bigm| & 0 \end{bmatrix}$	$\begin{bmatrix} 1 & m & \bigm| & n \\ 0 & 0 & \bigm| & p \end{bmatrix}$

m, n, p real numbers; $p \neq 0$

The process of solving systems of equations described in this section is referred to as **Gauss–Jordan elimination**. We will use this method to solve larger-scale systems in the next section, including systems where the number of equations and the number of variables are not the same.

Answers to Matched Problems

7. $x_1 = -2, x_2 = 3$ **8.** $x_1 = 2, x_2 = -1$

9. The system is dependent. For t any real number, $x_2 = t, x_1 = 3t - 3$ is a solution.

10. Inconsistent—no solution

Exercise 5-2 ■

A *Perform each of the indicated row operations on the following matrix.*

$$\begin{bmatrix} 1 & -3 & \bigm| & 2 \\ 4 & -6 & \bigm| & -8 \end{bmatrix}$$

1. $R_1 \leftrightarrow R_2$ **2.** $\frac{1}{2}R_2 \to R_2$

3. $-4R_1 \to R_1$ **4.** $-2R_1 \to R_1$

5. $2R_2 \to R_2$ **6.** $-1R_2 \to R_2$

7. $R_2 + (-4)R_1 \to R_2$ **8.** $R_1 + (-\frac{1}{2})R_2 \to R_1$

9. $R_2 + (-2)R_1 \to R_2$ **10.** $R_2 + (-3)R_1 \to R_2$

11. $R_2 + (-1)R_1 \to R_2$ **12.** $R_2 + (1)R_1 \to R_2$

Solve, using augmented matrix methods.

13. $x_1 + x_2 = 5$
$x_1 - x_2 = 1$

14. $x_1 - x_2 = 2$
$x_1 + x_2 = 6$

B *Solve, using augmented matrix methods.*

15. $x_1 - 2x_2 = 1$
$2x_1 - x_2 = 5$

16. $x_1 + 3x_2 = 1$
$3x_1 - 2x_2 = 14$

17. $x_1 - 4x_2 = -2$
$-2x_1 + x_2 = -3$

18. $x_1 - 3x_2 = -5$
$-3x_1 - x_2 = 5$

19. $3x_1 - x_2 = 2$
$x_1 + 2x_2 = 10$

20. $2x_1 + x_2 = 0$
$x_1 - 2x_2 = -5$

21. $x_1 + 2x_2 = 4$
$2x_1 + 4x_2 = -8$

22. $2x_1 - 3x_2 = -2$
$-4x_1 + 6x_2 = 7$

23. $2x_1 + x_2 = 6$
$x_1 - x_2 = -3$

24. $3x_1 - x_2 = -5$
$x_1 + 3x_2 = 5$

25. $3x_1 - 6x_2 = -9$
$-2x_1 + 4x_2 = 6$

26. $2x_1 - 4x_2 = -2$
$-3x_1 + 6x_2 = 3$

27. $4x_1 - 2x_2 = 2$
$-6x_1 + 3x_2 = -3$

28. $-6x_1 + 2x_2 = 4$
$3x_1 - x_2 = -2$

C *Solve, using augmented matrix methods.*

29. $3x_1 - x_2 = 7$
$2x_1 + 3x_2 = 1$

30. $2x_1 - 3x_2 = -8$
$5x_1 + 3x_2 = 1$

31. $3x_1 + 2x_2 = 4$
$2x_1 - x_2 = 5$

32. $4x_1 + 3x_2 = 26$
$3x_1 - 11x_2 = -7$

33. $0.2x_1 - 0.5x_2 = 0.07$
$0.8x_1 - 0.3x_2 = 0.79$

34. $0.3x_1 - 0.6x_2 = 0.18$
$0.5x_1 - 0.2x_2 = 0.54$

Section 5-3 Gauss–Jordan Elimination

■ Reduced Matrices
■ Solving Systems by Gauss–Jordan Elimination

Now that you have had some experience with row operations on simple augmented matrices, we will consider systems involving more than two variables. In addition, we will not require that a system have the same number of equations as variables.

■ **Reduced Matrices**

Our objective is to start with the augmented matrix of a linear system and transform it, using row operations from Theorem 2 in the preceding section, into a simple form where the solution can be read by inspection. The simple form we will obtain is called the *reduced form*, and we define it as follows:

Reduced Matrix

A matrix is in **reduced form** if:

1. Each row consisting entirely of 0's is below any row having at least one nonzero element.
2. The leftmost nonzero element in each row is 1.
3. The column containing the leftmost 1 of a given row has 0's above and below the 1.
4. The leftmost 1 in any row is to the right of the leftmost 1 in the preceding row.

EXAMPLE 11   The following matrices are in reduced form. Check each one carefully to convince yourself that the conditions in the definition are met.

$$\left[\begin{array}{cc|c} 1 & 0 & 2 \\ 0 & 1 & -3 \end{array}\right] \qquad \left[\begin{array}{ccc|c} 1 & 0 & 0 & 2 \\ 0 & 1 & 0 & -1 \\ 0 & 0 & 1 & 3 \end{array}\right] \qquad \left[\begin{array}{cc|c} 1 & 0 & 3 \\ 0 & 1 & -1 \\ 0 & 0 & 0 \end{array}\right]$$

$$\left[\begin{array}{cccc|c} 1 & 4 & 0 & 0 & -3 \\ 0 & 0 & 1 & 0 & 2 \\ 0 & 0 & 0 & 1 & 6 \end{array}\right] \qquad \left[\begin{array}{ccc|c} 1 & 0 & 4 & 0 \\ 0 & 1 & 3 & 0 \\ 0 & 0 & 0 & 1 \end{array}\right]$$

PROBLEM 11   The matrices below are not in reduced form. Indicate which condition in the definition is violated for each matrix.

(A) $\left[\begin{array}{cc|c} 1 & 0 & 2 \\ 0 & 3 & -6 \end{array}\right]$ (B) $\left[\begin{array}{ccc|c} 1 & 5 & 4 & 3 \\ 0 & 1 & 2 & -1 \\ 0 & 0 & 0 & 0 \end{array}\right]$

(C) $\left[\begin{array}{ccc|c} 0 & 1 & 2 & -3 \\ 1 & -2 & 3 & 0 \\ 0 & 0 & 1 & 2 \end{array}\right]$ (D) $\left[\begin{array}{ccc|c} 1 & 2 & 0 & 3 \\ 0 & 0 & 0 & 0 \\ 0 & 0 & 1 & 4 \end{array}\right]$

EXAMPLE 12   Write the linear system corresponding to each reduced augmented matrix and solve.

(A) $\left[\begin{array}{ccc|c} 1 & 0 & 0 & 2 \\ 0 & 1 & 0 & -1 \\ 0 & 0 & 1 & 3 \end{array}\right]$ (B) $\left[\begin{array}{ccc|c} 1 & 0 & 4 & 0 \\ 0 & 1 & 3 & 0 \\ 0 & 0 & 0 & 1 \end{array}\right]$

(C) $\left[\begin{array}{ccc|c} 1 & 0 & 2 & -3 \\ 0 & 1 & -1 & 8 \\ 0 & 0 & 0 & 0 \end{array}\right]$ (D) $\left[\begin{array}{cccc|c} 1 & 4 & 0 & 0 & 3 & -2 \\ 0 & 0 & 1 & 0 & -2 & 0 \\ 0 & 0 & 0 & 1 & 2 & 4 \end{array}\right]$

Solution   (A) $\begin{aligned} x_1 &&&= 2 \\ && x_2 &&= -1 \\ &&& x_3 = 3 \end{aligned}$

The solution is obvious: $x_1 = 2$, $x_2 = -1$, $x_3 = 3$.

(B) $\begin{aligned} x_1 && + 4x_3 &= 0 \\ & x_2 + 3x_3 &= 0 \\ 0x_1 + 0x_2 &+ 0x_3 &= 1 \end{aligned}$

The last equation implies $0 = 1$, which is a contradiction. Hence, the system is inconsistent and has no solution.

(C) $\begin{aligned} x_1 && + 2x_3 &= -3 \\ & x_2 - & x_3 &= 8 \end{aligned}$ We disregard the equation corresponding to the third row in the matrix, since it is satisfied by all values of x_1, x_2, and x_3.

When a reduced system (a system corresponding to a reduced augmented matrix) has more variables than equations, the system is dependent and has infinitely many solutions. To represent these solutions, it is useful to divide the variables into two types: **basic variables** and **nonbasic variables**. To represent the infinitely many solutions to the system, we solve for the basic variables in terms of the nonbasic variables. This can be accomplished very easily if *we choose as basic variables the first variable (with a nonzero coefficient) in each equation of the reduced system.* Since each of these variables occurs in exactly one equation, it is easy to solve for each in terms of the other variables, the nonbasic variables. Returning to our original system, we choose x_1 and x_2 (the first variable in each equation) as basic variables and x_3 as a nonbasic variable. We then solve for the basic variables x_1 and x_2 in terms of the nonbasic variable x_3:

$$x_1 = -2x_3 - 3$$
$$x_2 = x_3 + 8$$

If we let $x_3 = t$, then for any real number t,

$$x_1 = -2t - 3$$
$$x_2 = t + 8$$
$$x_3 = t$$

is a solution. For example,

If $t = 0$, then	If $t = -2$, then
$x_1 = -2(0) - 3 = -3$	$x_1 = -2(-2) - 3 = 1$
$x_2 = 0 + 8 = 8$	$x_2 = -2 + 8 = 6$
$x_3 = 0$	$x_3 = -2$
is a solution.	is a solution.

(D)
$$x_1 + 4x_2 \qquad\quad + 3x_5 = -2$$
$$x_3 \quad - 2x_5 = 0$$
$$x_4 + 2x_5 = 4$$

Solve for x_1, x_3, and x_4 (basic variables) in terms of x_2 and x_5 (nonbasic variables).

$$x_1 = -4x_2 - 3x_5 - 2$$
$$x_3 = 2x_5$$
$$x_4 = -2x_5 + 4$$

If we let $x_2 = s$ and $x_5 = t$, then for any real numbers s and t,

$$x_1 = -4s - 3t - 2$$
$$x_2 = s$$
$$x_3 = 2t$$
$$x_4 = -2t + 4$$
$$x_5 = t$$

is a solution. The system is dependent and has infinitely many solutions. Can you find two?

PROBLEM 12   Write the linear system corresponding to each reduced augmented matrix and solve.

(A) $\begin{bmatrix} 1 & 0 & 0 & | & -5 \\ 0 & 1 & 0 & | & 3 \\ 0 & 0 & 1 & | & 6 \end{bmatrix}$ (B) $\begin{bmatrix} 1 & 2 & -3 & | & 0 \\ 0 & 0 & 0 & | & 1 \\ 0 & 0 & 0 & | & 0 \end{bmatrix}$

(C) $\begin{bmatrix} 1 & 0 & -2 & | & 4 \\ 0 & 1 & 3 & | & -2 \\ 0 & 0 & 0 & | & 0 \end{bmatrix}$ (D) $\begin{bmatrix} 1 & 0 & 3 & 2 & | & 5 \\ 0 & 1 & -2 & -1 & | & 3 \\ 0 & 0 & 0 & 0 & | & 0 \end{bmatrix}$

■ Solving Systems by Gauss–Jordan Elimination

We are now ready to outline a step-by-step procedure for solving systems of linear equations called the *Gauss–Jordan elimination* method. The method provides us with a systematic way of transforming augmented matrices into a reduced form from which we can write the solution to the original system by inspection, if a solution exists. The method will also reveal when a solution fails to exist.

EXAMPLE 13   Solve by Gauss–Jordan elimination.

$$2x_1 - 2x_2 + x_3 = 3$$
$$3x_1 + x_2 - x_3 = 7$$
$$x_1 - 3x_2 + 2x_3 = 0$$

Solution   Write the augmented matrix and follow the steps indicated at the right.

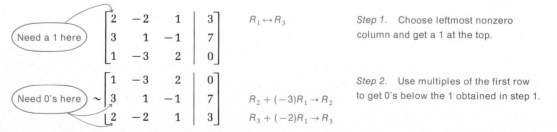

$\begin{bmatrix} 2 & -2 & 1 & | & 3 \\ 3 & 1 & -1 & | & 7 \\ 1 & -3 & 2 & | & 0 \end{bmatrix}$ $R_1 \leftrightarrow R_3$ *Step 1.* Choose leftmost nonzero column and get a 1 at the top.

Need a 1 here

$\sim \begin{bmatrix} 1 & -3 & 2 & | & 0 \\ 3 & 1 & -1 & | & 7 \\ 2 & -2 & 1 & | & 3 \end{bmatrix}$ $R_2 + (-3)R_1 \rightarrow R_2$ *Step 2.* Use multiples of the first row to get 0's below the 1 obtained in step 1.

Need 0's here $R_3 + (-2)R_1 \rightarrow R_3$

$$\boxed{\text{Need a 1 here}} \sim \begin{bmatrix} 1 & -3 & 2 & | & 0 \\ 0 & 10 & -7 & | & 7 \\ 0 & 4 & -3 & | & 3 \end{bmatrix} \quad \tfrac{1}{10}R_2 \to R_2$$

Step 3. Mentally delete R_1 and repeat steps 1 and 2 with the **submatrix** (the matrix that remains after deleting the top row and the first column). Continue the above process (steps 1–3) until it is not possible to go further; then proceed with step 4.

$$\boxed{\text{Need a 0 here}} \sim \begin{bmatrix} 1 & -3 & 2 & | & 0 \\ 0 & 1 & -\tfrac{7}{10} & | & \tfrac{7}{10} \\ 0 & 4 & -3 & | & 3 \end{bmatrix} \quad R_3 + (-4)R_2 \to R_3$$

$$\boxed{\text{Need a 1 here}} \sim \begin{bmatrix} 1 & -3 & 2 & | & 0 \\ 0 & 1 & -\tfrac{7}{10} & | & \tfrac{7}{10} \\ 0 & 0 & -\tfrac{1}{5} & | & \tfrac{1}{5} \end{bmatrix} \quad (-5)R_2 \to R_2$$

Mentally delete R_1 and R_2.

$$\boxed{\text{Need a 0 here}} \sim \begin{bmatrix} 1 & -3 & 2 & | & 0 \\ 0 & 1 & -\tfrac{7}{10} & | & \tfrac{7}{10} \\ 0 & 0 & 1 & | & -1 \end{bmatrix} \quad \begin{array}{l} R_1 + (-2)R_3 \to R_1 \\ R_2 + \tfrac{7}{10}R_3 \to R_2 \end{array}$$

Since steps 1–3 cannot be carried further, proceed to step 4.

Step 4. Return deleted rows. Begin with the bottom nonzero row and use appropriate multiples of it to get 0's above the leftmost 1. Continue the process, moving up row by row, until the matrix is in reduced form.

$$\boxed{\text{Need a 0 here}} \sim \begin{bmatrix} 1 & -3 & 0 & | & 2 \\ 0 & 1 & 0 & | & 0 \\ 0 & 0 & 1 & | & -1 \end{bmatrix} \quad R_1 + 3R_2 \to R_1$$

$$\sim \begin{bmatrix} 1 & 0 & 0 & | & 2 \\ 0 & 1 & 0 & | & 0 \\ 0 & 0 & 1 & | & -1 \end{bmatrix}$$

The matrix is in reduced form, and we can write the solution to the original system by inspection.

Solution: $x_1 = 2,\ x_2 = 0,\ x_3 = -1$. It is left to the reader to check this solution.

Steps 1–4 outlined in the solution of Example 13 are referred to as **Gauss–Jordan elimination**. The steps are summarized in the box for easy reference.

> **Gauss–Jordan Elimination**
>
> 1. Choose the leftmost nonzero column and use appropriate row operations to get a 1 at the top.
> 2. Use multiples of the first row to get 0's in all places below the 1 obtained in step 1.
> 3. Delete (mentally) the top row and first column of the matrix. Repeat steps 1 and 2 with the **submatrix** (the matrix that remains after deleting the top row and first column). Continue this process (steps 1–3) until it is not possible to go further.

4. Consider the whole matrix obtained after mentally returning all the rows to the matrix. Begin with the bottom nonzero row and use appropriate multiples of it to get 0's above the leftmost 1. Continue this process, moving up row by row, until the matrix is finally in reduced form.

[*Note:* If at any point in the above process we obtain a row with all 0's to the left of the vertical line and a nonzero number to the right, we can stop, since we will have a contradiction ($0 = n$, $n \neq 0$). We can then conclude that the system has no solution.]

PROBLEM 13   Solve by Gauss–Jordan elimination.

$$3x_1 + x_2 - 2x_3 = 2$$
$$x_1 - 2x_2 + x_3 = 3$$
$$2x_1 - x_2 - 3x_3 = 3$$

EXAMPLE 14   Solve by Gauss–Jordan elimination.

$$2x_1 - x_2 + 4x_3 = -2$$
$$3x_1 + 2x_2 - x_3 = 1$$

Solution

$$\left[\begin{array}{ccc|c} 2 & -1 & 4 & -2 \\ 3 & 2 & -1 & 1 \end{array}\right] \quad \tfrac{1}{2}R_1 \to R_1$$

(Need a 1 here)

$$\sim \left[\begin{array}{ccc|c} 1 & -\tfrac{1}{2} & 2 & -1 \\ 3 & 2 & -1 & 1 \end{array}\right] \quad R_2 + (-3)R_1 \to R_2$$

(Need a 0 here)

$$\sim \left[\begin{array}{ccc|c} 1 & -\tfrac{1}{2} & 2 & -1 \\ 0 & \tfrac{7}{2} & -7 & 4 \end{array}\right] \quad \tfrac{2}{7}R_2 \to R_2$$

(Need a 1 here)

$$\sim \left[\begin{array}{ccc|c} 1 & -\tfrac{1}{2} & 2 & -1 \\ 0 & 1 & -2 & \tfrac{8}{7} \end{array}\right] \quad R_1 + \tfrac{1}{2}R_2 \to R_1$$

(Need a 0 here)

$$\sim \left[\begin{array}{ccc|c} 1 & 0 & 1 & -\tfrac{3}{7} \\ 0 & 1 & -2 & \tfrac{8}{7} \end{array}\right]$$

The matrix is now in reduced form. Write the corresponding system and the solution.

$$x_1 \quad + x_3 = -\tfrac{3}{7}$$
$$x_2 - 2x_3 = \tfrac{8}{7}$$

Solve for the basic variables x_1 and x_2 in terms of the nonbasic variable x_3.

$$x_1 = -x_3 - \tfrac{3}{7}$$
$$x_2 = 2x_3 + \tfrac{8}{7}$$

If $x_3 = t$, then for t any real number,

$x_1 = -t - \frac{3}{7}$

$x_2 = 2t + \frac{8}{7}$

$x_3 = t$

is a solution.

Remark: In general, it can be proved that a system with more variables than equations cannot have a unique solution.

PROBLEM 14   Solve by Gauss–Jordan elimination.

$$3x_1 + 6x_2 - 3x_3 = 2$$
$$2x_1 - x_2 + 2x_3 = -1$$

EXAMPLE 15   Solve by Gauss–Jordan elimination.

$$2x_1 - x_2 = -4$$
$$x_1 + 2x_2 = 3$$
$$3x_1 - x_2 = -1$$

Solution

$$\begin{bmatrix} 2 & -1 & \bigm| & -4 \\ 1 & 2 & \bigm| & 3 \\ 3 & -1 & \bigm| & -1 \end{bmatrix} \quad R_1 \leftrightarrow R_2$$

$$\sim \begin{bmatrix} 1 & 2 & \bigm| & 3 \\ 2 & -1 & \bigm| & -4 \\ 3 & -1 & \bigm| & -1 \end{bmatrix} \quad \begin{array}{l} R_2 + (-2)R_1 \to R_2 \\ R_3 + (-3)R_1 \to R_3 \end{array}$$

$$\sim \begin{bmatrix} 1 & 2 & \bigm| & 3 \\ 0 & -5 & \bigm| & -10 \\ 0 & -7 & \bigm| & -10 \end{bmatrix} \quad -\tfrac{1}{5}R_2 \to R_2$$

$$\sim \begin{bmatrix} 1 & 2 & \bigm| & 3 \\ 0 & 1 & \bigm| & 2 \\ 0 & -7 & \bigm| & -10 \end{bmatrix} \quad R_3 + 7R_2 \to R_3$$

$$\sim \begin{bmatrix} 1 & 2 & \bigm| & 3 \\ 0 & 1 & \bigm| & 2 \\ 0 & 0 & \bigm| & 4 \end{bmatrix}$$

We stop the Gauss–Jordan elimination even though the matrix is not in reduced form, since the last row produces a contradiction.

The last row implies $0 = 4$, which is a contradiction; therefore, the system has no solution.

PROBLEM 15   Solve by Gauss–Jordan elimination.

$$3x_1 + x_2 = 5$$
$$2x_1 + 3x_2 = 1$$
$$x_1 - x_2 = 3 \ .$$

Answers to Matched Problems   **11.** (A) Condition 2 is violated: The 3 in the second row should be a 1.
(B) Condition 3 is violated: In the second column, the 5 should be a 0.
(C) Condition 4 is violated: The leftmost 1 in the second row is not to the right of the leftmost 1 in the first row.
(D) Condition 1 is violated: The all-zero second row should be at the bottom.

12. (A) $x_1 \qquad\quad = -5$
$\qquad x_2 \quad\; = 3$
$\qquad\qquad x_3 = 6$

Solution:
$x_1 = -5, x_2 = 3, x_3 = 6$

(B) $x_1 + 2x_2 - 3x_3 = 0$
$0x_1 + 0x_2 + 0x_3 = 1$
$0x_1 + 0x_2 + 0x_3 = 0$

Inconsistent; no solution.

(C) $x_1 \quad\;\; - 2x_3 = 4$
$\qquad x_2 + 3x_3 = -2$

Dependent: let $x_3 = t$. Then for any real t,
$x_1 = 2t + 4$
$x_2 = -3t - 2$
$x_3 = t$
is a solution.

(D) $x_1 \qquad\; + 3x_3 + 2x_4 = 5$
$\qquad x_2 - 2x_3 - \;\; x_4 = 3$

Dependent: let $x_3 = s$ and $x_4 = t$. Then for any real s and t,
$x_1 = -3s - 2t + 5$
$x_2 = 2s + t + 3$
$x_3 = s$
$x_4 = t$
is a solution.

13. $x_1 = 1, \quad x_2 = -1, \quad x_3 = 0$

14. $x_1 = -\frac{3}{5}t - \frac{4}{15}, \quad x_2 = \frac{4}{5}t + \frac{7}{15}, \quad x_3 = t, \quad t$ any real number

15. $x_1 = 2, \quad x_2 = -1$

Exercise 5-3 ▪ A   *Indicate whether each matrix is in reduced form.*

1. $\begin{bmatrix} 1 & 0 & | & 2 \\ 0 & 1 & | & -1 \end{bmatrix}$

2. $\begin{bmatrix} 0 & 1 & | & 2 \\ 1 & 0 & | & -1 \end{bmatrix}$

3. $\begin{bmatrix} 1 & 0 & 2 & | & 3 \\ 0 & 0 & 0 & | & 0 \\ 0 & 1 & -1 & | & 4 \end{bmatrix}$

4. $\begin{bmatrix} 1 & 0 & 0 & | & -2 \\ 0 & 1 & 0 & | & 0 \\ 0 & 0 & 1 & | & 1 \end{bmatrix}$

5. $\begin{bmatrix} 0 & 1 & 0 & | & 2 \\ 0 & 0 & 3 & | & -1 \\ 0 & 0 & 0 & | & 0 \end{bmatrix}$

6. $\begin{bmatrix} 1 & 3 & 0 & | & 0 \\ 0 & 0 & 1 & | & 0 \\ 0 & 0 & 0 & | & 1 \end{bmatrix}$

7. $\begin{bmatrix} 1 & 2 & 0 & 3 & | & 2 \\ 0 & 0 & 1 & -1 & | & 0 \end{bmatrix}$

8. $\begin{bmatrix} 0 & 1 & 2 & | & 1 \\ 1 & 0 & -3 & | & 2 \end{bmatrix}$

Write the linear system corresponding to each reduced augmented matrix and solve.

9. $\begin{bmatrix} 1 & 0 & 0 & | & -2 \\ 0 & 1 & 0 & | & 3 \\ 0 & 0 & 1 & | & 0 \end{bmatrix}$

10. $\begin{bmatrix} 1 & 0 & 0 & 0 & | & -2 \\ 0 & 1 & 0 & 0 & | & 0 \\ 0 & 0 & 1 & 0 & | & 1 \\ 0 & 0 & 0 & 1 & | & 3 \end{bmatrix}$

11. $\begin{bmatrix} 1 & 0 & -2 & | & 3 \\ 0 & 1 & 1 & | & -5 \\ 0 & 0 & 0 & | & 0 \end{bmatrix}$

12. $\begin{bmatrix} 1 & -2 & 0 & | & -3 \\ 0 & 0 & 1 & | & 5 \\ 0 & 0 & 0 & | & 0 \end{bmatrix}$

13. $\begin{bmatrix} 1 & 0 & | & 0 \\ 0 & 1 & | & 0 \\ 0 & 0 & | & 1 \end{bmatrix}$

14. $\begin{bmatrix} 1 & 0 & | & 5 \\ 0 & 1 & | & -3 \\ 0 & 0 & | & 0 \end{bmatrix}$

15. $\begin{bmatrix} 1 & -2 & 0 & -3 & | & -5 \\ 0 & 0 & 1 & 3 & | & 2 \end{bmatrix}$

16. $\begin{bmatrix} 1 & 0 & -2 & 3 & | & 4 \\ 0 & 1 & -1 & 2 & | & -1 \end{bmatrix}$

B *Use row operations to change each matrix to reduced form.*

17. $\begin{bmatrix} 1 & 2 & | & -1 \\ 0 & 1 & | & 3 \end{bmatrix}$

18. $\begin{bmatrix} 1 & 3 & | & 1 \\ 0 & 2 & | & -4 \end{bmatrix}$

19. $\begin{bmatrix} 1 & 0 & -3 & | & 1 \\ 0 & 1 & 2 & | & 0 \\ 0 & 0 & 3 & | & -6 \end{bmatrix}$

20. $\begin{bmatrix} 1 & 0 & 4 & | & 0 \\ 0 & 1 & -3 & | & -1 \\ 0 & 0 & -2 & | & 2 \end{bmatrix}$

21. $\begin{bmatrix} 1 & 2 & -2 & | & -1 \\ 0 & 3 & -6 & | & 1 \\ 0 & -1 & 2 & | & -\frac{1}{3} \end{bmatrix}$

22. $\begin{bmatrix} 0 & -2 & 8 & | & 1 \\ 2 & -2 & 6 & | & -4 \\ 0 & -1 & 4 & | & \frac{1}{2} \end{bmatrix}$

Solve, using Gauss–Jordan elimination.

23. $2x_1 + 4x_2 - 10x_3 = -2$
$3x_1 + 9x_2 - 21x_3 = 0$
$x_1 + 5x_2 - 12x_3 = 1$

24. $3x_1 + 5x_2 - x_3 = -7$
$x_1 + x_2 + x_3 = -1$
$2x_1 + 11x_3 = 7$

25. $3x_1 + 8x_2 - x_3 = -18$
$2x_1 + x_2 + 5x_3 = 8$
$2x_1 + 4x_2 + 2x_3 = -4$

26. $2x_1 + 7x_2 + 15x_3 = -12$
$4x_1 + 7x_2 + 13x_3 = -10$
$3x_1 + 6x_2 + 12x_3 = -9$

27. $2x_1 - x_2 - 3x_3 = 8$
$x_1 - 2x_2 = 7$

28. $2x_1 + 4x_2 - 6x_3 = 10$
$3x_1 + 3x_2 - 3x_3 = 6$

29. $2x_1 + 3x_2 - x_3 = 1$
$x_1 - 2x_2 + 2x_3 = -2$

30. $x_1 - 3x_2 + 2x_3 = -1$
$3x_1 + 2x_2 - x_3 = 2$

31. $2x_1 + 2x_2 = 2$
$x_1 + 2x_2 = 3$
$ - 3x_2 = -6$

32. $2x_1 - x_2 = 0$
$3x_1 + 2x_2 = 7$
$x_1 - x_2 = -1$

33. $\begin{aligned} 2x_1 - x_2 &= 0 \\ 3x_1 + 2x_2 &= 7 \\ x_1 - x_2 &= -2 \end{aligned}$ **34.** $\begin{aligned} x_1 - 3x_2 &= 5 \\ 2x_1 + x_2 &= 3 \\ x_1 - 2x_2 &= 5 \end{aligned}$

35. $\begin{aligned} 3x_1 - 4x_2 - x_3 &= 1 \\ 2x_1 - 3x_2 + x_3 &= 1 \\ x_1 - 2x_2 + 3x_3 &= 2 \end{aligned}$ **36.** $\begin{aligned} 3x_1 + 7x_2 - x_3 &= 11 \\ x_1 + 2x_2 - x_3 &= 3 \\ 2x_1 + 4x_2 - 2x_3 &= 10 \end{aligned}$

C   *Solve, using Gauss–Jordan elimination.*

37. $\begin{aligned} 2x_1 - 3x_2 + 3x_3 &= -15 \\ 3x_1 + 2x_2 - 5x_3 &= 19 \\ 5x_1 - 4x_2 - 2x_3 &= -2 \end{aligned}$ **38.** $\begin{aligned} 3x_1 - 2x_2 - 4x_3 &= -8 \\ 4x_1 + 3x_2 - 5x_3 &= -5 \\ 6x_1 - 5x_2 + 2x_3 &= -17 \end{aligned}$

39. $\begin{aligned} 5x_1 - 3x_2 + 2x_3 &= 13 \\ 2x_1 + 4x_2 - 3x_3 &= -9 \\ 4x_1 - 2x_2 + 5x_3 &= 13 \end{aligned}$ **40.** $\begin{aligned} 4x_1 - 2x_2 + 3x_3 &= 0 \\ 3x_1 - 5x_2 - 2x_3 &= -12 \\ 2x_1 + 4x_2 - 3x_3 &= -4 \end{aligned}$

41. $\begin{aligned} x_1 + 2x_2 - 4x_3 - x_4 &= 7 \\ 2x_1 + 5x_2 - 9x_3 - 4x_4 &= 16 \\ x_1 + 5x_2 - 7x_3 - 7x_4 &= 13 \end{aligned}$ **42.** $\begin{aligned} 2x_1 + 4x_2 + 5x_3 + 4x_4 &= 8 \\ x_1 + 2x_2 + 2x_3 + x_4 &= 3 \end{aligned}$

APPLICATIONS   *Solve all of the following problems using Gauss–Jordan elimination.*

43. *Production scheduling.* A small manufacturing plant makes three types of inflatable boats: one-person, two-person, and four-person models. Each boat requires the services of three departments, as listed in the table. The cutting, assembly, and packaging departments have available a maximum of 380, 330, and 120 work-hours per week, respectively. How many boats of each type must be produced each week for the plant to operate at full capacity?

	ONE-PERSON BOAT	TWO-PERSON BOAT	FOUR-PERSON BOAT
Cutting department	0.5 hr	1.0 hr	1.5 hr
Assembly department	0.6 hr	0.9 hr	1.2 hr
Packaging department	0.2 hr	0.3 hr	0.5 hr

44. *Production scheduling.* Repeat Problem 43 assuming the cutting, assembly, and packaging departments have available a maximum of 350, 330, and 115 work-hours per week, respectively.

45. *Production scheduling.* Work Problem 43 assuming the packaging department is no longer used.

46. *Production scheduling.* Work Problem 44 assuming the packaging department is no longer used.

47. *Production scheduling.* Work Problem 43 assuming the four-person boat is no longer produced.

48. *Production scheduling.* Work Problem 44 assuming the four-person boat is no longer produced.

49. *Nutrition.* A dietitian in a hospital is to arrange a special diet using three basic foods. The diet is to include exactly 340 units of calcium, 180 units of iron, and 220 units of vitamin A. The number of units per ounce of each special ingredient for each of the foods is indicated in the table. How many ounces of each food must be used to meet the diet requirements?

	UNITS PER OUNCE		
	Food A	Food B	Food C
Calcium	30	10	20
Iron	10	10	20
Vitamin A	10	30	20

50. *Nutrition.* Repeat Problem 49 if the diet is to include exactly 400 units of calcium, 160 units of iron, and 240 units of vitamin A.

51. *Nutrition.* Solve Problem 49 with the assumption that food C is no longer available.

52. *Nutrition.* Solve Problem 50 with the assumption that food C is no longer available.

53. *Nutrition.* Solve Problem 49 assuming the vitamin A requirement is deleted.

54. *Nutrition.* Solve Problem 50 assuming the vitamin A requirement is deleted.

55. *Sociology.* Two sociologists have grant money to study school busing in a particular city. They wish to conduct an opinion survey using 600 telephone contacts and 400 house contacts. Survey company A has personnel to do thirty telephone and ten house contacts per hour; survey company B can handle twenty telephone and twenty house contacts per hour. How many hours should be scheduled for each firm to produce exactly the number of contacts needed?

56. *Sociology.* Repeat Problem 55 if 650 telephone contacts and 350 house contacts are needed.

Section 5-4   Systems Involving Second-Degree Equations

■ Systems with First-Degree and Second-Degree Equations
■ Systems with Second-Degree Equations Only

In this section we will investigate systems of the form

$$4x^2 + y^2 = 25 \qquad x^2 - y^2 = 5 \qquad x^2 + 3xy + y^2 = 20$$
$$2x + y = 7 \qquad x^2 + 2y^2 = 17 \qquad xy - y^2 = 0$$

It can be shown that such systems have at most four solutions.

■ Systems with First-Degree and Second-Degree Equations

If a system involves a first-degree and a second-degree equation, the method of elimination using substitution is effective. We solve the first-degree equation for one variable in terms of the other, and substitute into the second-degree equation to obtain a quadratic in one variable. An example should make the process clear.

EXAMPLE 16   Solve the system.

$$4x^2 + y^2 = 25$$
$$2x + y = 7$$

Solution

$$2x + y = 7$$ Solve the first-degree equation for y in terms of x;
$$y = 7 - 2x$$ then substitute into the second-degree equation.

$$4x^2 + y^2 = 25$$ Second-degree equation.
$$4x^2 + (7 - 2x)^2 = 25$$ Simplify and write in standard quadratic form.

$$8x^2 - 28x + 24 = 0$$ Divide through by 4 to simplify further.
$$2x^2 - 7x + 6 = 0$$ Solve.
$$(2x - 3)(x - 2) = 0$$
$$x = \tfrac{3}{2}, 2$$

These values are substituted back into the linear equation to find the corresponding values for y. (Note that if we substitute these values back into the second-degree equations, we may obtain "extraneous" roots; try it and see why.)

For $x = \tfrac{3}{2}$,

$$2(\tfrac{3}{2}) + y = 7$$
$$y = 4$$

For x = 2,

$$2(2) + y = 7$$
$$y = 3$$

Thus $(\frac{3}{2}, 4)$ and $(2, 3)$ are solutions to the system, as can easily be checked.

PROBLEM 16   Solve.

$$2x^2 - y^2 = 1$$
$$3x + y = 2$$

■   Systems with Second-Degree Equations Only

We now look at a couple of systems where both equations are second degree.

EXAMPLE 17   Solve.

$$x^2 - y^2 = 5$$
$$x^2 + 2y^2 = 17$$

Solution   This type of system can be solved by elimination using addition. Multiply the second equation by -1 and add.

$$x^2 - y^2 = 5$$
$$\underline{-x^2 - 2y^2 = -17}$$
$$-3y^2 = -12$$
$$y^2 = 4$$
$$y = \pm 2$$

Now substitute $y = 2$ and $y = -2$ back into either original equation to find x.

For $y = 2$,

$$x^2 - (2)^2 = 5$$
$$x = \pm 3$$

For $y = -2$,

$$x^2 - (-2)^2 = 5$$
$$x = \pm 3$$

Thus, $(3, -2)$, $(3, 2)$, $(-3, -2)$, and $(-3, 2)$ are the four solutions to the system. The check of the solutions is left to the reader.

PROBLEM 17   Solve.

$$2x^2 - 3y^2 = 5$$
$$3x^2 + 4y^2 = 16$$

EXAMPLE 18   Solve.

$$x^2 + 3xy + y^2 = 20$$
$$xy - y^2 = 0$$

Solution   Factor the left side of the equation that has a 0 constant term.

$$xy - y^2 = 0$$
$$y(x - y) = 0$$
$$y = 0 \quad \text{or} \quad y = x$$

Thus, the original system is equivalent to the two systems:

$$y = 0 \qquad\qquad\qquad y = x$$
$$x^2 + 3xy + y^2 = 20 \qquad \text{or} \qquad x^2 + 3xy + y^2 = 20$$

These systems are solved as in Example 16 by substitution.

First system:

$$y = 0$$
$$x^2 + 3xy + y^2 = 20$$
$$x^2 + 3x(0) + (0)^2 = 20$$
$$x^2 = 20$$
$$x = \pm\sqrt{20} = \pm 2\sqrt{5}$$

Substitute $y = 0$ in the second equation and solve for x.

Second system:

$$y = x$$
$$x^2 + 3xy + y^2 = 20$$
$$x^2 + 3xx + x^2 = 20$$
$$5x^2 = 20$$
$$x^2 = 4$$
$$x = \pm 2$$

Substitute $y = x$ in the second equation and solve for x; then substitute these values back into $y = x$ to find y.

For $x = 2$, $y = 2$. For $x = -2$, $y = -2$.
The solutions for the original system are $(2\sqrt{5}, 0)$, $(-2\sqrt{5}, 0)$, $(2, 2)$, and $(-2, -2)$. The check of the solutions is left to the reader.

PROBLEM 18   Solve.

$$x^2 + xy - y^2 = 4$$
$$2x^2 - xy = 0$$

Example 18 is somewhat specialized; however, it suggests a procedure that is effective for some problems.

Answers to Matched Problems   **16.** $(1, -1), (\frac{5}{7}, -\frac{1}{7})$ **17.** $(2, 1), (2, -1), (-2, 1), (-2, -1)$

18. $(0, 2i), (0, -2i), (2i, 4i), (-2i, -4i)$

Exercise 5-4 ■   A *Solve each system.*

1. $x^2 + y^2 = 169$
$x = -12$

2. $x^2 + y^2 = 25$
$y = -4$

3. $8x^2 - y^2 = 16$
$y = 2x$

4. $y^2 = 2x$
$x = y - \frac{1}{2}$

5. $2x^2 - 3y^2 = 25$
$x + y = 0$

6. $x^2 + 4y^2 = 32$
$x + 2y = 0$

7. $y^2 = -x$
$x - 2y = 5$

8. $x^2 = 2y$
$3x = y + 5$

9. $2x^2 + y^2 = 24$
$x^2 - y^2 = -12$

10. $x^2 - y^2 = 3$
$x^2 + y^2 = 5$

11. $x^2 + y^2 = 10$
$16x^2 + y^2 = 25$

12. $x^2 - 2y^2 = 1$
$x^2 + 4y^2 = 25$

*solve for x
substitute*   B **13.** $xy = -4$
$y - x = 2$

14. $xy - 6 = 0$
$x - y = 4$

15. $x^2 - 2xy + y^2 = 1$
$x - 2y = 2$

16. $x^2 + xy - y^2 = -5$
$y - x = 3$

elim **17.** $2x^2 + 3y^2 = -4$
$4x^2 + 2y^2 = 8$

18. $2x^2 - 3y^2 = 10$
$x^2 + 4y^2 = -17$

(elim) either **19.** $x^2 - y^2 = 2$
$y^2 = x$

20. $x^2 + y^2 = 20$
$x^2 = y$

elim x's **21.** $x^2 + y^2 = 5$
$x^2 = 4(2 - y)$

22. $x^2 + y^2 = 16$
$y^2 = 4 - x$

23. $x^2 - y^2 = 3$
$xy = 2$ → *solve for x +
plug in*

24. $2x^2 + y^2 = 18$
$xy = 4$

C **25.** $2x^2 - xy + y^2 = 8$
$x^2 \quad\;\; - y^2 = 0$

26. $x^2 + 2xy + y^2 = 36$
$x^2 - \quad xy \quad\;\; = 0$

27. $x^2 + xy - 3y^2 = 3$
$x^2 + 4xy + 3y^2 = 0$

28. $x^2 - 2xy + 2y^2 = 16$
$x^2 \quad\quad\; - y^2 = 0$

stop

APPLICATIONS **29.** *Numbers.*   Find two numbers such that their sum is 1 and their product is 1.

30. *Numbers.*   Find two numbers such that their difference is 1 and their product is 1. (Let x be the larger number and y the smaller number.)

31. *Geometry.* Find the dimensions of a rectangle with an area of 60 square inches if its diagonal is 13 inches long.

32. *Geometry.* Find the dimensions of a rectangle with an area of 32 square meters if its perimeter is 36 meters long.

33. *Supply and demand.* The daily demand equation for a certain brand of ball-point pen in a given city is $dp = 1,000$, and the supply equation is $s = 5p - 50$, where d is the number of pen shoppers who are willing to buy at $p¢$ each, and s is the number of suppliers who are willing to sell at $p¢$ each. At what price will supply equal demand; that is, at what price will $s = d$?

Section 5-5   Systems of Linear Inequalities

- Single-Inequality Statements
- Systems of Inequality Statements
- Application

Single-Inequality Statements

We know how to graph first-degree equations such as

$$y = 2x - 3 \quad \text{and} \quad 2x - 3y = 5$$

but how do we graph first-degree inequalities such as

$$y \leq 2x - 3 \quad \text{and} \quad 2x - 3y > 5$$

We will find that graphing these inequalities is almost as easy as graphing the equalities. The following discussion leads to a simple solution of the problem.

A vertical line divides a plane into left and right **half-planes**; a non-vertical line divides a plane into upper and lower half-planes as indicated in Figure 1.

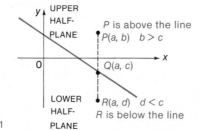

FIGURE 1

Now let us compare the graphs of the following:

$$y < 2x - 3 \qquad y = 2x - 3 \qquad y > 2x - 3$$

We start by graphing $y = 2x - 3$. For a fixed x, equality holds if a point is on the line. For the same x, if a point is below the line, then $y < 2x - 3$, and if a point is above the line, then $y > 2x - 3$. See Figure 2. Since the same results are obtained for each point on the x axis, we conclude that the graph of $y > 2x - 3$ is the upper half-plane determined by the line $y = 2x - 3$, and the graph of $y < 2x - 3$ is the lower half-plane determined by the same line.

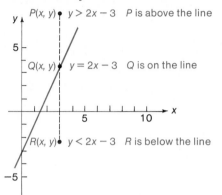

FIGURE 2

In graphing $y > 2x - 3$, we show the line $y = 2x - 3$ as a broken line, indicating that it is not part of the graph; in graphing $y \geq 2x - 3$, we show the line $y = 2x - 3$ as a solid line, indicating that it is part of the graph. Figure 3 illustrates four typical cases.

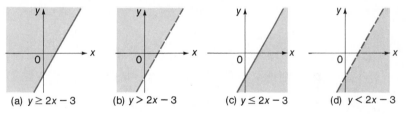

FIGURE 3   (a) $y \geq 2x - 3$ (b) $y > 2x - 3$ (c) $y \leq 2x - 3$ (d) $y < 2x - 3$

THEOREM 3

Graphs of Linear Inequalities

The graph of a linear inequality

$$Ax + By < C \quad \text{or} \quad Ax + By > C$$

with $B \neq 0$, is either the upper half-plane or the lower half-plane (but not both) determined by the line $Ax + By = C$. If $B = 0$, then the graph of

$$Ax < C \quad \text{or} \quad Ax > C$$

is either the left half-plane or the right half-plane (but not both) determined by the line $Ax = C$.

As a consequence of Theorem 3, we state a simple and fast mechanical procedure for graphing linear inequalities.

Procedure For Graphing Linear Inequalities

Step 1.   Graph $Ax + By = C$ as a broken line if equality is not included in the original statements, or as a solid line if equality is included.

Step 2.   Choose a test point anywhere in the plane not on the line [the origin $(0, 0)$ often requires the least computation] and substitute the coordinates into the inequality.

Step 3.   The graph of the original inequality includes the half-plane containing the test point if the inequality is satisfied by that point, or the half-plane not containing that point if the inequality is not satisfied by that point.

EXAMPLE 19   Graph: $3x - 4y \leq 12$

Solution   *Step 1.*   Graph $3x - 4y = 12$ as a solid line, since equality is included in the original statement.

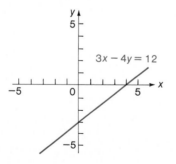

Step 2.   Pick a convenient test point above or below the line. The origin $(0, 0)$ requires the least computation. Substituting $(0, 0)$ into the inequality

$$3x - 4y \leq 12$$

$$3(0) - 4(0) = 0 \leq 12$$

produces a true statement; therefore, $(0, 0)$ is in the solution set.

Step 3.   The line $3x - 4y = 12$ and the half-plane containing the origin form the graph of $3x - 4y \leq 12$.

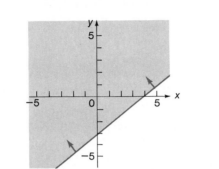

PROBLEM 19   Graph: $2x + 3y < 6$

EXAMPLE 20   Graph: (A) $y > -3$ (B) $2x \leq 5$

Solution   (A)   (B)

PROBLEM 20   Graph: (A) $y \leq 2$ (B) $3x > -8$

■ Systems of Inequality Statements

As in systems of linear equations in two variables, we say that the ordered pair of numbers (x_0, y_0) is a solution of a system of linear inequalities in two variables if the ordered pair satisfies each inequality in the system. Thus, **the graph of a system of linear inequalities is the intersection of the graphs of each inequality in the system**. In this book we will limit our investigation of solutions of systems of inequalities to graphical methods. An example will illustrate the process.

EXAMPLE 21   Solve the following linear system graphically.

$$0 \leq x \leq 8$$
$$0 \leq y \leq 4$$

Solution   This system is actually equivalent to the system

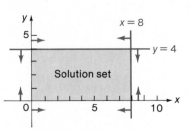

The solution to the system is the intersection of all four solution sets.

PROBLEM 21   Solve graphically.

$$2 \leq x \leq 6$$
$$1 \leq y \leq 3$$

EXAMPLE 22   Solve graphically.

$$3x + 5y \leq 60$$
$$4x + 2y \leq 40$$
$$x \geq 0$$
$$y \geq 0$$

Solution

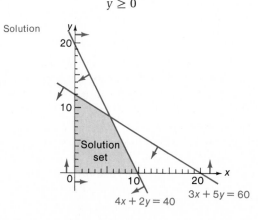

PROBLEM 22   Solve graphically.

$$x + 2y \geq 12$$
$$3x + 2y \geq 24$$
$$x \geq 0$$
$$y \geq 0$$

■ Application

EXAMPLE 23 *Production scheduling.* A manufacturer of surfboards makes a standard model and a competition model. The relevant manufacturing data are shown in the table. What combinations of boards can be produced each week so as not to exceed the number of work-hours available in each department per week?

	STANDARD MODEL (WORK-HOURS PER BOARD)	COMPETITION MODEL (WORK-HOURS PER BOARD)	MAXIMUM WORK-HOURS AVAILABLE PER WEEK
Fabricating	6	8	120
Finishing	1	3	30

Solution Let x and y be the respective number of standard and competition boards produced per week. These variables are restricted as follows:

$$6x + 8y \leq 120 \quad \text{Fabricating}$$
$$x + 3y \leq 30 \quad \text{Finishing}$$
$$x \geq 0$$
$$y \geq 0$$

The solution set of this system of inequalities is the shaded area in the figure and is referred to as the **feasible region**. Any point within the shaded area would represent a possible production schedule. Any point outside the shaded area would represent an impossible schedule. For example, it would be possible to produce 10 standard boards and five competition boards per week, but it would not be possible to produce thirteen standard boards and six competition boards per week.

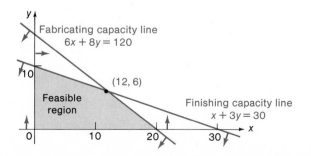

PROBLEM 23   Repeat Example 23 using 5 hours for fabricating a standard board in place of 6 hours, and a maximum of 27 work-hours for the finishing department.

Answers to Matched Problems   19.

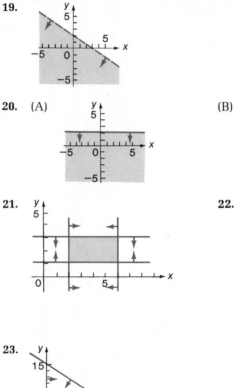

20.   (A)

(B)

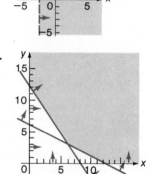

21.   22.

23.

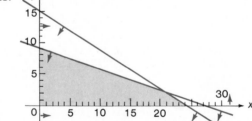

Exercise 5-5 ▪ A   *Graph each inequality.*

1. $2x - 3y < 6$ 2. $3x + 4y < 12$ 3. $3x + 2y \geq 18$

4. $3y - 2x \geq 24$ 5. $y \leq \tfrac{2}{3}x + 5$ 6. $y \geq \dfrac{x}{3} - 2$

7. $y < 8$ 8. $x > -5$ 9. $-3 \leq y < 2$

10. $-1 < x \leq 3$

B Find the solution set of each system graphically.

11. $-2 \leq x < 2$
$-1 < y \leq 6$

12. $-4 \leq x < -1$
$-2 < y \leq 5$

13. $2x + y \leq 8$
$0 \leq x \leq 3$
$0 \leq y \leq 5$

14. $x + 3y \leq 12$
$0 \leq x \leq 8$
$0 \leq y \leq 3$

15. $2x + y \leq 8$
$x + 3y \leq 12$
$x \geq 0$
$y \geq 0$

16. $x + 2y \leq 10$
$3x + y \leq 15$
$x \geq 0$
$y \geq 0$

17. $6x + 3y \leq 24$
$3x + 6y \leq 30$
$x \geq 0$
$y \geq 0$

18. $2x + y \leq 10$
$x + 2y \leq 8$
$x \geq 0$
$y \geq 0$

19. $3x + 4y \geq 8$
$4x + 3y \geq 24$
$x \geq 0$
$y \geq 0$

20. $x + 2y \geq 8$
$2x + y \geq 10$
$x \geq 0$
$y \geq 0$

C 21. $3x + 5y \geq 60$
$4x + 2y \geq 40$
$2 \leq x \leq 14$
$6 \leq y \leq 18$

22. $2x + y \geq 8$
$x + 2y \geq 10$
$1 \leq x \leq 7$
$3 \leq y \leq 9$

APPLICATIONS 23. *Manufacturing—resource allocation.* A manufacturing company makes two types of water skis: a trick ski and a slalom ski. The trick ski requires 6 work-hours for fabricating and 1 work-hour for finishing. The slalom ski requires 4 work-hours for fabricating and 1 work-hour for finishing. The maximum work-hours available per day for fabricating and finishing are 108 and 24, respectively. If x is the number of trick skis and y is the number of slalom skis produced per day, write a system of inequalities that indicates appropriate restraints on x and y. Find the set of feasible solutions graphically for the number of each type of ski that can be produced.

24. *Nutrition.* A dietitian in a hospital is to arrange a special diet using two foods. Each ounce of food M contains 30 units of calcium, 10 units of iron, and 10 units of vitamin A. Each ounce of food N contains 10 units of calcium, 10 units of iron, and 30 units of vitamin A. The minimum requirements in the diet are 360 units of calcium, 160 units of iron, and 240 units of vitamin A. If x is the number of ounces of food M used and y is the number of ounces of food N used, write a system of linear inequalities that reflects the conditions indicated. Find the set of feasible solutions graphically for the amount of each kind of food that can be used.

Section 5-6 Chapter Review

IMPORTANT TERMS
AND SYMBOLS
5-1 Systems of Linear Equations—A Review. Systems in two variables, solutions, solution set, equivalent systems, solution by graphing, solution by elimination using substitution, solution by elimination using addition, inconsistent systems, dependent systems, parameter, systems in three variables, solution, solution set, equivalent systems

$$a_1x + b_1y = k_1 \qquad a_1x + b_1y + c_1z = k_1$$
$$a_2x + b_2y = k_2 \qquad a_2x + b_2y + c_2z = k_2$$
$$a_3x + b_3y + c_3z = k_3$$

5-2 Systems and Augmented Matrices—An Introduction. Matrix, element, column, row, augmented matrix, row-equivalent matrices, row operations, Gauss–Jordan elimination

$$R_i \leftrightarrow R_j \qquad kR_i \to R_i \qquad R_i + kR_j \to R_i$$

5-3 Gauss–Jordan Elimination. Reduced form of a matrix, basic variables, nonbasic variables, submatrix, Gauss–Jordan elimination

5-4 Systems Involving Second-Degree Equations. First- and second-degree systems, solution by elimination using substitution, second-degree systems, solution by elimination using addition, solution by elimination using substitution

5-5 Systems of Linear Inequalities. Half-planes, left and right half-planes, upper and lower half-planes, solving single-inequality statements geometrically, solving systems of inequality statements geometrically

Exercise 5-6 Chapter Review

Work through all the problems in this chapter review and check answers in the back of the book. (Answers to all problems are there, and following each answer is a number in italics indicating the section in which that type of problem is discussed.) Where weaknesses show up, review appropriate sections in the text. When you are satisfied that you know the material, take the practice test following this review.

A *Solve by elimination using substitution or addition.*

1. $2x + \ y = 7$
    $3x - 2y = 0$

2. $3x - 6y = 5$
    $-2x + 4y = 1$

3. $4x - 3y = -8$
    $-2x + \frac{3}{2}y = 4$

4. $x + 2y + \ z = 3$
    $2x + 3y + 4z = 3$
    $x + 2y + 3z = 1$

5. $x^2 + y^2 = 2$
$2x - y = 3$

6. $3x^2 - y^2 = -6$
$2x^2 + 3y^2 = 29$

Solve by graphing.

7. $3x - 2y = 8$
$x + 3y = -1$

8. $2x + y \le 8$
$2x + 3y \le 12$
$x \ge 0$
$y \ge 0$

B *Solve using Gauss–Jordan elimination.*

9. $3x_1 + 2x_2 = 3$
$x_1 + 3x_2 = 8$

10. $x_1 + x_2 \quad\quad = 1$
$x_1 \quad\quad - x_3 = -2$
$x_2 + 2x_3 = 4$

11. $x_1 + 2x_2 + 3x_3 = 1$
$2x_1 + 3x_2 + 4x_3 = 3$
$x_1 + 2x_2 + x_3 = 3$

12. $x_1 + 2x_2 - x_3 = 2$
$2x_1 + 3x_2 + x_3 = -3$
$3x_1 + 5x_2 \quad\quad = -1$

13. $x_1 - 2x_2 = 1$
$2x_1 - x_2 = 0$
$x_1 - 3x_2 = -2$

14. $x_1 + 2x_2 - x_3 = 2$
$3x_1 - x_2 + 2x_3 = -3$

Solve Problems 15 and 16.

15. $x^2 - y^2 = 2$
$y^2 = x$

16. $2x^2 + xy + y^2 = 8$
$x^2 - y^2 = 0$

17. Solve graphically.

$2x + y \ge 8$
$x + 3y \ge 12$
$x \ge 0$
$y \ge 0$

C **18.** Solve using Gauss–Jordan elimination.

$x_1 + \quad x_2 + \quad x_3 = 7{,}000$
$0.04x_1 + 0.05x_2 + 0.06x_3 = 360$
$0.04x_1 + 0.05x_2 - 0.06x_3 = 120$

19. Solve.

$x^2 - xy + y^2 = 4$
$x^2 + xy - 2y^2 = 0$

APPLICATIONS **20.** *Business.* A container contains 120 packages. Some of the packages weigh $\frac{1}{2}$ pound each, and the rest weigh $\frac{1}{3}$ pound each. If the total contents of the container weigh 48 pounds, how many are there of each type of package? Solve using two-equations–two-unknowns methods.

21. *Geometry.* Find the dimensions of a rectangle with an area of 48 square meters and a perimeter of 28 meters. Solve using two-equations–two-unknowns methods.

22. *Diet.* A lab assistant wishes to obtain a food mix that contains, among other things, 27 grams of protein, 5.4 grams of fat, and 19 grams of moisture. He has available mixes of the compositions as listed in the table. How many grams of each mix should be used to get the desired diet mix? Set up a system of equations and solve using Gauss–Jordan elimination.

MIX	PROTEIN (%)	FAT (%)	MOISTURE (%)
A	30	3	10
B	20	5	20
C	10	4	10

Practice Test   Chapter 5

Take this practice test as if it were a graded test. Allow yourself up to 50 minutes. Work the problems without looking back in the chapter. Correct your work using the answers (keyed to appropriate sections) in the back of the book.

Solve by graphing.

1. $2x + y = 6$
$x + 4y = -4$

2. $3x - 4y \geq 24$

3. $x + 2y \leq 8$
$2x + y \leq 10$
$x \geq 0$
$y \geq 0$

Solve by elimination using substitution or addition.

4. $x^2 = y$
$y = 2x - 2$

5. $4x^2 - y^2 = -5$
$3x^2 + 2y^2 = 21$

6. $x^2 + 2xy + y^2 = 8$
$x^2 - xy = 0$

Solve using Gauss–Jordan elimination.

7. $2x_1 + 4x_2 = 2$
$3x_1 + 8x_2 = 1$

8. $2x_1 + x_2 + x_3 = 3$
$x_1 + 2x_2 - 4x_3 = -6$
$2x_1 + x_2 - x_3 = 1$

9. $4x_1 - 8x_2 = 8$
$3x_1 - 4x_2 = 9$
$2x_1 - 4x_2 = 2$

10. $x_1 + 2x_2 + 8x_3 = -1$
$3x_1 + 7x_2 + 29x_3 = -2$

Matrices and Determinants

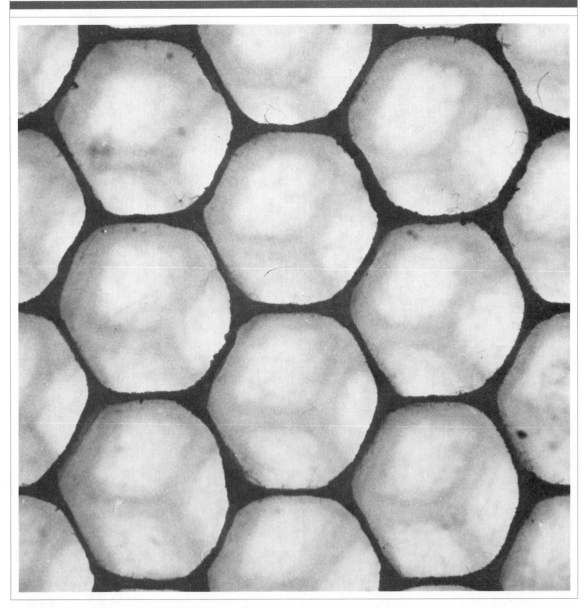

A natural design of mathematical interest. Can you guess the source? See the back of the book.

Chapter 6 ▪ Matrices and Determinants

Section 6-1 Matrix Addition; Multiplication of a Matrix by a Number

- ▪ Dimension of a Matrix
- ▪ Matrix Addition
- ▪ Multiplication of a Matrix by a Number
- ▪ Application

In the last chapter we introduced the important idea of matrix. In this and the following two sections, we will develop this concept further.

▪ Dimension of a Matrix

Recall that we defined a **matrix** as any rectangular array of numbers enclosed within brackets. The **size** or **dimension of a matrix** is important relative to operations on matrices. We define an **m × n matrix** (read "m by n matrix") to be one with m rows and n columns. It is important to note that the number of rows is always given first. If a matrix has the same number of rows and columns, it is called a **square matrix**. A matrix with only one column is called a **column matrix**, and one with only one row is called a **row matrix**. These definitions are illustrated by the following:

$$
\begin{array}{cccc}
3 \times 2 & 3 \times 3 & 4 \times 1 & 1 \times 4 \\[4pt]
\begin{bmatrix} -2 & 5 \\ 0 & -2 \\ 3 & 6 \end{bmatrix} &
\begin{bmatrix} 0.5 & 0.2 & 1.0 \\ 0.0 & 0.3 & 0.5 \\ 0.7 & 0.0 & 0.2 \end{bmatrix} &
\begin{bmatrix} 3 \\ -2 \\ 1 \\ 0 \end{bmatrix} &
[2 \quad \tfrac{1}{2} \quad 0 \quad -\tfrac{2}{3}] \\[6pt]
 & \textbf{Square matrix} & & \textbf{Row matrix} \\
 & & \textbf{Column matrix} &
\end{array}
$$

Two matrices are **equal** if they have the same dimension and their corresponding elements are equal. For example,

$$
\begin{array}{cc}
2 \times 3 & 2 \times 3 \\[4pt]
\begin{bmatrix} a & b & c \\ d & e & f \end{bmatrix} = \begin{bmatrix} u & v & w \\ x & y & z \end{bmatrix} &
\end{array}
\quad \text{if and only if} \quad
\begin{array}{ccc}
a = u & b = v & c = w \\
d = x & e = y & f = z
\end{array}
$$

■ Matrix Addition

The **sum of two matrices** of the same dimension is a matrix with elements that are the sums of the corresponding elements of the two given matrices. Addition is not defined for matrices with different dimensions.

EXAMPLE 1 (A) $\begin{bmatrix} a & b \\ c & d \end{bmatrix} + \begin{bmatrix} w & x \\ y & z \end{bmatrix} = \begin{bmatrix} (a+w) & (b+x) \\ (c+y) & (d+z) \end{bmatrix}$

(B) $\begin{bmatrix} 2 & -3 & 0 \\ 1 & 2 & -5 \end{bmatrix} + \begin{bmatrix} 3 & 1 & 2 \\ -3 & 2 & 5 \end{bmatrix} = \begin{bmatrix} 5 & -2 & 2 \\ -2 & 4 & 0 \end{bmatrix}$

PROBLEM 1 Add.

$$\begin{bmatrix} 3 & 2 \\ -1 & -1 \\ 0 & 3 \end{bmatrix} + \begin{bmatrix} -2 & 3 \\ 1 & -1 \\ 2 & -2 \end{bmatrix}$$

Because we add two matrices by adding their corresponding elements, it follows from the properties of real numbers that matrices of the same dimension are commutative and associative relative to addition. That is, if A, B, and C are matrices of the same dimension, then

$$A + B = B + A \qquad \text{Commutative}$$

$$(A + B) + C = A + (B + C) \quad \text{Associative}$$

A matrix with elements that are all 0's is called a **zero matrix**. For example,

$$[0 \quad 0 \quad 0] \qquad \begin{bmatrix} 0 & 0 \\ 0 & 0 \end{bmatrix} \qquad \begin{bmatrix} 0 \\ 0 \\ 0 \\ 0 \end{bmatrix} \qquad \begin{bmatrix} 0 & 0 & 0 & 0 \\ 0 & 0 & 0 & 0 \\ 0 & 0 & 0 & 0 \end{bmatrix}$$

are zero matrices of different dimensions. [*Note:* "0" may be used to denote the zero matrix of any dimension.] The **negative of a matrix M**, denoted by $-M$, is a matrix with elements that are the negative of the elements in M. Thus, if

$$M = \begin{bmatrix} a & b \\ c & d \end{bmatrix}$$

then

$$-M = \begin{bmatrix} -a & -b \\ -c & -d \end{bmatrix}$$

Note that $M + (-M) = 0$ (a zero matrix).

If A and B are matrices of the same dimension, then we define **subtraction** as follows:

$$A - B = A + (-B)$$

Thus, to subtract matrix B from matrix A, we simply subtract corresponding elements.

EXAMPLE 2 $\begin{bmatrix} 3 & -2 \\ 5 & 0 \end{bmatrix} - \begin{bmatrix} -2 & 2 \\ 3 & 4 \end{bmatrix} = \begin{bmatrix} 3 & -2 \\ 5 & 0 \end{bmatrix} + \begin{bmatrix} 2 & -2 \\ -3 & -4 \end{bmatrix} = \begin{bmatrix} 5 & -4 \\ 2 & -4 \end{bmatrix}$

PROBLEM 2   Subtract: $[2 \quad -3 \quad 5] - [3 \quad -2 \quad 1]$

■ Multiplication of a Matrix by a Number

Finally, the **product of a number k and a matrix M**, denoted by kM, is a matrix formed by multiplying each element of M by k. This definition is partly motivated by the fact that if M is a matrix, then we would like $M + M$ to equal $2M$.

EXAMPLE 3 $-2 \begin{bmatrix} 3 & -1 & 0 \\ -2 & 1 & 3 \\ 0 & -1 & -2 \end{bmatrix} = \begin{bmatrix} -6 & 2 & 0 \\ 4 & -2 & -6 \\ 0 & 2 & 4 \end{bmatrix}$

PROBLEM 3   Find: $10 \begin{bmatrix} 1.3 \\ 0.2 \\ 3.5 \end{bmatrix}$

We now consider an application that uses various operations.

■ Application

EXAMPLE 4   Ms. Smith and Mr. Jones are salespeople in a new-car agency that sells only two models. August was the last month for this year's models, and next year's models were introduced in September. Gross dollar sales for each month are given in the following matrices:

	AUGUST SALES			SEPTEMBER SALES	
	Compact	Luxury		Compact	Luxury
Ms. Smith	$18,000	$36,000		$72,000	$144,000
Mr. Jones	$36,000	0		$90,000	$108,000

where the August matrix $= A$ and the September matrix $= B$.

(For example, Ms. Smith had $18,000 in compact sales in August, and Mr. Jones had $108,000 in luxury car sales in September.)

(A) What was the combined dollar sales in August and September for each person and each model?

(B) What was the increase in dollar sales from August to September?

(C) If both salespeople receive 5% commissions on gross dollar sales, compute the commission for each person for each model sold in September.

Solution (A) $A + B =$

Compact   Luxury
$$\begin{bmatrix} \$\ 90{,}000 & \$180{,}000 \\ \$126{,}000 & \$108{,}000 \end{bmatrix} \begin{matrix} \text{Ms. Smith} \\ \text{Mr. Jones} \end{matrix}$$

(B) $B - A =$

Compact   Luxury
$$\begin{bmatrix} \$54{,}000 & \$108{,}000 \\ \$54{,}000 & \$108{,}000 \end{bmatrix} \begin{matrix} \text{Ms. Smith} \\ \text{Mr. Jones} \end{matrix}$$

(C) $0.05B =$

$$\begin{bmatrix} (0.05)(\$72{,}000) & (0.05)(\$144{,}000) \\ (0.05)(\$90{,}000) & (0.05)(\$108{,}000) \end{bmatrix}$$

Compact   Luxury
$$= \begin{bmatrix} \$3{,}600 & \$7{,}200 \\ \$4{,}500 & \$5{,}400 \end{bmatrix} \begin{matrix} \text{Ms. Smith} \\ \text{Mr. Jones} \end{matrix}$$

In Example 4 we chose a relatively simple example involving an agency with only two salespeople and two models. Consider the more realistic problem of an agency with nine models and perhaps seven salespeople—then you can begin to see the value of matrix methods.

PROBLEM 4   Repeat Example 4 with

$$A = \begin{bmatrix} \$36{,}000 & \$36{,}000 \\ \$18{,}000 & \$36{,}000 \end{bmatrix} \quad \text{and} \quad B = \begin{bmatrix} \$90{,}000 & \$108{,}000 \\ \$72{,}000 & \$108{,}000 \end{bmatrix}$$

Answers to Matched Problems

1. $\begin{bmatrix} 1 & 5 \\ 0 & -2 \\ 2 & 1 \end{bmatrix}$ **2.** $[-1 \quad -1 \quad 4]$ **3.** $\begin{bmatrix} 13 \\ 2 \\ 35 \end{bmatrix}$

4. (A) $\begin{bmatrix} \$126{,}000 & \$144{,}000 \\ \$\ 90{,}000 & \$144{,}000 \end{bmatrix}$ (B) $\begin{bmatrix} \$54{,}000 & \$72{,}000 \\ \$54{,}000 & \$72{,}000 \end{bmatrix}$

(C) $\begin{bmatrix} \$4{,}500 & \$5{,}400 \\ \$3{,}600 & \$5{,}400 \end{bmatrix}$

Exercise 6-1 ■ A *Problems 1–18 refer to the following matrices:*

$$A = \begin{bmatrix} 2 & -1 \\ 3 & 0 \end{bmatrix} \qquad B = \begin{bmatrix} -3 & 1 \\ 2 & -3 \end{bmatrix} \qquad C = \begin{bmatrix} 2 \\ -3 \\ 0 \end{bmatrix}$$

$$D = \begin{bmatrix} 1 \\ 3 \\ 5 \end{bmatrix} \qquad E = [-4 \quad 1 \quad 0 \quad -2] \qquad F = \begin{bmatrix} 2 & -3 \\ -2 & 0 \\ 1 & 2 \\ 3 & 5 \end{bmatrix}$$

1. What are the dimensions of B? Of E?
2. What are the dimensions of F? Of D?
3. What element is in the third row and second column of matrix F?
4. What element is in the second row and first column of matrix F?
5. Write a zero matrix of the same dimension as B.
6. Write a zero matrix of the same dimension as E.
7. Identify all column matrices.
8. Identify all row matrices.
9. Identify all square matrices.
10. How many additional columns would F have to have to be a square matrix?
11. Find $A + B$.
12. Find $C + D$.
13. Write the negative of matrix C.
14. Write the negative of matrix B.
15. Find $D - C$.
16. Find $A - A$.
17. Find $5B$.
18. Find $-2E$.

B   In Problems 19–24 perform the indicated operations.

19. $[-2 \quad 3 \quad 0] + 2[1 \quad -1 \quad 2]$

20. $\begin{bmatrix} 230 \\ 120 \end{bmatrix} + 3 \begin{bmatrix} 20 \\ 60 \end{bmatrix}$

21. $1{,}000 \begin{bmatrix} 0.25 & 0.36 \\ 0.04 & 0.35 \end{bmatrix}$

22. $100 \begin{bmatrix} 0.32 & 0.05 & 0.17 \\ 0.22 & 0.03 & 0.21 \end{bmatrix}$

23. $2 \begin{bmatrix} 1 & 2 \\ -1 & 3 \\ 0 & -2 \end{bmatrix} - \begin{bmatrix} 3 & 4 \\ -2 & 0 \\ 1 & -3 \end{bmatrix}$

24. $-2 \begin{bmatrix} 1 & 3 & 0 \\ -2 & -1 & 1 \end{bmatrix} - \begin{bmatrix} -3 & -1 & 1 \\ 0 & 2 & -1 \end{bmatrix}$

C   25. Find a, b, c, and d so that

$$\begin{bmatrix} a & b \\ c & d \end{bmatrix} + \begin{bmatrix} 2 & -3 \\ 0 & 1 \end{bmatrix} = \begin{bmatrix} 1 & -2 \\ 3 & -4 \end{bmatrix}$$

26. Find w, x, y, and z so that

$$\begin{bmatrix} 4 & -2 \\ -3 & 0 \end{bmatrix} + \begin{bmatrix} w & x \\ y & z \end{bmatrix} = \begin{bmatrix} 2 & -3 \\ 0 & 5 \end{bmatrix}$$

APPLICATIONS

27. *Cost analysis.* A company with two different plants manufactures guitars and banjos. Its production costs for each instrument are given in the following matrices:

PLANT X

	Guitar	Banjo	
Materials	$30	$25	= A
Labor	$60	$80	

PLANT Y

	Guitar	Banjo	
Materials	$36	$27	= B
Labor	$54	$74	

Find $\frac{1}{2}(A + B)$, the average cost of production for the two plants.

28. *Heredity.* Gregor Mendel (1822–1884), a Bavarian monk and botanist, made discoveries that revolutionized the science of heredity. In one experiment he crossed dihybrid yellow round peas (yellow and round are dominant characteristics; the peas also contained green and wrinkled as recessive genes) and obtained 560 peas of the types indicated in the matrix:

	Round	Wrinkled	
Yellow	319	101	= M
Green	108	32	

Suppose he carried out a second experiment of the same type and obtained 640 peas of the types indicated in this matrix:

	Round	Wrinkled	
Yellow	370	124	= N
Green	110	36	

If the results of the two experiments are combined, write the resulting matrix $M + N$. Compute the decimal fraction of the total number of peas (1,200) in each category of the combined results. [*Hint:* Compute $(1/1,200)(M + N)$.]

29. *Psychology.* Two psychologists independently carried out studies on the relationship between height and aggressive behavior in women over 18 years of age. The results of the studies are summarized in the following matrices:

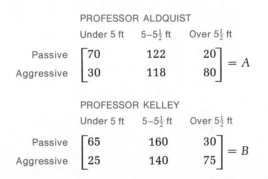

PROFESSOR ALDQUIST

	Under 5 ft	5–5½ ft	Over 5½ ft	
Passive	70	122	20	= A
Aggressive	30	118	80	

PROFESSOR KELLEY

	Under 5 ft	5–5½ ft	Over 5½ ft	
Passive	65	160	30	= B
Aggressive	25	140	75	

The two psychologists decided to combine their results and publish a joint paper. Write the matrix $A + B$ illustrating their combined results. If you have a hand calculator, compute the decimal fraction of the total sample in each category of the combined study. [*Hint:* Compute $(1/935)(A + B)$.]

Section 6-2 Matrix Multiplication

- Dot Product
- Matrix Product
- Multiplication Properties
- Application

In this section we are going to introduce two types of matrix multiplication that will at first seem rather strange. In spite of this apparent strangeness, these operations are well founded in the general theory of matrices and, as we will see, are extremely useful in practical problems.

- Dot Product

We start by defining the dot product of two special matrices.

Dot Product

The **dot product** of a $1 \times n$ row matrix and an $n \times 1$ column matrix is a real number given by:

$$\underset{1 \times n}{[a_1 \quad a_2 \cdots a_n]} \cdot \underset{n \times 1}{\begin{bmatrix} b_1 \\ b_2 \\ \vdots \\ b_n \end{bmatrix}} = a_1 b_1 + a_2 b_2 + \cdots + a_n b_n \quad \text{A real number}$$

The dot between the two matrices is important. If the dot is omitted, the multiplication is of another type, which we will consider later.

EXAMPLE 5
$$[2 \quad -3 \quad 0] \cdot \begin{bmatrix} -5 \\ 2 \\ -2 \end{bmatrix} = (2)(-5) + (-3)(2) + (0)(-2)$$

$$= -10 - 6 + 0 = -16$$

PROBLEM 5 $[-1 \quad 0 \quad 3 \quad 2] \cdot \begin{bmatrix} 2 \\ 3 \\ 4 \\ -1 \end{bmatrix} = ?$

EXAMPLE 6   A factory produces a slalom water ski that requires 4 work-hours in
the fabricating department and 1 work-hour in the finishing department. Fabricating personnel receive \$8 per hour and finishing person-
nel receive \$6 per hour. Total labor cost per ski is given by the dot
product:

$$[4 \quad 1] \cdot \begin{bmatrix} 8 \\ 6 \end{bmatrix} = (4)(8) + (1)(6) = 32 + 6 = \$38 \text{ per ski}$$

PROBLEM 6   If the factory in Example 6 also produces a trick water ski that requires
6 work-hours in the fabricating department and 1.5 work-hours in the
finishing department, write a dot product between appropriate row and
column matrices that will give the total labor cost for this ski. Compute
the cost.

■ Matrix Product

It is important to remember that the dot product of a row matrix and
a column matrix is a real number and not a matrix. We now define a
matrix product for certain matrices.

Matrix Product

The **product of two matrices** A and B is defined only on the assumption that the number of columns in A is equal to the number of rows
in B. If A is an $m \times p$ matrix and B is a $p \times n$ matrix, then the matrix
product of A and B, denoted by AB, is an $m \times n$ matrix whose element
in the ith row and jth column is the dot product of the ith row matrix
of A and the jth column matrix of B.

It is important to check dimensions before starting the multiplication
process. If matrix A has dimension $a \times b$ and matrix B has dimension
$c \times d$, then if $b = c$, the product AB will exist and will have dimension

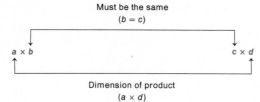

FIGURE 1

$a \times d$. This is shown schematically in Figure 1. The definition is not as complicated as it might first seem. An example should help to clarify the process. For

$$A = \begin{bmatrix} 2 & 3 & -1 \\ -2 & 1 & 2 \end{bmatrix} \quad \text{and} \quad B = \begin{bmatrix} 1 & 3 \\ 2 & 0 \\ -1 & 2 \end{bmatrix}$$

A is 2×3, B is 3×2, and AB will be 2×2. The four dot products used to produce the four elements in AB (usually calculated mentally for small whole numbers) are shown in the following large matrix.

$$
\underset{2 \times 3}{\begin{bmatrix} 2 & 3 & -1 \\ -2 & 1 & 2 \end{bmatrix}}
\underset{3 \times 2}{\begin{bmatrix} 1 & 3 \\ 2 & 0 \\ -1 & 2 \end{bmatrix}}
= \begin{bmatrix} [2 \quad 3 \quad -1] \cdot \begin{bmatrix} 1 \\ 2 \\ -1 \end{bmatrix} & [2 \quad 3 \quad -1] \cdot \begin{bmatrix} 3 \\ 0 \\ 2 \end{bmatrix} \\[2ex] [-2 \quad 1 \quad 2] \cdot \begin{bmatrix} 1 \\ 2 \\ -1 \end{bmatrix} & [-2 \quad 1 \quad 2] \cdot \begin{bmatrix} 3 \\ 0 \\ 2 \end{bmatrix} \end{bmatrix}
= \underset{2 \times 2}{\begin{bmatrix} 9 & 4 \\ -2 & -2 \end{bmatrix}}
$$

EXAMPLE 7

$$
\underset{3 \times 2}{\begin{bmatrix} 2 & 1 \\ 1 & 0 \\ -1 & 2 \end{bmatrix}}
\underset{2 \times 4}{\begin{bmatrix} 1 & -1 & 0 & 1 \\ 2 & 1 & 2 & 0 \end{bmatrix}}
= \underset{3 \times 4}{\begin{bmatrix} 4 & -1 & 2 & 2 \\ 1 & -1 & 0 & 1 \\ 3 & 3 & 4 & -1 \end{bmatrix}}
$$

PROBLEM 7   Find the product.

$$
\begin{bmatrix} 1 & -1 & 0 \\ 2 & 1 & 2 \end{bmatrix}
\begin{bmatrix} 2 & -1 \\ 1 & 0 \\ -1 & 2 \end{bmatrix}
$$

■ Multiplication Properties

In the arithmetic of real numbers it does not matter in which order we multiply; for example, $5 \times 7 = 7 \times 5$. In matrix multiplication it does make a difference; that is, MN does not always equal NM, even if both multiplications are defined (see Problems 11, 12, 25, 27, and 28 in Exercise 6-2). Also, MN may be zero with neither M nor N zero.

Matrices do, however, have other general properties. We state three important properties without proof. Assuming all products and sums are defined for the indicated matrices A, B, and C, then for k a real number:

1. $(AB)C = A(BC)$ Associative property
2. $A(B + C) = AB + AC$ Left-hand distributive property
3. $(B + C)A = BA + CA$ Right-hand distributive property
4. $k(AB) = (kA)B = A(kB)$

Since matrix multiplication is not commutative, properties 2 and 3 must be listed as distinct properties (see Problems 27 and 28 in Exercise 6-2).

■ Application

The next example illustrates the use of the dot and matrix product in a business application.

EXAMPLE 8   *Production scheduling.* Let us combine the time requirements discussed in Example 6 and Problem 6 into one matrix:

$$
\begin{array}{cc}
& \begin{array}{cc} \text{Fabricating} & \text{Finishing} \\ \text{department} & \text{department} \end{array} \\
\begin{array}{c} \text{Trick ski} \\ \text{Slalom ski} \end{array} &
\begin{bmatrix} 6 \text{ hr} & 1.5 \text{ hr} \\ 4 \text{ hr} & 1 \text{ hr} \end{bmatrix} = A
\end{array}
$$

Now suppose the company has two manufacturing plants X and Y in different parts of the country and that their hourly rates for each department are given in the following matrix:

$$
\begin{array}{cc}
& \begin{array}{cc} \text{Plant X} & \text{Plant Y} \end{array} \\
\begin{array}{c} \text{Fabricating department} \\ \text{Finishing department} \end{array} &
\begin{bmatrix} \$8 & \$7 \\ \$6 & \$4 \end{bmatrix} = B
\end{array}
$$

To find the total labor costs for each ski at each factory, we multiply A and B:

$$
AB = \begin{matrix} 2 \times 2 \\ \begin{bmatrix} 6 & 1.5 \\ 4 & 1 \end{bmatrix} \end{matrix} \begin{matrix} 2 \times 2 \\ \begin{bmatrix} 8 & 7 \\ 6 & 4 \end{bmatrix} \end{matrix} = \begin{matrix} \;\;X\quad\;\; Y \\ \begin{bmatrix} \$57 & \$48 \\ \$38 & \$32 \end{bmatrix} \end{matrix} \begin{matrix} \text{Trick ski} \\ \text{Slalom ski} \end{matrix}
$$

Notice that the dot product of the first row matrix of A and the first column matrix of B gives us the labor costs, $57, for a trick ski manufactured at plant X; the dot product of the second row matrix of A and the second column matrix of B gives us the labor costs, $32, for manufacturing a slalom ski at plant Y; and so on.

Example 8 is, of course, oversimplified. Companies that manufacture many different items in many different plants deal with matrices that have very large numbers of rows and columns.

PROBLEM 8   Repeat Example 8 with

$$A = \begin{bmatrix} 7\ hr & 2\ hr \\ 5\ hr & 1.5\ hr \end{bmatrix} \quad \text{and} \quad B = \begin{bmatrix} \$10 & \$8 \\ \$6 & \$4 \end{bmatrix}$$

Answers to Matched Problems

5. 8

6. $[6 \quad 1.5] \cdot \begin{bmatrix} 8 \\ 6 \end{bmatrix} = \$57$

7. $\begin{bmatrix} 1 & -1 \\ 3 & 2 \end{bmatrix}$

8.

	X	Y	
	$82	$64	Trick
	$59	$46	Slalom

Exercise 6-2 ■

A *Find the dot products.*

1. $[2 \quad 4] \cdot \begin{bmatrix} 3 \\ 1 \end{bmatrix}$

2. $[3 \quad 1] \cdot \begin{bmatrix} 2 \\ 4 \end{bmatrix}$

3. $[-3 \quad 2] \cdot \begin{bmatrix} -1 \\ -2 \end{bmatrix}$

4. $[3 \quad -2] \cdot \begin{bmatrix} -4 \\ -1 \end{bmatrix}$

Find the matrix products.

5. $[2 \quad 5] \begin{bmatrix} 1 & -1 \\ 2 & 3 \end{bmatrix}$

6. $[1 \quad 3] \begin{bmatrix} 2 & 3 \\ 1 & -4 \end{bmatrix}$

7. $\begin{bmatrix} 3 & 4 \\ -1 & -2 \end{bmatrix} \begin{bmatrix} -1 \\ 2 \end{bmatrix}$

8. $\begin{bmatrix} -1 & 1 \\ 2 & -3 \end{bmatrix} \begin{bmatrix} 4 \\ -2 \end{bmatrix}$

9. $\begin{bmatrix} 2 & -3 \\ 1 & 2 \end{bmatrix} \begin{bmatrix} 1 & -1 \\ 0 & -2 \end{bmatrix}$

10. $\begin{bmatrix} -3 & 2 \\ 4 & -1 \end{bmatrix} \begin{bmatrix} -2 & 5 \\ -1 & 3 \end{bmatrix}$

11. $\begin{bmatrix} -5 & -2 \\ 1 & -3 \end{bmatrix} \begin{bmatrix} -2 & 1 \\ 0 & -3 \end{bmatrix}$

12. $\begin{bmatrix} -2 & 1 \\ 0 & -3 \end{bmatrix} \begin{bmatrix} -5 & -2 \\ 1 & -3 \end{bmatrix}$

B *Find the dot products.*

13. $[-1 \quad -2 \quad 2] \cdot \begin{bmatrix} 2 \\ -1 \\ 3 \end{bmatrix}$

14. $[-2 \quad 4 \quad 0] \cdot \begin{bmatrix} -1 \\ -3 \\ 2 \end{bmatrix}$

15. $[-1 \quad -3 \quad 0 \quad 5] \cdot \begin{bmatrix} 4 \\ -3 \\ -1 \\ 2 \end{bmatrix}$ **16.** $[-1 \quad 2 \quad 3 \quad -2] \cdot \begin{bmatrix} 3 \\ -2 \\ 0 \\ 4 \end{bmatrix}$

Find the matrix products.

17. $\begin{bmatrix} 2 & -1 & 1 \\ 1 & 3 & -2 \end{bmatrix} \begin{bmatrix} 1 & 3 \\ 0 & -1 \\ -2 & 2 \end{bmatrix}$ **18.** $\begin{bmatrix} -1 & -4 & 3 \\ 2 & 0 & 1 \end{bmatrix} \begin{bmatrix} 2 & -3 \\ 1 & 2 \\ 0 & -1 \end{bmatrix}$

19. $\begin{bmatrix} 1 & 3 \\ 0 & -1 \\ -2 & 2 \end{bmatrix} \begin{bmatrix} 2 & -1 & 1 \\ 1 & 3 & -2 \end{bmatrix}$ **20.** $\begin{bmatrix} 2 & -3 \\ 1 & 2 \\ 0 & -1 \end{bmatrix} \begin{bmatrix} -1 & -4 & 3 \\ 2 & 0 & 1 \end{bmatrix}$

21. $[3 \quad -2 \quad -4] \begin{bmatrix} 1 \\ 2 \\ -3 \end{bmatrix}$ **22.** $[1 \quad -2 \quad 2] \begin{bmatrix} 2 \\ -1 \\ 1 \end{bmatrix}$

23. $\begin{bmatrix} 1 \\ 2 \\ -3 \end{bmatrix} [3 \quad -2 \quad -4]$ **24.** $\begin{bmatrix} 2 \\ -1 \\ 1 \end{bmatrix} [1 \quad -2 \quad 2]$

C   *In Problems 25–28 verify each statement using the following matrices:*

$$A = \begin{bmatrix} 1 & 2 \\ 0 & 1 \end{bmatrix} \qquad B = \begin{bmatrix} 1 & 1 \\ 2 & 3 \end{bmatrix} \qquad C = \begin{bmatrix} -3 & 1 \\ -1 & 2 \end{bmatrix}$$

25. $AB \neq BA$ **26.** $(AB)C = A(BC)$

27. $A(B + C) = AB + AC$ **28.** $(B + C)A = BA + CA$

APPLICATIONS **29.** *Labor costs.* A company with manufacturing plants located in different parts of the country has work-hour and wage requirements for the manufacturing of three types of inflatable boats as given in the following two matrices:

WORK-HOURS PER BOAT

	Cutting department	Assembly department	Packaging department	
$M = $	0.6 hr	0.6 hr	0.2 hr	One-person boat
	1.0 hr	0.9 hr	0.3 hr	Two-person boat
	1.5 hr	1.2 hr	0.4 hr	Four-person boat

HOURLY WAGES

	Plant I	Plant II	
$N = $	$6	$7	Cutting department
	$8	$10	Assembly department
	$3	$4	Packaging department

(A) Find the labor costs for a one-person boat manufactured at plant I; that is, find the dot product

$$[0.6 \quad 0.6 \quad 0.2] \cdot \begin{bmatrix} 6 \\ 8 \\ 3 \end{bmatrix}$$

(B) Find the labor costs for a four-person boat manufactured at plant II. Set up a dot product as in part (A) and multiply.

(C) What is the dimension of MN?

(D) Find MN and interpret.

30. *Nutrition.*   A nutritionist for a cereal company blends two cereals in different mixes. The amounts of protein, carbohydrate, and fat (in grams per ounce) in each cereal are given by matrix M. The amounts of each cereal used in the three mixes is given by matrix N.

	Cereal A	Cereal B	
$M =$	4 grams	2 grams	Protein
	20 grams	16 grams	Carbohydrate
	3 grams	1 gram	Fat

	Mix X	Mix Y	Mix Z	
$N =$	15 ounces	10 ounces	5 ounces	Cereal A
	5 ounces	10 ounces	15 ounces	Cereal B

(A) Find the amount of protein in mix X by computing the dot product

$$[4 \quad 2] \cdot \begin{bmatrix} 15 \\ 5 \end{bmatrix}$$

(B) Find the amount of fat in mix Z. Set up a dot product as in part (A) and multiply.

(C) What is the dimension of MN?

(D) Find MN and interpret.

(E) Find $\frac{1}{20} MN$ and interpret.

31. *Politics.*   In a local election a group hired a public relations firm to promote its candidate in three ways: telephone, house calls, and letters. The cost per contact is given in matrix M:

	Cost per contact	
$M =$	$0.40	Telephone
	$0.75	House call
	$0.25	Letter

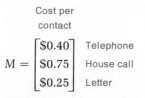

The number of contacts of each type made in two adjacent cities is given in matrix N:

$$N = \begin{array}{c} \\ \\ \end{array} \overset{\text{Telephone \quad House call \quad Letter}}{\begin{bmatrix} 1{,}000 & 500 & 5{,}000 \\ 2{,}000 & 800 & 8{,}000 \end{bmatrix}} \begin{array}{l} \text{Berkeley} \\ \text{Oakland} \end{array}$$

(A) Find the total amount spent in Berkeley by computing the dot product

$$[1{,}000 \quad 500 \quad 5{,}000] \cdot \begin{bmatrix} \$0.40 \\ \$0.75 \\ \$0.25 \end{bmatrix}$$

(B) Find the total amount spent in Oakland by computing the dot product of appropriate matrices.

(C) Compute NM and interpret.

(D) Multiply N by the matrix $[1 \quad 1]$ and interpret.

Section 6-3   Inverse of a Square Matrix; Matrix Equations

- Identity Matrix for Multiplication
- Inverse of a Square Matrix
- Matrix Equations
- Application

Identity Matrix for Multiplication

We know that

$$1a = a1 = a$$

for all real numbers a. The number 1 is called the **identity** for real number multiplication. Does the set of all matrices of a given dimension have an identity element for multiplication? The answer, in general, is no. However, the set of all **square matrices of order n** (dimension $n \times n$) does have an identity, and it is given as follows: The **identity element for multiplication** for the set of all square matrices of order n is the square matrix of order n, denoted by I, with 1's along the **main diagonal** (from upper left corner to lower right corner) and 0's elsewhere. For example,

$$\begin{bmatrix} 1 & 0 \\ 0 & 1 \end{bmatrix} \quad \text{and} \quad \begin{bmatrix} 1 & 0 & 0 \\ 0 & 1 & 0 \\ 0 & 0 & 1 \end{bmatrix}$$

are the identity matrices for square matrices of order 2 and 3, respectively.

EXAMPLE 9

$$\begin{bmatrix} 1 & 0 & 0 \\ 0 & 1 & 0 \\ 0 & 0 & 1 \end{bmatrix} \begin{bmatrix} a & b & c \\ d & e & f \\ g & h & i \end{bmatrix} = \begin{bmatrix} a & b & c \\ d & e & f \\ g & h & i \end{bmatrix}$$

$$= \begin{bmatrix} a & b & c \\ d & e & f \\ g & h & i \end{bmatrix} \begin{bmatrix} 1 & 0 & 0 \\ 0 & 1 & 0 \\ 0 & 0 & 1 \end{bmatrix}$$

PROBLEM 9   Multiply.

$$\begin{bmatrix} 1 & 0 \\ 0 & 1 \end{bmatrix} \begin{bmatrix} 2 & -3 \\ 5 & 7 \end{bmatrix} \quad \text{and} \quad \begin{bmatrix} 2 & -3 \\ 5 & 7 \end{bmatrix} \begin{bmatrix} 1 & 0 \\ 0 & 1 \end{bmatrix}$$

In general, we can show that if M is a square matrix of order n and I is the identity matrix of order n, then

$$\mathbf{IM = MI = M}$$

■ Inverse of a Square Matrix

In the set of real numbers we know that for each real number a (except 0) there exists a real number a^{-1} such that

$$a^{-1}a = 1$$

The number a^{-1} is called the **inverse** of the number a relative to multiplication, or the **multiplicative inverse** of a. For example, 2^{-1} is the multiplicative inverse of 2, since $2^{-1} \cdot 2 = 1$. For each square matrix M does there exist an inverse matrix M^{-1} such that the following relation is true?

$$M^{-1}M = MM^{-1} = I$$

If M^{-1} exists for a given matrix M, then M^{-1} is called the **inverse of M relative to multiplication**. Let us use this definition to find M^{-1} for

$$M = \begin{bmatrix} 2 & 3 \\ 1 & 2 \end{bmatrix}$$

We are looking for

$$M^{-1} = \begin{bmatrix} a & c \\ b & d \end{bmatrix}$$

such that

$$MM^{-1} = M^{-1}M = I$$

Thus, we write

$$\begin{matrix} M & M^{-1} & I \end{matrix}$$

$$\begin{bmatrix} 2 & 3 \\ 1 & 2 \end{bmatrix} \begin{bmatrix} a & c \\ b & d \end{bmatrix} = \begin{bmatrix} 1 & 0 \\ 0 & 1 \end{bmatrix}$$

and try to find a, b, c, and d so that the product of M and M^{-1} is the identity matrix I. Multiplying M and M^{-1} on the left side, we obtain

$$\begin{bmatrix} (2a + 3b) & (2c + 3d) \\ (a + 2b) & (c + 2d) \end{bmatrix} = \begin{bmatrix} 1 & 0 \\ 0 & 1 \end{bmatrix}$$

which is true only if

$$2a + 3b = 1 \qquad 2c + 3d = 0$$
$$a + 2b = 0 \qquad c + 2d = 1$$

Solving these two systems, we find that $a = 2$, $b = -1$, $c = -3$, and $d = 2$. Thus,

$$M^{-1} = \begin{bmatrix} 2 & -3 \\ -1 & 2 \end{bmatrix}.$$

as is easily checked:

$$\begin{matrix} M & M^{-1} & I & M^{-1} & M \end{matrix}$$

$$\begin{bmatrix} 2 & 3 \\ 1 & 2 \end{bmatrix} \begin{bmatrix} 2 & -3 \\ -1 & 2 \end{bmatrix} = \begin{bmatrix} 1 & 0 \\ 0 & 1 \end{bmatrix} = \begin{bmatrix} 2 & -3 \\ -1 & 2 \end{bmatrix} \begin{bmatrix} 2 & 3 \\ 1 & 2 \end{bmatrix}$$

Inverses do not always exist for square matrices. For example, if

$$M = \begin{bmatrix} 2 & 1 \\ 4 & 2 \end{bmatrix}$$

then, proceeding as before, we are led to the systems

$$2a + b = 1 \qquad 2c + d = 0$$
$$4a + 2b = 0 \qquad 4c + 2d = 1$$

These are both inconsistent and have no solution. Hence, M^{-1} does not exist.

Being able to find inverses, when they exist, leads to direct and simple solutions to many practical problems. At the end of this section, for example, we will show how inverses can be used to solve systems of linear equations.

The method outlined for finding M^{-1}, if it exists, gets very involved for matrices of order larger than 2. Now that we know what we are looking for, we can introduce the idea of the augmented matrix (considered in Sections 5-2 and 5-3) to make the process more efficient. For example, to find the inverse (if it exists) of

$$M = \begin{bmatrix} 1 & -1 & 1 \\ 0 & 2 & -1 \\ 2 & 3 & 0 \end{bmatrix}$$

we start as before and write

$$\underset{M}{\begin{bmatrix} 1 & -1 & 1 \\ 0 & 2 & -1 \\ 2 & 3 & 0 \end{bmatrix}} \underset{M^{-1}}{\begin{bmatrix} a & d & g \\ b & e & h \\ c & f & i \end{bmatrix}} = \underset{I}{\begin{bmatrix} 1 & 0 & 0 \\ 0 & 1 & 0 \\ 0 & 0 & 1 \end{bmatrix}}$$

which is true only if

$$
\begin{array}{lll}
a - b + c = 1 & d - e + f = 0 & g - h + i = 0 \\
2b - c = 0 & 2e - f = 1 & 2h - i = 0 \\
2a + 3b = 0 & 2d + 3e = 0 & 2g + 3h = 1
\end{array}
$$

Now we write augmented matrices for each of the three systems:

$$
\underset{\text{First}}{\left[\begin{array}{rrr|r} 1 & -1 & 1 & 1 \\ 0 & 2 & -1 & 0 \\ 2 & 3 & 0 & 0 \end{array}\right]}
\quad
\underset{\text{Second}}{\left[\begin{array}{rrr|r} 1 & -1 & 1 & 0 \\ 0 & 2 & -1 & 1 \\ 2 & 3 & 0 & 0 \end{array}\right]}
\quad
\underset{\text{Third}}{\left[\begin{array}{rrr|r} 1 & -1 & 1 & 0 \\ 0 & 2 & -1 & 0 \\ 2 & 3 & 0 & 1 \end{array}\right]}
$$

Since each matrix to the left of the vertical bar is the same, exactly the same row operations can be used on each total matrix to transform it into a reduced form. We can speed up the process substantially by combining all three augmented matrices into the single augmented matrix form

$$\left[\begin{array}{rrr|rrr} 1 & -1 & 1 & 1 & 0 & 0 \\ 0 & 2 & -1 & 0 & 1 & 0 \\ 2 & 3 & 0 & 0 & 0 & 1 \end{array}\right] = [M\,|\,I] \tag{1}$$

We now try to perform row operations on matrix (1) until we obtain a row-equivalent matrix that looks like matrix (2):

$$\left[\begin{array}{ccc|ccc} \overset{I}{1} & 0 & 0 & \overset{B}{a} & d & g \\ 0 & 1 & 0 & b & e & h \\ 0 & 0 & 1 & c & f & i \end{array}\right] \tag{2}$$

If this can be done, then the new matrix to the right of the vertical bar will be M^{-1}! Now let us try to transform (1) into a form like (2).

$$
\begin{array}{c}
\overset{M}{} \qquad\qquad \overset{I}{}
\end{array}
$$

$$
\left[\begin{array}{ccc|ccc}
1 & -1 & 1 & 1 & 0 & 0 \\
0 & 2 & -1 & 0 & 1 & 0 \\
2 & 3 & 0 & 0 & 0 & 1
\end{array}\right] \qquad R_3 + (-2)R_1 \to R_3
$$

$$
\sim \left[\begin{array}{ccc|ccc}
1 & -1 & 1 & 1 & 0 & 0 \\
0 & 2 & -1 & 0 & 1 & 0 \\
0 & 5 & -2 & -2 & 0 & 1
\end{array}\right] \qquad \tfrac{1}{2}R_2 \to R_2
$$

$$
\sim \left[\begin{array}{ccc|ccc}
1 & -1 & 1 & 1 & 0 & 0 \\
0 & 1 & -\tfrac{1}{2} & 0 & \tfrac{1}{2} & 0 \\
0 & 5 & -2 & -2 & 0 & 1
\end{array}\right] \qquad R_3 + (-5)R_2 \to R_3
$$

$$
\sim \left[\begin{array}{ccc|ccc}
1 & -1 & 1 & 1 & 0 & 0 \\
0 & 1 & -\tfrac{1}{2} & 0 & \tfrac{1}{2} & 0 \\
0 & 0 & \tfrac{1}{2} & -2 & -\tfrac{5}{2} & 1
\end{array}\right] \qquad 2R_3 \to R_3
$$

$$
\sim \left[\begin{array}{ccc|ccc}
1 & -1 & 1 & 1 & 0 & 0 \\
0 & 1 & -\tfrac{1}{2} & 0 & \tfrac{1}{2} & 0 \\
0 & 0 & 1 & -4 & -5 & 2
\end{array}\right] \qquad \begin{array}{l} R_1 + (-1)R_3 \to R_1 \\ R_2 + \tfrac{1}{2}R_3 \to R_2 \end{array}
$$

$$
\sim \left[\begin{array}{ccc|ccc}
1 & -1 & 0 & 5 & 5 & -2 \\
0 & 1 & 0 & -2 & -2 & 1 \\
0 & 0 & 1 & -4 & -5 & 2
\end{array}\right] \qquad R_1 + R_2 \to R_1
$$

$$
\begin{array}{c}
\overset{I}{} \qquad\qquad \overset{B}{}
\end{array}
$$

$$
\sim \left[\begin{array}{ccc|ccc}
1 & 0 & 0 & 3 & 3 & -1 \\
0 & 1 & 0 & -2 & -2 & 1 \\
0 & 0 & 1 & -4 & -5 & 2
\end{array}\right]
$$

Converting back to systems of equations equivalent to our three original systems, we have

$$
\begin{array}{lll}
a = 3 & d = 3 & g = -1 \\
b = -2 & e = -2 & h = 1 \\
c = -4 & f = -5 & i = 2
\end{array}
$$

And these are just the elements of M^{-1} that we are looking for! Hence,

$$
M^{-1} = \left[\begin{array}{ccc}
3 & 3 & -1 \\
-2 & -2 & 1 \\
-4 & -5 & 2
\end{array}\right]
$$

Note that this is the matrix to the right of the vertical line in the last augmented matrix. (You should check that $MM^{-1} = I$.)

Inverse of a Square Matrix M

If $[M|I]$ is transformed by row operations into $[I|B]$, then the resulting matrix B is M^{-1}. If, however, we obtain all 0's in one or more rows to the left of the vertical line, then M^{-1} will not exist.

EXAMPLE 10   Find M^{-1}, given

$$M = \begin{bmatrix} 3 & -1 \\ -4 & 2 \end{bmatrix}$$

Solution   Can you identify the row operations used in the transformations?

$$\begin{bmatrix} 3 & -1 & | & 1 & 0 \\ -4 & 2 & | & 0 & 1 \end{bmatrix} \sim \begin{bmatrix} 1 & -\frac{1}{3} & | & \frac{1}{3} & 0 \\ -4 & 2 & | & 0 & 1 \end{bmatrix}$$

$$\sim \begin{bmatrix} 1 & -\frac{1}{3} & | & \frac{1}{3} & 0 \\ 0 & \frac{2}{3} & | & \frac{4}{3} & 1 \end{bmatrix}$$

$$\sim \begin{bmatrix} 1 & -\frac{1}{3} & | & \frac{1}{3} & 0 \\ 0 & 1 & | & 2 & \frac{3}{2} \end{bmatrix}$$

$$\sim \begin{bmatrix} 1 & 0 & | & 1 & \frac{1}{2} \\ 0 & 1 & | & 2 & \frac{3}{2} \end{bmatrix}$$

Thus,

$$M^{-1} = \begin{bmatrix} 1 & \frac{1}{2} \\ 2 & \frac{3}{2} \end{bmatrix} \quad \text{Check by showing that } M^{-1}M = I.$$

PROBLEM 10   Find M^{-1}, given

$$M = \begin{bmatrix} 2 & -6 \\ 1 & -2 \end{bmatrix}$$

■ Matrix Equations

We will now show how systems of equations can be solved using inverses of square matrices.

EXAMPLE 11   Solve the system

$$\begin{aligned} 3x_1 - \ x_2 &= k_1 \\ -4x_1 + 2x_2 &= k_2 \end{aligned}$$ (3)

for:

(A) $k_1 = -5$, $k_2 = 8$ (B) $k_1 = 4$, $k_2 = -2$ (C) $k_1 = 0$, $k_2 = -4$

Solution   Once we obtain the inverse of the coefficient matrix,

$$A = \begin{bmatrix} 3 & -1 \\ -4 & 2 \end{bmatrix}$$

we will be able to solve parts (A)–(C) very easily. To see why, we convert system (3) into the following equivalent **matrix equation**:

$$\overset{A}{\begin{bmatrix} 3 & -1 \\ -4 & 2 \end{bmatrix}} \overset{X}{\begin{bmatrix} x_1 \\ x_2 \end{bmatrix}} = \overset{B}{\begin{bmatrix} k_1 \\ k_2 \end{bmatrix}} \tag{4}$$

You should check that matrix equation (4) is equivalent to system (3) by multiplying the left side and then equating corresponding elements on the left with those on the right.

We are now interested in finding a column matrix X that will satisfy the matrix equation

$$AX = B$$

To solve this equation we multiply both sides by A^{-1} (if it exists) to isolate X on the left side:

$$AX = B \qquad \text{Multiply both sides by } A^{-1}.$$
$$A^{-1}(AX) = A^{-1}B \qquad \text{Use the associative property.}$$
$$(A^{-1}A)X = A^{-1}B \qquad A^{-1}A = I$$
$$IX = A^{-1}B \qquad IX = X$$
$$X = A^{-1}B$$

The inverse of A was found in Example 10 to be

$$A^{-1} = \begin{bmatrix} 1 & \frac{1}{2} \\ 2 & \frac{3}{2} \end{bmatrix}$$

Thus,

$$\overset{X}{\begin{bmatrix} x_1 \\ x_2 \end{bmatrix}} = \overset{A^{-1}}{\begin{bmatrix} 1 & \frac{1}{2} \\ 2 & \frac{3}{2} \end{bmatrix}} \overset{B}{\begin{bmatrix} k_1 \\ k_2 \end{bmatrix}}$$

To solve parts (A)–(C) we simply replace k_1 and k_2 with appropriate values and multiply.

(A) $\begin{bmatrix} x_1 \\ x_2 \end{bmatrix} = \begin{bmatrix} 1 & \frac{1}{2} \\ 2 & \frac{3}{2} \end{bmatrix} \begin{bmatrix} -5 \\ 8 \end{bmatrix} = \begin{bmatrix} -1 \\ 2 \end{bmatrix}$

Thus, $x_1 = -1$ and $x_2 = 2$.

(B) $\begin{bmatrix} x_1 \\ x_2 \end{bmatrix} = \begin{bmatrix} 1 & \frac{1}{2} \\ 2 & \frac{3}{2} \end{bmatrix} \begin{bmatrix} 4 \\ -2 \end{bmatrix} = \begin{bmatrix} 3 \\ 5 \end{bmatrix}$

Thus, $x_1 = 3$ and $x_2 = 5$.

(C) $\begin{bmatrix} x_1 \\ x_2 \end{bmatrix} = \begin{bmatrix} 1 & \frac{1}{2} \\ 2 & \frac{3}{2} \end{bmatrix} \begin{bmatrix} 0 \\ -4 \end{bmatrix} = \begin{bmatrix} -2 \\ -6 \end{bmatrix}$

Thus, $x_1 = -2$ and $x_2 = -6$.

PROBLEM 11   Solve the system

$$2x_1 - 3x_2 = k_1$$
$$-x_1 + 2x_2 = k_2$$

using the inverse of the coefficient matrix, for:
(A) $k_1 = 10, \ k_2 = -6$ (B) $k_1 = 1, \ k_2 = 0$ (C) $k_1 = -8, \ k_2 = 5$

A great advantage of using an inverse to solve a system of linear equations is that once the inverse is found, it can be used to solve any new system formed through a change in the constant terms. This method is not suited, however, for cases where the numbers of equations and unknowns are not the same. (Why?)

■ Application

The following application will illustrate the usefulness of the inverse method.

EXAMPLE 12   An investment adviser currently has two types of investments available for clients: a conservative investment A that pays 10% per year and an investment B of higher risk that pays 20% per year. Clients may divide their investments between the two to achieve any total return desired between 10% and 20%. However, the higher the desired return, the higher the risk. How should each client listed in the table invest to achieve the indicated return?

	CLIENT			
	1	2	3	k
Total investment	$20,000	$50,000	$10,000	k_1
Annual return desired	$ 2,400	$ 7,500	$ 1,300	k_2
	(12%)	(15%)	(13%)	

Solution   We will solve the problem for an arbitrary client k, using inverses, and then apply the result to the three specific clients.

Let x_1 = Amount invested in A

x_2 = Amount invested in B

Then $x_1 + \quad x_2 = k_1$ Total invested

$0.1x_1 + 0.2x_2 = k_2$ Total annual return

Write as a matrix equation:

$$\underset{A}{\begin{bmatrix} 1 & 1 \\ 0.1 & 0.2 \end{bmatrix}} \underset{X}{\begin{bmatrix} x_1 \\ x_2 \end{bmatrix}} = \underset{B}{\begin{bmatrix} k_1 \\ k_2 \end{bmatrix}}$$

If A^{-1} exists, then

$X = A^{-1}B$

We now find A^{-1} by starting with $[A|I]$ and proceeding as discussed earlier in this section.

$$\begin{bmatrix} 1 & 1 & | & 1 & 0 \\ 0.1 & 0.2 & | & 0 & 1 \end{bmatrix} \quad 10R_2 \to R_2$$

$$\sim \begin{bmatrix} 1 & 1 & | & 1 & 0 \\ 1 & 2 & | & 0 & 10 \end{bmatrix} \quad R_2 + (-1)R_1 \to R_2$$

$$\sim \begin{bmatrix} 1 & 1 & | & 1 & 0 \\ 0 & 1 & | & -1 & 10 \end{bmatrix} \quad R_1 + (-1)R_2 \to R_1$$

$$\sim \begin{bmatrix} 1 & 0 & | & 2 & -10 \\ 0 & 1 & | & -1 & 10 \end{bmatrix}$$

Thus,

$$A^{-1} = \begin{bmatrix} 2 & -10 \\ -1 & 10 \end{bmatrix}$$

Check $$\underset{A^{-1}}{\begin{bmatrix} 2 & -10 \\ -1 & 10 \end{bmatrix}} \underset{A}{\begin{bmatrix} 1 & 1 \\ 0.1 & 0.2 \end{bmatrix}} = \underset{I}{\begin{bmatrix} 1 & 0 \\ 0 & 1 \end{bmatrix}}$$

and

$$\underset{X}{\begin{bmatrix} x_1 \\ x_2 \end{bmatrix}} = \underset{A^{-1}}{\begin{bmatrix} 2 & -10 \\ -1 & 10 \end{bmatrix}} \underset{B}{\begin{bmatrix} k_1 \\ k_2 \end{bmatrix}}$$

To solve each client's investment problem, we replace k_1 and k_2 with appropriate values from the table and multiply by A^{-1}:

Client 1

$$\begin{bmatrix} x_1 \\ x_2 \end{bmatrix} = \begin{bmatrix} 2 & -10 \\ -1 & 10 \end{bmatrix} \begin{bmatrix} 20{,}000 \\ 2{,}400 \end{bmatrix} = \begin{bmatrix} 16{,}000 \\ 4{,}000 \end{bmatrix}$$

Solution:　$x_1 = \$16{,}000$ in A,　$x_2 = \$4{,}000$ in B

Client 2

$$\begin{bmatrix} x_1 \\ x_2 \end{bmatrix} = \begin{bmatrix} 2 & -10 \\ -1 & 10 \end{bmatrix} \begin{bmatrix} 50{,}000 \\ 7{,}500 \end{bmatrix} = \begin{bmatrix} 25{,}000 \\ 25{,}000 \end{bmatrix}$$

Solution:　$x_1 = \$25{,}000$ in A,　$x_2 = \$25{,}000$ in B

Client 3

$$\begin{bmatrix} x_1 \\ x_2 \end{bmatrix} = \begin{bmatrix} 2 & -10 \\ -1 & 10 \end{bmatrix} \begin{bmatrix} 10{,}000 \\ 1{,}300 \end{bmatrix} = \begin{bmatrix} 7{,}000 \\ 3{,}000 \end{bmatrix}$$

Solution:　$x_1 = \$7{,}000$ in A,　$x_2 = \$3{,}000$ in B

PROBLEM 12　Repeat Example 12 with investment A paying 8% and investment B paying 24%.

Answers to Matched Problems　**9.** $\begin{bmatrix} 2 & -3 \\ 5 & 7 \end{bmatrix}$　　**10.** $\begin{bmatrix} -1 & 3 \\ -\frac{1}{2} & 1 \end{bmatrix}$

11. (A)　$x_1 = 2$,　$x_2 = -2$　　(B)　$x_1 = 2$,　$x_2 = 1$
(C)　$x_1 = -1$,　$x_2 = 2$

12.　$A^{-1} = \begin{bmatrix} 1.5 & -6.25 \\ -0.5 & 6.25 \end{bmatrix}$;

Client 1:　$\$15{,}000$ in A and $\$5{,}000$ in B;　Client 2:　$\$28{,}125$ in A and $\$21{,}875$ in B;　Client 3:　$\$6{,}875$ in A and $\$3{,}125$ in B

Exercise 6-3 ■ A　*Perform the indicated operations.*

1. $\begin{bmatrix} 1 & 0 \\ 0 & 1 \end{bmatrix}\begin{bmatrix} 2 & -3 \\ 4 & 5 \end{bmatrix}$　　**2.** $\begin{bmatrix} 2 & -3 \\ 4 & 5 \end{bmatrix}\begin{bmatrix} 1 & 0 \\ 0 & 1 \end{bmatrix}$

3. $\begin{bmatrix} 1 & 0 & 0 \\ 0 & 1 & 0 \\ 0 & 0 & 1 \end{bmatrix}\begin{bmatrix} -2 & 1 & 3 \\ 2 & 4 & -2 \\ 5 & 1 & 0 \end{bmatrix}$　　**4.** $\begin{bmatrix} -2 & 1 & 3 \\ 2 & 4 & -2 \\ 5 & 1 & 0 \end{bmatrix}\begin{bmatrix} 1 & 0 & 0 \\ 0 & 1 & 0 \\ 0 & 0 & 1 \end{bmatrix}$

For each problem show that the two matrices are inverses of each other by showing that their product is the identity matrix I.

5. $\begin{bmatrix} 3 & -4 \\ -2 & 3 \end{bmatrix}\begin{bmatrix} 3 & 4 \\ 2 & 3 \end{bmatrix}$　　**6.** $\begin{bmatrix} 5 & -7 \\ -2 & 3 \end{bmatrix}\begin{bmatrix} 3 & 7 \\ 2 & 5 \end{bmatrix}$

7. $\begin{bmatrix} 1 & -1 & 1 \\ 0 & 2 & -1 \\ 2 & 3 & 0 \end{bmatrix} \begin{bmatrix} 3 & 3 & -1 \\ -2 & -2 & 1 \\ -4 & -5 & 2 \end{bmatrix}$

8. $\begin{bmatrix} 3 & 3 & -1 \\ -2 & -2 & 1 \\ -4 & -5 & 2 \end{bmatrix} \begin{bmatrix} 1 & -1 & 1 \\ 0 & 2 & -1 \\ 2 & 3 & 0 \end{bmatrix}$

Find x_1 and x_2.

9. $\begin{bmatrix} x_1 \\ x_2 \end{bmatrix} = \begin{bmatrix} 3 & -2 \\ 1 & 4 \end{bmatrix} \begin{bmatrix} -2 \\ 1 \end{bmatrix}$
10. $\begin{bmatrix} x_1 \\ x_2 \end{bmatrix} = \begin{bmatrix} -2 & 1 \\ -1 & 2 \end{bmatrix} \begin{bmatrix} 3 \\ -2 \end{bmatrix}$

11. $\begin{bmatrix} x_1 \\ x_2 \end{bmatrix} = \begin{bmatrix} -2 & 3 \\ 2 & -1 \end{bmatrix} \begin{bmatrix} 3 \\ 2 \end{bmatrix}$
12. $\begin{bmatrix} x_1 \\ x_2 \end{bmatrix} = \begin{bmatrix} 3 & -1 \\ 0 & 2 \end{bmatrix} \begin{bmatrix} -2 \\ 1 \end{bmatrix}$

B Given M as indicated, find M^{-1} and show that $M^{-1}M = I$.

13. $\begin{bmatrix} 1 & 2 \\ 1 & 3 \end{bmatrix}$
14. $\begin{bmatrix} 2 & 1 \\ 5 & 3 \end{bmatrix}$

15. $\begin{bmatrix} 1 & 3 \\ 2 & 7 \end{bmatrix}$
16. $\begin{bmatrix} 2 & 1 \\ 1 & 1 \end{bmatrix}$

17. $\begin{bmatrix} 1 & -3 & 0 \\ 0 & 3 & 1 \\ 2 & -1 & 2 \end{bmatrix}$
18. $\begin{bmatrix} 2 & 9 & 0 \\ 1 & 2 & 3 \\ 0 & -1 & 1 \end{bmatrix}$

19. $\begin{bmatrix} 1 & 1 & 0 \\ 0 & 3 & -1 \\ 1 & 0 & 1 \end{bmatrix}$
20. $\begin{bmatrix} 1 & 0 & -1 \\ 2 & -1 & 0 \\ 1 & 1 & 1 \end{bmatrix}$

Write each system as a matrix equation and solve using inverses.
[Note: The inverses were found in Problems 13–18.]

21. $x_1 + 2x_2 = k_1$
$x_1 + 3x_2 = k_2$

(A) $k_1 = 1$, $k_2 = 3$
(B) $k_1 = 3$, $k_2 = 5$
(C) $k_1 = -2$, $k_2 = 1$

22. $2x_1 + x_2 = k_1$
$5x_1 + 3x_2 = k_2$

(A) $k_1 = 2$, $k_2 = 13$
(B) $k_1 = -2$, $k_2 = 4$
(C) $k_1 = 1$, $k_2 = -3$

23. $x_1 + 3x_2 = k_1$
$2x_1 + 7x_2 = k_2$

(A) $k_1 = 2$, $k_2 = -1$
(B) $k_1 = 1$, $k_2 = 0$
(C) $k_1 = 3$, $k_2 = -1$

24. $2x_1 + x_2 = k_1$
$x_1 + x_2 = k_2$

(A) $k_1 = -1$, $k_2 = -2$
(B) $k_1 = 2$, $k_2 = 3$
(C) $k_1 = 2$, $k_2 = 0$

25. $\begin{aligned} x_1 - 3x_2 \qquad &= k_1 \\ 3x_2 + x_3 &= k_2 \\ 2x_1 - x_2 + 2x_3 &= k_3 \end{aligned}$

(A) $k_1 = 1, \quad k_2 = 0, \quad k_3 = 2$
(B) $k_1 = -1, \quad k_2 = 1, \quad k_3 = 0$
(C) $k_1 = 2, \quad k_2 = -2, \quad k_3 = 1$

26. $\begin{aligned} 2x_1 + 9x_2 \qquad &= k_1 \\ x_1 + 2x_2 + 3x_3 &= k_2 \\ - x_2 + x_3 &= k_3 \end{aligned}$

(A) $k_1 = 0, \quad k_2 = 2, \quad k_3 = 1$
(B) $k_1 = -2, \quad k_2 = 0, \quad k_3 = 1$
(C) $k_1 = 3, \quad k_2 = 1, \quad k_3 = 0$

C   Write each system as a matrix equation and solve using inverses.
[Note: The inverses were found in Problems 19 and 20.]

27. $\begin{aligned} x_1 + x_2 \qquad &= k_1 \\ 3x_2 - x_3 &= k_2 \\ x_1 \qquad + x_3 &= k_3 \end{aligned}$

(A) $k_1 = 2, \quad k_2 = 0, \quad k_3 = 4$
(B) $k_1 = 0, \quad k_2 = 4, \quad k_3 = -2$
(C) $k_1 = 4, \quad k_2 = 2, \quad k_3 = 0$

28. $\begin{aligned} x_1 \qquad - x_3 &= k_1 \\ 2x_1 - x_2 \qquad &= k_2 \\ x_1 + x_2 + x_3 &= k_3 \end{aligned}$

(A) $k_1 = 4, \quad k_2 = 8, \quad k_3 = 0$
(B) $k_1 = 4, \quad k_2 = 0, \quad k_3 = -4$
(C) $k_1 = 0, \quad k_2 = 8, \quad k_3 = -8$

Show that the inverses of the following matrices do not exist.

29. $\begin{bmatrix} 3 & 9 \\ 2 & 6 \end{bmatrix}$

30. $\begin{bmatrix} 2 & -4 \\ -3 & 6 \end{bmatrix}$

31. $\begin{bmatrix} 2 & 1 & 1 \\ 1 & 1 & 0 \\ -1 & -1 & 0 \end{bmatrix}$

32. $\begin{bmatrix} 1 & -1 & 0 \\ 2 & -1 & 1 \\ 0 & 1 & 1 \end{bmatrix}$

33. Show that $(A^{-1})^{-1} = A$ for

$$A = \begin{bmatrix} 3 & 4 \\ 2 & 3 \end{bmatrix}$$

34. Show that $(AB)^{-1} = B^{-1}A^{-1}$ for

$$A = \begin{bmatrix} 3 & 4 \\ 2 & 3 \end{bmatrix} \quad \text{and} \quad B = \begin{bmatrix} 3 & 7 \\ 2 & 5 \end{bmatrix}$$

APPLICATIONS   *Solve using systems of equations and inverses.*

35. *Resource allocation.* A concert hall has 10,000 seats. If tickets are $4 and $8, how many of each type of ticket should be sold (assuming all seats can be sold) to bring in each of the returns indicated in the table? Use decimals in computing the inverse.

	CONCERT		
	1	2	3
Tickets sold	10,000	10,000	10,000
Return required	$56,000	$60,000	$68,000

36. *Production scheduling.* Labor and material costs for manufacturing two guitar models are given in the table below.

GUITAR MODEL	LABOR COST	MATERIAL COST
A	$30	$20
B	$40	$30

If a total of $3,000 a week is allowed for labor and material, how many of each model should be produced each week to exactly use each of the allocations of the $3,000 indicated in the following table? Use decimals in computing the inverse.

	WEEKLY ALLOCATION		
	1	2	3
Labor	$1,800	$1,750	$1,720
Material	$1,200	$1,250	$1,280

37. *Diets.* A biologist has available two commercial food mixes with the following percentages of protein and fat:

MIX	PROTEIN (%)	FAT (%)
A	20	2
B	10	6

How many ounces of each mix should be used to prepare each of the diets listed in the following table?

	DIET		
	1	2	3
Protein	20 ounces	10 ounces	10 ounces
Fat	6 ounces	4 ounces	6 ounces

Section 6-4　Determinant Functions

- Determinant Functions
- Second-Order Determinants
- Third-Order Determinants
- Remarks

Determinant Functions

In this section we are going to introduce a new function, called a **determinant function**. Its domain is the set of all square matrices with real elements, and its range is the set of all real numbers. If A is a square matrix, then the determinant of A is denoted by **det A** or simply by writing the array of elements in A using vertical lines in place of square brackets. For example,

$$\det \begin{bmatrix} 2 & -3 \\ 5 & 1 \end{bmatrix} = \begin{vmatrix} 2 & -3 \\ 5 & 1 \end{vmatrix}$$

$$\det \begin{bmatrix} 1 & -2 & 3 \\ 0 & 5 & -7 \\ -2 & 1 & 6 \end{bmatrix} = \begin{vmatrix} 1 & -2 & 3 \\ 0 & 5 & -7 \\ -2 & 1 & 6 \end{vmatrix}$$

The expressions on the right are often referred to simply as **determinants**.

A determinant of **order n** is one with n rows and n columns. In this section we will concentrate most of our attention on determining the value of determinants of orders 2 and 3. Many of the results and procedures we will discuss generalize completely to determinants of order n.

Second-Order Determinants

In general, we can symbolize a **second-order determinant** as follows:

$$\begin{vmatrix} a_{11} & a_{12} \\ a_{21} & a_{22} \end{vmatrix}$$

where we use a single letter with a **double subscript** to facilitate generalization to higher-order determinants. The first subscript number indicates the row in which the element lies, and the second subscript number indicates the column. Thus a_{21} is the element in the second row and first column, and a_{12} is the element in the first row and second column. Each second-order determinant represents a real number given by the following formula:

Value of a Second-Order Determinant

$$\begin{vmatrix} a_{11} & a_{12} \\ a_{21} & a_{22} \end{vmatrix} = a_{11}a_{22} - a_{21}a_{12} \tag{1}$$

Formula (1) is easily remembered if you notice that the expression on the right is the product of the principal diagonal (from upper left to lower right) minus the product of the secondary diagonal (from lower left to upper right).

EXAMPLE 13 $\begin{vmatrix} -1 & 2 \\ -3 & 4 \end{vmatrix} = (-1)(-4) - (-3)(2) = 4 - (-6) = 10$

PROBLEM 13   Find: $\begin{vmatrix} 3 & -5 \\ 4 & -2 \end{vmatrix}$

■ **Third-Order Determinants**

A determinant of order 3 is a square array of nine elements and represents a real number given by the following formula:

Value of a Third-Order Determinant

$$\begin{vmatrix} a_{11} & a_{12} & a_{13} \\ a_{21} & a_{22} & a_{23} \\ a_{31} & a_{32} & a_{33} \end{vmatrix} = a_{11}a_{22}a_{33} - a_{11}a_{32}a_{23} + a_{21}a_{32}a_{13} - a_{21}a_{12}a_{33} + a_{31}a_{12}a_{23} - a_{31}a_{22}a_{13} \tag{2}$$

Note that each term in the expansion on the right of equation (2) contains exactly one element from each row and each column. Don't panic! You do not need to memorize formula (2). After we introduce the ideas of "minor" and "cofactor," we will state a theorem that can be used to obtain the same result with much less memory strain.

The **minor of an element** in a third-order determinant is a second-order determinant obtained by deleting the row and column that contains the element. For example, in the determinant in formula (2),

$$\text{Minor of } a_{23} = \begin{vmatrix} a_{11} & a_{12} & a_{13} \\ a_{21} & a_{22} & a_{23} \\ a_{31} & a_{32} & a_{33} \end{vmatrix} = \begin{vmatrix} a_{11} & a_{12} \\ a_{31} & a_{32} \end{vmatrix}$$

$$\text{Minor of } a_{32} = \begin{vmatrix} a_{11} & a_{12} & a_{13} \\ a_{21} & a_{22} & a_{23} \\ a_{31} & a_{32} & a_{33} \end{vmatrix} = \begin{vmatrix} a_{11} & a_{13} \\ a_{21} & a_{23} \end{vmatrix}$$

A quantity closely associated with the minor of an element is the cofactor of an element. The **cofactor of an element** a_{ij} (from the ith row and jth column) is the product of the minor of a_{ij} and $(-1)^{i+j}$.

Cofactor

$$\text{Cofactor of } a_{ij} = (-1)^{i+j}(\text{Minor of } a_{ij})$$

Thus, a cofactor of an element is nothing more than a signed minor. The sign is determined by raising -1 to a power that is the sum of the numbers indicating the row and column in which the element lies. Note that $(-1)^{i+j}$ is -1 if $i+j$ is odd and 1 if $i+j$ is even. Thus, if we are given the determinant

$$\begin{vmatrix} a_{11} & a_{12} & a_{13} \\ a_{21} & a_{22} & a_{23} \\ a_{31} & a_{32} & a_{33} \end{vmatrix}$$

then

$$\text{Cofactor of } a_{23} = (-1)^{2+3} \begin{vmatrix} a_{11} & a_{12} \\ a_{31} & a_{32} \end{vmatrix} = -\begin{vmatrix} a_{11} & a_{12} \\ a_{31} & a_{32} \end{vmatrix}$$

$$\text{Cofactor of } a_{11} = (-1)^{1+1} \begin{vmatrix} a_{22} & a_{23} \\ a_{32} & a_{33} \end{vmatrix} = \begin{vmatrix} a_{22} & a_{23} \\ a_{32} & a_{33} \end{vmatrix}$$

EXAMPLE 14   Find the cofactors of -2 and 5 in the determinant

$$\begin{vmatrix} -2 & 0 & 3 \\ 1 & -6 & 5 \\ -1 & 2 & 0 \end{vmatrix}$$

Solution$\quad$ Cofactor of $-2 = (-1)^{1+1} \begin{vmatrix} -6 & 5 \\ 2 & 0 \end{vmatrix} = \begin{vmatrix} -6 & 5 \\ 2 & 0 \end{vmatrix}$

$$= (-6)(0) - (2)(5) = -10$$

$$\text{Cofactor of } 5 = (-1)^{2+3} \begin{vmatrix} -2 & 0 \\ -1 & 2 \end{vmatrix} = - \begin{vmatrix} -2 & 0 \\ -1 & 2 \end{vmatrix}$$

$$= -[(-2)(2) - (-1)(0)] = 4$$

PROBLEM 14$\quad$ Find the cofactors of 2 and 3 in the determinant in Example 14.

[*Note:* The sign in front of the minor, $(-1)^{i+j}$, can be determined rather mechanically by using a checkerboard pattern of $+$ and $-$ signs over the determinant, starting with $+$ in the upper left-hand corner:

$$\begin{matrix} + & - & + \\ - & + & - \\ + & - & + \end{matrix}$$

Use either the checkerboard or the exponent method, whichever is easier for you, to determine the sign in front of the minor.]

Now we are ready for the central theorem of this section, Theorem 1. It will provide us with an efficient means of evaluating third-order determinants.

THEOREM 1

Value of a Third-Order Determinant
The value of a determinant of order 3 is the sum of three products obtained by multiplying each element of any one row (or each element of any one column) by its cofactor.

To prove this theorem we must show that the expansions indicated by the theorem for any row or any column (six cases) produce the expression on the right of formula (2). Proofs of special cases of this theorem are left to the C problems in Exercise 6-4.

EXAMPLE 15$\quad$ Evaluate by expanding by (A) the first row, and (B) the second column.

$$\begin{vmatrix} 2 & -2 & 0 \\ -3 & 1 & 2 \\ 1 & -3 & -1 \end{vmatrix}$$

Solution   (A) $\begin{vmatrix} 2 & -2 & 0 \\ -3 & 1 & 2 \\ 1 & -3 & -1 \end{vmatrix}$

$$= a_{11}\left(\begin{array}{c}\text{Cofactor} \\ \text{of } a_{11}\end{array}\right) + a_{12}\left(\begin{array}{c}\text{Cofactor} \\ \text{of } a_{12}\end{array}\right) + a_{13}\left(\begin{array}{c}\text{Cofactor} \\ \text{of } a_{13}\end{array}\right)$$

$$= 2\left[(-1)^{1+1}\begin{vmatrix} 1 & 2 \\ -3 & -1 \end{vmatrix}\right] + (-2)\left[(-1)^{1+2}\begin{vmatrix} -3 & 2 \\ 1 & -1 \end{vmatrix}\right] + 0$$

$$= (2)(1)[(1)(-1) - (-3)(2)] + (-2)(-1)[(-3)(-1) - (1)(2)]$$

$$= (2)(5) + (2)(1) = 12$$

(B) $\begin{vmatrix} 2 & -2 & 0 \\ -3 & 1 & 2 \\ 1 & -3 & -1 \end{vmatrix}$

$$= a_{12}\left(\begin{array}{c}\text{Cofactor} \\ \text{of } a_{12}\end{array}\right) + a_{22}\left(\begin{array}{c}\text{Cofactor} \\ \text{of } a_{22}\end{array}\right) + a_{32}\left(\begin{array}{c}\text{Cofactor} \\ \text{of } a_{32}\end{array}\right)$$

$$= (-2)\left[(-1)^{1+2}\begin{vmatrix} -3 & 2 \\ 1 & -1 \end{vmatrix}\right] + (1)\left[(-1)^{2+2}\begin{vmatrix} 2 & 0 \\ 1 & -1 \end{vmatrix}\right]$$

$$+ (-3)\left[(-1)^{3+2}\begin{vmatrix} 2 & 0 \\ -3 & 2 \end{vmatrix}\right]$$

$$= (-2)(-1)[(-3)(-1) - (1)(2)] + (1)(1)[(2)(-1) - (1)(0)]$$

$$+ (-3)(-1)[(2)(2) - (-3)(0)]$$

$$= (2)(1) + (1)(-2) + (3)(4)$$

$$= 12$$

PROBLEM 15   Evaluate by expanding by (A) the first row, and (B) the third column.

$$\begin{vmatrix} 2 & 1 & -1 \\ -2 & -3 & 0 \\ -1 & 2 & 1 \end{vmatrix}$$

■ Remarks

1. It should now be apparent that we can greatly reduce the work involved in evaluating a determinant by choosing to expand (using Theorem 1) by the row or column with the greatest number of 0's.
2. Theorem 1 and the definitions of minors and cofactors generalize completely for determinants of arbitrary order.

3. Where are determinants used? Many equations and formulas have particularly simple and compact representations in determinant form that are easily remembered (see Problems 44–48 in Exercise 6-5). In addition, determinants are involved in theoretical work. For example, one can show that the inverse of a square matrix exists if and only if its determinant is not 0.

Answers to Matched Problems   **13.**   14   **14.**   13, −4

15.   (A)   3   (B)   3

Exercise 6-4 ▪ A   *Evaluate each second-order determinant.*

1. $\begin{vmatrix} 2 & 2 \\ -3 & 1 \end{vmatrix}$ **2.** $\begin{vmatrix} 2 & 4 \\ 3 & -1 \end{vmatrix}$ **3.** $\begin{vmatrix} 6 & -2 \\ -1 & -3 \end{vmatrix}$

4. $\begin{vmatrix} 5 & -4 \\ -2 & 2 \end{vmatrix}$ **5.** $\begin{vmatrix} 1.8 & -1.6 \\ -1.9 & 1.2 \end{vmatrix}$ **6.** $\begin{vmatrix} 0.5 & -3.2 \\ 1.4 & -6.7 \end{vmatrix}$

Given the determinant

$$\begin{vmatrix} a_{11} & a_{12} & a_{13} \\ a_{21} & a_{22} & a_{23} \\ a_{31} & a_{32} & a_{33} \end{vmatrix}$$

write the minor of each of the following elements.

7. a_{11} **8.** a_{33} **9.** a_{23} **10.** a_{22}

Write the cofactor of each of the following elements.

11. a_{11} **12.** a_{33} **13.** a_{23} **14.** a_{22}

Problems 15–22 pertain to the determinant.

$$\begin{vmatrix} -2 & 3 & 0 \\ 5 & 1 & -2 \\ 7 & -4 & 8 \end{vmatrix}$$

Write the minor of each of the following elements. (Leave the answer in determinant form.)

15. a_{11} **16.** a_{22} **17.** a_{32} **18.** a_{21}

Write the cofactor of each of the following elements and evaluate each.

19. a_{11} **20.** a_{22} **21.** a_{32} **22.** a_{21}

Evaluate Problems 23–28 using cofactors.

23. $\begin{vmatrix} 1 & 0 & 0 \\ -2 & 4 & 3 \\ 5 & -2 & 1 \end{vmatrix}$ 24. $\begin{vmatrix} 2 & -3 & 5 \\ 0 & -3 & 1 \\ 0 & 6 & 2 \end{vmatrix}$

25. $\begin{vmatrix} 0 & 1 & 5 \\ 3 & -7 & 6 \\ 0 & -2 & -3 \end{vmatrix}$ 26. $\begin{vmatrix} 4 & -2 & 0 \\ 9 & 5 & 4 \\ 1 & 2 & 0 \end{vmatrix}$

27. $\begin{vmatrix} -1 & 2 & -3 \\ -2 & 0 & -6 \\ 4 & -3 & 2 \end{vmatrix}$ 28. $\begin{vmatrix} 0 & 2 & -1 \\ -6 & 3 & 1 \\ 7 & -9 & -2 \end{vmatrix}$

B   *Given the determinant*

$$\begin{vmatrix} a_{11} & a_{12} & a_{13} & a_{14} \\ a_{21} & a_{22} & a_{23} & a_{24} \\ a_{31} & a_{32} & a_{33} & a_{34} \\ a_{41} & a_{42} & a_{43} & a_{44} \end{vmatrix}$$

write the cofactor (in determinant form) of each of the following elements.

29. a_{11} 30. a_{44} 31. a_{43} 32. a_{23}

Evaluate each of the following determinants using cofactors.

33. $\begin{vmatrix} 3 & -2 & -8 \\ -2 & 0 & -3 \\ 1 & 0 & -4 \end{vmatrix}$ 34. $\begin{vmatrix} 4 & -4 & 6 \\ 2 & 8 & -3 \\ 0 & -5 & 0 \end{vmatrix}$

35. $\begin{vmatrix} 1 & 4 & 1 \\ 1 & 1 & -2 \\ 2 & 1 & -1 \end{vmatrix}$ 36. $\begin{vmatrix} 3 & 2 & 1 \\ -1 & 5 & 1 \\ 2 & 3 & 1 \end{vmatrix}$

37. $\begin{vmatrix} 1 & 4 & 3 \\ 2 & 1 & 6 \\ 3 & -2 & 9 \end{vmatrix}$ 38. $\begin{vmatrix} 4 & -6 & 3 \\ -1 & 4 & 1 \\ 5 & -6 & 3 \end{vmatrix}$

39. $\begin{vmatrix} 2 & 6 & 1 & 7 \\ 0 & 3 & 0 & 0 \\ 3 & 4 & 2 & 5 \\ 0 & 9 & 0 & 2 \end{vmatrix}$ 40. $\begin{vmatrix} 0 & 1 & 0 & 1 \\ 2 & 4 & 7 & 6 \\ 0 & 3 & 0 & 1 \\ 0 & 6 & 2 & 5 \end{vmatrix}$

C 41. $\begin{vmatrix} -2 & 0 & 0 & 0 & 0 \\ 9 & -1 & 0 & 0 & 0 \\ 2 & 1 & 3 & 0 & 0 \\ -1 & 4 & 2 & 2 & 0 \\ 7 & -2 & 3 & 5 & 5 \end{vmatrix}$ 42. $\begin{vmatrix} 2 & 0 & 0 & 0 & 0 \\ 0 & 3 & 0 & 0 & 0 \\ 0 & 0 & 2 & 0 & 0 \\ 0 & 0 & 0 & 1 & 0 \\ 0 & 0 & 0 & 0 & 4 \end{vmatrix}$

If all the letters in Problems 43–46 represent real numbers, show that each statement is true.

43. $\begin{vmatrix} a & b \\ ka & kb \end{vmatrix} = 0$

44. $\begin{vmatrix} a & b \\ c & d \end{vmatrix} = - \begin{vmatrix} b & a \\ d & c \end{vmatrix}$

45. $\begin{vmatrix} a & b \\ c & d \end{vmatrix} = \begin{vmatrix} a & c \\ b & d \end{vmatrix}$

46. $\begin{vmatrix} ka & kb \\ c & d \end{vmatrix} = k \begin{vmatrix} a & b \\ c & d \end{vmatrix}$

47. Show that the expansion of the determinant

$$\begin{vmatrix} a_{11} & a_{12} & a_{13} \\ a_{21} & a_{22} & a_{23} \\ a_{31} & a_{32} & a_{33} \end{vmatrix}$$

by the first column is the same as its expansion by the third row.

48. Repeat Problem 47, using the second row and the third column.

49. If

$$A = \begin{bmatrix} 2 & 3 \\ 1 & -2 \end{bmatrix} \quad \text{and} \quad B = \begin{bmatrix} -1 & 3 \\ 2 & 1 \end{bmatrix}$$

show that $\det(AB) = \det A \cdot \det B$.

50. If

$$A = \begin{bmatrix} a & b \\ c & d \end{bmatrix} \quad \text{and} \quad B = \begin{bmatrix} w & x \\ y & z \end{bmatrix}$$

show that $\det(AB) = \det A \cdot \det B$.

Section 6-5   Properties of Determinants

The following theorems greatly facilitate the task of evaluating determinants of order 3 or greater. Because the proofs for the general case are involved and notationally difficult, we will sketch only informal proofs for determinants of order 3.

THEOREM 2 | If each element of any row (or column) of a determinant is multiplied by a constant k, the new determinant is k times the original.

Partial Proof   Let C_{ij} be the cofactor of a_{ij}. Then

$$\begin{vmatrix} ka_{11} & ka_{12} & ka_{13} \\ a_{21} & a_{22} & a_{23} \\ a_{31} & a_{32} & a_{33} \end{vmatrix} = ka_{11}C_{11} + ka_{12}C_{12} + ka_{13}C_{13}$$

$$= k(a_{11}C_{11} + a_{12}C_{12} + a_{13}C_{13})$$

$$= k\begin{vmatrix} a_{11} & a_{12} & a_{13} \\ a_{21} & a_{22} & a_{23} \\ a_{31} & a_{32} & a_{33} \end{vmatrix}$$

Theorem 2 also states that a factor common to all elements of a row (or column) can be taken out as a factor of the determinant.

EXAMPLE 16 $$\begin{vmatrix} 6 & 1 & 3 \\ -2 & 7 & -2 \\ 4 & 5 & 0 \end{vmatrix} = 2\begin{vmatrix} 3 & 1 & 3 \\ -1 & 7 & -2 \\ 2 & 5 & 0 \end{vmatrix}$$

where 2 is a common factor of the first column.

PROBLEM 16   Take out factors common to any row or any column.

$$\begin{vmatrix} 3 & 2 & 1 \\ 6 & 3 & -9 \\ 1 & 0 & -5 \end{vmatrix}$$

THEOREM 3   If every element in a row (or column) is 0, the value of the determinant is zero.

Theorem 3 is an immediate consequence of Theorem 2, and its proof is left as an exercise. It is illustrated in the following example:

$$\begin{vmatrix} 3 & -2 & 5 \\ 0 & 0 & 0 \\ -1 & 4 & 9 \end{vmatrix} = 0$$

THEOREM 4   If two rows (or two columns) of a determinant are interchanged, the new determinant is the negative of the old.

A proof of Theorem 4 even for a determinant of order 3 is notationally involved. We suggest that you partially prove the theorem by direct expansion of the determinants before and after the interchange of two rows (or columns). The theorem is illustrated by the following example where the second and third columns are interchanged:

$$\begin{vmatrix} 1 & 0 & 9 \\ -2 & 1 & 5 \\ 3 & 0 & 7 \end{vmatrix} = -\begin{vmatrix} 1 & 9 & 0 \\ -2 & 5 & 1 \\ 3 & 7 & 0 \end{vmatrix}$$

THEOREM 5

> If the corresponding elements are equal in two rows (or columns), the value of the determinant is zero.

Proof The general proof of Theorem 5 is easy, making direct use of Theorem 4. If we start with a determinant D that has two rows (or columns) equal and we interchange the equal rows (or columns), the new determinant will be the same as the old. But by Theorem 4,

$$D = -D$$

hence,

$$2D = 0$$

and

$$D = 0$$

THEOREM 6

> If a multiple of any row (or column) of a determinant is added to any other row (or column), the value of the determinant is not changed.

Partial Proof If, in a general third-order determinant, we add a k multiple of the second column to the first, we obtain (where C_{ij} is the cofactor of a_{ij} in the original determinant)

$$\begin{vmatrix} a_{11} + ka_{12} & a_{12} & a_{13} \\ a_{21} + ka_{22} & a_{22} & a_{23} \\ a_{31} + ka_{32} & a_{32} & a_{33} \end{vmatrix} = (a_{11} + ka_{12})C_{11} + (a_{21} + ka_{22})C_{21} + (a_{31} + ka_{32})C_{31}$$

$$= (a_{11}C_{11} + a_{21}C_{21} + a_{31}C_{31}) + k(a_{12}C_{11} + a_{22}C_{21} + a_{32}C_{31})$$

$$= \begin{vmatrix} a_{11} & a_{12} & a_{13} \\ a_{21} & a_{22} & a_{23} \\ a_{31} & a_{32} & a_{33} \end{vmatrix} + k\begin{vmatrix} a_{12} & a_{12} & a_{13} \\ a_{22} & a_{22} & a_{23} \\ a_{32} & a_{32} & a_{33} \end{vmatrix} = \begin{vmatrix} a_{11} & a_{12} & a_{13} \\ a_{21} & a_{22} & a_{23} \\ a_{31} & a_{32} & a_{33} \end{vmatrix}$$

since the determinant following k is zero. (Why?)

Note the similarity in the process described in Theorem 6 to that used to obtain row-equivalent matrices. We use this theorem to transform a determinant without 0 elements into one that contains a row or column with all elements 0 but one. The determinant can then be easily expanded by this row (or column). An example best illustrates the process.

EXAMPLE 17   Evaluate the determinant.

$$\begin{vmatrix} 3 & -1 & 2 \\ -2 & 4 & -3 \\ 4 & -2 & 5 \end{vmatrix}$$

We use Theorem 6 to obtain two 0's in the first row, and then expand the determinant by this row. To start, we replace the third column with the sum of it and 2 times the second column to obtain a 0 in the a_{13} position.

$$\begin{vmatrix} 3 & -1 & 2 \\ -2 & 4 & -3 \\ 4 & -2 & 5 \end{vmatrix} = \begin{vmatrix} 3 & -1 & 0 \\ -2 & 4 & 5 \\ 4 & -2 & 1 \end{vmatrix} \qquad C_3 + 2C_2 \to C_3{}^*$$

Next, to obtain a 0 in the a_{11} position, we replace the first column with the sum of it and 3 times the second column.

$$\begin{vmatrix} 3 & -1 & 0 \\ -2 & 4 & 5 \\ 4 & -2 & 1 \end{vmatrix} = \begin{vmatrix} 0 & -1 & 0 \\ 10 & 4 & 5 \\ -2 & -2 & 1 \end{vmatrix} \qquad C_1 + 3C_2 \to C_1$$

Now it is an easy matter to expand this last determinant by the first row to obtain

$$0 + (-1)\left[(-1)^{1+2} \begin{vmatrix} 10 & 5 \\ -2 & 1 \end{vmatrix} \right] + 0 = 20$$

PROBLEM 17   Evaluate the following determinant by first using Theorem 6 to obtain 0's in the a_{11} and a_{31} positions, and then expand by the first column.

$$\begin{vmatrix} 3 & 10 & -5 \\ 1 & 6 & -3 \\ 2 & 3 & 4 \end{vmatrix}$$

Answers to Matched Problems   **16.** $3\begin{vmatrix} 3 & 2 & 1 \\ 2 & 1 & -3 \\ 1 & 0 & -5 \end{vmatrix}$ **17.** 44

* C_1, C_2, and C_3 represent columns 1, 2, and 3, respectively.

Exercise 6-5 ■

A *For each statement, identify the theorem from this section that justifies it. Do not evaluate.*

1. $\begin{vmatrix} 16 & 8 \\ 0 & -1 \end{vmatrix} = 8 \begin{vmatrix} 2 & 1 \\ 0 & -1 \end{vmatrix}$

2. $\begin{vmatrix} 1 & -9 \\ 0 & -6 \end{vmatrix} = -3 \begin{vmatrix} 1 & 3 \\ 0 & 2 \end{vmatrix}$

3. $-2 \begin{vmatrix} 2 & 1 \\ -3 & 4 \end{vmatrix} = \begin{vmatrix} -4 & 1 \\ 12 & 4 \end{vmatrix}$

4. $4 \begin{vmatrix} -1 & 3 \\ 2 & 1 \end{vmatrix} = \begin{vmatrix} -4 & 12 \\ 2 & 1 \end{vmatrix}$

5. $\begin{vmatrix} 3 & 0 \\ -2 & 0 \end{vmatrix} = 0$

6. $\begin{vmatrix} 5 & -7 \\ 0 & 0 \end{vmatrix} = 0$

7. $\begin{vmatrix} 5 & -1 \\ 8 & 0 \end{vmatrix} = - \begin{vmatrix} -1 & 5 \\ 0 & 8 \end{vmatrix}$

8. $\begin{vmatrix} 6 & 9 \\ 0 & 1 \end{vmatrix} = - \begin{vmatrix} 0 & 1 \\ 6 & 9 \end{vmatrix}$

9. $\begin{vmatrix} 4 & 3 \\ 1 & 2 \end{vmatrix} = \begin{vmatrix} 4-4 & 3-8 \\ 1 & 2 \end{vmatrix}$

10. $\begin{vmatrix} 3 & 2 \\ 5 & 1 \end{vmatrix} = \begin{vmatrix} 3+4 & 2 \\ 5+2 & 1 \end{vmatrix}$

Theorem 6 was used to transform the determinant on the left to that on the right. Replace each letter x with an appropriate numeral to complete the transformation.

11. $\begin{vmatrix} -1 & 3 \\ 2 & -4 \end{vmatrix} = \begin{vmatrix} -1 & x \\ 2 & 2 \end{vmatrix}$

12. $\begin{vmatrix} -1 & 3 \\ 5 & -2 \end{vmatrix} = \begin{vmatrix} -1 & 3 \\ x & 13 \end{vmatrix}$

13. $\begin{vmatrix} -1 & 2 & 3 \\ 2 & 1 & 4 \\ 1 & 3 & 2 \end{vmatrix} = \begin{vmatrix} -1 & 2 & 0 \\ 2 & 1 & 10 \\ 1 & 3 & x \end{vmatrix}$

14. $\begin{vmatrix} -1 & 2 & 3 \\ 2 & 1 & 4 \\ 1 & 3 & 2 \end{vmatrix} = \begin{vmatrix} -1 & 0 & 3 \\ 2 & x & 4 \\ 1 & 5 & 2 \end{vmatrix}$

Use Theorem 6 to transform each determinant into one that contains a row (or column) with all elements 0 but one (if possible); then expand the transformed determinant by this row (or column).

15. $\begin{vmatrix} -1 & 0 & 3 \\ 2 & 5 & 4 \\ 1 & 5 & 2 \end{vmatrix}$

16. $\begin{vmatrix} -1 & 2 & 0 \\ 2 & 1 & 10 \\ 1 & 3 & 5 \end{vmatrix}$

17. $\begin{vmatrix} 3 & 5 & 0 \\ 1 & 1 & -2 \\ 2 & 1 & -1 \end{vmatrix}$

18. $\begin{vmatrix} 2 & 0 & 1 \\ -1 & -3 & 4 \\ 1 & 2 & 3 \end{vmatrix}$

B *For each statement, identify the theorem from this section that justifies it.*

19. $-2 \begin{vmatrix} 1 & 0 & 2 \\ 3 & -2 & 4 \\ 0 & 1 & 1 \end{vmatrix} = \begin{vmatrix} 1 & 0 & 2 \\ -6 & 4 & -8 \\ 0 & 1 & 1 \end{vmatrix}$

20. $\begin{vmatrix} 8 & 0 & 1 \\ 12 & -1 & 0 \\ 4 & 3 & 2 \end{vmatrix} = 4 \begin{vmatrix} 2 & 0 & 1 \\ 3 & -1 & 0 \\ 1 & 3 & 2 \end{vmatrix}$

21. $\begin{vmatrix} 1 & 2 & 0 \\ -1 & 3 & 0 \\ 0 & 1 & 0 \end{vmatrix} = 0$

22. $\begin{vmatrix} -2 & 5 & 13 \\ 1 & 7 & 12 \\ 0 & 8 & 15 \end{vmatrix} = - \begin{vmatrix} 5 & -2 & 13 \\ 7 & 1 & 12 \\ 8 & 0 & 15 \end{vmatrix}$

23. $\begin{vmatrix} 4 & 2 & -1 \\ 2 & 0 & 2 \\ -3 & 5 & -2 \end{vmatrix} = \begin{vmatrix} 4-4 & 2 & -1 \\ 2+8 & 0 & 2 \\ -3-8 & 5 & -2 \end{vmatrix}$

24. $\begin{vmatrix} 7 & 7 & 1 \\ -3 & -3 & 11 \\ 2 & 2 & 0 \end{vmatrix} = 0$

Theorem 6 was used to transform the determinant on the left to that on the right. Replace each letter with an appropriate numeral to complete the transformation.

25. $\begin{vmatrix} 2 & 1 & -1 \\ 3 & 4 & 1 \\ 1 & 2 & -2 \end{vmatrix} = \begin{vmatrix} 0 & 0 & -1 \\ x & 5 & 1 \\ -3 & y & -2 \end{vmatrix}$

26. $\begin{vmatrix} 3 & -1 & 1 \\ -2 & 4 & 3 \\ 1 & 5 & 2 \end{vmatrix} = \begin{vmatrix} 0 & -1 & 0 \\ 10 & 4 & 7 \\ x & 5 & y \end{vmatrix}$

27. $\begin{vmatrix} 7 & 9 & 4 \\ 2 & 3 & 1 \\ 3 & 4 & -2 \end{vmatrix} = \begin{vmatrix} -1 & x & 0 \\ 2 & 3 & 1 \\ 7 & y & 0 \end{vmatrix}$

28. $\begin{vmatrix} 5 & 2 & 3 \\ 3 & 1 & 2 \\ -4 & -3 & 5 \end{vmatrix} = \begin{vmatrix} x & 0 & -1 \\ 3 & 1 & 2 \\ 5 & 0 & y \end{vmatrix}$

Use Theorem 6 to transform each determinant into one that contains a row (or column) with all elements 0 but one (if possible); then expand the transformed determinant by this row (or column).

29. $\begin{vmatrix} 1 & 5 & 3 \\ 4 & 2 & 1 \\ 3 & 1 & 2 \end{vmatrix}$

30. $\begin{vmatrix} -1 & 5 & 1 \\ 2 & 3 & 1 \\ 3 & 2 & 1 \end{vmatrix}$

31. $\begin{vmatrix} 5 & 2 & -3 \\ -2 & 4 & 4 \\ 1 & -1 & 3 \end{vmatrix}$

32. $\begin{vmatrix} 5 & 3 & -6 \\ -1 & 1 & 4 \\ 4 & 3 & -6 \end{vmatrix}$

33. $\begin{vmatrix} 3 & -4 & 1 \\ 6 & -1 & 2 \\ 9 & 2 & 3 \end{vmatrix}$

34. $\begin{vmatrix} 2 & 3 & -1 \\ 5 & 4 & 7 \\ -4 & -6 & 2 \end{vmatrix}$

35. $\begin{vmatrix} 0 & 1 & 0 & 1 \\ 1 & -2 & 4 & 3 \\ 2 & 1 & 5 & 4 \\ 1 & 2 & 1 & 2 \end{vmatrix}$

36. $\begin{vmatrix} 2 & 3 & 1 & -1 \\ 3 & 1 & 2 & 1 \\ 0 & 5 & 4 & 0 \\ -1 & 2 & 3 & 0 \end{vmatrix}$

C 37. $\begin{vmatrix} 3 & 2 & 3 & 1 \\ 3 & -2 & 8 & 5 \\ 2 & 1 & 3 & 1 \\ 4 & 5 & 4 & -3 \end{vmatrix}$

38. $\begin{vmatrix} -1 & 4 & 2 & 1 \\ 5 & -1 & -3 & -1 \\ 2 & -1 & -2 & 3 \\ -3 & 3 & 3 & 3 \end{vmatrix}$

Prove each of the following statements.

39. $\begin{vmatrix} a & b & a \\ d & e & d \\ g & h & g \end{vmatrix} = 0$

40. $\begin{vmatrix} a & b & c \\ kd & ke & kf \\ g & h & i \end{vmatrix} = k \begin{vmatrix} a & b & c \\ d & e & f \\ g & h & i \end{vmatrix}$

41. $\begin{vmatrix} a_1 & b_1 & c_1 \\ a_2 & b_2 & c_2 \\ a_3 & b_3 & c_3 \end{vmatrix} = - \begin{vmatrix} b_1 & a_1 & c_1 \\ b_2 & a_2 & c_2 \\ b_3 & a_3 & c_3 \end{vmatrix}$

42. $\begin{vmatrix} a_1 & b_1 & c_1 \\ a_2 & b_2 & c_2 \\ a_3 & b_3 & c_3 \end{vmatrix} = \begin{vmatrix} a_1 + kc_1 & b_1 & c_1 \\ a_2 + kc_2 & b_2 & c_2 \\ a_3 + kc_3 & b_3 & c_3 \end{vmatrix}$

43. Show, without expanding, that $(2, 5)$ and $(-3, 4)$ satisfy the equation

$$\begin{vmatrix} x & y & 1 \\ 2 & 5 & 1 \\ -3 & 4 & 1 \end{vmatrix} = 0$$

44. Show that

$$\begin{vmatrix} x & y & 1 \\ 2 & 3 & 1 \\ -1 & 2 & 1 \end{vmatrix} = 0$$

is the equation of a line that passes through $(2, 3)$ and $(-1, 2)$.

45. Show that

$$\begin{vmatrix} x & y & 1 \\ x_1 & y_1 & 1 \\ x_2 & y_2 & 1 \end{vmatrix} = 0$$

is the equation of a line that passes through (x_1, y_1) and (x_2, y_2).

46. In analytic geometry it is shown that the area of a triangle with vertices (x_1, y_1), (x_2, y_2), and (x_3, y_3) is the absolute value of

$$\frac{1}{2} \begin{vmatrix} x_1 & y_1 & 1 \\ x_2 & y_2 & 1 \\ x_3 & y_3 & 1 \end{vmatrix}$$

Use this result to find the area of a triangle with vertices $(-1, 4)$, $(4, 8)$, and $(1, 1)$.

47. What can we say about the three points (x_1, y_1), (x_2, y_2), and (x_3, y_3) if

$$\begin{vmatrix} x_1 & y_1 & 1 \\ x_2 & y_2 & 1 \\ x_3 & y_3 & 1 \end{vmatrix} = 0$$

[*Hint*: See Problem 46.]

48. If the three points (x_1, y_1), (x_2, y_2), and (x_3, y_3) are all on the same line, what can we say about the value of the determinant

$$\begin{vmatrix} x_1 & y_1 & 1 \\ x_2 & y_2 & 1 \\ x_3 & y_3 & 1 \end{vmatrix}$$

Section 6-6   Cramer's Rule

■   Two-Equations–Two-Unknowns
■   Three-Equations–Three-Unknowns

Now let us see how determinants arise rather naturally in the process of solving systems of linear equations. We will start by investigating two equations and two unknowns, and then extend any results to three equations and three unknowns.

■ Two-Equations—Two-Unknowns

Instead of thinking of each system of linear equations in two unknowns as a different problem, let us see what happens when we attempt to solve the general system

$$a_{11}x + a_{12}y = k_1 \qquad \text{(1A)}$$

$$a_{21}x + a_{22}y = k_2 \qquad \text{(1B)}$$

once and for all in terms of the unspecified real constants a_{11}, a_{12}, a_{21}, a_{22}, k_1, and k_2.

We proceed by multiplying equations (1A) and (1B) by suitable constants so that when the resulting equations are added, left side to left side and right side to right side, one of the variables drops out. Suppose we choose to eliminate y; what constant should we use to make the coefficients of y the same except for the signs? Multiply equation (1A) by a_{22} and (1B) by $-a_{12}$; then add.

$$
\begin{array}{ll}
a_{22}(1A): & a_{11}a_{22}x + a_{12}a_{22}y = k_1a_{22} \\
-a_{12}(1B): & \underline{-a_{21}a_{12}x - a_{12}a_{22}y = -k_2a_{12}} \\
& a_{11}a_{22}x - a_{21}a_{12}x + 0y = k_1a_{22} - k_2a_{12}
\end{array}
$$

$$(a_{11}a_{22} - a_{21}a_{12})x = k_1a_{22} - k_2a_{12}$$

$$x = \frac{k_1a_{22} - k_2a_{12}}{a_{11}a_{22} - a_{21}a_{12}}$$

$$a_{11}a_{22} - a_{21}a_{12} \neq 0$$

What do the numerator and denominator remind you of? From your experience with determinants in the last two sections, you should recognize these expressions as

$$x = \frac{\begin{vmatrix} k_1 & a_{12} \\ k_2 & a_{22} \end{vmatrix}}{\begin{vmatrix} a_{11} & a_{12} \\ a_{21} & a_{22} \end{vmatrix}}$$

Similarly, starting with system (1) and eliminating x (this is left as an exercise), we obtain

$$y = \frac{\begin{vmatrix} a_{11} & k_1 \\ a_{21} & k_2 \end{vmatrix}}{\begin{vmatrix} a_{11} & a_{12} \\ a_{21} & a_{22} \end{vmatrix}}$$

These results are summarized in Theorem 7, which is named after the Swiss mathematician, G. Cramer (1704–1752).

THEOREM 7 | **Cramer's Rule for Two Equations and Two Unknowns**

Given the system

$$a_{11}x + a_{12}y = k_1$$
$$a_{21}x + a_{22}y = k_2$$

with

$$D = \begin{vmatrix} a_{11} & a_{12} \\ a_{21} & a_{22} \end{vmatrix} \neq 0$$

then

$$x = \frac{\begin{vmatrix} k_1 & a_{12} \\ k_2 & a_{22} \end{vmatrix}}{D} \quad \text{and} \quad y = \frac{\begin{vmatrix} a_{11} & k_1 \\ a_{21} & k_2 \end{vmatrix}}{D}$$

The determinant D is called the **coefficient determinant**. If $D \neq 0$, then the system has exactly one solution, which is given by Cramer's rule. If, on the other hand, $D = 0$, then it can be shown that the system is either inconsistent or dependent; that is, the system either has no solutions or has an infinite number of solutions.

EXAMPLE 18   Solve using Cramer's rule.

$$2x - 3y = 7$$
$$-3x + y = -7$$

Solution $$D = \begin{vmatrix} 2 & -3 \\ -3 & 1 \end{vmatrix} = -7$$

$$x = \frac{\begin{vmatrix} 7 & -3 \\ -7 & 1 \end{vmatrix}}{-7} = \frac{-14}{-7} = 2$$

$$y = \frac{\begin{vmatrix} 2 & 7 \\ -3 & -7 \end{vmatrix}}{-7} = \frac{7}{-7} = -1$$

PROBLEM 18   Solve using Cramer's rule.

$$3x + 2y = -3$$
$$-4x + 3y = -13$$

■ Three-Equations–Three-Unknowns

Cramer's rule generalizes completely for any size linear system that has the same number of unknowns as equations. We state without proof in Theorem 8 the rule for three equations and three unknowns. (Augmented matrix methods can be used to prove this rule.)

THEOREM 8

Cramer's Rule for Three Equations and Three Unknowns

Given the system

$$a_{11}x + a_{12}y + a_{13}z = k_1$$
$$a_{21}x + a_{22}y + a_{23}z = k_2$$
$$a_{31}x + a_{32}y + a_{33}z = k_3$$

with

$$D = \begin{vmatrix} a_{11} & a_{12} & a_{13} \\ a_{21} & a_{22} & a_{23} \\ a_{31} & a_{32} & a_{33} \end{vmatrix} \neq 0$$

then

$$x = \frac{\begin{vmatrix} k_1 & a_{12} & a_{13} \\ k_2 & a_{22} & a_{23} \\ k_3 & a_{32} & a_{33} \end{vmatrix}}{D} \qquad y = \frac{\begin{vmatrix} a_{11} & k_1 & a_{13} \\ a_{21} & k_2 & a_{23} \\ a_{31} & k_3 & a_{33} \end{vmatrix}}{D} \qquad z = \frac{\begin{vmatrix} a_{11} & a_{12} & k_1 \\ a_{21} & a_{22} & k_2 \\ a_{31} & a_{32} & k_3 \end{vmatrix}}{D}$$

It is easy to remember these determinant formulas for x, y, and z if one observes the following:

1. Determinant D is formed from the coefficients of x, y, and z, keeping the same relative position in the determinant as found in the system.
2. Determinant D appears in the denominators for x, y, and z.
3. The numerator for x can be obtained from D by replacing the coefficients of x—a_{11}, a_{21}, and a_{31}—with the constants k_1, k_2, and k_3, respectively. Similar statements can be made for the numerators for y and z.

EXAMPLE 19   Use Cramer's rule to solve.

$$x + y \qquad = \quad 1$$
$$3y - z = -4$$
$$x \qquad + z = \quad 3$$

Solution $D = \begin{vmatrix} 1 & 1 & 0 \\ 0 & 3 & -1 \\ 1 & 0 & 1 \end{vmatrix} = 2$

$$x = \frac{\begin{vmatrix} 1 & 1 & 0 \\ -4 & 3 & -1 \\ 3 & 0 & 1 \end{vmatrix}}{2} = \frac{4}{2} = 2 \qquad y = \frac{\begin{vmatrix} 1 & 1 & 0 \\ 0 & -4 & -1 \\ 1 & 3 & 1 \end{vmatrix}}{2} = \frac{-2}{2} = -1$$

$$z = \frac{\begin{vmatrix} 1 & 1 & 1 \\ 0 & 3 & -4 \\ 1 & 0 & 3 \end{vmatrix}}{2} = \frac{2}{2} = 1$$

PROBLEM 19   Use Cramer's rule to solve.

$$3x \quad\quad - z = 5$$
$$x - y + z = 0$$
$$x + y \quad\quad = 0$$

In practice, Cramer's rule is rarely used to solve systems of order higher than 2 or 3; more efficient methods are available, including the methods discussed in Chapter 5. Cramer's rule is, however, a valuable tool in theoretical mathematics.

Answers to Matched Problems   **18.** $x = 1, \quad y = -3$ **19.** $x = 1, \quad y = -1, \quad z = -2$

Exercise 6-6 ■   *Solve, using Cramer's rule.*

A   **1.** $x + 2y = 1$ **2.** $x + 2y = 3$
$x + 3y = -1$ $x + 3y = 5$

3. $2x + y = 1$ **4.** $x + 3y = 1$
$5x + 3y = 2$ $2x + 8y = 0$

5. $2x - y = -3$ **6.** $2x + y = 1$
$-x + 3y = 4$ $5x + 3y = 2$

B   **7.** $x + y \quad = 0$ **8.** $x + y \quad = -4$
$2y + z = -5$ $2y + z = 0$
$-x + \quad z = -3$ $-x + \quad z = 5$

9. $x + y \quad = 1$ **10.** $x + y \quad = -4$
$2y + z = 0$ $2y + z = 3$
$-x + \quad z = 0$ $-x + \quad z = 7$

11.
$$y + z = -4$$
$$x + 2z = 0$$
$$x - y = 5$$

12.
$$x - z = 2$$
$$2x - y = 8$$
$$x + y + z = 2$$

13.
$$2y - z = -4$$
$$x - y - z = 0$$
$$x - y + 2z = 6$$

14.
$$2x + y = 2$$
$$x - y + z = -1$$
$$x + y + z = -1$$

C It is clear that $x = 0$, $y = 0$, $z = 0$ is a solution to each of the following systems. Use Cramer's rule to determine whether this solution is unique. [*Hint:* If $D \neq 0$, what can you conclude? If $D = 0$, what can you conclude?]

15.
$$x - 4y + 9z = 0$$
$$4x - y + 6z = 0$$
$$x - y + 3z = 0$$

16.
$$3x - y + 3z = 0$$
$$5x + 5y - 9z = 0$$
$$-2x + y - 3z = 0$$

17. Prove Theorem 7 for y.

Section 6-7 Chapter Review

IMPORTANT TERMS AND SYMBOLS

6-1 Matrix Addition; Multiplication of a Matrix by a Number. Dimension of a matrix, $m \times n$ matrix, square matrix, column matrix, row matrix, equal matrices, matrix addition, addition properties, zero matrix, negative of a matrix, matrix subtraction, multiplication of a matrix by a number, $A + B$, $-B$, $A - B$, kA

6-2 Matrix Multiplication. Dot product, matrix product, multiplication properties, $A \cdot B$, AB

6-3 Inverse of a Square Matrix; Matrix Equations. Identity matrix for multiplication, main diagonal, inverse of a matrix relative to multiplication, matrix equation, I, M^{-1}, $AX = B$

6-4 Determinant Functions. Determinant function, determinant, second-order determinant, third-order determinant, det M

6-5 Properties of Determinants. Row operations, column operations

6-6 Cramer's Rule. Two-equations–two-unknowns, three-equations–three-unknowns, coefficient determinant

Exercise 6-7 Chapter Review

Work through all the problems in this chapter review and check answers in the back of the book. (Answers to all problems are there, and following each answer is a number in italics indicating the section in which that type of problem is discussed.) Where weaknesses show up, review appropriate sections in the text. When you are satisfied that you know the material, take the practice test following this review.

A   In Problems 1–9 perform the operations that are defined, given the following matrices:

$$A = \begin{bmatrix} 1 & 2 \\ 3 & 1 \end{bmatrix} \qquad B = \begin{bmatrix} 2 & 1 \\ 1 & 1 \end{bmatrix} \qquad C = [2 \quad 3] \qquad D = \begin{bmatrix} 1 \\ 2 \end{bmatrix}$$

1. $A + B$ 2. $B + D$ 3. $A - 2B$

4. AB 5. AC 6. AD

7. DC 8. $C \cdot D$ 9. $C + D$

10.   Find the inverse of

$$A = \begin{bmatrix} 3 & 2 \\ 4 & 3 \end{bmatrix}$$

by appropriate row operations on $[A\,|\,I]$. Show that $A^{-1}A = I$.

11.   Write the system

$$3x_1 + 2x_2 = k_1$$
$$4x_1 + 3x_2 = k_2$$

as a matrix equation and solve using the inverse found in Problem 10 for

(A) $k_1 = 3, \quad k_2 = 5$ (B) $k_1 = 7, \quad k_2 = 10$
(C) $k_1 = 4, \quad k_2 = 2$

Evaluate Problems 12 and 13.

12. $\begin{vmatrix} 2 & -3 \\ -5 & -1 \end{vmatrix}$ 13. $\begin{vmatrix} 2 & 3 & -4 \\ 0 & 5 & 0 \\ 1 & -4 & -2 \end{vmatrix}$

14.   Solve the system using Cramer's rule.

$$3x - 2y = 8$$
$$x + 3y = -1$$

B   In Problems 15–20 perform the operations that are defined, given the following matrices:

$$A = \begin{bmatrix} 2 & -2 \\ 1 & 0 \\ 3 & 2 \end{bmatrix} \qquad B = \begin{bmatrix} -1 \\ 2 \\ 3 \end{bmatrix} \qquad C = [2 \quad 1 \quad 3]$$

$$D = \begin{bmatrix} 3 & -2 & 1 \\ -1 & 1 & 2 \end{bmatrix} \qquad E = \begin{bmatrix} 3 & -4 \\ -1 & 0 \end{bmatrix}$$

15. $A + D$ 16. $E + DA$ 17. $DA - 3E$

18. $C \cdot B$ 19. CB 20. $AD - BC$

21. Find the inverse of

$$A = \begin{bmatrix} 1 & 2 & 3 \\ 2 & 3 & 4 \\ 1 & 2 & 1 \end{bmatrix}$$

by appropriate row operations on $[A|I]$. Show that $A^{-1}A = I$.

22. Write the system

$$x_1 + 2x_2 + 3x_3 = k_1$$
$$2x_1 + 3x_2 + 4x_3 = k_2$$
$$x_1 + 2x_2 + x_3 = k_3$$

as a matrix equation and solve using the inverse found in Problem 21 for

(A) $k_1 = 1$, $k_2 = 3$, $k_3 = 3$ (B) $k_1 = 0$, $k_2 = 0$, $k_3 = -2$
(C) $k_1 = -3$, $k_2 = -4$, $k_3 = 1$

Evaluate Problems 23 and 24.

23. $\begin{vmatrix} -\frac{1}{4} & \frac{3}{2} \\ \frac{1}{2} & \frac{2}{3} \end{vmatrix}$ 24. $\begin{vmatrix} 2 & -1 & 1 \\ -3. & 5 & 2 \\ 1 & -2 & 4 \end{vmatrix}$

25. Solve for y only using Cramer's rule.

$$x - 2y + z = -6$$
$$y - z = 4$$
$$2x + 2y + z = 2$$

Find the numerator and denominator first; then reduce.

C 26. Find the inverse of

$$A = \begin{bmatrix} 4 & 5 & 6 \\ 4 & 5 & -6 \\ 1 & 1 & 1 \end{bmatrix}$$

Show that $A^{-1}A = I$.

27. Clear the decimals in the system

$$0.04x_1 + 0.05x_2 + 0.06x_3 = 360$$
$$0.04x_1 + 0.05x_2 - 0.06x_3 = 120$$
$$x_1 + x_2 + x_3 = 7{,}000$$

by multiplying the first two equations by 100; then write the resulting system as a matrix equation and solve using the inverse found in Problem 26.

28. $\begin{vmatrix} -1 & 4 & 1 & 1 \\ 5 & -1 & 2 & -1 \\ 2 & -1 & 0 & 3 \\ -3 & 3 & 0 & 3 \end{vmatrix} = ?$

29. Show that

$$\begin{vmatrix} u & v \\ w & x \end{vmatrix} = \begin{vmatrix} u + kv & v \\ w + kx & x \end{vmatrix}$$

Practice Test   Chapter 6

Take this practice test as if it were a graded test. Allow yourself up to 50 minutes. Work the problems without looking back in the chapter. Correct your work using the answers (keyed to appropriate sections) in the back of the book.

In Problems 1–5 perform the indicated operations (if possible) given the following matrices:

$$A = \begin{bmatrix} 2 & -1 & 3 \\ -1 & 2 & 0 \end{bmatrix} \qquad B = [1 \quad -2 \quad -3] \qquad C = \begin{bmatrix} 4 \\ -1 \\ 2 \end{bmatrix}$$

$$D = \begin{bmatrix} 2 & -3 \\ -1 & 2 \end{bmatrix} \qquad E = \begin{bmatrix} 1 & -2 & 0 \\ 3 & 1 & 2 \\ -1 & 0 & 1 \end{bmatrix}$$

1. $B \cdot C$ **2.** DA **3.** AD

4. $CB + 2E$ **5.** $A + E$

6. Find the inverse for

$$A = \begin{bmatrix} 2 & -3 \\ 3 & -4 \end{bmatrix}$$

using row operations on $[A \mid I]$. Check by showing that $A^{-1}A = I$.

7. Write the system

$$2x_1 - 3x_2 = k_1$$
$$3x_1 - 4x_2 = k_2$$

as a matrix equation and solve using the inverse in Problem 6 for

(A) $\quad k_1 = 2, \quad k_2 = -3$ (B) $\quad k_1 = -2, \quad k_2 = 1$

8.   Evaluate.

$$\begin{vmatrix} 3 & -1 & 4 \\ -1 & 2 & -3 \\ 1 & 3 & 2 \end{vmatrix}$$

9.   Solve the system for z only, using Cramer's rule.

$$2x - y + z = 9$$
$$x \qquad - 2z = -8$$
$$-x + 3y + z = 3$$

10. It is clear that (0, 0, 0) is a solution to the system

$$x_1 - x_2 + x_3 = 0$$
$$-x_1 - 3x_2 - x_3 = 0$$
$$x_1 + x_2 + x_3 = 0$$

Evaluate the coefficient determinant and determine whether (0, 0, 0) is the only solution or whether there are infinitely many others.

Sequences and Series

A natural design of mathematical interest. Can you guess the source? See the back of the book.

Chapter 7 ▪ Sequences and Series

In this chapter we are going to consider functions whose domains are special subsets of the set of integers; that is, subsets whose members are successive integers. These special functions, called **sequences**, are encountered with increased frequency as one progresses in mathematics.

Section 7-1 Sequences and Series

▪ Sequences
▪ Series

▪ Sequences

Consider the function f given by

$$f(n) = 2n - 1 \tag{1}$$

where the domain of f is the set of natural numbers N. The function f is an example of a sequence; however, one hardly ever sees sequences represented in this way. A special notation for sequences has evolved, which we now discuss.

To start, the range value $f(n)$ is usually symbolized more compactly with a symbol such as a_n. Thus, in place of equation (1) we would write

$$a_n = 2n - 1$$

and the domain would be understood to be the set of natural numbers N unless something was said to the contrary or the context indicated otherwise. The elements in the range are called **terms of the sequence**; a_1 is the first term, a_2 the second term, and a_n the nth term.

$a_1 = 2(1) - 1 = 1$ First term
$a_2 = 2(2) - 1 = 3$ Second term
$a_3 = 2(3) - 1 = 5$ Third term
$\vdots$ $\vdots$

When the terms in a sequence are written in their natural order with respect to domain values

$$a_1, a_2, a_3, \ldots, a_n, \ldots$$

or

$$1, 3, 5, \ldots, 2n - 1, \ldots$$

this ordered list of elements is often informally referred to as a sequence. A sequence is also represented in the abbreviated form $\{a_n\}$ where a symbol for the nth term is placed between braces. For example, we could refer to the sequence

$$1, 3, 5, \ldots, 2n - 1, \ldots$$

as the sequence $\{2n - 1\}$.

If the domain of a function is a finite set of successive integers, then the sequence is called a **finite sequence**; if the domain is an infinite set of successive integers, then the sequence is called an **infinite sequence**. The sequence $\{2n - 1\}$ is an infinite sequence. We now illustrate a finite sequence and another general way of specifying a sequence by use of a *recursion formula*.

EXAMPLE 1   List the terms of the sequence specified by

$$a_1 = 5$$
$$a_n = a_{n-1} + 2 \qquad n \in \{2, 3, 4\}$$

Solution $a_1 \quad = 5$

$$a_2 \;\big|\; = a_{2-1} + 2 = a_1 + 2 = 5 + 2 \;\big|\; = 7$$
$$a_3 \;\big|\; = a_{3-1} + 2 = a_2 + 2 = 7 + 2 \;\big|\; = 9$$
$$a_4 \;\big|\; = a_{4-1} + 2 = a_3 + 2 = 9 + 2 \;\big|\; = 11$$

The formula $a_n = a_{n-1} + 2$ is called a **recursion formula** and is used to generate the terms of a sequence in terms of preceding terms. Of course, a starting term must be provided in order to use the formula. Recursion formulas are particularly suitable for use with calculators and computers (see Problems 55 and 56 in Exercise 7-1).

PROBLEM 1   Find the first five terms of a sequence specified by

$$a_1 = 4$$
$$a_n = \tfrac{1}{2}a_{n-1} \qquad n \geq 2$$

Now let us look at the problem in reverse; that is, given the first few terms of a sequence (assuming the terms of the sequence continue in the indicated pattern), find a_n in terms of n.

EXAMPLE 2   Find a_n in terms of n for the sequences whose first four terms are

(A) 5, 6, 7, 8, . . .   (B) 2, -4, 8, -16, . . .

Solution   (A) $a_n = n + 4$ (B) $a_n = (-1)^{n+1} 2^n$

[*Note:* These representations are not unique. Also, since it is not stated to the contrary, the domain of each sequence is assumed to be the set of natural numbers N.]

PROBLEM 2   Find a_n in terms of n for

(A) 2, 4, 6, 8, . . .   (B) 1, $-\frac{1}{2}$, $\frac{1}{4}$, $-\frac{1}{8}$, . . .

■ Series

The indicated sum of the terms of a sequence is called a **series**. If the sequence is finite, the corresponding series is a **finite series**; if the sequence is infinite, the corresponding series is an **infinite series**. We will restrict our discussion to finite series in this section. For example,

1, 2, 4, 8, 16   Finite sequence

$1 + 2 + 4 + 8 + 16$ Finite series

Series are often represented in a compact form using summation notation. Consider the following examples:

$$\sum_{k=1}^{4} a_k = a_1 + a_2 + a_3 + a_4$$

$$\sum_{k=3}^{7} b_k = b_3 + b_4 + b_5 + b_6 + b_7$$

$$\sum_{k=0}^{n} c_k = c_0 + c_1 + c_2 + \cdots + c_n$$

The terms on the right are obtained from the left expression by successively replacing the **summing index** k with integers, starting with the first number indicated below $\sum$ and ending with the number that appears above $\sum$. Thus, for example, if we are given the sequence

$$\frac{1}{2}, \frac{1}{4}, \frac{1}{8}, \ldots, \frac{1}{2^n}$$

the corresponding series is

$$\frac{1}{2} + \frac{1}{4} + \frac{1}{8} + \cdots + \frac{1}{2^n}$$

or, more compactly,

$$\sum_{k=1}^{n} \frac{1}{2^k}$$

EXAMPLE 3   Write $\displaystyle\sum_{k=1}^{5} \frac{k-1}{k}$ without summation notation.

Solution $\displaystyle\sum_{k=1}^{5} \frac{k-1}{k} = \frac{1-1}{1} + \frac{2-1}{2} + \frac{3-1}{3} + \frac{4-1}{4} + \frac{5-1}{5}$

$$= 0 + \frac{1}{2} + \frac{2}{3} + \frac{3}{4} + \frac{4}{5}$$

PROBLEM 3   Write $\displaystyle\sum_{k=0}^{5} \frac{(-1)^k}{2k+1}$ without summation notation.

EXAMPLE 4   Write the following series using summation notation.

$$1 - \frac{1}{2} + \frac{1}{3} - \frac{1}{4} + \frac{1}{5} - \frac{1}{6}$$

(A) Start the summing index at $k = 1$.
(B) Start the summing index at $k = 0$.

Solution   (A) $(-1)^{k+1}$ provides the alternation of sign.
$1/k$ provides the other part of each term.
Thus, we can write

$$\sum_{k=1}^{6} \frac{(-1)^{k+1}}{k}$$

as can be easily checked.

(B) $(-1)^k$ provides the alternation of sign.
$1/(k+1)$ provides the other part of each term.
Thus, we can write

$$\sum_{k=0}^{5} \frac{(-1)^k}{k+1}$$

as can be checked.

PROBLEM 4   Write the following series using summation notation.

$$1 - \frac{2}{3} + \frac{4}{9} - \frac{8}{27} + \frac{16}{81}$$

(A) Start with $k = 1$.   (B) Start with $k = 0$.

1. $4, 2, 1, \frac{1}{2}, \frac{1}{4}$

2.   (A) $a_n = 2n$ (B) $a_n = (-1)^{n+1} 2^{1-n}$

3. $1 - \dfrac{1}{3} + \dfrac{1}{5} - \dfrac{1}{7} + \dfrac{1}{9} - \dfrac{1}{11}$

4.   (A) $\displaystyle\sum_{k=1}^{5} \left(-\dfrac{2}{3}\right)^{k-1}$ (B) $\displaystyle\sum_{k=0}^{4} \left(-\dfrac{2}{3}\right)^{k}$

Exercise 7-1 ■

A   Write the first four terms for each sequence.

1. $a_n = n - 2$ 2. $a_n = n + 3$ 3. $a_n = \dfrac{n-1}{n+1}$

4. $a_n = \left(1 + \dfrac{1}{n}\right)^n$ 5. $a_n = (-2)^{n+1}$ 6. $a_n = \dfrac{(-1)^{n+1}}{n^2}$

7.   Write the eighth term in the sequence in Problem 1.

8.   Write the tenth term in the sequence in Problem 2.

9.   Write the one-hundredth term in the sequence in Problem 3.

10.   Write the two-hundredth term in the sequence in Problem 4.

Write each series in expanded form without summation notation.

11. $\displaystyle\sum_{k=1}^{5} k$ 12. $\displaystyle\sum_{k=1}^{4} k^2$ 13. $\displaystyle\sum_{k=1}^{3} \dfrac{1}{10^k}$

14. $\displaystyle\sum_{k=1}^{5} \left(\dfrac{1}{3}\right)^k$ 15. $\displaystyle\sum_{k=1}^{4} (-1)^k$ 16. $\displaystyle\sum_{k=1}^{6} (-1)^{k+1} k$

B   Write the first five terms of each sequence.

17. $a_n = (-1)^{n+1} n^2$ 18. $a_n = (-1)^{n+1}\left(\dfrac{1}{2^n}\right)$

19. $a_n = \dfrac{1}{3}\left(1 - \dfrac{1}{10^n}\right)$ 20. $a_n = n[1 - (-1)^n]$

21. $a_n = \left(-\dfrac{1}{2}\right)^{n-1}$ 22. $a_n = \left(-\dfrac{3}{2}\right)^{n-1}$

23. $a_1 = 7; \quad a_n = a_{n-1} - 4, \, n \geq 2$

24. $a_1 = a_2 = 1; \quad a_n = a_{n-1} + a_{n-2}, \, n \geq 3$

25. $a_1 = 4; \quad a_n = \frac{1}{4} a_{n-1}, \, n \geq 2$ 26. $a_1 = 2; \quad a_n = 2a_{n-1}, \, n \geq 2$

Find a_n in terms of n.

27. $4, 5, 6, 7, \ldots$ 28. $-2, -1, 0, 1, \ldots$

29. $3, 6, 9, 12, \ldots$ 30. $-2, -4, -6, -8, \ldots$

31. $\frac{1}{2}, \frac{2}{3}, \frac{3}{4}, \frac{4}{5}, \ldots$ **32.** $\frac{1}{2}, \frac{3}{4}, \frac{5}{6}, \frac{7}{8}, \ldots$

33. $1, -1, 1, -1, \ldots$ **34.** $1, -2, 3, -4, \ldots$

35. $-2, 4, -8, 16, \ldots$ **36.** $1, -3, 5, -7, \ldots$

37. $x, \dfrac{x^2}{2}, \dfrac{x^3}{3}, \dfrac{x^4}{4}, \ldots$ **38.** $x, -x^3, x^5, -x^7, \ldots$

Write each series in expanded form without summation notation.

39. $\displaystyle\sum_{k=1}^{4} \frac{(-2)^{k+1}}{k}$ **40.** $\displaystyle\sum_{k=1}^{5} (-1)^{k+1}(2k-1)^2$

41. $\displaystyle\sum_{k=1}^{3} \frac{1}{k} x^{k+1}$ **42.** $\displaystyle\sum_{k=1}^{5} x^{k-1}$

43. $\displaystyle\sum_{k=1}^{5} \frac{(-1)^{k+1}}{k} x^k$ **44.** $\displaystyle\sum_{k=0}^{4} \frac{(-1)^k x^{2k+1}}{2k+1}$

Write each series using summation notation.

45. $S_4 = 1^2 + 2^2 + 3^2 + 4^2$ **46.** $S_5 = 2 + 3 + 4 + 5 + 6$

47. $S_5 = \dfrac{1}{2} + \dfrac{1}{2^2} + \dfrac{1}{2^3} + \dfrac{1}{2^4} + \dfrac{1}{2^5}$ **48.** $S_4 = 1 - \frac{1}{2} + \frac{1}{3} - \frac{1}{4}$

49. $S_n = 1 + \dfrac{1}{2^2} + \dfrac{1}{3^2} + \cdots + \dfrac{1}{n^2}$

50. $S_n = 2 + \frac{2}{3} + \frac{4}{3} + \cdots + \dfrac{n+1}{n}$

51. $S_n = 1 - 4 + 9 - \cdots + (-1)^{n+1} n^2$

52. $S_n = \frac{1}{2} + \frac{1}{4} + \frac{1}{8} + \cdots + \dfrac{(-1)^{n+1}}{2^n}$

C **53.** Show that: $\displaystyle\sum_{k=1}^{n} ca_k = c \sum_{k=1}^{n} a_k$

54. Show that: $\displaystyle\sum_{k=1}^{n} (a_k + b_k) = \sum_{k=1}^{n} a_k + \sum_{k=1}^{n} b_k$

CALCULATOR PROBLEMS   *The sequence*

$$a_n = \frac{a_{n-1}^2 + M}{2a_{n-1}} \qquad n \geq 2; \quad M \text{ a positive real number}$$

can be used to find $\sqrt{M}$ *to any decimal-place accuracy desired. To start the sequence, choose* a_1 *arbitrarily from the positive real numbers.*

55. (A)   Find the first four terms of the sequence

$$a_1 = 3 \qquad a_n = \frac{a_{n-1}^2 + 2}{2a_{n-1}} \qquad n \geq 2$$

(B)   Compare the terms with $\sqrt{2}$ from a calculator or a table.

(C)   Repeat parts (A) and (B) by letting a_1 be any other positive number, say 1.

56. (A)   Find the first four terms of the sequence

$$a_1 = 2 \qquad a_n = \frac{a_{n-1}^2 + 5}{2a_{n-1}} \qquad n \geq 2$$

(B)   Find $\sqrt{5}$ in a table and compare with part (A).

(C)   Repeat parts (A) and (B) by letting a_1 be any other positive number, say 3.

In calculus, it can be shown that

$$e^x = \sum_{k=0}^{\infty} \frac{x^k}{k!} \approx 1 + \frac{x}{1!} + \frac{x^2}{2!} + \frac{x^3}{3!} + \cdots + \frac{x^n}{n!}$$

the larger n, the better the approximation. Note that $0! = 1$ and $n! = 1 \cdot 2 \cdot 3 \cdots \cdot n$ for $n \in N$. Problems 57 and 58 refer to this series.

57. Approximate $e^{0.2}$ using the first five terms of the series. Compare this approximation with your calculator evaluation of $e^{0.2}$.

58. Approximate $e^{-0.5}$ using the first five terms of the series. Compare this approximation with your calculator evaluation of $e^{-0.5}$.

Section 7-2   Mathematical Induction

- Introduction
- Mathematical Induction
- Three Famous Problems

■ Introduction

In common usage the word **induction** means the generalization from particular cases or facts. The ability to formulate general hypotheses from a limited number of facts is a distinguishing characteristic of a creative mathematician. The creative process does not stop here, however; these hypotheses must then be proved or disproved. In mathematics, we have a special method of proof called **mathematical induction** that ranks among the most important basic tools in a mathematician's tool box. This method of proof, using deductive reasoning, enters frequently into the second part of the process described above.

We illustrate the first part of the process by an example. Suppose we write the sums of consecutive odd integers as follows:

$$1 = 1$$
$$1 + 3 = 4$$
$$1 + 3 + 5 = 9$$
$$1 + 3 + 5 + 7 = 16$$
$$1 + 3 + 5 + 7 + 9 = 25$$

Is there something very regular about 1, 4, 9, 16, and 25? You no doubt guessed that each is a perfect square and, perhaps, even guessed that each is the square of the number of terms being added. Have we "discovered" a general property of integers? What does this property appear to be?

Conjecture P: The sum of the first n odd integers is n^2 for all positive integers n [that is, $1 + 3 + 5 + \cdots + (2n - 1) = n^2$ for $n \in N$].*

Thus far we have used ordinary induction to arrive at Conjecture P. But how do we prove that Conjecture P is true for all positive integers? Continuing by one-by-one testing will never accomplish a general proof—not in your lifetime or all of your descendants' lifetimes. Mathematical induction is the answer to this dilemma. Before we discuss this method of proof, let us consider another conjecture.

Conjecture Q: For each positive integer n, the number $n^2 - n + 41$ is a prime number.

It is important to recognize that a conjecture can be proved false if it fails for only one case, called a **counterexample**. Let us check the conjecture for a few particular cases:

n	$n^2 - n + 41$	Prime?
1	41	Yes
2	43	Yes
3	47	Yes
4	53	Yes
5	61	Yes

* The equation $1 + 3 + 5 + \cdots + (2n - 1) = n^2$ is a symbolic way of representing the statement, "The sum of the first n odd integers is n^2." On the left side of the equation, we start at 1 and add successive odd integers until we reach $2n - 1$, for a given n. If $n = 1$, then $2n - 1 = 2(1) - 1 = 1$, and we start at 1 and stop there! Thus, for $n = 1$, the equation becomes $1 = 1^2$. If $n = 2$, then $2n - 1 = 2(2) - 1 = 3$, and we start at 1 and stop at 3. Thus, for $n = 2$, the equation becomes $1 + 3 = 2^2$. For $n = 3$, then $2n - 1 = 2(3) - 1 = 5$, and we start at 1 and stop at 5. Thus, for $n = 3$, the equation becomes $1 + 3 + 5 = 3^2$. And so on.

It certainly appears that Conjecture Q has a good chance of being true. The reader may want to check a few more cases, and if she or he persists, it will be found that Conjecture Q is true for n up to 41. What happens at $n = 41$?

$$41^2 - 41 + 41 = 41^2$$

which is not prime. Thus, Conjecture Q is false; $n = 41$ provides a counterexample. Here we see the danger of generalizing without proof from a few special cases. This example was discovered by Euler (1707–1783).

■ Mathematical Induction

Now to discuss mathematical induction. To start, we state a rather obvious property of the integers as an axiom.

AXIOM

Axiom—Well-Ordering Principle
Let S be any set that contains one or more positive integers; then there must be a positive integer in S that is smaller than each of the others.

Sets in which we have a particular interest are sets of integers that are closed under the addition of 1; that is, if k is in the set, then $k + 1$ is in the set. We will refer to such sets as **inductive sets**. Now we state the important theorem of this section.

THEOREM 1

Principle of Mathematical Induction
If p is a positive integer and S is a set of integers such that **1.** $p \in S$ Hypothesis 1 **2.** S is inductive   Hypothesis 2 then S contains all integers greater than or equal to p.

Theorem 1 certainly seems reasonable, since if $p \in S$, then by Hypothesis 2, $p + 1 \in S$; if $p + 1 \in S$, then by Hypothesis 2, $p + 2 \in S$; and so on. Clearly, all integers greater than or equal to p are in S. The only catch to this "proof" is in the use of "and so on" and "clearly." We proceed now to a rigorous proof of this important theorem.

Proof (by Contradiction)   Assume, under the hypothesis of the theorem, S does not contain all integers greater than or equal to p. Let G be the set of all integers greater than p not in S. From the well-ordering axiom, G has a least element, say r, that is not p, since, by Hypothesis 1, $p \in S$. Thus $r - 1$, the integer preceding r, is in S. But by Hypothesis 2, if $r - 1 \in S$, then $(r - 1) + 1 = r \in S$, which is a contradiction, since r is in G. Our assumption must be false, and we conclude that S contains all integers from p on.

Let us now use Theorem 1 to prove that Conjecture P is true, as well as several other conjectures. To facilitate the writing of induction proofs in a more concise way, we introduce two special symbols:

SYMBOL	MEANING
$\therefore$	Therefore
$\Rightarrow$	Implies ("$p \Rightarrow q$" is read "p implies q" or, equivalently, "If p then q.")

EXAMPLE 5   Prove: $1 + 3 + 5 + \cdots + (2n - 1) = n^2, \quad n \in N$

Proof   Write

$$P_n: \quad 1 + 3 + 5 + \cdots + (2n - 1) = n^2$$
$$S = \{n \in N \mid P_n \text{ is true}\}$$

where S is the truth set for the open statement P_n. To show that $S = N$, we must establish both parts of Theorem 1.

Part 1.   Show that $1 \in S$.

$$1 = 1^2$$
$$\therefore 1 \in S$$

Part 2.   Show that S is inductive (that is, prove generally that $k \in S \Rightarrow k + 1 \in S$).

We want to show that if P_n is true for $n = k$, it follows logically that P_n is true for $n = k + 1$. We write P_k and P_{k+1} first to obtain an idea of where we must start and where we must finish:

$$P_k: \quad 1 + 3 + 5 + \cdots + (2k - 1) = k^2$$
$$P_{k+1}: \quad 1 + 3 + 5 + \cdots + (2k - 1) + (2k + 1) = (k + 1)^2$$

Starting with P_k, we can add $2k + 1$ to both members, and after simplifying the right member we note that we have obtained P_{k+1} as a logical consequence of P_k:

$$1 + 3 + 5 + \cdots + (2k - 1) = k^2 \qquad\qquad P_k$$
$$1 + 3 + 5 + \cdots + (2k - 1) + (2k + 1) = k^2 + (2k + 1) \quad \text{Equality property}$$
$$= (k + 1)^2 \qquad\qquad P_{k+1}$$

Thus, $k \in S \Rightarrow k + 1 \in S$, and S is inductive.

Conclusion. $S = N$; that is, P_n is true for all natural numbers n.

PROBLEM 5   Prove: $1 + 2 + 3 + \cdots + n = \dfrac{n(n + 1)}{2}, \quad n \in N$

We are now in a position to prove the laws of exponents for natural numbers n. First, we redefine a^n, $n \in N$, using a recursion formula:

Definition of a^n

$$a^1 = a$$
$$a^{n+1} = a^n a \qquad n \in N$$

Thus,

$$a^4 = a^3 a = (a^2 a)a = [(a^1 a)a]a = [(aa)a]a$$

EXAMPLE 6   Prove that $(xy)^n = x^n y^n$ for all positive integers n.

Proof   Write

$$P_n: \quad (xy)^n = x^n y^n \quad \text{and} \quad S = \{n \in N \,|\, P_n \text{ is true}\}$$

Part 1. Show that $1 \in S$.

$$(xy)^1 = xy \qquad \text{Definition}$$
$$= x^1 y^1 \quad \text{Definition}$$

$$\therefore 1 \in S$$

Part 2. Show that S is inductive.

$$P_k: \qquad (xy)^k = x^k y^k$$
$$P_{k+1}: \quad (xy)^{k+1} = x^{k+1} y^{k+1}$$

Here we start with the left member of P_{k+1} and use P_k to find the right member of P_{k+1}.

$$(xy)^{k+1} = (xy)^k(xy) \qquad \text{Definition}$$
$$= x^k y^k xy \qquad \text{Use of } P_k$$
$$= (x^k x)(y^k y) \qquad \text{Property of real numbers}$$
$$= x^{k+1} y^{k+1} \qquad \text{Definition}$$

Thus, $k \in S \Rightarrow k + 1 \in S$, and S is inductive.

Conclusion. $S = N$

PROBLEM 6   Prove that $(x/y)^n = x^n/y^n$ for all positive integers n.

We consider one last example. (Before we start, recall that integer p is **divisible** by integer q if $p = qr$ for some integer r.)

EXAMPLE 7   Prove that $4^{2n} - 1$ is divisible by 5 for all positive integers n.

Proof   Write

$$P_n: \quad 4^{2n} - 1 \text{ is divisible by 5} \quad \text{and} \quad S = \{n \in N \mid P_n \text{ is true}\}$$

Part 1.   Show that $1 \in S$.

$$4^{2 \cdot 1} - 1 = 15 \quad \text{Divisible by 5}$$
$$\therefore 1 \in S$$

Part 2.   Show that S is inductive.

$$P_k: \qquad 4^{2k} - 1 = 5r \qquad \text{for some integer } r$$
$$P_{k+1}: \quad 4^{2(k+1)} - 1 = 5s \qquad \text{for some integer } s$$
$$4^{2k} - 1 = 5r \qquad\qquad P_k$$
$$4^2(4^{2k} - 1) = 4^2(5r) \qquad\qquad \text{Property of equality (multiply each side by } 4^2)$$
$$4^{2k+2} - 4^2 = 4^2(5r) \qquad\qquad \text{Property of a real number}$$
$$4^{2(k+1)} - 1 = 15 + 4^2(5r) \qquad \text{Property of equality (add 15 to each side)}$$
$$= 5(3 + 16r) \qquad \text{Property of a real number (factor out 5)}$$
$$= 5s \qquad \text{where } s = (3 + 16r), \text{ an integer}$$

Thus, $k \in S \Rightarrow k + 1 \in S$, and S is inductive.

Conclusion. $S = N$

PROBLEM 7   Prove that $8^n - 1$ is divisible by 7 for all positive integers n.

■   Three Famous Problems

We conclude this section by stating three famous problems. Instant worldwide fame awaits anyone who can prove or disprove either of the first two; neither has been proved or disproved to date.

1. **Goldbach's problem, 1742:** Every positive even integer greater than 2 is the sum of two prime numbers.
2. **Fermat's last theorem, 1637:** For $n > 2$, $x^n + y^n = z^n$ does not have solutions in the natural numbers.
3. **Each positive integer can be expressed as the sum of four or fewer squares of positive integers. (Considered by the early Greeks and finally proved in 1772 by Lagrange.)**

Answers to Matched Problems

5. Sketch of proof: Write

$$P_n: \quad 1 + 2 + 3 + \cdots + n = \frac{n(n+1)}{2}$$

and

$$S = \{n \in N \mid P_n \text{ is true}\}$$

Part 1. Show that $1 \in S$.

$$1 = \frac{1(1+1)}{2}$$

$$= 1$$

$$\therefore 1 \in S$$

Part 2. Show that S is inductive. (Supply reasons.)

$$1 + 2 + 3 + \cdots + k = \frac{k(k+1)}{2} \qquad P_k$$

$$1 + 2 + 3 + \cdots + k + (k+1) = \frac{k(k+1)}{2} + (k+1)$$

$$= \frac{(k+1)(k+2)}{2} \qquad P_{k+1}$$

Thus, $k \in S \Rightarrow k + 1 \in S$, and S is inductive.

Conclusion. $S = N$

6. Sketch of proof: Write

$$P_n: \quad \left(\frac{x}{y}\right)^n = \frac{x^n}{y^n} \quad \text{and} \quad S = \{n \in N \mid P_n \text{ is true}\}$$

Part 1. Show that $1 \in S$. (Supply reasons.)

$$\left(\frac{x}{y}\right)^1 = \frac{x}{y}$$

$$= \frac{x^1}{y^1}$$

$$\therefore 1 \in S$$

Part 2. Show that S is inductive. (Supply reasons.)

$$\left(\frac{x}{y}\right)^{k+1} = \left(\frac{x}{y}\right)^k \frac{x}{y}$$

$$= \frac{x^k}{y^k} \frac{x}{y}$$

$$= \frac{x^k x}{y^k y}$$

$$= \frac{x^{k+1}}{y^{k+1}}$$

Thus, $k \in S \Rightarrow k + 1 \in S$, and S is inductive.

Conclusion. $S = N$

7. Sketch of proof: Write

P_n: $8^n - 1$ is divisible by 7 and $S = \{n \in N \,|\, P_n$ is true$\}$

Part 1. Show that $1 \in S$.

$8^1 - 1 = 7$

$\therefore 1 \in S$

Part 2. Show that S is inductive. (Supply reasons.)

$$8^k - 1 = 7r$$

$$8(8^k - 1) = 8(7r)$$

$$8^{k+1} - 1 = 7 + 8(7r)$$

$$= 7(1 + 8r)$$

Thus, $k \in S \Rightarrow k + 1 \in S$, and S is inductive.

Conclusion. $S = N$

Exercise 7-2 ■ A *Find the first positive integer n that causes the statement to fail.*

all
1. $(3 + 5)^n = 3^n + 5^n$ 2. $n < 10$
3. $n^2 = 3n - 2$ 4. $n^3 + 11n = 6n^2 + 6$

Verify each open statement P_n for $n = 1, 2,$ and 3.

5. P_n: $2 + 6 + 10 + \cdots + (4n - 2) = 2n^2$
6. P_n: $4 + 8 + 12 + \cdots + 4n = 2n(n + 1)$
7. P_n: $a^5 a^n = a^{5+n}$
8. P_n: $(a^5)^n = a^{5n}$

9. P_n: $9^n - 1$ is divisible by 4

10. P_n: $4^n - 1$ is divisible by 3

Write P_k and P_{k+1} for each of the following.

11. P_n in Problem 5   12. P_n in Problem 6

13. P_n in Problem 7   14. P_n in Problem 8

15. P_n in Problem 9   16. P_n in Problem 10

Use mathematical induction to prove that each P_n holds for all positive integers n.

17. P_n in Problem 5   18. P_n in Problem 6

19. P_n in Problem 7   20. P_n in Problem 8

21. P_n in Problem 9   22. P_n in Problem 10

B   *Use mathematical induction to prove each of the following propositions for all positive integers n, unless restricted otherwise.*

23. $2 + 2^2 + 2^3 + \cdots + 2^n = 2^{n+1} - 2$

24. $\dfrac{1}{2} + \dfrac{1}{4} + \dfrac{1}{8} + \cdots + \dfrac{1}{2^n} = 1 - \left(\dfrac{1}{2}\right)^n$

25. $1^2 + 3^2 + 5^2 + \cdots + (2n - 1)^2 = \frac{1}{3}(4n^3 - n)$

26. $1 + 8 + 16 + \cdots + 8(n - 1) = (2n - 1)^2$

27. $1^2 + 2^2 + 3^2 + \cdots + n^2 = \dfrac{n(n + 1)(2n + 1)}{6}$

28. $1 \cdot 2 + 2 \cdot 3 + 3 \cdot 4 + \cdots + n(n + 1) = \dfrac{n(n + 1)(n + 2)}{3}$

29. $\dfrac{a^n}{a^3} = a^{n-3}$; $n > 3$

30. $\dfrac{a^5}{a^n} = \dfrac{1}{a^{n-5}}$; $n > 5$

31. $a^m a^n = a^{m+n}$; $m, n \in N$
     [*Hint:* Choose m as an arbitrary element of N, and then use induction on n.]

32. $(a^n)^m = a^{mn}$; $m, n \in N$

33. $x^n - 1$ is divisible by $x - 1$, $x \neq 1$
     [*Hint:* Divisible means that $x^n - 1 = (x - 1)Q(x)$ for some polynomial $Q(x)$.]

34. $x^n - y^n$ is divisible by $x - y$, $x \neq y$

35. $x^{2n} - 1$ is divisible by $x - 1$, $x \neq 1$

36. $x^{2n} - 1$ is divisible by $x + 1$, $x \neq -1$

37. $1^3 + 2^3 + 3^3 + \cdots + n^3 = (1 + 2 + 3 + \cdots + n)^2$

[*Hint:* See Problem 5 following Example 5.]

38. $\dfrac{1}{1 \cdot 2 \cdot 3} + \dfrac{1}{2 \cdot 3 \cdot 4} + \dfrac{1}{3 \cdot 4 \cdot 5} + \cdots + \dfrac{1}{n(n + 1)(n + 2)}$

$$= \dfrac{n(n + 3)}{4(n + 1)(n + 2)}$$

C *Discover a formula for each of the following, and prove your hypothesis using mathematical induction, $n \in N$.*

39. $2 + 4 + 6 + \cdots + 2n$

40. $\dfrac{1}{1 \cdot 2} + \dfrac{1}{2 \cdot 3} + \dfrac{1}{3 \cdot 4} + \cdots + \dfrac{1}{n(n + 1)}$

41. The number of lines determined by n points in a plane, no three of which are collinear

42. The number of diagonals in a polygon with n sides

Prove Problems 43–46 true for all integers n as specified.

43. $a > 1 \Rightarrow a^n > 1, \quad n \in N$

44. $0 < a < 1 \Rightarrow 0 < a^n < 1, \quad n \in N$

45. $n^2 > 2n, \quad n \geq 3$ **46.** $2^n > n^2, \quad n \geq 5$

47. Prove or disprove the generalization of the following two facts:

$$3^2 + 4^2 = 5^2$$
$$3^3 + 4^3 + 5^3 = 6^3$$

48. Prove or disprove: $n^2 + 21n + 1$ is a prime number for all natural numbers n.

If $\{a_n\}$ and $\{b_n\}$ are two sequences, we write $\{a_n\} = \{b_n\}$ if and only if $a_n = b_n$, $n \in N$. Use mathematical induction to show that $\{a_n\} = \{b_n\}$ where

49. $a_1 = 1, \quad a_n = a_{n-1} + 2; \quad b_n = 2n - 1$

50. $a_1 = 2, \quad a_n = a_{n-1} + 2; \quad b_n = 2n$

51. $a_1 = 2, \quad a_n = 2^2 a_{n-1}; \quad b_n = 2^{2n-1}$

52. $a_1 = 2, \quad a_n = 3a_{n-1}; \quad b_n = 2 \cdot 3^{n-1}$

Section 7-3 Arithmetic Sequences and Series

- Arithmetic Sequences
- *n*th-Term Formula
- Finite Arithmetic Series

■ Arithmetic Sequences

Consider the sequence

  5, 9, 13, 17, . . .

Can you guess what the fifth term is? If you guessed 21, you have observed that each term after the first can be obtained from the preceding one by adding 4 to it. This is an example of an arithmetic sequence.

Arithmetic Sequence

A sequence

  $a_1, a_2, a_3, \ldots, a_n, \ldots$

is called an **arithmetic sequence** (or **arithmetic progression**) if there exists a constant d, called the **common difference**, such that

  $a_n - a_{n-1} = d$

That is,

  $a_n = a_{n-1} + d$ for every $n > 1$

EXAMPLE 8 Which sequence is an arithmetic sequence and what is its common difference?

  (A) 1, 2, 3, 5, . . .   (B) 3, 5, 7, 9, . . .

Solution Sequence (B) is an arithmetic sequence with $d = 2$.

PROBLEM 8 Repeat Example 8 with (A) $-4, -1, 2, 5, \ldots$, and (B) 2, 4, 8, 16, ...

■ *n*th-Term Formula

Arithmetic sequences have several convenient properties. For example, one can derive formulas for the nth term in terms of n and the sum of any number of consecutive terms. To obtain an nth-term formula, we note that if $\{a_n\}$ is an arithmetic sequence, then

  $a_2 = a_1 + d$
  $a_3 = a_2 + d = a_1 + 2d$
  $a_4 = a_3 + d = a_1 + 3d$

which suggests

$$a_n = a_1 + (n - 1)d \qquad \text{for every } n > 1$$

We have arrived at this formula by ordinary induction; its proof requires mathematical induction, which we leave as an exercise (see Problem 31 in Exercise 7-3).

EXAMPLE 9   If the first and tenth terms of an arithmetic sequence are 3 and 30, respectively, find the fiftieth term of the sequence.

Solution   First find d:

$$a_n = a_1 + (n - 1)d$$
$$a_{10} = a_1 + (10 - 1)d$$
$$30 = 3 + 9d$$
$$d = 3$$

Now find a_{50}:

$$a_{50} = a_1 + (50 - 1)3$$
$$= 3 + 49 \cdot 3$$
$$= 150$$

PROBLEM 9   If the first and fifteenth terms of an arithmetic sequence are -5 and 23, respectively, find the seventy-third term of the sequence.

■ Finite Arithmetic Series

The sum of the terms of an arithmetic sequence is called an **arithmetic series**. We will derive two simple and very useful formulas for finding the sum of an arithmetic series. Let

$$S_n = a_1 + (a_1 + d) + \cdots + [a_1 + (n - 2)d] + [a_1 + (n - 1)d]$$

which is the sum of the first n terms of an arithmetic sequence. Reversing the order of the sum, we obtain

$$S_n = [a_1 + (n - 1)d] + [a_1 + (n - 2)d] + \cdots + (a_1 + d) + a_1$$

Adding left members and corresponding elements of the right members of the two equations, we see that

$$2S_n = [2a_1 + (n - 1)d] + [2a_1 + (n - 1)d] + \cdots + [2a_1 + (n - 1)d]$$
$$= n[2a_1 + (n - 1)d]$$

or

$$S_n = \frac{n}{2}[2a_1 + (n-1)d]$$

By replacing $a_1 + (n-1)d$ with a_n, we obtain a second useful formula for the sum:

$$S_n = \frac{n}{2}(a_1 + a_n)$$

The proof of the first sum formula by mathematical induction is left as an exercise (see Problem 32 in Exercise 7-3).

EXAMPLE 10   Find the sum of the first twenty-six terms of an arithmetic series if the first term is -7 and $d = 3$.

Solution $$S_n = \frac{n}{2}[2a_1 + (n-1)d]$$

$$S_{26} = \frac{26}{2}[2(-7) + (26-1)3]$$
$$= 793$$

PROBLEM 10   Find the sum of the first fifty-two terms of an arithmetic series if the first term is 23 and $d = -2$.

EXAMPLE 11   Find the sum of all the odd numbers between 51 and 99, inclusive.

Solution   First find n:

$$a_n = a_1 + (n-1)d$$
$$99 = 51 + (n-1)2$$
$$n = 25$$

Now find S_{25}:

$$S_n = \frac{n}{2}(a_1 + a_n)$$

$$S_{25} = \frac{25}{2}(51 + 99)$$
$$= 1{,}875$$

PROBLEM 11   Find the sum of all the even numbers between -22 and 52, inclusive.

8. Sequence (A) with $d = 3$ **9.**   139

10. $-1,456$ **11.**   570

Exercise 7-3 ■ A

1. Determine which of the following are arithmetic sequences. Find d and the next two terms for those that are.
   (A)   2, 4, 8, . . .   (B)   7, 6.5, 6, . . .
   (C) $-11, -16, -21, . . .$ (D) $\frac{1}{2}, \frac{1}{6}, \frac{1}{18}, . . .$

2. Repeat Problem 1 for
   (A)   5, $-1, -7, . . .$ (B)   12, 4, $\frac{4}{3}, . . .$
   (C) $\frac{1}{2}, \frac{2}{3}, \frac{3}{4}, . . .$ (D)   16, 48, 80, . . .

Let $a_1, a_2, a_3, \ldots, a_n, \ldots$ be an arithmetic sequence. In Problems 3–18 find the indicated quantities.

3. $a_1 = -5,$ $d = 4$; $a_2 = ?,$ $a_3 = ?,$ $a_4 = ?$

4. $a_1 = -18,$ $d = 3$; $a_2 = ?,$ $a_3 = ?,$ $a_4 = ?$

5. $a_1 = -3,$ $d = 5$; $a_{15} = ?,$ $S_{11} = ?$

6. $a_1 = 3,$ $d = 4$; $a_{22} = ?,$ $S_{21} = ?$

7. $a_1 = 1,$ $a_2 = 5$; $S_{21} = ?$

8. $a_1 = 5,$ $a_2 = 11$; $S_{11} = ?$

9. $a_1 = 7,$ $a_2 = 5$; $a_{15} = ?$

10. $a_1 = -3,$ $d = -4$; $a_{10} = ?$

B

11. $a_1 = 3,$ $a_{20} = 117$; $d = ?,$ $a_{101} = ?$

12. $a_1 = 7,$ $a_8 = 28$; $d = ?,$ $a_{25} = ?$

13. $a_1 = -12,$ $a_{40} = 22$; $S_{40} = ?$

14. $a_1 = 24,$ $a_{24} = -28$; $S_{24} = ?$

15. $a_1 = \frac{1}{3},$ $a_2 = \frac{1}{2}$; $a_{11} = ?,$ $S_{11} = ?$

16. $a_1 = \frac{1}{6},$ $a_2 = \frac{1}{4}$; $a_{19} = ?,$ $S_{19} = ?$

17. $a_3 = 13,$ $a_{10} = 55$; $a_1 = ?$

18. $a_9 = -12,$ $a_{13} = 3$; $a_1 = ?$

19. $S_{51} = \sum_{k=1}^{51} (3k + 3) = ?$ **20.** $S_{40} = \sum_{k=1}^{40} (2k - 3) = ?$

21. Find $g(1) + g(2) + g(3) + \cdots + g(51)$ if $g(t) = 5 - t$.

22. Find $f(1) + f(2) + f(3) + \cdots + f(20)$ if $f(x) = 2x - 5$.

23. Find the sum of all the even integers between 21 and 135.

24. Find the sum of all the odd integers between 100 and 500.

25. Show that the sum of the first n odd natural numbers is n^2, using appropriate formulas from this section.

26. Show that the sum of the first n even natural numbers is $n + n^2$, using appropriate formulas from this section.

27. For a given sequence in which $a_1 = -3$ and $a_n = a_{n-1} + 3, n > 1$, find a_n in terms of n.

28. For the sequence in Problem 27 find $S_n = \sum_{k=1}^{n} a_k$ in terms of n.

29. An object falling from rest in a vacuum near the surface of the earth falls 16 feet during the first second, 48 feet during the second second, 80 feet during the third second, and so on.
(A) How far will the object fall during the eleventh second?
(B) How far will the object fall in 11 seconds?
(C) How far will the object fall in t seconds?

30. In investigating different job opportunities, you find that firm A will start you at $25,000 per year and guarantee you a raise of $1,200 each year, while firm B will start you at $28,000 per year but will guarantee you a raise of only $800 each year. Over a 15-year period how much would you receive from each firm?

C **31.** Prove, using mathematical induction, that if $\{a_n\}$ is an arithmetic sequence, then

$$a_n = a_1 + (n - 1)d$$

32. Prove, using mathematical induction, that if $\{a_n\}$ is an arithmetic sequence, then

$$S_n = \frac{n}{2}[2a_1 + (n - 1)d]$$

33. Show that $(x^2 + xy + y^2)$, $(z^2 + xz + x^2)$, and $(y^2 + yz + z^2)$ are consecutive terms of an arithmetic progression if x, y, and z form an arithmetic progression. (From USSR Mathematical Olympiads, 1955–1956, Grade 9.)

34. Take 121 terms of each arithmetic progression 2, 7, 12, ... and 2, 5, 8, ... How many numbers will there be in common? (From USSR Mathematical Olympiads, 1955–1956, Grade 9.)

35. Given the system of equations

$$ax + by = c$$
$$dx + ey = f$$

where a, b, c, d, e, f is any arithmetic progression with a nonzero constant difference, show that the system has a unique solution.

Section 7-4   Geometric Sequences and Series

- ■ Geometric Sequences
- ■ nth-Term Formula
- ■ Finite Geometric Series
- ■ Infinite Geometric Series

■ **Geometric Sequences**

Consider the sequence

$$2, -4, 8, -16, \ldots$$

Can you guess what the fifth and sixth terms are? If you guessed 32 and -64, respectively, you have observed that each term after the first can be obtained from the preceding one by multiplying it by -2. This is an example of a *geometric sequence*.

Geometric Sequence

A sequence

$$a_1, a_2, a_3, \ldots, a_n, \ldots$$

is called a **geometric sequence** (or a **geometric progression**) if there exists a nonzero constant r, called the **common ratio**, such that

$$\frac{a_n}{a_{n-1}} = r$$

That is,

$$a_n = ra_{n-1} \qquad \text{for every } n > 1$$

EXAMPLE 12   Which sequence is a geometric sequence and what is its common ratio?

(A) $2, 6, 8, 10, \ldots$ (B) $-1, 3, -9, 27, \ldots$

Solution   Sequence (B) is a geometric sequence with $r = -3$.

PROBLEM 12   Repeat Example 12 with (A) $\frac{1}{4}, \frac{1}{2}, 1, 2, \ldots$, and (B) $\frac{1}{2}, \frac{1}{4}, \frac{1}{16}, \frac{1}{256}, \ldots$.

■ nth-Term Formula

Just as with arithmetic sequences, geometric sequences have several convenient properties. It is easy to derive formulas for the nth term in terms of n and the sum of any number of consecutive terms. To obtain an nth-term formula, we note that if $\{a_n\}$ is a geometric sequence, then

$$a_2 = ra_1$$
$$a_3 = ra_2 = r^2a_1$$
$$a_4 = ra_3 = r^3a_1$$

which suggests that

$$a_n = a_1r^{n-1} \qquad \text{for every } n > 1$$

We have arrived at this formula using ordinary induction; its proof requires mathematical induction, which we leave as an exercise.

EXAMPLE 13   Find the seventh term of the geometric sequence $1, \frac{1}{2}, \frac{1}{4}, \ldots$.

Solution
$$r = \frac{1}{2}$$
$$a_n = a_1r^{n-1}$$
$$a_7 = 1(\tfrac{1}{2})^{7-1} = \tfrac{1}{64}$$

PROBLEM 13   Find the eighth term of the geometric sequence $\frac{1}{64}, -\frac{1}{32}, \frac{1}{16}, \ldots$.

EXAMPLE 14   If the first and tenth terms of a geometric sequence are 1 and 2, respectively, find the common ratio r to two decimal places.

Solution
$$a_n = a_1r^{n-1}$$
$$2 = 1r^{10-1}$$
$$r = 2^{1/9} = 1.08 \qquad \text{Calculation by calculator using } y^x \text{ button.}$$

PROBLEM 14   If the first and eighth terms of a geometric sequence are 2 and 16, respectively, find the common ratio r to three decimal places.

■ Finite Geometric Series

The sum of the terms of a geometric sequence is called a **geometric series**. As was the case with an arithmetic series, we can derive two simple and very useful formulas for finding the **sum of a geometric series**.

Let

$$S_n = a_1 + a_1 r + a_1 r^2 + a_1 r^3 + \cdots + a_1 r^{n-2} + a_1 r^{n-1}$$

which is the sum of the first n terms of a geometric sequence. Multiply both members by r to obtain

$$r S_n = a_1 r + a_1 r^2 + a_1 r^3 + \cdots + a_1 r^{n-1} + a_1 r^n$$

Now subtract the left member of the second equation from the left member of the first, and the right member of the second equation from the right member of the first to obtain

$$S_n - r S_n = a_1 - a_1 r^n$$
$$S_n(1 - r) = a_1 - a_1 r^n$$

Thus,

$$S_n = \frac{a_1 - a_1 r^n}{1 - r} \qquad r \neq 1$$

Since $a_n = a_1 r^{n-1}$, or $r a_n = a_1 r^n$, the sum formula can also be written in the form

$$S_n = \frac{a_1 - r a_n}{1 - r} \qquad r \neq 1$$

The proof of the first sum formula by mathematical induction is left as an exercise.

If $r = 1$, then

$$S_n = a_1 + a_1 1 + a_1 1^2 + \cdots + a_1 1^{n-1} = n a_1$$

EXAMPLE 15   Find the sum of the first twenty terms of a geometric series if the first term is 1 and $r = 2$.

Solution $$S_n = \frac{a_1 - a_1 r^n}{1 - r}$$

$$= \frac{1 - 1 \cdot 2^{20}}{1 - 2} \approx 1{,}050{,}000 \qquad \text{Calculation using a calculator}$$

PROBLEM 15   Find the sum (to two decimal places) of the first fourteen terms of a geometric series if the first term is $\frac{1}{64}$ and $r = -2$.

■ Infinite Geometric Series

Consider a geometric series with $a_1 = 5$ and $r = \frac{1}{2}$. What happens to the sum S_n as n increases? To answer this question, we first write the sum formula in the more convenient form

$$S_n = \frac{a_1 - a_1 r^n}{1 - r} = \frac{a_1}{1 - r} - \frac{a_1 r^n}{1 - r} \tag{1}$$

For $a_1 = 5$ and $r = \frac{1}{2}$,

$$S_n = 10 - 10\left(\frac{1}{2}\right)^n$$

Thus,

$$S_2 = 10 - 10\left(\frac{1}{4}\right)$$

$$S_4 = 10 - 10\left(\frac{1}{16}\right)$$

$$S_{10} = 10 - 10\left(\frac{1}{1,024}\right)$$

$$S_{20} = 10 - 10\left(\frac{1}{1,048,576}\right)$$

It appears that $(\frac{1}{2})^n$ becomes smaller and smaller as n increases, and that the sum gets closer and closer to 10.

In general, it is possible to show that, if $|r| < 1$ (that is, $-1 < r < 1$), then r^n will tend to 0 as n increases. Thus,

$$\frac{a_1 r^n}{1 - r}$$

in equation (1) will tend to 0 as n increases, and S_n will tend to

$$\frac{a_1}{1 - r}$$

In other words, if $|r| < 1$, then S_n can be made as close to

$$\frac{a_1}{1 - r}$$

as we wish by taking n sufficiently large. Thus, we define

$$S_\infty = \frac{a_1}{1 - r} \qquad |r| < 1$$

and call this the **sum of an infinite geometric series**. If $|r| \geq 1$, an infinite geometric series has no sum.

EXAMPLE 16   Represent the repeating decimal $0.45\overline{45}$ as the quotient of two integers. (Recall that a repeating decimal names a rational number, and that any rational number can be represented as the quotient of two integers.)

Solution $0.45\overline{45} = 0.45 + 0.0045 + 0.000045 + \cdots$

The right member of the equation is an infinite geometric series with $a_1 = 0.45$ and $r = 0.01$. Thus,

$$S_\infty = \frac{a_1}{1 - r} = \frac{0.45}{1 - 0.01} = \frac{0.45}{0.99} = \frac{5}{11}$$

Hence, $0.45\overline{45}$ and $\frac{5}{11}$ name the same rational number. Check the result by dividing 5 by 11.

PROBLEM 16   Repeat Example 16 for $0.8181\overline{81}$.

Answers to Matched Problems   **12.**   Sequence (A)   with $r = 2$ **13.** -2

14. $r = 1.346$ **15.** -85.33

16. $\frac{9}{11}$

Exercise 7-4 ■   *A calculator will be useful in some problems.*

A   **1.**   Determine which of the following are geometric sequences. Find r and the next two terms for those that are.
   (A) $2, -4, 8, \ldots$ (B) $7, 6.5, 6, \ldots$
   (C) $-11, -16, -21, \ldots$ (D) $\frac{1}{2}, \frac{1}{6}, \frac{1}{18}, \ldots$

2.   Repeat Problem 1 for
   (A) $5, -1, -7, \ldots$ (B) $12, 4, \frac{4}{3}, \ldots$
   (C) $\frac{1}{2}, \frac{2}{3}, \frac{3}{4}, \ldots$ (D) $16, 48, 80, \ldots$

Let $a_1, a_2, a_3, \ldots, a_n, \ldots$ be a geometric sequence. Find each of the indicated quantities in Problems 3–16.

3. $a_1 = -6$, $r = -\frac{1}{2}$; $a_2 = ?$, $a_3 = ?$, $a_4 = ?$

4. $a_1 = 12$, $r = \frac{2}{3}$; $a_2 = ?$, $a_3 = ?$, $a_4 = ?$

5. $a_1 = 81$, $r = \frac{1}{3}$; $a_{10} = ?$

6. $a_1 = 64$, $r = \frac{1}{2}$; $a_{13} = ?$

7. $a_1 = 3$, $a_7 = 2,187$, $r = 3$; $S_7 = ?$

8. $a_1 = 1$, $a_7 = 729$, $r = -3$; $S_7 = ?$

B   **9.** $a_1 = 100,\quad a_6 = 1;\quad r = ?$ **10.** $a_1 = 10,\quad a_{10} = 30;\quad r = ?$

11. $a_1 = 5,\quad r = -2;\quad S_{10} = ?$ **12.** $a_1 = 3,\quad r = 2;\quad S_{10} = ?$

13. $a_1 = 9,\quad a_4 = \frac{8}{3};\quad a_2 = ?,\quad a_3 = ?$

14. $a_1 = 12,\quad a_4 = -\frac{4}{9};\quad a_2 = ?,\quad a_3 = ?$

15. $S_7 = \sum\limits_{k=1}^{7} (-3)^{k-1} = ?$ **16.** $S_7 = \sum\limits_{k=1}^{7} 3^k = ?$

17. Find $g(1) + g(2) + \cdots + g(10)$ if $g(x) = (\frac{1}{2})^x$.

18. Find $f(1) + f(2) + \cdots + f(10)$ if $f(x) = 2^x$.

19. Find a positive number x so that $-2 + x - 6$ is a three-term geometric series.

20. Find a positive number x so that $6 + x + 8$ is a three-term geometric series.

Find the sum of each infinite geometric series that has a sum.

21. $3 + 1 + \frac{1}{3} + \cdots$ **22.** $16 + 4 + 1 + \cdots$

23. $2 + 4 + 8 + \cdots$ **24.** $4 + 6 + 9 + \cdots$

25. $2 - \frac{1}{2} + \frac{1}{8} - \cdots$ **26.** $21 - 3 + \frac{3}{7} - \cdots$

Represent each repeating decimal fraction as the quotient of two integers.

27. $0.777\overline{7}$ **28.** $0.555\overline{5}$ **29.** $0.5454\overline{54}$

30. $0.2727\overline{27}$ **31.** $3.216216\overline{216}$ **32.** $5.6363\overline{63}$

33. *Business.* If $P is invested at r% compounded annually, the amount A present after n years forms a geometric progression with a common ratio $(1 + r)$. Write a formula for the amount present after n years. How long will it take a sum of money P to double if invested at 6% interest compounded annually?

34. *Population growth.* If a population of A_0 people grows at the constant rate of r% per year, the population after t years forms a geometric progression with a common ratio $(1 + r)$. Write a formula for the total population after t years. If the world's population is increasing at the rate of 2% per year, how long will it take to double?

35. *Engineering.*   A rotating flywheel coming to rest rotates 300 revolutions the first minute. If in each subsequent minute it rotates two-thirds as many times as in the preceding minute, how many revolutions will the wheel make before coming to rest?

36. *Physics.*   The first swing of a bob on a pendulum is 10 inches. If on each subsequent swing it travels 0.9 as far as on the preceding swing, how far will the bob travel before coming to rest?

C   **37.**   If in a given sequence, $a_1 = -2$ and $a_n = -3a_{n-1}$, $n > 1$, find a_n in terms of n.

38.   For the sequence in Problem 37 find $S_n = \sum\limits_{k=1}^{n} a_k$ in terms of n.

39.   Prove, using mathematical induction, that if $\{a_n\}$ is a geometric sequence, then

$$a_n = a_1 r^{n-1} \qquad n \in N$$

40.   Prove, using mathematical induction, that if $\{a_n\}$ is a geometric sequence, then

$$S_n = \frac{a_1 - a_1 r^n}{1 - r} \qquad n \in N$$

41.   *Economics.*   The government, through a subsidy program, distributes $1,000,000. If we assume that each individual or agency spends 0.8 of what is received, and 0.8 of this is spent, and so on, how much total increase in spending results from this government action? (Let $a_1 = \$800,000$.)

42.   *Zeno's paradox.*   Visualize a hypothetical 440-yard oval racetrack that has tapes stretched across the track at the halfway point and at each point that marks the halfway point of each remaining distance thereafter. A runner running around the track has to break the first tape before the second, the second before the third, and so on. From this point of view it appears that he will never finish the race. (This famous paradox is attributed to the Greek philosopher, Zeno, 495–435 B.C.) If we assume the runner runs at 440 yards/minute, the times between tape breakings form an infinite geometric progression. What is the sum of this progression?

Section 7-5   Additional Applications

This section includes additional applications involving progressions (mainly geometric) from many different fields. The problems are self-contained and require no previous knowledge of the subjects concerned.

Exercise 7-5 ∎

APPLICATIONS   *Difficult problems are double-starred (★★), moderately difficult problems are single-starred (★), and the easier problems are not marked.*

   A calculator will be helpful in solving some of these problems.

Business and Economics   **1.**   If you received $7,000 a year 11 years ago and now receive $14,000 a year, and if your salary has been increased the same amount each year, what is that yearly increase and how much money have you received from the company over the 11 years?

2. Let us suppose the government has reduced taxes so that you have $600 more in spendable income. What is the net effect of this extra $600 on the economy? According to the "multiplier" doctrine in economics, the effect of the $600 is multiplied. Let us assume that you spend 0.7 of the $600 on consumer goods, that the producers of these goods in turn spend 0.7 of what they receive on consumer goods, and that this chain continues indefinitely, forming a geometric progression. What is the total amount spent on consumer goods if the process continues indefinitely? (Let $a_1 = \$420$.)

Earth Sciences   ★3. If atmospheric pressure decreases (roughly) by a factor of 10 for each 10-mile increase in altitude up to 60 miles, and if the pressure is 15 pounds/square inch at sea level, what will the pressure be 40 miles up?

4. As dry air moves upward it expands, and in so doing cools at the rate of about 5°F for each 1,000-foot rise. This is known as the **adiabatic process**.

   (A) Temperatures at altitudes that are multiples of 1,000 form what kind of a sequence?

   (B) If the ground temperature is 80°F, write a formula for the temperature T_n in terms of n, if n is in thousands of feet.

Life Sciences—Ecology   5. A plant is eaten by an insect, an insect by a trout, a trout by a salmon, a salmon by a bear, and the bear is eaten by you. If only 20% of the energy is transformed from one stage to the next, how many calories must be supplied by plant food to provide you with 2,000 calories from the bear meat?

★6. If there are 30 years in a generation, how many direct ancestors did each of us have 600 years ago? (By *direct* ancestors we mean parents, grandparents, great-grandparents, and so on.)

★7. A single cholera bacterium divides every $\frac{1}{2}$ hour to produce two complete cholera bacteria. If we start with a colony of A_0 bacteria, in t hours (assuming adequate food supply) how many bacteria will we have?

★8. One leukemic cell injected into a healthy mouse will divide into two cells in about $\frac{1}{2}$ day; at the end of the day these two cells will divide again, with the doubling process continuing each half-day until there are 1 billion cells, at which time the mouse dies. On which day after the experiment is started does this happen?

Astronomy ★★9. Ever since the time of the Greek astronomer Hipparchus (second century B.C.), the brightness of stars has been measured in terms of magnitude. The brightest stars (excluding the sun) are classed as magnitude 1, and the dimmest visible to the eye are classed as magnitude 6. In 1856, the English astronomer N. R. Pogson showed that first-magnitude stars are 100 times brighter than sixth-magnitude stars. If the ratio of brightness between consecutive magnitudes is constant, find this ratio. [Hint: If b_n is the brightness of an nth-magnitude star, find r for the geometric progression $b_1, b_2, b_3, \ldots$, given $b_1 = 100b_6$.]

Music ★10. The notes on a piano, as measured in cycles per second, form a geometric progression.

(A) If A is 400 cycles per second and A′, 12 notes higher, is 800 cycles per second, find the constant ratio r.

(B) Find the cycles per second for C, three notes higher than A.

Geometry 11. If the midpoints of the sides of an equilateral triangle are joined by straight lines, the new figure will be an equilateral triangle with a perimeter half the old. If we start with an equilateral triangle with perimeter 1, and form a sequence of "nested" equilateral triangles proceeding as above, what will be the total perimeter of all the triangles that can be formed in this way?

Photography 12. The shutter speeds and f-stops on a camera are given as follows:

Shutter speeds: $1, \frac{1}{2}, \frac{1}{4}, \frac{1}{8}, \frac{1}{15}, \frac{1}{30}, \frac{1}{60}, \frac{1}{125}, \frac{1}{250}, \frac{1}{500}$

f-stops: 1.4, 2, 2.8, 4, 5.6, 8, 10.3, 16

These are very close to being geometric progressions. Estimate their common ratios.

Puzzles 13. If you place 1¢ on the first square of a chessboard, 2¢ on the second square, 4¢ on the third, and so on, continuing to double the amount until all sixty-four squares are covered, how much money will be on the sixty-fourth square? How much money will there be on the whole board?

★14. If a sheet of very thin paper 0.001 inch thick is torn in half, and each half is again torn in half, and this process is repeated for a total of thirty-two times, how high will the stack of paper be if the pieces are placed one on top of the other? Give the answer to the nearest mile. (5,280 feet = 1 mile)

Section 7-6   Binomial Formula

- Factorial
- Binomial Formula

The binomial form

$$(a + b)^n$$

where n is a natural number, appears more frequently than you might expect. The coefficients in the expansion play an important role in probability studies. The binomial formula, which we will derive below, enables us to expand $(a + b)^n$ directly for n any natural number. Since the formula involves factorials, we digress for a moment to introduce this important concept.

- Factorial

For n a natural number, **n factorial**—denoted by $n!$—is the product of the first n natural numbers. **Zero factorial** is defined to be 1. Symbolically,

---

n Factorial

$$n! = n(n - 1) \cdot \cdots \cdot 2 \cdot 1$$
$$1! = 1$$
$$0! = 1$$

It is also useful to note that

$$n! = n \cdot (n - 1)!$$

EXAMPLE 17   (A) $4! = 4 \cdot 3! = 4 \cdot 3 \cdot 2! = 4 \cdot 3 \cdot 2 \cdot 1! = 4 \cdot 3 \cdot 2 \cdot 1 = 24$

(B) $5! = 5 \cdot 4 \cdot 3 \cdot 2 \cdot 1 = 120$

(C) $\dfrac{7!}{6!} = \dfrac{7 \cdot 6!}{6!} = 7$

(D) $\dfrac{8!}{5!} = \dfrac{8 \cdot 7 \cdot 6 \cdot 5!}{5!} = 336$

PROBLEM 17   Find: (A) 6!   (B) 6!/5!   (C) 9!/6!

The symbol $\dbinom{n}{r}$ is frequently used in probability studies and will be used by us shortly. It is called the **combinatorial symbol** and is defined for nonnegative r and n, as follows:

Combinatorial Symbol

For nonnegative integers r and n, $0 \le r \le n$,

$$\binom{n}{r} = \frac{n!}{r!(n-r)!}$$

$$= \frac{n(n-1)(n-2) \cdots \cdots (n-r+1)}{r(r-1) \cdots \cdots 2 \cdot 1}$$

EXAMPLE 18   (A) $\dbinom{8}{3} = \dfrac{8!}{3!(8-3)!} = \dfrac{8!}{3!5!} = \dfrac{8 \cdot 7 \cdot 6 \cdot 5!}{3 \cdot 2 \cdot 1 \cdot 5!} = 56$

(B) $\dbinom{7}{0} = \dfrac{7!}{0!(7-0)!} = \dfrac{7!}{7!} = 1$

PROBLEM 18   Find: (A) $\dbinom{9}{2}$ (B) $\dbinom{5}{5}$

■ Binomial Formula

We are now ready to try to discover a formula for the expansion of $(a + b)^n$ using ordinary induction; that is, we will look at a few special cases and try to postulate a general formula from them. If successful, we will try to prove that the formula holds for all natural numbers, using mathematical induction. To start, let us calculate directly the first five natural number powers of $(a + b)^n$:

$$(a + b)^1 = a + b$$
$$(a + b)^2 = a^2 + 2ab + b^2$$
$$(a + b)^3 = a^3 + 3a^2b + 3ab^2 + b^3$$
$$(a + b)^4 = a^4 + 4a^3b + 6a^2b^2 + 4ab^3 + b^4$$
$$(a + b)^5 = a^5 + 5a^4b + 10a^3b^2 + 10a^2b^3 + 5ab^4 + b^5$$

Observations

---

1. The expansion of $(a + b)^n$ has $(n + 1)$ terms.
2. The power of a decreases by 1 for each term as we move from left to right.
3. The power of b increases by 1 for each term as we move from left to right.
4. In each term the sum of the powers of a and b always adds up to n.
5. Starting with a given term, we can get the coefficient of the next term by multiplying the coefficient of the given term by the exponent of a and dividing by the number that represents the position of the term in the series of terms. For example, in the expansion of $(a + b)^4$, the coefficient of the third term is found from the second term by multiplying 4 and 3 and then dividing by 2 [that is, the coefficient of the third term $= (4 \cdot 3)/2 = 6$].

We now postulate the properties for the general case:

$$(a + b)^n = a^n + \frac{n}{1} a^{n-1}b + \frac{n(n-1)}{1 \cdot 2} a^{n-2}b^2$$

$$+ \frac{n(n-1)(n-2)}{1 \cdot 2 \cdot 3} a^{n-3}b^3 + \cdots + b^n$$

$$= \frac{n!}{0!(n-0)!} a^n + \frac{n!}{1!(n-1)!} a^{n-1}b + \frac{n!}{2!(n-2)!} a^{n-2}b^2$$

$$+ \frac{n!}{3!(n-3)!} a^{n-3}b^3 + \cdots + \frac{n!}{n!(n-n)!} b^n$$

$$= \binom{n}{0} a_n + \binom{n}{1} a^{n-1}b + \binom{n}{2} a^{n-2}b^2$$

$$+ \binom{n}{3} a^{n-3}b^3 + \cdots + \binom{n}{n} b^n$$

Thus, it appears that

Binomial Formula

$$(a + b)^n = \sum_{k=0}^{n} \binom{n}{k} a^{n-k} b^k \qquad n \geq 1$$

This result is known as the **binomial formula**, and we now proceed to prove that it holds for all natural numbers n.

Proof   Write

$$P_n: \quad (a + b)^n = \sum_{j=0}^{n} \binom{n}{j} a^{n-j} b^j$$

$$S = \{n \in N \mid P_n \text{ is true}\}$$

Part 1. Show that $1 \in S$.

$$\sum_{j=0}^{1} \binom{1}{j} a^{1-j} b^j = \binom{1}{0} a + \binom{1}{1} b = a + b = (a + b)^1$$

$$\therefore \ 1 \in S$$

Part 2. Show that $k \in S \Rightarrow k + 1 \in S$.

$$P_k: \quad (a + b)^k = \sum_{j=0}^{k} \binom{k}{j} a^{k-j} b^j$$

$$P_{k+1}: \quad (a + b)^{k+1} = \sum_{j=0}^{k+1} \binom{k+1}{j} a^{k+1-j} b^j$$

Starting with P_k, we multiply both members by $a + b$ and try to obtain P_{k+1}:

$$(a + b)^k (a + b) = \left[\sum_{j=0}^{k} \binom{k}{j} a^{k-j} b^j \right] (a + b)$$

$$= \left[\binom{k}{0} a^k + \binom{k}{1} a^{k-1} b + \binom{k}{2} a^{k-2} b^2 + \cdots + \binom{k}{k} b^k \right] (a + b)$$

$$= \left[\binom{k}{0} a^{k+1} + \binom{k}{1} a^k b + \binom{k}{2} a^{k-1} b^2 + \cdots + \binom{k}{k} ab^k \right]$$

$$+ \left[\binom{k}{0} a^k b + \binom{k}{1} a^{k-1} b^2 + \cdots + \binom{k}{k-1} ab^k + \binom{k}{k} b^{k+1} \right]$$

$$(a + b)^k(a + b) = \binom{k}{0}a^{k+1} + \left[\binom{k}{0} + \binom{k}{1}\right]a^kb + \left[\binom{k}{1} + \binom{k}{2}\right]a^{k-1}b^2 + \cdots$$

$$+ \left[\binom{k}{k-1} + \binom{k}{k}\right]ab^k + \binom{k}{k}b^{k+1}$$

We now use the facts (the proofs left as exercises) that

$$\binom{k}{r-1} + \binom{k}{r} = \binom{k+1}{r} \qquad \binom{k}{0} = \binom{k+1}{0} \qquad \binom{k}{k} = \binom{k+1}{k+1}$$

to rewrite the right side as

$$\binom{k+1}{0}a^{k+1} + \binom{k+1}{1}a^kb + \binom{k+1}{2}a^{k-1}b^2 + \cdots + \binom{k+1}{k}ab^k + \binom{k+1}{k+1}b^{k+1}$$

$$= \sum_{j=0}^{k+1} \binom{k+1}{j}a^{k+1-j}b^j$$

Thus, $k \in S \Rightarrow k + 1 \in S$, and S is inductive.

Conclusion. $S = N$

EXAMPLE 19   Use the binomial formula to expand $(x + y)^6$.

Solution $$(x + y)^6 = \sum_{k=0}^{6} \binom{6}{k}x^{6-k}y^k$$

$$= \binom{6}{0}x^6 + \binom{6}{1}x^5y + \binom{6}{2}x^4y^2 + \binom{6}{3}x^3y^3$$

$$+ \binom{6}{4}x^2y^4 + \binom{6}{5}xy^5 + \binom{6}{6}y^6$$

$$= x^6 + 6x^5y + 15x^4y^2 + 20x^3y^3 + 15x^2y^4 + 6xy^5 + y^6$$

PROBLEM 19   Use the binomial formula to expand $(x + 1)^5$.

EXAMPLE 20   Use the binomial formula to find the fourth term in the expansion of $(x - 2)^{20}$.

Solution   Fourth term $= \binom{20}{3}x^{17}(-2)^3$ In the expansion of $(a + b)^n$, the exponent of b in the rth term is $r - 1$ and the exponent of a is $n - (r - 1)$.

$$= \frac{20 \cdot 19 \cdot 18}{3 \cdot 2 \cdot 1}x^{17}(-8)$$

$$= -9,120x^{17}$$

PROBLEM 20   Use the binomial formula to find the fifth term in the expansion of $(u - 1)^{18}$.

17. (A)   720   (B)   6   (C) 504

18. (A)   36   (B)   1

19. $x^5 + 5x^4 + 10x^3 + 10x^2 + 5x + 1$

20. $3{,}060u^{14}$

Exercise 7-6 ■ A *Evaluate.*

1. $6!$

2. $4!$

3. $\dfrac{20!}{19!}$

4. $\dfrac{5!}{4!}$

5. $\dfrac{10!}{7!}$

6. $\dfrac{9!}{6!}$

7. $\dfrac{6!}{4!\,2!}$

8. $\dfrac{5!}{2!\,3!}$

9. $\dfrac{9!}{0!(9-0)!}$

10. $\dfrac{8!}{8!(8-8)!}$

11. $\dfrac{8!}{2!(8-2)!}$

12. $\dfrac{7!}{3!(7-3)!}$

Write as the quotient of two factorials.

13. 9

14. 12

15. $6 \cdot 7 \cdot 8$

16. $9 \cdot 10 \cdot 11 \cdot 12$

B *Evaluate.*

17. $\dbinom{9}{5}$

18. $\dbinom{5}{2}$

19. $\dbinom{6}{5}$

20. $\dbinom{7}{1}$

21. $\dbinom{9}{9}$

22. $\dbinom{5}{0}$

23. $\dbinom{17}{13}$

24. $\dbinom{20}{16}$

Expand, using the binomial formula.

25. $(m + n)^3$

26. $(x + 2)^3$

27. $(2x - 3y)^3$

28. $(3u + 2v)^3$

29. $(x - 2)^4$

30. $(x - y)^4$

31. $(m + 3n)^4$

32. $(3p - q)^4$

33. $(2x - y)^5$

34. $(2x - 1)^5$

35. $(m + 2n)^6$

36. $(2x - y)^6$

Find the indicated term in each expansion.

37. $(u + v)^{15}$;   seventh term

38. $(a + b)^{12}$;   fifth term

39. $(2m + n)^{12}$;   eleventh term

40. $(x + 2y)^{20}$;   third term

41. $[(w/2) - 2]^{12}$;   seventh term

42. $(x - 3)^{10}$;   fourth term

43. $(3x - 2y)^8$;   sixth term

44. $(2p - 3q)^7$;   fourth term

C **45.** Evaluate $(1.01)^{10}$ to four decimal places, using the binomial formula.

[*Hint:*   Let $1.01 = 1 + 0.01$.]

46. Evaluate $(0.99)^6$ to four decimal places, using the binomial formula.

47. Show that

$$\binom{n}{r} = \binom{n}{n-r}$$

48. Show that

$$\binom{n}{0} = \binom{n}{n}$$

49. Show that

$$\binom{k}{r-1} + \binom{k}{r} = \binom{k+1}{r}$$

50. Show that

$$\binom{k}{0} = \binom{k+1}{0}$$

51. Show that

$$\binom{k}{k} = \binom{k+1}{k+1}$$

52. Show that

$$\binom{n}{r}$$

is given by the recursion formula

$$\binom{n}{r} = \frac{n-r+1}{r}\binom{n}{r-1}$$

where $\binom{n}{0} = 1$

53. Write $2^n = (1+1)^n$ and expand, using the binomial formula, to obtain

$$2^n = \binom{n}{0} + \binom{n}{1} + \binom{n}{2} + \cdots + \binom{n}{n}$$

54. Can you guess what the next two rows in **Pascal's triangle** are? Compare the numbers in the triangle with the binomial coefficients obtained with the binomial formula.

```
        1
       1 1
      1 2 1
     1 3 3 1
    1 4 6 4 1
```

Section 7-7   Chapter Review

7-1 Sequences and Series.   Sequence, terms of a sequence, finite sequence, infinite sequence, recursion formula, series, finite series, infinite series, summation notation, summing index

$$a_1, a_2, \ldots, a_n, \ldots \qquad \{a_n\}$$

$$a_1 + a_2 + \cdots + a_n \qquad \sum_{k=1}^{n} a_k$$

7-2 Mathematical Induction.   Well-ordering principle, inductive sets, principle of mathematical induction

7-3 Arithmetic Sequences and Series.   Arithmetic sequence, arithmetic progression, common difference, nth-term formula, finite arithmetic series, sum formulas

$$a_n = a_1 + (n-1)d$$

$$S_n = \frac{n}{2}[2a_1 + (n-1)d]$$

$$S_n = \frac{n}{2}(a_1 + a_n)$$

7-4 Geometric Sequences and Series.   Geometric sequence, geometric progression, common ratio, nth-term formula, finite geometric series, infinite geometric series, sum formulas

$$a_n = a_1 r^{n-1}$$

$$S_n = \frac{a_1 - a_1 r^n}{1 - r} \qquad r \neq 1$$

$$S_n = \frac{a_1 - r a_n}{1 - r} \qquad r \neq 1$$

$$S_\infty = \frac{a_1}{1 - r} \qquad |r| < 1$$

7-5 Additional Applications

7-6 Binomial Formula.   Factorial, combinatorial symbol, binomial formula

$$n! = n(n-1)(n-2) \cdot \ldots \cdot 1 \qquad \binom{n}{r} = \frac{n!}{r!(n-r)!}$$

$$(a + b)^n = \sum_{k=0}^{n} \binom{n}{k} a^{n-k} b^k$$

Exercise 7-7   Chapter Review

Work through all the problems in this chapter review and check answers in the back of the book. (Answers to all problems are there, and following each answer is a number in italics indicating the section in which that type of problem is discussed.) Where weaknesses show up, review appropriate sections in the text. When you are satisfied that you know the material, take the practice test following this review.

A   1.   Determine whether the sequence is geometric, arithmetic, or neither.
(A) $16, -8, 4, \ldots$ (B) $5, 7, 9, \ldots$
(C) $-8, -5, -2, \ldots$ (D) $2, 3, 5, 8, \ldots$
(E) $-1, 2, -4, \ldots$

(A) Write the first four terms of each sequence, (B) find a_{10}, and (C) find S_{10}.

2. $a_n = 2n + 3$ 3. $a_n = 32(\tfrac{1}{2})^n$

4. $a_1 = -8; \quad a_n = a_{n-1} + 3, \quad n \geq 2$

5. $a_1 = -1; \quad a_n = (-2)a_{n-1}, \quad n \geq 2$

6.   Find S_∞ in Problem 3.

Evaluate.

7. $6!$ 8. $\dfrac{22!}{19!}$ 9. $\dfrac{7!}{2!(7-2)!}$

Verify for n = 1, 2, and 3.

10. P_n: $5 + 7 + 9 + \cdots + (2n + 3) = n^2 + 4n$

11. P_n: $2 + 4 + 8 + \cdots + 2^n = 2^{n+1} - 2$

12. P_n: $49^n - 1$ is divisible by 6

Write P_k and P_{k+1}.

13.   For P_n in Problem 10   14.   For P_n in Problem 11

15.   For P_n in Problem 12

B   *Write without summation notation and find the sum.*

16. $S_{10} = \displaystyle\sum_{k=1}^{10} (2k - 8)$ 17. $S_7 = \displaystyle\sum_{k=1}^{7} \dfrac{16}{2^k}$

18. $S_\infty = 27 - 18 + 12 + \cdots = ?$

19.   Write $S_n = \tfrac{1}{3} - \tfrac{1}{9} + \tfrac{1}{27} + \cdots + \dfrac{(-1)^{n+1}}{3^n}$ using summation notation, and find S_∞.

20. If in an arithmetic sequence $a_1 = 13$ and $a_7 = 31$, find the common difference d and the fifth term a_5.

21. Write $0.72727\overline{2}$ as the quotient of two integers.

Evaluate.

22. $\dfrac{20!}{18!(20-18)!}$ **23.** $\dbinom{16}{12}$ **24.** $\dbinom{11}{11}$

25. Expand $(x - y)^5$ using the binomial formula.

26. Find the tenth term in the expansion of $(2x - y)^{12}$.

Establish each statement for all natural numbers, using mathematical induction.

27. P_n in Problem 10   **28.** P_n in Problem 11

29. P_n in Problem 12

C   **30.** A free-falling body travels $g/2$ feet in the first second, $3g/2$ feet during the next second, $5g/2$ feet the next, and so on. Find the distance fallen during the twenty-fifth second, and the total distance fallen from the start to the end of the twenty-fifth second.

31. Expand $(x + i)^6$, i the complex unit, using the binomial formula.

Prove that each of the following statements holds for all positive integers, using mathematical induction.

32. $\displaystyle\sum_{k=1}^{n} k^3 = \left(\sum_{k=1}^{n} k\right)^2$

33. $x^{2n} - y^{2n}$ is divisible by $x - y$, $x \neq y$

34. $\dfrac{a^n}{a^m} = a^{n-m}$; $n > m$, $n, m \in N$

35. $\{a_n\} = \{b_n\}$ where $a_n = a_{n-1} + 2$, $a_1 = -3$, $b_n = -5 + 2n$

36. $(1!)1 + (2!)2 + (3!)3 + \cdots + (n!)n = (n + 1)! - 1$ (From USSR Mathematical Olympiad, 1955–1956, Grade 10.)

Practice Test   Chapter 7

Take this practice test as if it were a graded test. Allow yourself up to 50 minutes. Work the problems without looking back in the chapter. Correct your work using the answers (keyed to appropriate sections) in the back of the book.

1. Determine whether the sequence is geometric, arithmetic, or neither. If geometric, find the common ratio r; if arithmetic, find the common difference d.

(A) $2, 5, 7, \ldots$ (B) $8, -2, \frac{1}{2}, \ldots$ (C) $7, 3, -1, \ldots$

2. Given the sequence $a_1 = 64$, $a_n = a_{n-1} - 4$, $n \geq 2$, find the first four terms, a_{51}, and S_{31}.

3. Given the sequence $\{64/2^n\}$, find the first four terms, a_{10}, and S_{10}.

4. Write $\displaystyle\sum_{k=1}^{4} k^2 x^{3k-1}$ without summation notation.

5. Write $\frac{2}{3} - \frac{4}{9} + \frac{8}{27} - \frac{16}{81}$ using summation notation starting the summing index at $k = 1$.

6. Write $0.018\overline{018}$ as the quotient of integers using the sum of an appropriate geometric series.

7. Use mathematical induction to prove the following proposition for all positive integers:

$$3 + 3^2 + 3^3 + \cdots + 3^n = \frac{3^{n+1} - 3}{2}$$

8. Evaluate.   (A) $\dfrac{15!}{3!12!}$ (B) $\dbinom{23}{21}$

9. Expand using the binomial formula: $(3x - y)^4$

10. Write the sixth term in the expansion of $(x - 2)^9$.

An Introduction to Probability

■ 8

A natural design of mathematical interest. Can you guess the source? See the back of the book.

Chapter 8 ■ An Introduction to Probability

Section 8-1   Introduction

Probability can be thought of as the science of uncertainty. If, for example, a single die is rolled, it is uncertain which number will turn up. But if a die is rolled many times, a particular number, say 2, will occur over the long run with a relative frequency that is approximately predictable. Probability theory is concerned with determining the long-run relative frequency of the occurrence of a given event.

How are probabilities assigned to events? There are two basic approaches to this problem, one theoretical and the other empirical. An example will illustrate the difference between the two approaches.

Returning to our original example, suppose you were asked, "What is the probability of obtaining a 2 on a single throw of a die?" Using a **theoretical approach**, we would reason as follows: Since there are six equally likely ways the die can turn up (assuming it is fair) and there is only one way a 2 can turn up, then the probability of obtaining a 2 is one-sixth. Here we have arrived at a probability assignment using certain assumptions and a reasoning process. What does the result have to do with reality? We would, of course, expect that in the long run the 2 would appear approximately one-sixth of the time. With the **empirical approach**, we make no assumption about the equally likely ways the die can turn up. We simply set up an experiment and roll the die a large number of times. Then we compute the percentage of times the 2 appears, and use this number as an estimate of the probability of obtaining a 2 on a single roll of the die.

We will start our study by considering the theoretical approach and develop procedures that will lead to the solution of a variety of interesting problems. The procedures we will look at require the counting of the number of ways certain events can occur, and this is not always easy. However, there are effective mathematical tools that can assist us in this counting task. The development of these tools is the subject matter of the next two sections.

Section 8-2   The Fundamental Principle of Counting

■ Tree Diagrams
■ Fundamental Principle of Counting

- **Tree Diagrams**

The best way to start the discussion is with an example that uses an effective visual device called a *tree diagram*.

EXAMPLE 1 Suppose we flip a coin and then throw a single die. What are the possible combined outcomes?

Solution To solve this problem, let us use a tree diagram.

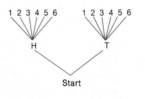

Thus there are twelve possible combined outcomes (there are two ways in which the coin can come up followed by six ways in which the die can come up).

PROBLEM 1 Use a tree diagram to determine the number of possible outcomes of throwing a single die followed by flipping a coin.

- **Fundamental Principle of Counting**

Now suppose you are asked how many 5-card hands are possible from a 52-card deck? To try to count the possibilities using a tree diagram, or any other one-by-one process, would be extremely time-consuming and tedious to say the least. The fundamental principle of counting will enable us to solve this problem easily, and, in addition, it forms the basis for several other counting devices that are developed in the next section.

Fundamental Principle of Counting

1. If two operations are performed in order, with N_1 possible outcomes for the first and N_2 possible outcomes for the second, then there are

$$N_1 \cdot N_2$$

possible combined outcomes of the first operation followed by the second.

> **2.** In general, if n operations $O_1, O_2, \ldots, O_n$ are performed in order, with possible number of outcomes $N_1, N_2, \ldots, N_n$, respectively, then there are
>
> $$N_1 \cdot N_2 \cdot \cdots \cdot N_n$$
>
> possible combined outcomes of the operations performed in the given order.

In Example 1, we see that there are two possible outcomes of flipping a coin (first operation) and six possible outcomes from throwing a die (second operation); hence, by the fundamental principle of counting, there are $2 \cdot 6 = 12$ possible combined operations. Use the fundamental principle to solve Problem 1. [*Answer:* $6 \cdot 2 = 12$]

EXAMPLE 2   A pizza parlor produces pizzas with two special ingredients, one different from the second, and only this type. For example, on the menu one finds mushroom and sausage pizzas, olive and pepperoni pizzas, and so on. If the parlor has twelve different types of special ingredients that can go together, how many different types of pizzas can it produce?

Solution   Operation 1:   Choose the first ingredient—12 ways

    Operation 2:   Choose the second ingredient after the first is chosen— 11 ways

Total number of different pizzas $= 12 \cdot 11 = 132$

PROBLEM 2   A snack bar has three kinds of bread and four kinds of meat. How many different kinds of sandwiches are possible?

EXAMPLE 3   How many four-letter code words are possible using the first ten letters of the alphabet if:

(A) No letter can be repeated.
(B) Letters can be repeated.
(C) Adjacent letters cannot be alike.

Solution   To form a four-letter code word with the ten letters available, we put a letter in the first position, one in the second position, one in the third position, and finally one in the fourth position—four operations. The table illustrates the different possibilities under the three conditions of the problem.

OPERATION	NUMBER OF WAYS OF COMPLETING OPERATION UNDER CONDITION		
	A	B	C
O_1: Fill first position	10	10	10
O_2: Fill second position after O_1	9	10	9
O_3: Fill third position after O_1 and O_2	8	10	9
O_4: Fill fourth position after O_1, O_2, and O_3	7	10	9

(A) $10 \cdot 9 \cdot 8 \cdot 7 = 5{,}040$ (B) $10 \cdot 10 \cdot 10 \cdot 10 = 10{,}000$
(C) $10 \cdot 9 \cdot 9 \cdot 9 = 7{,}290$

PROBLEM 3   How many three-letter code words are possible using the first eight letters of the alphabet? Answer the question under the three conditions stated in Example 3.

Answers to Matched Problems   **1.** 12

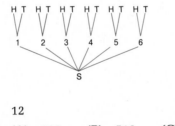

2. 12

3. (A)   336   (B)   512   (C)   392

Exercise 8-2 ■

A   *Solve using tree diagrams.*

1. In how many ways can two coins turn up? [*Note:* (H, T) is to be distinguished from (T, H).]

2. How many two-letter "words" can be formed from the first four letters of the alphabet with no letter being used more than once?

3. In how many ways can three coins turn up?

4. How many three-letter "words" can be formed from the first four letters of the alphabet with no letter being used more than once?

Solve, using the fundamental principle of counting.

5. Problem 1   6. Problem 2

7. Problem 3   8. Problem 4

9. In how many ways can two dice turn up?

10. In how many ways can two coins and a single die turn up?

11. How many four-letter "words" are possible from the first six letters of the alphabet with no letters repeated? Allowing letters to repeat?

12. How many five-letter "words" are possible from the first seven letters of the alphabet with no letters repeated? Allowing letters to repeat?

B **13.** On a particularly hectic day, a student has to go to the dentist, see a counselor, and pick up her brother at the airport. The dentist has openings at 9, 11, and 2 o'clock; the counselor at 8, 9, 1, and 3 o'clock; and her brother's plane can come in at 1 or 5 o'clock. Use a tree diagram to determine the number of choices available.

14. A student plans to take French, psychology, and mathematics. French classes are available at 10, 11, and 2 o'clock; psychology at 10 and 1 o'clock; and mathematics at 11, 1, and 3 o'clock. Use a tree diagram to determine the number of schedules possible.

15. In how many ways can a chairperson and a vice chairperson be selected from a committee of eight people? Assume one person cannot hold more than one position.

16. In how many ways can a chairperson, a vice chairperson, and a secretary be selected from a committee of ten people? Assume one person cannot hold more than one position.

17. How many different license plates can be made using three letters followed by three numbers? For example, one plate could be SZA 051.

18. How many seven-digit telephone numbers are possible if 0 cannot be used as the first digit?

19. Three signal flags are to be arranged on a flagpole, one below the other. How many three-flag signals are possible out of ten different flags? Out of twenty different flags? Out of n different flags?

20. A man packs three shirts, five ties, two pairs of slacks, and two sport coats for a 1-week trip. How many different outfits are possible if it is assumed that one object from each category is selected for one complete outfit?

21. How many three-letter code words are possible out of the alphabet if
(A) No letter can be used more than once?
(B) Letters can be repeated?
(C) Adjacent letters cannot be alike?

22. Repeat Problem 21 for four-letter code words.

23. Given five chairs in a row, how many seating arrangements are possible with five people? With ten chairs and ten people? With n chairs and n people?

24. Each of two countries sends five delegates to a negotiating conference. A rectangular table is used with five chairs on each side. If each country is assigned one side of the table (operation 1), how many seating arrangements are possible?

C 25. At a family reunion, a family portrait is to be made. The children are to sit on the floor in the front row, and the adults are to sit on chairs in a second row behind the children. How many different arrangements of the family are possible if there are twelve adults and fourteen children? (Use a calculator to estimate the answer.)

26. Two cards are drawn in succession from a deck of 52 cards.
    (A) In how many ways can one draw a heart followed by a spade if the first card is retained? If the first card is returned to the deck before the second draw?
    (B) Repeat part (A) for a heart followed by a heart.

27. Given a row of four chairs, how many distinguishable seating arrangements are possible with two people?
    (A) Solve the problem using a tree diagram.
    (B) Solve the problem using the fundamental principle of counting.

28. How many distinguishable seating arrangements are possible in a row of ten chairs with six people? Thirty chairs and ten people? (Use a calculator for the second part.)

29. Use the fundamental principle of counting to find the number of subsets of a set with five elements. [Hint: Think of selecting a subset from $\{a, b, c, d, e\}$ as a combined operation involving five individual operations. If a is in a subset, we write Y (yes); if not, we write N (no). Thus, YYNNY represents the subset $\{a, b, e\}$.]

30. Use the fundamental principle of counting to find the number of subsets of a set with n elements.

Section 8-3   Permutations, Combinations, and Set Partitioning

- Permutations
- Combinations
- Set Partitioning
- Distinguishable Permutations

The fundamental principle of counting studied in the last section can be used to develop three additional devices for counting that are extremely useful in more complicated counting problems.

■ Permutations

Suppose four pictures are to be arranged from left to right on one wall of an art gallery. How many arrangements are possible? Using the fundamental principle of counting, there are four ways of selecting the first picture; after the first picture is selected, there are three ways of selecting the second picture; after the first two pictures are selected, there are two ways of selecting the third picture; and after the first three pictures are selected, there remains but one way to select the fourth. Thus, the number of arrangements possible for the four pictures is

$$4 \cdot 3 \cdot 2 \cdot 1 = 4! \quad \text{or} \quad 24$$

In general, we refer to a particular arrangement or ordering of n objects as a **permutation** of the n objects. How many orderings (permutations) of n objects are there? Reasoning as above, there are n ways in which the first object is chosen, and there are $n - 1$ ways in which the second object can be chosen, and so on. Using the fundamental principle of counting, we have

---

Permutations of n Objects

Number of permutations of n objects $= n(n - 1) \cdot \cdots \cdot 2 \cdot 1 = n!$

Now suppose the museum director decides to use only two of the four available pictures on the wall arranged from left to right. How many arrangements of two pictures can be formed from the four? There are four ways the first picture can be selected; after selecting the first picture, there are three ways the second picture can be selected. Thus, the number of arrangements of two pictures from four pictures, denoted by $P_{4,2}$, is given by

$$P_{4,2} = 4 \cdot 3$$

or in terms of factorials, multiplying $4 \cdot 3$ by $2!/2!$, we have

$$P_{4,2} = \frac{4 \cdot 3 \cdot 2!}{2!} = \frac{4!}{2!}$$

We write this last form for purposes of generalization. Reasoning in the same way as in the example, we find **the number of permutations of n objects taken r at a time**, $0 \le r \le n$, denoted by $P_{n,r}$, to be given by

$$P_{n,r} = n(n - 1)(n - 2) \cdot \cdots \cdot (n - r + 1)$$

Multiplying the right side by 1 in the form $(n - r)!/(n - r)!$, we obtain a factorial form for $P_{n,r}$:

$$P_{n,r} = n(n - 1)(n - 2) \cdot \cdots \cdot (n - r + 1) \frac{(n - r)!}{(n - r)!}$$

But $n(n - 1)(n - 2) \cdot \cdots \cdot (n - r + 1)(n - r)! = n!$; hence,

---

Permutation of n Objects Taken r at a Time

The number of permutations of n objects taken r at a time is given by*

$$P_{n,r} = \frac{n!}{(n - r)!} \qquad 0 \le r \le n$$

Note: $\quad P_{n,n} = \dfrac{n!}{(n - n)!} = \dfrac{n!}{0!} = n!$ Permutations of n objects

---

EXAMPLE 4   From a committee of eight people, in how many ways can we choose a chairperson and a vice chairperson, assuming one person cannot hold more than one position?

Solution   We are actually asking for the number of permutations of eight objects taken two at a time—that is, $P_{8,2}$.

$$P_{8,2} = \frac{8!}{(8 - 2)!} = \frac{8!}{6!} = \frac{8 \cdot 7 \cdot 6!}{6!} = 56$$

PROBLEM 4   From a committee of ten people, in how many ways can we choose a chairperson, a vice chairperson, and a secretary?

Many hand calculators have an n! button, and some even have a $P_{n,r}$ button. The use of such a calculator will greatly facilitate many of the calculations in this and the next section.

---

* In place of the symbol $P_{n,r}$, one will also see P_r^n, $_nP_r$, and $P(n, r)$.

EXAMPLE 5   Find the number of permutations of twenty-five objects taken eight at a time. Compute the answer using a calculator to four significant digits.

Solution $P_{25,8} = \dfrac{25!}{(25-8)!} = \dfrac{25!}{17!} = 4.361 \times 10^{10}$

PROBLEM 5   Find the number of permutations of thirty objects taken four at a time. Compute the answer exactly using a calculator.

■ Combinations

Now suppose that an art museum owns eight paintings by a given artist and another art museum wishes to borrow three of these paintings for a special show. How many ways can three paintings be selected out of the eight available? Here the order does not matter. What we are actually interested in is how many three-object subsets can be formed from a set of eight objects. We call such a subset a **combination** of eight objects taken three at a time. The total number of such subsets (combinations) is denoted by the symbol

$$C_{8,3} \quad \text{or} \quad \binom{8}{3}$$

To find the number of combinations of eight objects taken three at a time, $C_{8,3}$, we make use of the formula for $P_{n,r}$ and the fundamental principle of counting. We know that the number of permutations of eight objects taken three at a time is given by $P_{8,3}$, and we have a formula for computing this. Now suppose we think of $P_{8,3}$ in terms of two operations:

O_1: Select a subset of three objects (paintings); $C_{8,3}$ ways

O_2: Arrange the subset in a given order; 3! ways

The combined operation, O_1 followed by O_2, produces a permutation of eight objects taken three at a time. Thus,

$$P_{8,3} = C_{8,3} \cdot 3!$$

To find $C_{8,3}$, the number of combinations of eight objects taken three at a time, we replace $P_{8,3}$ with $8!/(8-3)!$ and solve for $C_{8,3}$.

$$\frac{8!}{(8-3)!} = C_{8,3} \cdot 3!$$

$$C_{8,3} = \frac{8!}{3!(8-3)!} = 56$$

Thus, there are fifty-six choices the museum can make in selecting three paintings from the eight available.

In general, reasoning in the same way as in the example, the number of combinations of n objects taken r at a time, $0 \leq r \leq n$, denoted by $C_{n,r}$, can be obtained by replacing $P_{n,r}$ with $n!/(n-r)!$ and solving $C_{n,r}$ in the relationship.

$$P_{n,r} = C_{n,r} \cdot r!$$

$$\frac{n!}{(n-r)!} = C_{n,r} \cdot r!$$

$$C_{n,r} = \frac{n!}{r!(n-r)!}$$

In summary,

Combinations of n Objects Taken r at a Time

The number of combinations of n objects taken r at a time is given by*

$$C_{n,r} = \binom{n}{r} = \frac{n!}{r!(n-r)!} \qquad 0 \leq r \leq n$$

If n and r are other than small numbers, a calculator with an $n!$ button will simplify the computation, and one with a $C_{n,r}$ button will simplify the computation even further.

EXAMPLE 6   From a committee of eight people, in how many ways can a subcommittee of two people be chosen?

Solution   Notice how this example differs from Example 4, where we asked in how many ways we can choose a chairperson and a vice chairperson from a committee of eight people. In Example 4 ordering mattered; in choosing a two-person subcommittee, the ordering does not matter. Thus, we are actually asking how many combinations there are of eight objects taken two at a time. The number is given by

$$C_{8,2} = \binom{8}{2} = \frac{8!}{2!(8-2)!} = 28$$

---

* In place of the symbols $C_{n,r}$ and $\binom{n}{r}$, one will also see C_r^n, $_nC_r$, and $C(n, r)$.

PROBLEM 6   How many three-person subcommittees can be chosen from a committee of eight people?

EXAMPLE 7   Find the number of combinations of twenty-five objects taken eight at a time. Compute the answer using a calculator to four significant digits.

Solution
$$C_{25,8} = \binom{25}{8} = \frac{25!}{8!(25-8)!} = 1.082 \times 10^6$$

Compare this result with that obtained in Example 5.

PROBLEM 7   Find the number of combinations of thirty objects taken four at a time. Compute the answer exactly using a calculator.

REMEMBER: **In a permutation order counts.**
In a combination order does not count.

■   Set Partitioning

The combination of n objects taken r at a time can be thought of in another way, a way that generalizes into something very useful. Let us return to the art museum that agreed to lend three paintings from the eight by a given artist. In choosing the three, they are actually **partitioning** (dividing) the set of eight paintings into two subsets: the subset containing the three paintings to be loaned and the subset containing the five paintings that will remain. Now let us suppose that in addition to the one museum wishing to borrow three paintings, a second museum wishes to borrow two paintings by the same artist. The museum that owns the eight paintings is now confronted with the following problem: how to partition (divide) the set of eight paintings into three subsets, one containing three paintings, one containing two paintings, and one containing three paintings (the three left over). We now ask: In how many ways can a set of eight objects be partitioned into three subsets with the first containing three objects, the second containing two objects, and the third containing three objects? Again we call on the fundamental principle of counting. Think of the problem in terms of the following three operations:

O_1: Select a subset with three paintings from eight paintings; $C_{8,3}$ ways.

O_2: Select a subset with two paintings from the five paintings left; $C_{5,2}$ ways.

O_3: Select a subset with three paintings from the three paintings left; $C_{3,3}$ ways.

The combined operation—O_1, O_2, and O_3—produces a *partition* of the set of eight paintings into three subsets as desired. Thus, the number of such partitions, denoted by $\begin{pmatrix} 8 \\ 3,\ 2,\ 3 \end{pmatrix}$, is given by

$$\begin{pmatrix} 8 \\ 3,\ 2,\ 3 \end{pmatrix} = C_{8,3} \cdot C_{5,2} \cdot C_{3,3}$$

$$= \frac{8!}{3!(8-3)!} \cdot \frac{5!}{2!(5-2)!} \cdot \frac{3!}{3!(3-3)!} = \frac{8!}{3!5!} \cdot \frac{5!}{2!3!} \cdot \frac{3!}{3!0!}$$

$$= \frac{8!}{3!2!3!} \quad \text{Compare the arrangement with the arrangement in } \begin{pmatrix} 8 \\ 3,\ 2,\ 3 \end{pmatrix}.$$

$$= 560$$

In general, to partition a set of n elements into k subsets such that r_1 elements are in the first subset, r_2 elements are in the second subset, ..., and r_k elements are in the kth subset, $r_1 + r_2 + \cdots + r_k = n$, we can think of the problem in terms of k operations as above and apply the fundamental principle of counting to obtain

$$\begin{pmatrix} n \\ r_1,\ r_2,\ r_3,\ \ldots,\ r_k \end{pmatrix} = C_{n,r_1} \cdot C_{n-r_1,r_2} \cdot C_{n-r_1-r_2,r_3} \cdot \cdots \cdot C_{n-r_1-r_2-\cdots-r_{k-1},r_k}$$

$$= \frac{n!}{r_1!(n-r_1)!} \cdot \frac{(n-r_1)!}{r_2!(n-r_1-r_2)!} \cdot \frac{(n-r_1-r_2)!}{r_3!(n-r_1-r_2-r_3)!} \cdot \cdots$$

$$\cdot \frac{(n-r_1-r_2-\cdots-r_{k-1})!}{r_k!(n-r_1-r_2-\cdots-r_k)!} = \frac{n!}{r_1!r_2!\cdots r_k!}$$

Note: $(n - r_1 - r_2 - \cdots - r_k)! = 0! = 1$

Partition of n Elements into k Subsets

The number of partitions of a set with n elements into k subsets is given by

$$\begin{pmatrix} n \\ r_1,\ r_2,\ \ldots,\ r_k \end{pmatrix} = \frac{n!}{r_1!r_2!\cdot \cdots \cdot r_k!}$$

where r_i elements are in the ith subset and

$$r_1 + r_2 + \cdots + r_k = n$$

Note: $\begin{pmatrix} n \\ r_1,\ r_2 \end{pmatrix} = C_{n,r_1} = C_{n,r_2}$

EXAMPLE 8   If four people are playing poker, how many deals are possible in which each person receives five cards?

Solution   This is a partition problem. We are actually dividing (partitioning) the deck (set of 52 cards) into five subsets: four subsets correspond to the hands for four players and the fifth subset is what is left over after dealing the four hands. We use a calculator to compute the following to four significant digits:

$$\binom{52}{5, 5, 5, 5, 32} = \frac{52!}{5!\,5!\,5!\,5!\,32!}$$

$$= \frac{52!}{(5!)^4 32!} \approx 1.478 \times 10^{24} \text{ deals}$$

PROBLEM 8   If three people are playing cards and each is dealt 7 cards from a 52-card deck, how many deals are possible? Compute the answer to four significant digits using a calculator.

■ Distinguishable Permutations

We finish this section by considering a variation on the partitioning problem.

EXAMPLE 9   We know that there are 7! permutations of the first seven letters of the alphabet, but how many distinguishable permutations are there of the seven letters in the word "alababa"?

Solution   In "alababa" there are repeats of certain letters. Thus, for example, a rearrangement of the a's, keeping the other letters fixed, would not produce a distinguishable permutation. Let us appeal to the fundamental principle of counting directly in terms of the following four operations:

O_1:   Form a distinguishable arrangement; ? ways

O_2:   Arrange all the a's in the arrangement in a particular order; 4! ways

O_3:   After O_1 and O_2, arrange all the b's in a particular order; 2! ways

O_4:   After completing O_1, O_2, and O_3, arrange the l in a particular order; 1! way

The combined operation of O_1 through O_4 is the number of permutations of seven elements—that is, 7! Thus,

7! = (Number of distinguishable arrangements) · 4! · 2! · 1!

or

$$\text{Number of distinguishable arrangements} = \frac{7!}{4!2!1!}$$

$$= \binom{7}{4,\, 2,\, 1}$$

$$= 105$$

Note: $\binom{7}{4,\, 2,\, 1}$ is the same symbol that is used to represent the number of partitions of a set of seven elements into three subsets, one with four elements, one with two elements, and one with one element.

Reasoning in the same way as in Example 9, we can establish the following:

Distinguishable Permutations

If a set has n elements with r_1 elements of one kind, r_2 elements of a second kind, . . . , and r_k elements of a kth kind, $r_1 + r_2 + \cdots + r_k = n$, then the number of distinguishable permutations of the n elements is given by

$$\binom{n}{r_1,\, r_2,\, \ldots,\, r_k} = \frac{n!}{r_1!r_2! \cdot \cdots \cdot r_k!}$$

Note: The formula is exactly the same as the partition formula.

PROBLEM 9   How many distinguishable permutations are there of the letters aabccc?

Answers to Matched Problems

4. $P_{10,3} = \dfrac{10!}{(10 - 3)!} = 720$ **5.** $P_{30,4} = \dfrac{30!}{(30 - 4)!} = 657{,}720$

6. $C_{8,3} = \dfrac{8!}{3!(8 - 3)!} = 56$ **7.** $C_{30,4} = \dfrac{30!}{4!(30 - 4)!} = 27{,}405$

8. $\dbinom{52}{7,\, 7,\, 7,\, 31} = \dfrac{52!}{7!7!7!31!} \approx 7.662 \times 10^{22}$

9. $\dbinom{6}{2,\, 1,\, 3} = \dfrac{6!}{2!1!3!} = 60$

Exercise 8-3 ■   ▦   *A calculator with n! will be useful in many problems.*

A   *Evaluate Problems 1–12.*

1. $P_{4,2}$ 2. $P_{5,3}$ 3. $P_{52,2}$

4. $P_{52,4}$ 5. $C_{4,2}$ 6. $C_{5,3}$

7. $C_{52,2}$ 8. $C_{52,4}$ 9. $\begin{pmatrix} 8 \\ 5,\ 3 \end{pmatrix}$

10. $\begin{pmatrix} 7 \\ 2,\ 5 \end{pmatrix}$ 11. $\begin{pmatrix} 10 \\ 2,\ 7,\ 1 \end{pmatrix}$ 12. $\begin{pmatrix} 12 \\ 3,\ 1,\ 1,\ 7 \end{pmatrix}$

13. A bookshelf has space for three books. Out of six different books available, how many arrangements can be made on the shelf?

14. A small combination lock on a suitcase has ten positions. How many three-number opening combinations are possible, assuming the same position cannot be used more than once?

15. An ice cream parlor has twenty-five different flavors of ice cream. How many different two-scoop cones can they offer if two different flavors are to be used and the order of the scoops does not matter?

16. There are ten teams in a conference. If each team is to play every other team exactly once, how many games must be scheduled?

B 17. In how many different ways can six candidates for an office be listed on a ballot?

18. How many three-digit numbers can be obtained from the digits 3, 4, and 5 if no digit can be used more than once?

19. How many three-letter code words can be formed from the letters in the word "agent"? How many five-letter code words? (No letter can be used more than once.)

20. How many three-digit numbers can be formed from the digits 3, 4, 5, 6, and 7? How many five-digit numbers? (Assume no digit can be used more than once.)

21. How many 13-card bridge hands are possible from a standard 52-card deck?

22. Given seven points, no three of which are on a straight line, how many lines can be drawn joining two points at a time?

23. If four people are playing cards and each is dealt 13 cards, how many different deals are possible from a 52-card deck?

24. If six people are playing cards and each is dealt 7 cards, how many different deals are possible from a 52-card deck?

25. How many distinguishable five-digit numbers can be formed by permuting the digits in the number 35355?

26. How many distinguishable permutations can be obtained from all the letters in the word "Mississippi"?

27. Suppose the Senate has sixty Democrats and forty Republicans. How many committees consisting of five Democrats and five Republicans can be formed? (Use logarithms or a calculator to estimate the answer.)

28. A student council consists of eighteen boys and sixteen girls.
(A) How many six-student committees can be formed?
(B) How many committees consisting of three boys and three girls are possible?

C 29. Show that: $C_{n,r} = \begin{pmatrix} n \\ r, n-r \end{pmatrix}$

30. Show that: $C_{n,n-r} = \begin{pmatrix} n \\ r, n-r \end{pmatrix}$

31. In how many ways can three people be seated around a circular table with three chairs? Two seating arrangements are different only if at least one person is sitting next to a new person on his left or right.

32. Repeat Problem 31 for five people and five chairs; for n people and n chairs.

33. How many distinguishable arrangements are possible for the eight major chess pieces along one row?

34. How many seating arrangements are possible with five people and a row of five chairs? Six chairs? Nine chairs?

35. *Binomial formula revisited.* Interpret the coefficients in the right member of

$$(a + b)^n = \sum_{k=0}^{n} \binom{n}{k} a^{n-k} b^k$$

in terms of the number of subsets of k elements taken from a set of n elements.
[*Hint:* Write $(a + b)^n = (a + b)(a + b) \cdots \cdots (a + b)$ and consider selecting b from k factors and a from $n - k$ factors for a term involving $a^{n-k} b^k$.]

Section 8-4 Experiments, Sample Spaces, and Probability Functions

- ■ Experiments
- ■ Sample Spaces
- ■ Events
- ■ Probability Functions
- ■ Equally Likely Assumptions

■ Experiments

In science certain experiments produce the same results when performed under exactly the same conditions. Experiments of this type are called **deterministic**—the conditions of the experiment determine the outcome. There are also experiments that do not yield the same results no matter how carefully they are repeated under the same conditions. These experiments are called **random experiments**. Familiar examples of the latter are rolling dice, flipping coins, observing the weather, and observing the frequency of automobile accidents on a given day of the year. Probability theory is a branch of mathematics that has been developed to deal with random experiments, both real and conceptual.

In the developments that follow, the word *experiment* will be used to mean a random experiment. It is also important to keep in mind that by experiment we also mean any act or activity, real or conceptual, that leads to an observation or measurement. We may observe controlled experiments or uncontrolled natural phenomena.

■ Sample Spaces

Consider the experiment "a car accident occurs." What can we observe about the accident? We might be interested in whether it was light or dark; whether it was raining or not raining; whether the driver was sober or intoxicated; the day of the week; and so on. The list of possible outcomes of the experiment appears to be endless. There is no unique way of analyzing all possible outcomes of an experiment. Therefore, before conducting an experiment, it is important to decide just what outcomes are of interest.

In the accident experiment, suppose we limit our interest to questions concerning days of the week on which an accident occurs. Having decided what to observe, we make a list of outcomes of the experiment such that on each trial of the experiment one and only one of the results on the list will occur. Thus,

$$S_1 = \{M, T, W, Th, F, S, Su\}$$

is an appropriate list for our interests (M represents the outcome "the accident occurred on Monday," and so on). Note that each accident will correspond to exactly one element in S_1. The set of outcomes S_1 is called a *sample space* for the experiment. In general, a **sample space** for an experiment is a set of all possible outcomes S such that any performance of the experiment results in one and only one element in S. Each element in S is called a **simple outcome** or **sample point**.

Notice that we cannot add an event such as "the accident occurred on a weekend" to our list S_1, since this outcome would occur if either

S occurs or Su occurs, violating the condition that one and only one of the outcomes in the sample space occurs on a given trial. The outcome "the accident occurred on the weekend" is called a *compound outcome*. In general, C is a **compound outcome** relative to a sample space S if there exist at least two simple outcomes in S that imply the occurrence of C. The outcome "the accident was on a weekday" is a compound outcome relative to the sample space S_1, since this outcome will occur if any of the simple outcomes in the set {M, T, W, Th, F} occurs. Of course, none of the outcomes in a sample space are compound outcomes relative to that space; that is why they are called *simple outcomes*.

Suppose we are interested in whether it is light or dark as well as the day of the week on which an accident occurs. Then we must refine the sample space S_1 further. A suitable new sample space for the experiment "an accident occurs," reflecting our new interest, is

$$S_2 = \{\text{M-L, M-D, } \ldots \text{, Su-L, Su-D}\}$$

where M-L is the outcome "the accident occurred on Monday while it was light," and so on. Note that each sample outcome in S_1 is now a compound outcome in S_2; that is, we will know that Su occurred if we know that either Su-L or Su-D has occurred.

EXAMPLE 10   An experiment consists of recording the boy-girl composition of a two-child family. How shall we identify a sample space for this experiment? There are a number of possibilities, depending on our interest. We will consider three.

(A) If we are interested in the sex of each child in the order of their birth, then, using a tree diagram, we can easily determine an appropriate sample space for the experiment:

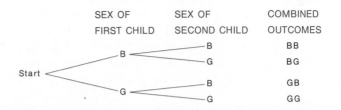

Thus, $S_1 = \{\text{BB, BG, GB, GG}\}$; there are four points in the sample space.

(B) If we are only interested in the number of girls in the family, then we can let

$$S_2 = \{0, 1, 2\}$$

and there are three points in the sample space.

(C)   If we are only interested in whether the sexes are alike (a) or different (d), then we can let

$$S_3 = \{a, d\}$$

and there are only two points in the sample space.

Important Observations

1. The earlier discussion on the accident experiment and Example 10 illustrate an important point: An experiment may have more than one sample space. Consequently, we generally refer to "a" sample space and not "the" sample space for an experiment. In a given experiment we may need a very fine or a very rough analysis of outcomes, depending on our interests, and this causes the sample space to have many or few elements.

2. In Example 10, sample space S_1 contains more information than either S_2 or S_3. If we know which outcome has occurred in S_1, then we know which outcome has occurred in S_2 and S_3. However, the reverse is not true. In this sense, we say that S_1 is a more fundamental sample space than either S_2 or S_3. In general, when specifying a sample space for an experiment, we include as much detail as is necessary to answer all questions of interest regarding the outcomes of an experiment. We will, however, limit our investigation to experiments with finite sample spaces.

PROBLEM 10   An experiment consists of tossing a dime and a quarter.

(A)   What is an appropriate sample space if we are interested in a head (H) or a tail (T) turning up on each coin? Draw a tree diagram.

(B)   What is an appropriate sample space if we are only interested in the number of heads that appear on a single toss of the two coins?

(C)   What is an appropriate sample space if we are only interested in whether the coins match (M) or do not match (D)?

(D)   What is an appropriate sample space for all three interests expressed above?

■   Events

To facilitate the building of a mathematical model of an experiment, we will restate a number of ideas introduced in set language. Consider Example 10 on the boy-girl composition of a two-child family, and the sample space.

$$S_1 = \{BB, BG, GB, GG\}$$

Suppose we are interested in the compound outcome "exactly one girl is in a two-child family." Looking at S_1, we find that it will occur if either of the two simple outcomes BG or GB occurs. Thus, to say that the compound outcome "exactly one girl is in a two-child family" occurs is the same as saying the experiment results in an outcome in the set

$$E = \{BG, GB\}$$

which is a subset of the sample space S_1. We will call the set E an **event**.

In general, given a sample space S for an experiment, we define an **event E** to be *any* subset of S. We say that an **event E occurs** if any of the simple outcomes in E occurs. If an event E has only one element in it, it is called a **simple event**; if it has more than one element, it is called a **compound event**.

EXAMPLE 11   If in the two-child experiment in Example 10 we use the sample space

$$S_1 = \{BB, BG, GB, GG\}$$

then:

(A) The event E_1 corresponding to "the family has at least one boy" is

$$E_1 = \{BB, BG, GB\}$$

(B) The event E_2 corresponding to "both children are the same sex" is

$$E_2 = \{BB, GG\}$$

PROBLEM 11   For the sample space in Example 11, write the subset of S that represents each of the following events:

(A) The family has exactly one boy.
(B) The family has at least one girl.

■   Probability Functions

We are now ready to introduce the concept of a **probability function**. This function will assign to each simple event in a sample space a real number between 0 and 1. Since an arbitrary event relative to a sample space S can be thought of as the union of simple events in S, we start by defining a probability function relative to simple events. We will then use these results as building blocks in assigning probabilities for compound events.

Probability Function

Given a finite sample space

$$S = \{e_1, e_2, \ldots, e_n\}$$

If to each simple event* e_i we assign a number, denoted by $P(e_i)$, such that

1. $0 \le P(e_i) \le 1$
2. $P(e_1) + P(e_2) + \cdots + P(e_n) = 1$

then the function P is called a **probability function** and $P(e_i)$ is called the **probability of the simple event e_i**. Any probability assignment to simple events in S that meets conditions 1 and 2 is called an **acceptable assignment**.

Any function P that satisfies conditions 1 and 2 is called a **probability function**. How probabilities are assigned to simple events is a question our mathematical theory does not answer. These assignments, however, are generally based on expected or actual long-run relative frequencies of the occurrences of the various simple events for a given experiment.

Now we are ready to define the probability of an arbitrary event E:

Probability of an Arbitrary Event E

Given a sample space S and a probability function defined on S, then the **probability of an arbitrary event E** relative to S, denoted by $P(E)$, is determined as follows:

1. If E is the empty set, then $P(E) = 0$.
2. If E is a simple event, then $P(E)$ has already been assigned.
3. If E is the union of two or more simple events, then $P(E)$ is the sum of the probabilities of the simple events whose union is E.

---

* Technically, we should write $\{e_i\}$, since there is a logical distinction between an element of a set and a subset consisting only of that element. But we will just keep this in mind, and drop the braces for simple events to simplify the notation.

Let us summarize the key steps in determining probabilities of events:

Steps for Finding Probabilities of Events

1. Set up an appropriate sample space S for the experiment.
2. Assign acceptable probabilities to the simple events of S.
3. To obtain the probability of an event E, add the probabilities of the simple events whose union is E.

EXAMPLE 12   Let a sample space for the roll of a single die (*dice* plural) be

$$S = \{1, 2, 3, 4, 5, 6\}$$

Suppose the die is loaded and

$P(1) = .15$ $P(2) = .21$ *Note:* .15 + .21 + .25 + .16 + .08 + .15 = 1.
$P(3) = .25$ $P(4) = .16$ Hence, P is a legitimate probability function.
$P(5) = .08$ $P(6) = .15$

What is the probability of getting an odd number on a single roll of the die?

Solution   The event "getting an odd number on a single roll of the die" is

$$E = \{1, 3, 5\}$$

Thus,

$$P(E) = P(1) + P(3) + P(5)$$
$$= .15 + .25 + .08 = .48$$

PROBLEM 12   Using the die in Example 13, what is the probability of getting a number larger than 3?

■ Equally Likely Assumptions

If we assume the die in Example 12 is not loaded—that is, it is a fair die—then in a large number of throws of the die we would expect each face to turn up with the same relative frequency—approximately one-sixth of the time for each, since there are six faces. Thus, a reasonable assignment of the probability of obtaining a given face is $\frac{1}{6}$. In general:

Equally Likely Simple Events

Given the sample space

$$S = \{e_1, e_2, \ldots, e_n\}$$

If each simple event has the same probability of occurring as any other, we write

$$P(e_1) = P(e_2) = \cdots = P(e_n) = \frac{1}{n}$$

and say that the simple events are **equally likely**.

EXAMPLE 13   What is the probability of getting a tail on a single toss of a fair coin?

Solution   We choose $S = \{H, T\}$ and assume H is as likely to occur as T; thus,

$$P(T) = \tfrac{1}{2}$$

PROBLEM 13   What is the probability of drawing the 9 of clubs from a standard 52-card deck, assuming each card is as likely to be drawn as any other?

Now let us look at a compound event related to the rolling of a fair die with sample space

$$S = \{1, 2, 3, 4, 5, 6\}$$

What is the probability of rolling a number divisible by 3? We are talking about the compound event

$$E = \{3, 6\}$$

From the definition of the probability of an event, we sum the probabilities whose union is E:

$$P(E) = P(3) + P(6)$$

$$= \frac{1}{6} + \frac{1}{6}$$

$$= \frac{2}{6} = \frac{\text{Number of elements in } E}{\text{Number of elements in } S}$$

In general, under the assumption that the occurrence of each simple event is as likely to occur as any other, the computation of the probability of the occurrence of any event E relative to S is completed as follows: We count the number of elements in E and divide by the number of elements in the sample space S.

Probability of an Event *E* Assuming Equally Likely Simple Events

Given the sample space S and event E, if each simple event in S is as likely to occur as any other, then

$$P(E) = \frac{\text{Number of elements in } E}{\text{Number of elements in } S} = \frac{n(E)}{n(S)}$$

EXAMPLE 14   The following pertain to the sexual composition of a three-child family, excluding multiple births.

    (A) Under the assumption that a girl is as likely as a boy at each birth, select a sample space S such that all simple events can be assumed equally likely to occur.

    (B) What is the probability of having three boys?

    (C) What is the probability of having two girls and a boy—in that order?

    (D) What is the probability of having two girls and a boy—in any order?

Solution   (A) A tree diagram is helpful in selecting a sample space S:

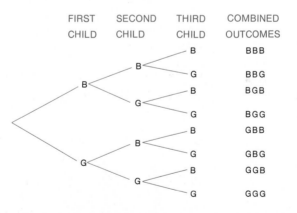

Under the assumption that a boy is as likely as a girl at each birth, each branch at the end is as likely as any other; hence, each combined outcome is as likely as any other. Thus, we choose the sample space

    $S = \{$BBB, BBG, BGB, BGG, GBB, GBG, GGB, GGG$\}$

    (B) The event of having three boys is the simple event

    $E = \{$BBB$\}$

thus,

$$P(E) = \frac{n(E)}{n(S)} = \frac{1}{8}$$

(C) The event of having two girls and a boy (in that order) is the simple event

$$E = \{GGB\}$$

thus,

$$P(E) = \frac{n(E)}{n(S)} = \frac{1}{8}$$

(D) The event of having two girls and a boy in any order is

$$E = \{BGG, GBG, GGB\}$$

thus,

$$P(E) = \frac{n(E)}{n(S)} = \frac{3}{8}$$

PROBLEM 14   Using the sample space in Example 14, find the probability of having

(A) Three girls   (B) At least one girl

EXAMPLE 15   In drawing 7 cards from a 52-card deck without replacement (without replacing a drawn card before selecting the next card), what is the probability of getting 7 clubs? Approximate the answer to three significant digits using a calculator with $n!$.

Solution   Let the sample space S be the set of all 7-card hands from a 52-card deck. Since the order in a hand does not matter, $n(S) = C_{52,7}$. The event E is the set of all 7-card club hands from 13 clubs. Again, the order does not matter and $n(E) = C_{13,7}$. Thus, assuming each 7-card hand is as likely as any other,

$$P(E) = \frac{n(E)}{n(S)} = \frac{C_{13,7}}{C_{52,7}} \approx .000\ 128$$

PROBLEM 15   In drawing 5 cards from a 52-card deck without replacement, what is the probability of getting 5 hearts? Approximate the answer to three significant digits using a calculator with $n!$.

Answers to Matched Problems   **10.** (A)

	DIME OUTCOMES	QUARTER OUTCOMES	COMBINED OUTCOMES
		H	HH
	H	T	HT
Start		H	TH
	T	T	TT

Thus, $S_1 = \{HH, HT, TH, TT\}$

(B) $S_2 = \{0, 1, 2\}$ (C) $S_3 = \{M, D\}$

(D)   The sample space in part (A)

11.   (A) $\{BG, GB\}$ (B) $\{BG, GB, GG\}$

12.   .39   13. $\frac{1}{52}$

14.   (A) $\frac{1}{8}$ (B) $\frac{7}{8}$ 13. $C_{13,5}/C_{52,5} \approx .000\ 495$

Exercise 8-4 ■ ▦   *A calculator with n! will be useful in many problems.*

A   An eight-sided die is numbered from 1 to 8, and each number is as likely to turn up as any other. An experiment consists of rolling the die once. Problems 1–6 refer to this experiment.

1.   Find a sample space S for the experiment composed of equally likely simple events.

2.   Find $n(S)$.

3.   What is the probability of rolling a 2?

4.   What is the probability of rolling an 8?

5.   What is the event associated with rolling an odd number? What is the probability of rolling an odd number?

6.   What is the event associated with rolling a number exactly divisible by 4? What is the probability of rolling a number exactly divisible by 4?

7.   What is the probability of having at least one girl in a two-child family? (See Example 10 in the text.)

8.   What is the probability of getting exactly one head in tossing two coins? (See Problem 10 in the text.)

B   9.   What would $P(E) = 0$ mean?

10.   What would $P(E) = 1$ mean?

An experiment consists of flipping a penny three times in succession. Problems 11–14 pertain to this experiment.

11.   Find a sample space S for the experiment consisting of equally likely simple events. (See Example 14 in the text.)

12.   Find the event E associated with exactly two tails occurring in three tosses of the penny. What is the probability of getting exactly two tails in three tosses?

13.   Find the event E associated with at least two tails occurring. What is the probability of at least two tails occurring?

14. Find the event E associated with at least one head occurring. What is the probability of at least one head occurring?

15. A spinner card is divided so that a pointer will land on yellow (Y), blue (B), or red (R), but not all with equal likelihood. Which of the following probability assignments for the simple events Y, B, and R would have to be rejected and why?
(A) $P(Y) = .42,$ $P(B) = -.13,$ $P(R) = 1.03$
(B) $P(Y) = .38,$ $P(B) = .21,$ $P(R) = .50$
(C) $P(Y) = .27,$ $P(B) = .43,$ $P(R) = .30$

16. Using the probability assignment in part (C) of Problem 15, what is the probability that the spinner will not land on red?

17. Using the probability assignment in part (C) of Problem 15, what is the probability that the spinner lands on yellow or red?

18. Using the probability assignment in part (C) of Problem 15, what is the probability that the spinner lands on green?

For a single roll of two fair dice a sample space S of equally likely simple events is given in the accompanying table. There are thirty-six equally likely outcomes; thus, S has thirty-six elements. [It is important to distinguish between (1, 2) and (2, 1), for example, even though both add up to 3.] Find the probability of the sum of the dots on the two faces:

SECOND DIE

(1, 1)	(1, 2)	(1, 3)	(1, 4)	(1, 5)	(1, 6)
(2, 1)	(2, 2)	(2, 3)	(2, 4)	(2, 5)	(2, 6)
(3, 1)	(3, 2)	(3, 3)	(3, 4)	(3, 5)	(3, 6)
(4, 1)	(4, 2)	(4, 3)	(4, 4)	(4, 5)	(4, 6)
(5, 1)	(5, 2)	(5, 3)	(5, 4)	(5, 5)	(5, 6)
(6, 1)	(6, 2)	(6, 3)	(6, 4)	(6, 5)	(6, 6)

FIRST DIE

19. Being 7

20. Being 12

21. Being 11

22. Being less than 4

23. Being prime

24. Being divisible by 3

25. Not being 12

26. Not being 3

C *An experiment consists of dealing 5 cards from a standard 52-card deck. In Problems 27–34, what is the probability of being dealt:*

27. 5 face cards, jacks through aces?

28. 5 nonface cards, 2 through 10?

29. 4 aces?

30. 4 of a kind?

31. Straight flush, ace high?

32. Straight flush, starting with 2?

33. 2 aces and 3 queens?

34. 2 kings and 3 aces?

APPLICATIONS

35. *Politics.* A two-person delegation is to be chosen at random from a six-person executive committee of a student council. What is the probability that Mary Smith and John Jones (both members of the executive committee) will be chosen?

36. *Politics.* A steering committee of five people at a convention decides to form a subcommittee of two people to be responsible for delegate checking. If the subcommittee is chosen at random, what is the probability that two particular people on the steering committee will be chosen?

37. *Consumer testing.* From six known brands of beer, three are chosen at random for a consumer to identify in a blindfold tasting test. What is the probability that the three could be identified by just guessing?

38. *Medicine.* An applicant for a laboratory technician job is given three different blood samples chosen at random from eight possible blood types. What is the probability that the applicant could identify these by just guessing?

Section 8-5   Empirical Probability

- ■ Expected Frequency
- ■ Empirical Probability
- ■ Summary

■ Expected Frequency

In flipping a penny, we reason that if the coin is fair, then a head (H) is as likely to turn up as a tail (T). And since there are only two faces, we assign a probability of $\frac{1}{2}$ to the occurrence of a head and $\frac{1}{2}$ to the occurrence of a tail.

We are naturally interested in whether these theoretical probability assignments have anything to do with reality. If they do, then we would expect out of, say, 1,000 flips of the coin, a head should turn up about 500 times. (We multiply the probability of the occurrence of event E by the total number of trials n to obtain the **expected frequency** of occurrence of E in n trials.)

■ Empirical Probability

The equally likely assumption behind many theoretical probability determinations may not be warranted in reality. Or a problem may be so complex that it is extremely difficult to calculate theoretical probabilities even if the equally likely assumption is warranted. We have another way of assigning probabilities to events that is based on experience.

Suppose we actually flip a penny 1,000 times and find the following frequency for each face:

Heads: 423
Tails: 577

It would then seem reasonable to say that

$$P(H) = \frac{423}{1,000} = .423$$

$$P(T) = \frac{577}{1,000} = .577$$

[*Note:* $P(H) + P(T) = 1$]

As we increase the number of flips of the coin, our confidence in the probability assignments would likely increase, and we would tend to trust this type of probability assignment more than a theoretical one, since we know that no coin can be perfectly fair.

In general, if we conduct an experiment n times and an event E occurs with frequency $f(E)$, then the ratio $f(E)/n$ is called the **relative frequency** of the occurrence of event E in n trials. We define the **empirical probability** of E, denoted by $P(E)$, by the number (if it exists) that the relative frequencies $f(E)/n$ approach as n gets larger and larger. Of course, for any particular n, the relative frequency $f(E)/n$ is generally only approximately equal to $P(E)$, but as n increases in size we would expect the approximation to improve.

Empirical Probability Approximation

$$P(E) \approx \frac{\text{Frequency of occurrence of } E}{\text{Total number of trials}} = \frac{f(E)}{n}$$

(The larger n, the better the approximation.)

If equally likely assumptions used to obtain theoretical probability assignments are actually warranted, then we would also expect corresponding approximate empirical probabilities to approach the theoretical ones as the number of trials n of actual experiments becomes very large.

EXAMPLE 16   One thousand randomly chosen two-child families are surveyed (twins are excluded) with the following frequencies of each type of family:

Two girls: 235

One girl: 544

No girls: 221

We calculate approximate empirical probabilities for each of the three family types as follows:

$$P(2 \text{ girls}) \approx \frac{235}{1,000} = .235$$

$$P(1 \text{ girl}) \approx \frac{544}{1,000} = .544$$

$$P(0 \text{ girls}) \approx \frac{221}{1,000} = .221$$

PROBLEM 16   One die is rolled 1,000 times with the following frequencies of outcomes:

1	167	2	141
3	190	4	120
5	205	6	177

(A) Calculate approximate empirical probabilities for each indicated outcome.

(B) Does it appear that the indicated outcomes are equally likely?

(C) Assuming the indicated outcomes are equally likely, compute their theoretical probabilities.

EXAMPLE 17   An insurance company selected 1,000 drivers at random in a particular city to determine a relationship between age and moving traffic citations. The data obtained are listed in the table. Compute the approximate empirical probabilities for a driver chosen at random in the city.

AGE, YEARS	MOVING CITATIONS IN ONE YEAR				
	0	1	2	3	Over 3
Under 20	18	30	45	33	14
20–29	28	54	60	30	18
30–39	36	64	40	23	17
40–49	57	54	30	21	8
50–59	46	48	28	17	11
Over 59	32	57	52	20	9

E_1: Being under 20 years old and having exactly three citations in one year

E_2: Being between 40 and 60 years old and having no citations in one year

E_3: Having over three citations in one year

Solution $P(E_1) \approx \dfrac{33}{1,000} = .033$

$P(E_2) \approx \dfrac{57 + 46}{1,000} = \dfrac{103}{1,000} = .103$

$P(E_3) \approx \dfrac{14 + 18 + 17 + 8 + 11 + 9}{1,000} = \dfrac{77}{1,000} = .077$

Notice that in this type of problem, which is typical of many realistic problems, approximate empirical probabilities are the only type we can compute.

PROBLEM 17   Referring to the results of the survey in Example 17, compute each of the following approximate empirical probabilities for a driver chosen at random from the city:

E_1: Being under 20 years old with no citations in one year

E_2: Being 20 to 40 years old with less than two citations in one year

E_3: Having less than two citations in one year

■ Summary

Approximate empirical probabilities are often used to test theoretical probabilities. As we said before, assumptions such as equally likely may not be justified in reality. In addition to this use, there are many situations in which it is either very difficult or not possible to compute the theoretical probabilities for given events. For example, insurance companies use past experience to establish approximate empirical probabilities to predict the future; baseball teams use batting averages (approximate empirical probabilities based on past experience) to predict the future performance of a player, and pollsters use approximate empirical probabilities to predict outcomes of elections.

Answers to Matched Problems   **16.** (A) $P(1) \approx .167,$ $P(2) \approx .141,$ $P(3) \approx .190,$ $P(4) \approx .120,$
          $P(5) \approx .205,$ $P(6) \approx .177$
        (B)   No   (C) $\frac{1}{6} \approx .167$ for each

17. $P(E_1) \approx .018,$ $P(E_2) \approx .182,$ $P(E_3) \approx .524$

Exercise 8-5 ■ ▦   *A calculator will be useful in some of the problems.*

A   **1.**   Out of 500 times at bat a baseball player gets 176 hits. What is the approximate empirical probability that the player will get a hit the next time at bat?

2.   A pole vaulter has vaulted over 15 feet in sixteen out of seventy-two attempts. What is the approximate empirical probability of the vaulter vaulting over 15 feet on his next attempt?

3.   A new drug is found to help 3,820 out of 5,000 people with a particular disease. What is the approximate empirical probability that a new patient with this disease will be helped by the drug?

4.   It is found that out of 10,000 people contracting a certain disease, 8,240 recover. What is the approximate empirical probability that a person contracting this disease will recover?

B   **5.**   A thumbtack is tossed 500 times with the following outcome frequencies:

Point down: 196
Point up: 304

Compute the approximate empirical probability for each outcome. Does each outcome appear to be equally likely?

6.   Actually toss a thumbtack fifty times, letting it fall on a table or the floor. What is the approximate empirical probability of the tack landing point down? Point up? (To speed up the process, you can toss fifty tacks once, twenty-five tacks twice, ten tacks five times, or five tacks ten times.)

7. Three coins are flipped 1,000 times with the following frequencies of outcomes:

   Three tails: 128

   Two tails: 371

   One tail: 380

   No tails: 121

   (A) Compute the approximate empirical probabilities for each outcome.

   (B) Compute the theoretical probability for each outcome, assuming fair coins.

   (C) Compute the expected frequency for each outcome assuming fair coins. [Use the theoretical probabilities from part (B).]

8. Toss three coins fifty times and compute the approximate empirical probability for three tails, two tails, one tail, and no tails, respectively.

C 9. Toss three thumbtacks fifty times and compute the approximate empirical probabilities for each of the following outcomes:

   E_1: Three points up

   E_2: Two points up

   E_3: One point up

   E_4: No points up

10. Shuffle a standard 52-card deck. Count out 22 cards and then discard them without looking at any of the cards.

    (A) With the remaining 30-card deck, deal two cards, face up, and record whether both are face cards (including aces) or not, and then return the cards to the 30-card deck and shuffle. Repeat this experiment fifty times and use the results to determine an approximate empirical probability for dealing two face cards from the 30-card deck.

    (B) Now check the results in part (A) by actually computing the theoretical probability of dealing two face cards from the 30-card deck by actually looking at the composition of the deck and using counting techniques studied in Sections 8-2 and 8-3.

Section 8-6   Chapter Review

IMPORTANT TERMS
AND SYMBOLS

8-1 Introduction.   Theoretical approach, empirical approach

8-2 The Fundamental Principle of Counting.   Tree diagrams, fundamental principle of counting

8-3 Permutations, Combinations, and Set Partitioning.   Permutation of n objects, permutation of n objects taken r at a time, combination of n objects taken r at a time, partition of n elements into k subsets, distinguishable permutations

$$P_{n,r} = \frac{n!}{(n-r)!} \qquad C_{n,r} = \frac{n!}{r!(n-r)!}$$

$$\binom{n}{r} = C_{n,r} \qquad \binom{n}{r_1, r_2, \ldots, r_k} = \frac{n!}{r_1!r_2!\cdots\cdot r_k!}$$

8-4 Experiments, Sample Spaces, and Probability Functions.   Experiments, sample space, simple outcome, sample point, compound outcome, event, simple event, compound event, probability function, probability of an event, equally likely simple events, probability of an event E assuming equally likely simple events

$$P(E) = \frac{n(E)}{n(S)}$$

8-5 Empirical Probability.   Expected frequency, relative frequency, empirical probability, empirical probability approximation

$$P(E) \approx \frac{\text{Frequency of occurrence of } E}{\text{Total number of trials}} = \frac{f(E)}{n}$$

Exercise 8-6   Chapter Review

▦ *A calculator with n! will be useful in some of the problems.*

Work through all the problems in this chapter review and check answers in the back of the book. (Answers to all problems are there, and following each answer is a number in italics indicating the section in which that type of problem is discussed.) Where weaknesses show up, review appropriate sections in the text. When you are satisfied that you know the material, take the practice test following this review.

A   1.   A spinner can land on yellow or blue and a die can turn up on any of the integers from 1 to 6. How many combined outcomes are there? Answer the question using
    (A)   A tree diagram
    (B)   The fundamental principle of counting

2. $C_{7,3} = ?$ 3. $P_{7,3} = ?$

4.   How many three-card hands can be dealt from a deck of thirteen different cards?

5. How many signals can be made from eight different flags, if one signal is made by hanging two flags on a vertical pole?

6. In the game of Scrabble, letters are generally printed on flat wooden squares. If a player draws seven different letters, how many different arrangements of the seven letters are possible if all seven letters are used?

7. In a single roll of a die, what is the probability of getting a prime number? Of not getting a composite number? (Remember, 1 is neither prime nor composite.)

8. If 750 students out of 1,000 randomly chosen students are helped by "flu shots," what is the approximate empirical probability that you will be helped by these shots?

B   9. A student plans to take history, English, and mathematics. History classes are given at 9, 10, and 2 o'clock; English classes at 10, 11, and 2; and math classes at 9, 2, and 3. How many schedules are possible? (Use a tree diagram.)

10. In how many ways can a coin and two dice turn up? (Use the fundamental principle of counting.)

11. In preparing for a party, a person wants to have five records stacked in a record player. Out of ten favorites, how many choices are possible?
    (A) Assuming order does not matter
    (B) Assuming order does matter

12. Given $S = \{e_1, e_2, e_3\}$ with $P(e_1) = -.3$, $P(e_2) = .7$, and $P(e_3) = .6$. Is P a probability function? Explain.

13. A single die is rolled twice. If the sample space is chosen as the set of all ordered pairs of integers taken from $\{1, 2, 3, 4, 5, 6\}$, what is the event E associated with having the combined number of dots add up to 4? What is the probability of this event?

14. George and Pat belong to a ski club with twelve members. If two people are chosen at random from the club to serve as president and vice president, what is the probability that Pat will be chosen as president and George as vice president?

15. If in Problem 14 two people are selected by lottery to cook dinner for a weekend trip, what is the probability that Pat and George will be chosen?

16. Flip two thumbtacks fifty times and determine approximate empirical probabilities for the events:

    E_1: Two points up

    E_2: One point up

    E_3: No points up

17. Flip five thumbtacks thirty times and determine empirical probabilities for the events:

E_1: Five points up E_4: Two points up

E_2: Four points up E_5: One point up

E_3: Three points up E_6: No points up

C 18. A club has fourteen men and twelve women. A committee is to be selected containing two men and two women. In how many ways can this be done?

19. What is the probability of being dealt a 5-card hand of 2 clubs and 3 hearts from a standard 52-card deck?

20. $\begin{pmatrix} 8 \\ 3,\,5 \end{pmatrix} = ?$

21. Show that: $C_{n,r} = \begin{pmatrix} n \\ n,\, n - r \end{pmatrix}$

22. In a four-person card game, each person is dealt 13 cards from a standard 52-card deck. How many deals are possible?

23. Eight books are placed on a shelf. Included are three copies of one title, two copies of a second title, two copies of a third title, and one copy of a fourth title. How many distinguishable arrangements are possible?

Practice Test   Chapter 8

Take this practice test as if it were a graded test. Allow yourself up to 50 minutes. Work the problems without looking back in the chapter. Correct your work using the answers (keyed to the appropriate sections) in the back of the book.

1. Consider two boxes where box 1 contains a white (W) and a green (G) ball, and box 2 contains a red (R), a yellow (Y), a white (W), and a green (G) ball. Suppose that the balls are identical except for color and that one ball is drawn from box 1; then one from box 2. Use a tree diagram to illustrate all possible outcomes of this experiment. What is the probability of drawing:
(A) Two green balls?   (B) At least one green ball?

Evaluate Problems 2–4.

2. $\begin{pmatrix} 10 \\ 7,\,3 \end{pmatrix}$ 3. $P_{12,2}$ 4. $C_{12,2}$

5. How many three-letter code words can be formed using the letters in the word "numbers"?
    (A) If no letter is repeated   (B) If letters can be repeated

6. How many three-digit numbers greater than 400 can be formed from the digits 1, 2, 3, 4, 5, 6, and 7?
    (A) If digits are not repeated
    (B) If adjacent digits cannot be alike

7. Jan and Mike are members of a five-person steering committee. If two names are drawn in succession without replacement, the first for chairperson and the second for recorder, what is the probability that Jan will be the chairperson and Mike the recorder?

8. In Problem 7 what is the probability that Jan and Mike will be on a two-person subcommittee if two names are drawn at random from the five-member steering committee?

9. An alumni fund-raising committee of three has nine potentially large contributors to contact personally. If one member can contact two contributors, another member three, and the third member four, in how many ways can the list be divided to accomplish this task?

10. How many distinguishable numbers can be formed using all the digits in the number 1010101?

For Problems 11 and 12 consider a box with one red, three green, and six yellow identically sized balls.

11. Suppose two balls are drawn from the box without replacement. Find the probability of obtaining two yellow balls. Use combination forms and compute the answer to two significant digits.

12. Suppose a ball is drawn at random from the box 400 times, with replacement each time, and the following results are recorded: red, 52; green, 116; yellow, 232. (Compute answers to two significant digits.)
    (A) Find the approximate empirical probabilities for drawing each color.
    (B) Find the theoretical probabilities for drawing each color.

Appendix A ▪ Significant Digits

Most calculations involving problems of the real world deal with figures that are only approximate. It would therefore seem reasonable to assume that a final answer could not be any more accurate than the least accurate figure used in the calculation. This is an important point, since calculators tend to give the impression that greater accuracy is achieved than is warranted.

Suppose we wish to compute the length of the diagonal of a rectangular field from measurements of its sides of 237.8 meters and 61.3 meters. Using the Pythagorean theorem and a calculator, we find

$$d = \sqrt{237.8^2 + 61.3^2}$$
$$= 245.573\ 878 \ldots$$

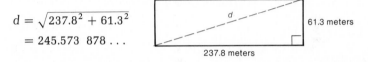

61.3 meters

237.8 meters

The calculator answer suggests an accuracy that is not justified. What accuracy is justified? To answer this question, we introduce the idea of **significant digits**.

The measurement 61.3 meters indicates that the measurement was made to the nearest tenth of a meter; that is, the actual width is between 61.25 and 61.35 meters. The number 61.3 has three significant digits. If we had written, instead, 61.30 meters as the width, then the actual width would be between 61.295 and 61.305 meters, and our measurement, 61.30 meters, would have four significant digits.

The number of significant digits in a number is found by counting the digits from left to right, starting with the first nonzero digit and ending with the last digit present.

The significant digits in the following numbers are underlined:

719.37   82,395   5.600   0.000 830   0.000 08

The definition takes care of all cases except one. Consider, for example, the number 7,800. It is not clear whether the number has been rounded to the hundreds place, the tens place, or the units place. This ambiguity can be resolved by writing this type of number in scientific notation. Thus,

7.8×10^3 has two significant digits

7.80×10^3 has three significant digits

7.800×10^3 has four significant digits

All three are equal to 7,800 when written without powers of 10.

In calculations involving multiplication, division, powers, and roots, we adopt the following convention:

> We will round off the answer to match the number of significant digits in the number with the least number of significant digits used in the calculation.

Thus, in computing the length of the diagonal of the field, we would write the answer to three significant digits because the width, the least accurate of the two numbers involved, has three significant digits:

$$d = 246 \text{ meters} \quad \text{Three significant digits}$$

One final note: In rounding a number that is exactly halfway between a larger and a smaller number, we will use the convention of making the final result even.

EXAMPLE 1   Round each number to three significant digits.

(A) 43.0690   (B) 48.05   (C) 48.15   (D) $8.017\ 632 \times 10^{-3}$

Solution   (A) 43.1

(B) 48.0⎱ Use the convention of making the digit before the
(C) 48.2⎰ 5 even if it is odd, or leaving it alone if it is even.

(D) 8.02×10^{-3}

PROBLEM 1   Round each number to three significant digits:

(A) 3.1495   (B) 0.004 135   (C) 32,450
(D) $4.314\ 764\ 09 \times 10^{12}$

Answers to matched problem   **1.** (A) 3.15   (B) 0.004 14   (C) 32,400   (D) 4.31×10^{12}

◼ Tables

TABLE I   Values of e^x and e^{-x} (0.00 to 3.00)

x	e^x	e^{-x}	x	e^x	e^{-x}	x	e^x	e^{-x}
0.00	1.0000	1.000 00	0.50	1.6487	0.606 53	1.00	2.7183	0.367 88
0.01	1.0101	0.990 05	0.51	1.6653	0.600 50	1.01	2.7456	0.364 22
0.02	1.0202	0.980 20	0.52	1.6820	0.594 52	1.02	2.7732	0.360 59
0.03	1.0305	0.970 45	0.53	1.6989	0.588 60	1.03	2.8011	0.357 01
0.04	1.0408	0.960 79	0.54	1.7160	0.582 75	1.04	2.8292	0.353 45
0.05	1.0513	0.951 23	0.55	1.7333	0.576 95	1.05	2.8577	0.349 94
0.06	1.0618	0.941 76	0.56	1.7507	0.571 21	1.06	2.8864	0.346 46
0.07	1.0725	0.932 39	0.57	1.7683	0.565 53	1.07	2.9154	0.343 01
0.08	1.0833	0.923 12	0.58	1.7860	0.559 90	1.08	2.9447	0.339 60
0.09	1.0942	0.913 93	0.59	1.8040	0.554 33	1.09	2.9743	0.336 22
0.10	1.1052	0.904 84	0.60	1.8221	0.548 81	1.10	3.0042	0.332 87
0.11	1.1163	0.895 83	0.61	1.8404	0.543 35	1.11	3.0344	0.329 56
0.12	1.1275	0.886 92	0.62	1.8589	0.537 94	1.12	3.0649	0.326 28
0.13	1.1388	0.878 10	0.63	1.8776	0.532 59	1.13	3.0957	0.323 03
0.14	1.1503	0.869 36	0.64	1.8965	0.527 29	1.14	3.1268	0.319 82
0.15	1.1618	0.860 71	0.65	1.9155	0.522 05	1.15	3.1582	0.316 64
0.16	1.1735	0.852 14	0.66	1.9348	0.516 85	1.16	3.1899	0.313 49
0.17	1.1853	0.843 66	0.67	1.9542	0.511 71	1.17	3.2220	0.310 37
0.18	1.1972	0.835 27	0.68	1.9739	0.506 62	1.18	3.2544	0.307 28
0.19	1.2092	0.826 96	0.69	1.9937	0.501 58	1.19	3.2871	0.304 22
0.20	1.2214	0.818 73	0.70	2.0138	0.496 59	1.20	3.3201	0.301 19
0.21	1.2337	0.810 58	0.71	2.0340	0.491 64	1.21	3.3535	0.298 20
0.22	1.2461	0.802 52	0.72	2.0544	0.486 75	1.22	3.3872	0.295 23
0.23	1.2586	0.794 53	0.73	2.0751	0.481 91	1.23	3.4212	0.292 29
0.24	1.2712	0.786 63	0.74	2.0959	0.477 11	1.24	3.4556	0.289 38
0.25	1.2840	0.778 80	0.75	2.1170	0.472 37	1.25	3.4903	0.286 50
0.26	1.2969	0.771 05	0.76	2.1383	0.467 67	1.26	3.5254	0.283 65
0.27	1.3100	0.763 38	0.77	2.1598	0.463 01	1.27	3.5609	0.280 83
0.28	1.3231	0.755 78	0.78	2.1815	0.458 41	1.28	3.5966	0.278 04
0.29	1.3364	0.748 26	0.79	2.2034	0.453 84	1.29	3.6328	0.275 27
0.30	1.3499	0.740 82	0.80	2.2255	0.449 33	1.30	3.6693	0.272 53
0.31	1.3634	0.733 45	0.81	2.2479	0.444 86	1.31	3.7062	0.269 82
0.32	1.3771	0.726 15	0.82	2.2705	0.440 43	1.32	3.7434	0.267 14
0.33	1.3910	0.718 92	0.83	2.2933	0.436 05	1.33	3.7810	0.264 48
0.34	1.4049	0.711 77	0.84	2.3164	0.431 71	1.34	3.8190	0.261 85
0.35	1.4191	0.704 69	0.85	2.3396	0.427 41	1.35	3.8574	0.259 24
0.36	1.4333	0.697 68	0.86	2.3632	0.423 16	1.36	3.8962	0.256 66
0.37	1.4477	0.690 73	0.87	2.3869	0.418 95	1.37	3.9354	0.254 11
0.38	1.4623	0.683 86	0.88	2.4109	0.414 78	1.38	3.9749	0.251 58
0.39	1.4770	0.677 06	0.89	2.4351	0.410 66	1.39	4.0149	0.249 08
0.40	1.4918	0.670 32	0.90	2.4596	0.406 57	1.40	4.0552	0.246 60
0.41	1.5068	0.663 65	0.91	2.4843	0.402 52	1.41	4.0960	0.244 14
0.42	1.5220	0.657 05	0.92	2.5093	0.398 52	1.42	4.1371	0.241 71
0.43	1.5373	0.650 51	0.93	2.5345	0.394 55	1.43	4.1787	0.239 31
0.44	1.5527	0.644 04	0.94	2.5600	0.390 63	1.44	4.2207	0.236 93
0.45	1.5683	0.637 63	0.95	2.5857	0.386 74	1.45	4.2631	0.234 57
0.46	1.5841	0.631 28	0.96	2.6117	0.382 89	1.46	4.3060	0.232 24
0.47	1.6000	0.625 00	0.97	2.6379	0.379 08	1.47	4.3492	0.229 93
0.48	1.6161	0.618 78	0.98	2.6645	0.375 31	1.48	4.3939	0.227 64
0.49	1.6323	0.612 63	0.99	2.6912	0.371 58	1.49	4.4371	0.225 37
0.50	1.6487	0.606 53	1.00	2.7183	0.367 88	1.50	4.4817	0.223 13

x	e^x	e^{-x}	x	e^x	e^{-x}	x	e^x	e^{-x}
1.50	4.4817	0.223 13	2.00	7.3891	0.135 34	2.50	12.182	0.082 085
1.51	4.5267	0.220 91	2.01	7.4633	0.133 99	2.51	12.305	0.081 268
1.52	4.5722	0.218 71	2.02	7.5383	0.132 66	2.52	12.429	0.080 460
1.53	4.6182	0.216 54	2.03	7.6141	0.131 34	2.53	12.554	0.079 659
1.54	4.6646	0.214 38	2.04	7.6906	0.130 03	2.54	12.680	0.078 866
1.55	4.7115	0.212 25	2.05	7.7679	0.128 73	2.55	12.807	0.078 082
1.56	4.7588	0.210 14	2.06	7.8460	0.127 45	2.56	12.936	0.077 305
1.57	4.8066	0.208 05	2.07	7.9248	0.126 19	2.57	13.066	0.076 536
1.58	4.8550	0.205 98	2.08	8.0045	0.124 93	2.58	13.197	0.075 774
1.59	4.9037	0.203 93	2.09	8.0849	0.123 69	2.59	13.330	0.075 020
1.60	4.9530	0.201 90	2.10	8.1662	0.122 46	2.60	13.464	0.074 274
1.61	5.0028	0.199 89	2.11	8.2482	0.121 24	2.61	13.599	0.073 535
1.62	5.0531	0.197 90	2.12	8.3311	0.120 03	2.62	13.736	0.072 803
1.63	5.1039	0.195 93	2.13	8.4149	0.118 84	2.63	13.874	0.072 078
1.64	5.1552	0.193 98	2.14	8.4994	0.117 65	2.64	14.013	0.071 361
1.65	5.2070	0.192 05	2.15	8.5849	0.116 48	2.65	14.154	0.070 651
1.66	5.2593	0.190 14	2.16	8.6711	0.115 33	2.66	14.296	0.069 948
1.67	5.3122	0.188 25	2.17	8.7583	0.114 18	2.67	14.440	0.069 252
1.68	5.3656	0.186 37	2.18	8.8463	0.113 04	2.68	14.585	0.068 563
1.69	5.4195	0.184 52	2.19	8.9352	0.111 92	2.69	14.732	0.067 881
1.70	5.4739	0.182 68	2.20	9.0250	0.110 80	2.70	14.880	0.067 206
1.71	5.5290	0.180 87	2.21	9.1157	0.109 70	2.71	15.029	0.066 537
1.72	5.5845	0.179 07	2.22	9.2073	0.108 61	2.72	15.180	0.065 875
1.73	5.6407	0.177 28	2.23	9.2999	0.107 53	2.73	15.333	0.065 219
1.74	5.6973	0.175 52	2.24	9.3933	0.106 46	2.74	15.487	0.064 570
1.75	5.7546	0.173 77	2.25	9.4877	0.105 40	2.75	15.643	0.063 928
1.76	5.8124	0.172 04	2.26	9.5831	0.104 35	2.76	15.800	0.063 292
1.77	5.8709	0.170 33	2.27	9.6794	0.103 31	2.77	15.959	0.062 662
1.78	5.9299	0.168 64	2.28	9.7767	0.102 28	2.78	16.119	0.062 039
1.79	5.9895	0.166 96	2.29	9.8749	0.101 27	2.79	16.281	0.061 421
1.80	6.0496	0.165 30	2.30	9.9742	0.100 26	2.80	16.445	0.060 810
1.81	6.1104	0.163 65	2.31	10.074	0.099 261	2.81	16.610	0.060 205
1.82	6.1719	0.162 03	2.32	10.176	0.098 274	2.82	16.777	0.059 606
1.83	6.2339	0.160 41	2.33	10.278	0.097 296	2.83	16.945	0.059 013
1.84	6.2965	0.158 82	2.34	10.381	0.096 328	2.84	17.116	0.058 426
1.85	6.3598	0.157 24	2.35	10.486	0.095 369	2.85	17.288	0.057 844
1.86	6.4237	0.155 67	2.36	10.591	0.094 420	2.86	17.462	0.057 269
1.87	6.4883	0.154 12	2.37	10.697	0.093 481	2.87	17.637	0.056 699
1.88	6.5535	0.152 59	2.38	10.805	0.092 551	2.88	17.814	0.056 135
1.89	6.6194	0.151 07	2.39	10.913	0.091 630	2.89	17.993	0.055 576
1.90	6.6859	0.149 57	2.40	11.023	0.090 718	2.90	18.174	0.055 023
1.91	6.7531	0.148 08	2.41	11.134	0.089 815	2.91	18.357	0.054 476
1.92	6.8210	0.146 61	2.42	11.246	0.088 922	2.92	18.541	0.053 934
1.93	6.8895	0.145 15	2.43	11.359	0.088 037	2.93	18.728	0.053 397
1.94	6.9588	0.143 70	2.44	11.473	0.087 161	2.94	18.916	0.052 866
1.95	7.0287	0.142 27	2.45	11.588	0.086 294	2.95	19.106	0.052 340
1.96	7.0993	0.140 86	2.46	11.705	0.085 435	2.96	19.298	0.051 819
1.97	7.1707	0.139 46	2.47	11.822	0.084 585	2.97	19.492	0.051 303
1.98	7.2427	0.138 07	2.48	11.941	0.083 743	2.98	19.688	0.050 793
1.99	7.3155	0.136 70	2.49	12.061	0.082 910	2.99	19.886	0.050 287
2.00	7.3891	0.135 34	2.50	12.182	0.082 085	3.00	20.086	0.049 787

TABLE II Common Logarithms

x	0	1	2	3	4	5	6	7	8	9
1.0	0.0000	0.004321	0.008600	0.01284	0.01703	0.02119	0.02531	0.02938	0.03342	0.03743
1.1	0.04139	0.04532	0.04922	0.05308	0.05690	0.06070	0.06446	0.06819	0.07188	0.07555
1.2	0.07918	0.08279	0.08636	0.08991	0.09342	0.09691	0.1004	0.1038	0.1072	0.1106
1.3	0.1139	0.1173	0.1206	0.1239	0.1271	0.1303	0.1335	0.1367	0.1399	0.1430
1.4	0.1461	0.1492	0.1523	0.1553	0.1584	0.1614	0.1644	0.1673	0.1703	0.1732
1.5	0.1761	0.1790	0.1818	0.1847	0.1875	0.1903	0.1931	0.1959	0.1987	0.2014
1.6	0.2041	0.2068	0.2095	0.2122	0.2148	0.2175	0.2201	0.2227	0.2253	0.2279
1.7	0.2304	0.2330	0.2355	0.2380	0.2405	0.2430	0.2455	0.2480	0.2504	0.2529
1.8	0.2553	0.2577	0.2601	0 2625	0.2648	0.2673	0.2695	0.2718	0.2742	0.2765
1.9	0.2788	0.2810	0.2833	0.2856	0.2878	0.2900	0.2923	0.2945	0.2967	0.2989
2.0	0.3010	0.3032	0.3054	0.3075	0.3096	0.3118	0.3139	0.3160	0.3181	0.3201
2.1	0.3222	0.3243	0.3263	0.3284	0.3304	0.3324	0.3345	0.3365	0.3385	0.3404
2.2	0.3424	0.3444	0.3464	0.3483	0.3502	0.3522	0.3541	0.3560	0.3579	0.3598
2.3	0.3617	0.3636	0.3655	0.3674	0.3692	0.3711	0.3729	0.3747	0.3766	0.3784
2.4	0.3802	0.3820	0.3838	0.3856	0.3874	0.3892	0.3909	0.3927	0.3945	0.3962
2.5	0.3979	0.3997	0.4014	0.4031	0.4048	0.4065	0.4082	0.4099	0.4116	0.4133
2.6	0.4150	0.4166	0.4183	0.4200	0.4216	0.4232	0.4249	0.4265	0.4281	0.4298
2.7	0.4314	0.4330	0.4346	0.4362	0.4378	0.4393	0.4409	0.4425	0.4440	0.4456
2.8	0.4472	0.4487	0.4502	0.4518	0.4533	0.4548	0.4564	0.4579	0.4594	0.4609
2.9	0.4624	0.4639	0.4654	0.4669	0.4683	0.4698	0.4713	0.4728	0.4742	0.4757
3.0	0.4771	0.4786	0.4800	0.4814	0.4829	0.4843	0.4857	0.4871	0.4886	0.4900
3.1	0.4914	0.4928	0.4942	0.4955	0.4969	0.4983	0.4997	0.5011	0.5024	0.5038
3.2	0.5051	0.5065	0.5079	0.5092	0.5105	0.5119	0.5132	0.5145	0.5159	0.5172
3.3	0.5185	0.5198	0.5211	0.5224	0.5237	0.5250	0.5263	0.5276	0.5289	0.5302
3.4	0.5315	0.5328	0.5340	0.5353	0.5366	0.5378	0.5391	0.5403	0.5416	0.5428
3.5	0.5441	0.5453	0.5465	0.5478	0.5490	0.5502	0.5514	0.5527	0.5539	0.5551
3.6	0.5563	0.5575	0.5587	0.5599	0.5611	0.5623	0.5635	0.5647	0.5658	0.5670
3.7	0.5682	0.5694	0.5705	0.5717	0.5729	0.5740	0.5752	0.5763	0.5775	0.5786
3.8	0.5798	0.5809	0.5821	0.5832	0.5843	0.5855	0.5866	0.5877	0.5888	0.5899
3.9	0.5911	0.5922	0.5933	0.5944	0.5955	0.5966	0.5977	0.5988	0.5999	0.6010
4.0	0.6021	0.6031	0.6042	0.6053	0.6064	0.6075	0.6085	0.6096	0.6107	0.6117
4.1	0.6128	0.6138	0.6149	0.6160	0.6170	0.6180	0.6191	0.6201	0.6212	0.6222
4.2	0.6232	0.6243	0.6253	0.6263	0.6274	0.6284	0.6294	0.6304	0.6314	0.6325
4.3	0.6335	0.6345	0.6355	0.6365	0.6375	0.6385	0.6395	0.6405	0.6415	0.6425
4.4	0.6435	0.6444	0.6454	0.6464	0.6474	0.6484	0.6493	0.6503	0.6513	0.6522
4.5	0.6532	0.6542	0.6551	0.6561	0.6571	0.6580	0.6590	0.6599	0.6609	0.6618
4.6	0.6628	0.6637	0.6646	0.6656	0.6665	0.6675	0.6684	0.6693	0.6702	0.6712
4.7	0.6721	0.6730	0.6739	0.6749	0.6758	0.6767	0.6776	0.6785	0.6794	0.6803
4.8	0.6812	0.6821	0.6830	0.6839	0.6848	0.6857	0.6866	0.6875	0.6884	0.6893
4.9	0.6902	0.6911	0.6920	0.6928	0.6937	0.6946	0.6955	0.6964	0.6972	0.6981
5.0	0.6990	0.6998	0.7007	0.7016	0.7024	0.7033	0.7042	0.7050	0.7059	0.7067
5.1	0.7076	0.7084	0.7093	0.7101	0.7110	0.7118	0.7126	0.7135	0.7143	0.7152
5.2	0.7160	0.7168	0.7177	0.7185	0.7193	0.7202	0.7210	0.7218	0.7226	0.7235
5.3	0.7243	0.7251	0.7259	0.7267	0.7275	0.7284	0.7292	0.7300	0.7308	0.7316
5.4	0.7324	0.7332	0.7340	0.7348	0.7356	0.7364	0.7372	0.7380	0.7388	0.7396

x	0	1	2	3	4	5	6	7	8	9
5.5	0.7404	0.7412	0.7419	0.7427	0.7435	0.7443	0.7451	0.7459	0.7466	0.7474
5.6	0.7482	0.7490	0.7497	0.7505	0.7513	0.7520	0.7528	0.7536	0.7543	0.7551
5.7	0.7559	0.7566	0.7574	0.7582	0.7589	0.7597	0.7604	0.7612	0.7619	0.7627
5.8	0.7634	0.7642	0.7649	0.7657	0.7664	0.7672	0.7679	0.7686	0.7694	0.7701
5.9	0.7709	0.7716	0.7723	0.7731	0.7738	0.7745	0.7752	0.7760	0.7767	0.7774
6.0	0.7782	0.7789	0.7796	0.7803	0.7810	0.7818	0.7825	0.7832	0.7839	0.7846
6.1	0.7853	0.7860	0.7868	0.7875	0.7882	0.7889	0.7896	0.7903	0.7910	0.7917
6.2	0.7924	0.7931	0.7938	0.7945	0.7952	0.7959	0.7966	0.7973	0.7980	0.7987
6.3	0.7993	0.8000	0.8007	0.8014	0.8021	0.8028	0.8035	0.8041	0.8048	0.8055
6.4	0.8062	0.8069	0.8075	0.8082	0.8089	0.8096	0.8102	0.8109	0.8116	0.8122
6.5	0.8129	0.8136	0.8142	0.8149	0.8156	0.8162	0.8169	0.8176	0.8182	0.8189
6.6	0.8195	0.8202	0.8209	0.8215	0.8222	0.8228	0.8235	0.8241	0.8248	0.8254
6.7	0.8261	0.8267	0.8274	0.8280	0.8287	0.8293	0.8299	0.8306	0.8312	0.8319
6.8	0.8325	0.8331	0.8338	0.8344	0.8351	0.8357	0.8363	0.8370	0.8376	0.8382
6.9	0.8388	0.8395	0.8401	0.8407	0.8414	0.8420	0.8426	0.8432	0.8439	0.8445
7.0	0.8451	0.8457	0.8463	0.8470	0.8476	0.8482	0.8488	0.8494	0.8500	0.8506
7.1	0.8513	0.8519	0.8525	0.8531	0.8537	0.8543	0.8549	0.8555	0.8561	0.8567
7.2	0.8573	0.8579	0.8585	0.8591	0.8597	0.8603	0.8609	0.8615	0.8621	0.8627
7.3	0.8633	0.8639	0.8645	0.8651	0.8657	0.8663	0.8669	0.8675	0.8681	0.8686
7.4	0.8692	0.8698	0.8704	0.8710	0.8716	0.8722	0.8727	0.8733	0.8739	0.8745
7.5	0.8751	0.8756	0.8762	0.8768	0.8774	0.8779	0.8785	0.8791	0.8797	0.8802
7.6	0.8808	0.8814	0.8820	0.8825	0.8831	0.8837	0.8842	0.8848	0.8854	0.8859
7.7	0.8865	0.8871	0.8876	0.8882	0.8887	0.8893	0.8899	0.8904	0.8910	0.8915
7.8	0.8921	0.8927	0.8932	0.8938	0.8943	0.8949	0.8954	0.8960	0.8965	0.8971
7.9	0.8976	0.8982	0.8987	0.8993	0.8998	0.9004	0.9009	0.9015	0.9020	0.9025
8.0	0.9031	0.9036	0.9042	0.9047	0.9053	0.9058	0.9063	0.9069	0.9074	0.9079
8.1	0.9085	0.9090	0.9096	0.9101	0.9106	0.9112	0.9117	0.9122	0.9128	0.9133
8.2	0.9138	0.9143	0.9149	0.9154	0.9159	0.9165	0.9170	0.9175	0.9180	0.9186
8.3	0.9191	0.9196	0.9201	0.9206	0.9212	0.9217	0.9222	0.9227	0.9232	0.9238
8.4	0.9243	0.9248	0.9253	0.9258	0.9263	0.9269	0.9274	0.9279	0.9284	0.9289
8.5	0.9294	0.9299	0.9304	0.9309	0.9315	0.9320	0.9325	0.9330	0.9335	0.9340
8.6	0.9345	0.9350	0.9355	0.9360	0.9365	0.9370	0.9375	0.9380	0.9385	0.9390
8.7	0.9395	0.9400	0.9405	0.9410	0.9415	0.9420	0.9425	0.9430	0.9435	0.9440
8.8	0.9445	0.9450	0.9455	0.9460	0.9465	0.9469	0.9474	0.9479	0.9484	0.9489
8.9	0 9494	0.9499	0.9504	0.9509	0.9513	0.9518	0.9523	0.9528	0.9533	0.9538
9.0	0.9542	0.9547	0.9552	0.9557	0.9562	0.9566	0.9571	0.9576	0.9581	0.9586
9.1	0.9590	0.9595	0.9600	0.9605	0.9609	0.9614	0.9619	0.9624	0.9628	0.9633
9.2	0.9638	0.9643	0.9647	0.9652	0.9657	0.9661	0.9666	0.9671	0.9675	0.9680
9.3	0.9685	0.9689	0.9694	0.9699	0.9703	0.9708	0.9713	0.9717	0.9722	0.9727
9.4	0.9731	0.9736	0.9741	0.9745	0.9750	0.9754	0.9759	0.9763	0.9768	0.9773
9.5	0.9777	0.9782	0.9786	0.9791	0.9795	0.9800	0.9805	0.9809	0.9814	0.9818
9.6	0.9823	0.9827	0.9832	0.9836	0.9841	0.9845	0.9850	0.9854	0.9859	0.9863
9.7	0.9868	0.9872	0.9877	0.9881	0.9886	0.9890	0.9894	0.9899	0.9903	0.9908
9.8	0.9912	0.9917	0.9921	0.9926	0.9930	0.9934	0.9939	0.9943	0.9948	0.9952
9.9	0.9956	0.9961	0.9965	0.9969	0.9974	0.9978	0.9983	0.9987	0.9991	0.9996

TABLE III Natural Logarithms (ln x = log$_e$ x)

	ln 10 = 2.3026	5 ln 10 = 11.5130	9 ln 10 = 20.7233
	2 ln 10 = 4.6052	6 ln 10 = 13.8155	10 ln 10 = 23.0259
	3 ln 10 = 6.9078	7 ln 10 = 16.1181	
	4 ln 10 = 9.2103	8 ln 10 = 18.4207	

x	.00	.01	.02	.03	.04	.05	.06	.07	.08	.09
1.0	0.0000	0.0100	0.0198	0.0296	0.0392	0.0488	0.0583	0.0677	0.0770	0.0862
1.1	0.0953	0.1044	0.1133	0.1222	0.1310	0.1398	0.1484	0.1570	0.1655	0.1740
1.2	0.1823	0.1906	0.1989	0.2070	0.2151	0.2231	0.2311	0.2390	0.2469	0.2546
1.3	0.2624	0.2700	0.2776	0.2852	0.2927	0.3001	0.3075	0.3148	0.3221	0.3293
1.4	0.3365	0.3436	0.3507	0.3577	0.3646	0.3716	0.3784	0.3853	0.3920	0.3988
1.5	0.4055	0.4121	0.4187	0.4253	0.4318	0.4383	0.4447	0.4511	0.4574	0.4637
1.6	0.4700	0.4762	0.4824	0.4886	0.4947	0.5008	0.5068	0.5128	0.5188	0.5247
1.7	0.5306	0.5365	0.5423	0.5481	0.5539	0.5596	0.5653	0.5710	0.5766	0.5822
1.8	0.5878	0.5933	0.5988	0.6043	0.6098	0.6152	0.6206	0.6259	0.6313	0.6366
1.9	0.6419	0.6471	0.6523	0.6575	0.6627	0.6678	0.6729	0.6780	0.6831	0.6881
2.0	0.6931	0.6981	0.7031	0.7080	0.7129	0.7178	0.7227	0.7275	0.7324	0.7372
2.1	0.7419	0.7467	0.7514	0.7561	0.7608	0.7655	0.7701	0.7747	0.7793	0.7839
2.2	0.7885	0.7930	0.7975	0.8020	0.8065	0.8109	0.8154	0.8198	0.8242	0.8286
2.3	0.8329	0.8372	0.8416	0.8459	0.8502	0.8544	0.8587	0.8629	0.8671	0.8713
2.4	0.8755	0.8796	0.8838	0.8879	0.8920	0.8961	0.9002	0.9042	0.9083	0.9123
2.5	0.9163	0.9203	0.9243	0.9282	0.9322	0.9361	0.9400	0.9439	0.9478	0.9517
2.6	0.9555	0.9594	0.9632	0.9670	0.9708	0.9746	0.9783	0.9821	0.9858	0.9895
2.7	0.9933	0.9969	1.0006	1.0043	1.0080	1.0116	1.0152	1.0188	1.0225	1.0260
2.8	1.0296	1.0332	1.0367	1.0403	1.0438	1.0473	1.0508	1.0543	1.0578	1.0613
2.9	1.0647	1.0682	1.0716	1.0750	1.0784	1.0818	1.0852	1.0886	1.0919	1.0953
3.0	1.0986	1.1019	1.1053	1.1086	1.1119	1.1151	1.1184	1.1217	1.1249	1.1282
3.1	1.1314	1.1346	1.1378	1.1410	1.1442	1.1474	1.1506	1.1537	1.1569	1.1600
3.2	1.1632	1.1663	1.1694	1.1725	1.1756	1.1787	1.1817	1.1848	1.1878	1.1909
3.3	1.1939	1.1969	1.2000	1.2030	1.2060	1.2090	1.2119	1.2149	1.2179	1.2208
3.4	1.2238	1.2267	1.2296	1.2326	1.2355	1.2384	1.2413	1.2442	1.2470	1.2499
3.5	1.2528	1.2556	1.2585	1.2613	1.2641	1.2669	1.2698	1.2726	1.2754	1.2782
3.6	1.2809	1.2837	1.2865	1.2892	1.2920	1.2947	1.2975	1.3002	1.3029	1.3056
3.7	1.3083	1.3110	1.3137	1.3164	1.3191	1.3218	1.3244	1.3271	1.3297	1.3324
3.8	1.3350	1.3376	1.3403	1.3429	1.3455	1.3481	1.3507	1.3533	1.3558	1.3584
3.9	1.3610	1.3635	1.3661	1.3686	1.3712	1.3737	1.3762	1.3788	1.3813	1.3838
4.0	1.3863	1.3888	1.3913	1.3938	1.3962	1.3987	1.4012	1.4036	1.4061	1.4085
4.1	1.4110	1.4134	1.4159	1.4183	1.4207	1.4231	1.4255	1.4279	1.4303	1.4327
4.2	1.4351	1.4375	1.4398	1.4422	1.4446	1.4469	1.4493	1.4516	1.4540	1.4563
4.3	1.4586	1.4609	1.4633	1.4656	1.4679	1.4702	1.4725	1.4748	1.4770	1.4793
4.4	1.4816	1.4839	1.4861	1.4884	1.4907	1.4929	1.4951	1.4974	1.4996	1.5019
4.5	1.5041	1.5063	1.5085	1.5107	1.5129	1.5151	1.5173	1.5195	1.5217	1.5239
4.6	1.5261	1.5282	1.5304	1.5326	1.5347	1.5369	1.5390	1.5412	1.5433	1.5454
4.7	1.5476	1.5497	1.5518	1.5539	1.5560	1.5581	1.5602	1.5623	1.5644	1.5665
4.8	1.5686	1.5707	1.5728	1.5748	1.5769	1.5790	1.5810	1.5831	1.5851	1.5872
4.9	1.5892	1.5913	1.5933	1.5953	1.5974	1.5994	1.6014	1.6034	1.6054	1.6074
5.0	1.6094	1.6114	1.6134	1.6154	1.6174	1.6194	1.6214	1.6233	1.6253	1.6273
5.1	1.6292	1.6312	1.6332	1.6351	1.6371	1.6390	1.6409	1.6429	1.6448	1.6467
5.2	1.6487	1.6506	1.6525	1.6544	1.6563	1.6582	1.6601	1.6620	1.6639	1.6658
5.3	1.6677	1.6696	1.6715	1.6734	1.6752	1.6771	1.6790	1.6808	1.6827	1.6845
5.4	1.6864	1.6882	1.6901	1.6919	1.6938	1.6956	1.6974	1.6993	1.7011	1.7029

x	.00	.01	.02	.03	.04	.05	.06	.07	.08	.09
5.5	1.7047	1.7066	1.7084	1.7102	1.7120	1.7138	1.7156	1.7174	1.7192	1.7210
5.6	1.7228	1.7246	1.7263	1.7281	1.7299	1.7317	1.7334	1.7352	1.7370	1.7387
5.7	1.7405	1.7422	1.7440	1.7457	1.7475	1.7492	1.7509	1.7527	1.7544	1.7561
5.8	1.7579	1.7596	1.7613	1.7630	1.7647	1.7664	1.7681	1.7699	1.7716	1.7733
5.9	1.7750	1.7766	1.7783	1.7800	1.7817	1.7834	1.7851	1.7867	1.7884	1.7901
6.0	1.7918	1.7934	1.7951	1.7967	1.7984	1.8001	1.8017	1.8034	1.8050	1.8066
6.1	1.8083	1.8099	1.8116	1.8132	1.8148	1.8165	1.8181	1.8197	1.8213	1.8229
6.2	1.8245	1.8262	1.8278	1.8294	1.8310	1.8326	1.8342	1.8358	1.8374	1.8390
6.3	1.8405	1.8421	1.8437	1.8453	1.8469	1.8485	1.8500	1.8516	1.8532	1.8547
6.4	1.8563	1.8579	1.8594	1.8610	1.8625	1.8641	1.8656	1.8672	1.8687	1.8703
6.5	1.8718	1.8733	1.8749	1.8764	1.8779	1.8795	1.8810	1.8825	1.8840	1.8856
6.6	1.8871	1.8886	1.8901	1.8916	1.8931	1.8946	1.8961	1.8976	1.8991	1.9006
6.7	1.9021	1.9036	1.9051	1.9066	1.9081	1.9095	1.9110	1.9125	1.9140	1.9155
6.8	1.9169	1.9184	1.9199	1.9213	1.9228	1.9242	1.9257	1.9272	1.9286	1.9301
6.9	1.9315	1.9330	1.9344	1.9359	1.9373	1.9387	1.9402	1.9416	1.9430	1.9445
7.0	1.9459	1.9473	1.9488	1.9502	1.9516	1.9530	1.9544	1.9559	1.9573	1.9587
7.1	1.9601	1.9615	1.9629	1.9643	1.9657	1.9671	1.9685	1.9699	1.9713	1.9727
7.2	1.9741	1.9755	1.9769	1.9782	1.9796	1.9810	1.9824	1.9838	1.9851	1.9865
7.3	1.9879	1.9892	1.9906	1.9920	1.9933	1.9947	1.9961	1.9974	1.9988	2.0001
7.4	2.0015	2.0028	2.0042	2.0055	2.0069	2.0082	2.0096	2.0109	2.0122	2.0136
7.5	2.0149	2.0162	2.0176	2.0189	2.0202	2.0215	2.0229	2.0242	2.0255	2.0268
7.6	2.0281	2.0295	2.0308	2.0321	2.0334	2.0347	2.0360	2.0373	2.0386	2.0399
7.7	2.0412	2.0425	2.0438	2.0451	2.0464	2.0477	2.0490	2.0503	2.0516	2.0528
7.8	2.0541	2.0554	2.0567	2.0580	2.0592	2.0605	2.0618	2.0631	2.0643	2.0656
7.9	2.0669	2.0681	2.0694	2.0707	2.0719	2.0732	2.0744	2.0757	2.0769	2.0782
8.0	2.0794	2.0807	2.0819	2.0832	2.0844	2.0857	2.0869	2.0882	2.0894	2.0906
8.1	2.0919	2.0931	2.0943	2.0956	2.0968	2.0980	2.0992	2.1005	2.1017	2.1029
8.2	2.1041	2.1054	2.1066	2.1078	2.1090	2.1102	2.1114	2.1126	2.1138	2.1150
8.3	2.1163	2.1175	2.1187	2.1199	2.1211	2.1223	2.1235	2.1247	2.1258	2.1270
8.4	2.1282	2.1294	2.1306	2.1318	2.1330	2.1342	2.1353	2.1365	2.1377	2.1389
8.5	2.1401	2.1412	2.1424	2.1436	2.1448	2.1459	2.1471	2.1483	2.1494	2.1506
8.6	2.1518	2.1529	2.1541	2.1552	2.1564	2.1576	2.1587	2.1599	2.1610	2.1622
8.7	2.1633	2.1645	2.1656	2.1668	2.1679	2.1691	2.1702	2.1713	2.1725	2.1736
8.8	2.1748	2.1759	2.1770	2.1782	2.1793	2.1804	2.1815	2.1827	2.1838	2.1849
8.9	2.1861	2.1872	2.1883	2.1894	2.1905	2.1917	2.1928	2.1939	2.1950	2.1961
9.0	2.1972	2.1983	2.1994	2.2006	2.2017	2.2028	2.2039	2.2050	2.2061	2.2072
9.1	2.2083	2.2094	2.2105	2.2116	2.2127	2.2138	2.2148	2.2159	2.2170	2.2181
9.2	2.2192	2.2203	2.2214	2.2225	2.2235	2.2246	2.2257	2.2268	2.2279	2.2289
9.3	2.2300	2.2311	2.2322	2.2332	2.2343	2.2354	2.2364	2.2375	2.2386	2.2396
9.4	2.2407	2.2418	2.2428	2.2439	2.2450	2.2460	2.2471	2.2481	2.2492	2.2502
9.5	2.2513	2.2523	2.2534	2.2544	2.2555	2.2565	2.2576	2.2586	2.2597	2.2607
9.6	2.2618	2.2628	2.2638	2.2649	2.2659	2.2670	2.2680	2.2690	2.2701	2.2711
9.7	2.2721	2.2732	2.2742	2.2752	2.2762	2.2773	2.2783	2.2793	2.2803	2.2814
9.8	2.2824	2.2834	2.2844	2.2854	2.2865	2.2875	2.2885	2.2895	2.2905	2.2915
9.9	2.2925	2.2935	2.2946	2.2956	2.2966	2.2976	2.2986	2.2996	2.3006	2.3016

TABLE IV   Logarithms of Factorial *n*

n	log n!	n	log n!	n	log n!	n	log n!
		50	64.483 07	100	157.970 00	150	262.756 89
1	0.000 00	51	66.190 64	101	159.974 32	151	264.935 87
2	0.301 03	52	67.906 65	102	161.982 93	152	267.117 71
3	0.778 15	53	69.630 92	103	163.995 76	153	269.302 41
4	1.380 21	54	71.363 32	104	166.012 80	154	271.489 93
5	2.079 18	55	73.103 68	105	168.033 99	155	273.680 26
6	2.857 33	56	74.851 87	106	170.059 29	156	275.873 38
7	3.702 43	57	76.607 74	107	172.086 67	157	278.069 28
8	4.605 52	58	78.371 17	108	174.122 10	158	280.267 94
9	5.559 76	59	80.142 02	109	176.159 52	159	282.469 34
10	6.559 76	60	81.920 17	110	178.200 92	160	284.673 46
11	7.601 16	61	83.705 50	111	180.246 24	161	286.880 28
12	8.680 34	62	85.497 90	112	182.295 46	162	289.089 80
13	9.794 28	63	87.297 24	113	184.348 54	163	291.301 98
14	10.940 41	64	89.103 42	114	186.405 44	164	293.516 83
15	12.116 50	65	90.916 33	115	188.466 14	165	295.734 31
16	13.320 62	66	92.735 87	116	190.530 60	166	297.954 42
17	14.551 07	67	94.561 95	117	192.598 78	167	300.177 14
18	15.806 34	68	96.394 46	118	194.670 67	168	302.402 45
19	17.085 09	69	98.233 31	119	196.746 21	169	304.630 33
20	18.386 12	70	100.078 40	120	198.825 39	170	306.860 78
21	19.708 34	71	101.929 66	121	200.908 18	171	309.093 78
22	21.050 77	72	103.787 00	122	202.994 54	172	311.329 31
23	22.412 49	73	105.650 32	123	205.084 44	173	313.567 35
24	23.792 71	74	107.519 55	124	207.177 87	174	315.807 90
25	25.190 65	75	109.394 61	125	209.274 78	175	318.050 94
26	26.605 62	76	111.275 43	126	211.375 15	176	320.296 45
27	28.036 98	77	113.161 92	127	213.478 95	177	322.544 43
28	29.484 14	78	115.054 01	128	215.586 16	178	324.794 85
29	30.946 54	79	116.951 64	129	217.696 75	179	327.047 70
30	32.423 66	80	118.854 73	130	219.810 69	180	329.302 97
31	33.915 02	81	120.763 21	131	221.927 96	181	331.560 65
32	35.420 17	82	122.677 03	132	224.048 54	182	333.820 72
33	36.938 69	83	124.596 10	133	226.172 39	183	336.083 17
34	38.470 16	84	126.520 38	134	228.299 49	184	338.347 99
35	40.014 23	85	128.449 80	135	230.429 83	185	340.615 16
36	41.570 54	86	130.384 30	136	232.563 37	186	342.884 68
37	43.138 74	87	132.323 82	137	234.700 09	187	345.156 52
38	44.718 52	88	134.268 30	138	236.839 97	188	347.430 67
39	46.309 59	89	136.217 69	139	238.982 98	189	349.707 14
40	47.911 65	90	138.171 94	140	241.129 11	190	351.985 89
41	49.524 43	91	140.130 98	141	242.278 33	191	354.266 92
42	51.147 68	92	142.094 76	142	245.430 62	192	356.550 22
43	52.781 15	93	144.063 25	143	247.585 95	193	358.835 78
44	54.424 60	94	146.036 38	144	249.744 32	194	361.123 58
45	56.077 81	95	148.014 10	145	251.905 68	195	363.413 62
46	57.740 57	96	149.996 37	146	254.070 04	196	365.705 87
47	59.412 67	97	151.983 14	147	256.237 35	197	368.000 34
48	61.093 91	98	153.974 37	148	258.407 62	198	370.297 01
49	62.784 10	99	155.970 00	149	260.580 80	199	372.595 86

▪ Answers

Chapter 0 — Exercise 0-1

1. Commutative   **3.** Associative   **5.** Distributive   **7.** Negatives   **9.** Def. subtraction   **11.** Negatives
13. Identity   **15.** Def. division   **17.** Inverse   **19.** Distributive   **21.** Distributive   **23.** Zero
25. Commutative   **27.** Associative   **29.** Distributive   **31.** Distributive   **33.** Negatives   **35.** Yes
37. (A) T; (B) F; (C) T   **39.** $\frac{3}{5}$ and -1.43 are two examples of infinitely many.
41. (A) Z, Q, R; (B) Q, R; (C) R; (D) Q, R
43. (B) is false, since, for example, $5 - 3 \neq 3 - 5$; (D) is false, since, for example, $9 \div 3 \neq 3 \div 9$.   **45.** $\frac{1}{11}$
47.
$$
\begin{array}{l}
23 \\
\underline{12} \\
46 \\
\underline{230} \\
276
\end{array}
\qquad
\begin{aligned}
23 \cdot 12 &= 23(2 + 10) \\
&= 23 \cdot 2 + 23 \cdot 10 \\
&= 46 + 230 \\
&= 276
\end{aligned}
$$
49. (A) $0.888\ 888\ 88\ldots$; (B) $0.272\ 727\ 27\ldots$; (C) $2.236\ 067\ 97\ldots$; (D) $1.375\ 000\ 00\ldots$

Exercise 0-2

1. 1   **3.** $6x^9$ **5.** $9x^6/y^4$ **7.** $(a^4 b^{12})/(c^8 d^4)$ **9.** 10^{17} **11.** $(2x)/y^2$ **13.** n^2 **15.** 4×10^8
17. 3.225×10^7 **19.** 8.5×10^{-2} **21.** 7.29×10^{-8} **23.** 0.005   **25.** 26,900,000   **27.** 0.000 000 000 59
29. $3y^4/2$ **31.** x^{12}/y^8 **33.** $4x^8/y^6$ **35.** $w^{12}/(u^{20} v^4)$ **37.** $(27y^3)/(2x^3)$ **39.** $1/(x+y)^2$ **41.** $2(a + 2b)^5$
43. $5/(u - v + w)^3$ **45.** 1.3×10^{25} pounds
47. 10^8 or 100 million, 10^{10} or 10 billion; 6×10^9 or 6 billion, 6×10^{11} or 600 billion   **49.** 6.65×10^{-17}
51. 1.54×10^{12} **53.** 1.0295×10^{11} **55.** -4.3647×10^{-18} **57.** 9.4697×10^{29}

Exercise 0-3

1. 4   **3.** 64   **5.** -6 **7.** Not a real number   **9.** $\frac{8}{125}$ **11.** $\frac{1}{27}$ **13.** $y^{3/5}$ **15.** $d^{1/3}$ **17.** $1/y^{1/2}$
19. $2x/y^2$ **21.** $1/(a^{1/4} b^{1/3})$ **23.** $xy^2/2$ **25.** $2b^2/3a^2$ **27.** $125y^{1/2}/27x^3$ **29.** $2/3x^{7/12}$ **31.** $a^{1/3}/b^2$
33. $3x^{1/2}$ **35.** $1/25y^{2/9}$ **37.** $a^{1/n} b^{1/m}$ **39.** $1/(x^{3m} y^{4n})$ **41.** (A) $x = -2$, for example; (B) $x = 2$, for example
43. 29.52   **45.** 0.03093   **47.** 5.421   **49.** 107.6

Exercise 0-4

1. $\sqrt[3]{m^2}$ or $\left(\sqrt[3]{m}\right)^2$ (first preferred)   **3.** $6\sqrt[5]{x^3}$ $\left(\text{not } \sqrt[5]{6x^3}\right)$ **5.** $\sqrt[5]{(4xy^3)^2}$ **7.** $\sqrt{x+y}$ **9.** $b^{1/5}$ **11.** $5x^{3/4}$
13. $(2x^2 y)^{3/5}$ **15.** $x^{1/3} + y^{1/3}$ **17.** -2 **19.** $3x^4 y^2$ **21.** $2mn^2$ **23.** $2ab^2\sqrt{2ab}$ **25.** $2xy^2\sqrt[3]{2xy}$
27. $\sqrt{m}$ **29.** $\sqrt[12]{xy}$ **31.** $3x\sqrt[3]{3}$ **33.** $\sqrt{5}/5$ **35.** $2\sqrt[3]{9}$ **37.** $\sqrt{42xy}/7y$ or $(1/7y)\sqrt{42xy}$ **39.** $3x^2 y^2\sqrt[3]{3x^2 y}$
41. $n\sqrt[4]{2m^3 n}$ **43.** $\sqrt[3]{a^2(b-a)}$ **45.** $\sqrt[4]{a^3 b}$ **47.** $x^2 y\sqrt[3]{6xy^2}$ **49.** $\sqrt{2}/2$ or $\frac{1}{2}\sqrt{2}$ **51.** $2a^2 b\sqrt[3]{4a^2 b}$
53. $\sqrt[4]{12xy^3}/2x$ or $(1/2x)\sqrt[4]{12xy^3}$ **55.** $\left(a\sqrt[3]{6ab}\right)/3b^2$ or $\left(a/3b^2\right)\sqrt[3]{6ab}$ **57.** $x/2\sqrt{2x}$ **59.** $1/m\sqrt[3]{2m^2}$ **61.** $2x/\sqrt[4]{2x}$

63. $\sqrt[6]{2a^3b^2}$ **65.** $2x\sqrt[3]{4y^2}$ **67.** $2a^4b^3\sqrt[6]{2^5a^5b^5}$ **69.** $\sqrt[6]{2ab}$ **71.** $\sqrt[n]{x/y} = (x/y)^{1/n} = (x^{1/n})/(y^{1/n}) = \sqrt[n]{x}/\sqrt[n]{y}$
73. (A) 6x; (B) 2x   **75.** (A) 2x; (B) 0   **77.** (A) 5x; (B) x   **79.** 0.2222   **81.** 1.934   **83.** 0.069 79
85. 2.073   **87.** Both are 1.059.   **89.** Both are 0.6300.

Exercise 0-5

1. 3   **3.** $2x^3 - x^2 + 2x + 4$ **5.** $2x^3 - 5x^2 + 6$ **7.** $6x^4 - 13x^3 + 9x^2 + 13x - 10$ **9.** $4x - 6$
11. $6y^2 - 16y$ **13.** $4\sqrt{5} - 2\sqrt{3}$ **15.** $5\sqrt[3]{a} - \sqrt[4]{a}$ **17.** $9x^2 - 4y^2$ **19.** $c - d$ **21.** $16x^2 - 8xy + y^2$
23. $x + 2x^{1/2}y^{1/2} + y$ **25.** $c - 2\sqrt{cd} + d$ **27.** $a^3 + b^3$ **29.** $-x + 27$ **31.** $2x^4 - 5x^3 + 5x^2 + 11x - 10$
33. $-5x^2 - 4x + 5$ **35.** $8m^3 - 12m^2n + 6mn^2 - n^3$ **37.** $6\sqrt{2} - 2\sqrt{5}$ **39.** $-4\sqrt{3x}$ **41.** $2a^2 - 5b^2$
43. $9x - 6\sqrt{xy} + y$ **45.** $2x + 3x^{1/2}y^{1/2} + y$ **47.** $6x - 2x^{19/3}$ **49.** 1   **51.** $a + b$ **53.** h **55.** 0
57. -4 **59.** $1/x - 2/(x^{1/2}y^{1/2}) + 1/y$ **61.** $t - x$ **63.** $6x + 12 - 5(x + 3)^{1/2}$ **65.** $x + 2\sqrt[3]{xy} - \sqrt[3]{x^2y^2} - 2y$
67. Perimeter $= 2x + 2(x - 5) = 4x - 10$ **69.** Value $= 5x + 10(x - 5) + 25(x - 3) = 40x - 125$ **71.** -2.000 **73.** 0

Exercise 0-6

1. $2x^2(3x^2 - 4x - 1)$ **3.** $5xy(2x^2 + 4xy - 3y^2)$ **5.** $(5x - 3)(x + 1)$ **7.** $(2w - x)(y - 2z)$ **9.** $(x + 3)(x - 2)$
11. $(2m - 1)(3m + 5)$ **13.** $(2x - 3y)(x - 2y)$ **15.** $(4a - 3b)(2c - d)$ **17.** $(2x - 1)(x + 3)$ **19.** $(x - 6y)(x + 2y)$
21. Prime   **23.** $(5m - 4n)(5m + 4n)$ **25.** $(x + 5y)^2$ **27.** Prime   **29.** $6(x + 2)(x + 6)$ **31.** $2y(y - 3)(y - 8)$
33. $y(4x - 1)^2$ **35.** $(3s - t)(2s + 3t)$ **37.** $xy(x - 3y)(x + 3y)$ **39.** $3m(m^2 - 2m + 5)$
41. $(m + n)(m^2 - mn + n^2)$ **43.** $(c - 1)(c^2 + c + 1)$ **45.** $[(a - b) - 2(c - d)][(a - b) + 2(c - d)]$
47. $(2m - 3n)(a + b)$ **49.** Prime   **51.** $(x + 3)(x - 3)^2$ **53.** $(a - 2)(a + 1)(a - 1)$
55. $[4(A + B) + 3][(A + B) - 2]$ **57.** $(m - n)(m + n)(m^2 + n^2)$ **59.** $st(st - 2)(s^2t^2 + 2st + 4)$
61. $(m + n)(m + n - 1)$ **63.** $(2/x - 3/y)(2/x + 3/y)$ **65.** $(x + \frac{3}{2})^2$ **67.** $(1/x - 2)(1/x^2 + 2/x + 4)$
69. $(x^{1/3} - 2)(x^{1/3} - 1)$ **71.** $(2x^{-1} - y^{-2})(2x^{-1} + y^{-2})$ **73.** $2a[3a - 2(x + 4)][3a + 2(x + 4)]$
75. $(x^2 - x + 1)(x^2 + x + 1)$ **77.** $(4x^2 + 2x + 1)(4x^2 - 2x + 1)$
79. $(x^3 - 1)(x^3 + 1) = (x - 1)(x^2 + x + 1)(x + 1)(x^2 - x + 1)$ **81.** $(x^2 - 2xy + 5y^2)(x^2 + 2xy + 5y^2)$

Exercise 0-7

1. $a^2/2$ **3.** $(22y + 9)/252$ **5.** $(x^2 + 8)/8x^3$ **7.** $1/(2x - 1)$ **9.** $1/m$ **11.** $2a/[(a + b)^2(a - b)]$
13. $(m^2 - 6m + 7)/(m - 2)$ **15.** $7/(x - 3)$ **17.** $3/(y + 3)$ **19.** $(x + y)/x$ **21.** $1/m$
23. $(y + 3)/[(y - 2)(y + 7)]$ **25.** -1 **27.** $(7y - 9x)/[xy(a - b)]$ **29.** $(x^2 - x + 1)/[2(x - 9)]$
31. $[(x - y)^2]/[y^2(x + y)]$ **33.** $1/(x - 4)$ **35.** $1/xy$ **37.** $(x - 3)/(x - 1)$ **39.** $(x^2 + xy + y^2)/xy$
41. $[-x(x + y)]/y$ **43.** $(a - 2)/a$ **45.** c/a **47.** $(6 + \sqrt{a} - a)/(a - 4)$ **49.** $(62 - 19\sqrt{10})/117$
51. $\sqrt{x^2 + 9} + 3$ **53.** $1/(\sqrt{t} + \sqrt{x})$ **55.** $1/(\sqrt{x + h} + \sqrt{x})$ **57.** $\frac{5}{6}\sqrt{6xy}$ **59.** $\frac{11}{2}\sqrt{2}$ **61.** $(y - x)/y$
63. $(x - 1)/x$ **65.** $1/(\sqrt[3]{t^2} + \sqrt[3]{tx} + \sqrt[3]{x^2})$

Exercise 0-8   Chapter Review

1. $6x^5y^{15}$ $(0\text{-}2)$ **2.** $3u^4/v^2$ $(0\text{-}2)$ **3.** 6×10^2 $(0\text{-}2)$ **4.** x^6/y^4 $(0\text{-}2)$ **5.** $u^{7/3}$ $(0\text{-}3)$ **6.** $3a^2/b$ $(0\text{-}3)$
7. $3\sqrt[5]{x^2}$ $(0\text{-}4)$ **8.** $-3(xy)^{2/3}$ $(0\text{-}4)$ **9.** $3x^2y\sqrt[3]{x^2y}$ $(0\text{-}4)$ **10.** $6x^2y^3\sqrt{xy}$ $(0\text{-}4)$ **11.** $2b\sqrt{3a}$ $(0\text{-}4)$
12. $\sqrt{7} - 2\sqrt{3}$ $(0\text{-}4)$ **13.** $(3\sqrt{5} + 5)/4$ $(0\text{-}4)$ **14.** $\sqrt[4]{y^3}$ $(0\text{-}4)$ **15.** $x^3 + 3x^2 + 5x - 2$ $(0\text{-}5)$
16. $x^3 - 3x^2 - 3x + 22$ $(0\text{-}5)$ **17.** $3x^5 + x^4 - 8x^3 + 24x^2 + 8x - 64$ $(0\text{-}5)$ **18.** $(3x - 2)^2$ $(0\text{-}6)$
19. Prime $(0\text{-}6)$ **20.** $3n(2n - 5)(n + 1)$ $(0\text{-}6)$ **21.** $(12a^3b - 40b^2 - 5a)/(30a^3b^2)$ $(0\text{-}7)$
22. $(7x - 4)/[6x(x - 4)]$ $(0\text{-}7)$ **23.** $(y + 2)/[y(y - 2)]$ $(0\text{-}7)$ **24.** u $(0\text{-}7)$ **25.** Def. subtraction $(0\text{-}1)$
26. Commutative $(0\text{-}1)$ **27.** Distributive $(0\text{-}1)$ **28.** Associative $(0\text{-}1)$ **29.** Negatives $(0\text{-}1)$
30. Identity $(0\text{-}1)$ **31.** (A) T; (B) F $(0\text{-}1)$ **32.** 0 and -3 are two examples of infinitely many. $(0\text{-}1)$

33. (A) a and d; (B) None (0-5) **34.** $2x^3 - 4x^2 + 12x$ (0-5) **35.** $(x - y)(7x - y)$ (0-6) **36.** Prime (0-6)

37. $3xy(2x^2 + 4xy - 5y^2)$ (0-6) **38.** $(y - b)(y - b - 1)$ (0-6) **39.** $3(x + 2y)(x^2 - 2xy + 4y^2)$ (0-6)

40. $(y - 2)(y + 2)^2$ (0-6) **41.** $2m/[(m + 2)(m - 2)^2]$ (0-7) **42.** y^2/x (0-7) **43.** $(x - y)/(x + y)$ (0-7)

44. $(-ab)/(a^2 + ab + b^2)$ (0-7) **45.** $\frac{1}{4}$ (0-2) **46.** $\frac{5}{9}$ (0-2) **47.** $3x^2/2y^2$ (0-2, 0-3) **48.** $27a^{1/6}/b^{1/2}$ (0-3)

49. $x + 2x^{1/2}y^{1/2} + y$ (0-3) **50.** $x - y$ (0-2, 0-3) **51.** 2×10^{-7} (0-2) **52.** $-6x^2y^2\sqrt[5]{3x^2y}$ (0-4)

53. $x\sqrt[3]{2x^2}$ (0-4) **54.** $\sqrt[5]{12x^3y^2}/2x$ (0-4) **55.** $y\sqrt[3]{2x^2y}$ (0-4) **56.** $-\frac{2}{3}\sqrt[3]{3}$ (0-4) **57.** $2x - 3\sqrt{xy} - 5y$ (0-4)

58. $\sqrt{y^2 + 4} + 2$ (0-4) **59.** $\left(6x + 3\sqrt{xy}\right)/(4x - y)$ (0-4) **60.** $\sqrt[3]{2x^2}$ (0-4) **61.** 0 (0-4, 0-5) **62.** $\frac{6}{11}$ (0-1)

63. $[(x - 2)(x + 1)]/2x$ (0-5) **64.** $\frac{2}{3}(x - 2)(x + 3)^4$ (0-2) **65.** $(a^2b^2)/(a^3 + b^3)$ (0-2, 0-7) **66.** $x - y$ (0-3, 0-5)

67. x^{m-1} (0-2, 0-5, 0-7) **68.** $2xy\sqrt[6]{2xy}$ (0-4) **69.** $\sqrt[6]{2xy}$ (0-4) **70.** $4\sqrt[12]{xy}$ (0-4) **71.** x^{n+1} (0-2, 0-3, 0-4, 0-6)

72. (A) $2x$; (B) $6x$ (0-4) **73.** $1/\left(\sqrt{5 + h} + \sqrt{5}\right)$ (0-4) **74.** $(2x^{-1/5} - 5y^{1/3})(2x^{-1/5} + 5y^{1/3})$ (0-3, 0-6)

75. $[5xy^2 - 2(3x + 2y)][5xy^2 + 2(3x + 2y)]$ (0-6)

Practice Test: Chapter 0

1. (A) Commutative; (B) Distributive; (C) Def. subtraction; (D) Identity; (E) Associative (0-1)

2. (A) T; (B) F (0-1) **3.** $-15x^2 + 23x - 6$ (0-5) **4.** $(4x - 13)/[(x - 3)(x + 4)]$ (0-7)

5. $x^4/[7(x - 3)]$ (0-6, 0-7) **6.** $x - 3$ (0-7) **7.** $(2a - 3b)(4x - y)$ (0-6) **8.** $(2x + y)^2(2x - y)^2$ (0-6)

9. $(x - 5)(x + 5)^2$ (0-6) **10.** $4y^{5/3}/3x^3$ (0-2, 0-3) **11.** $4x^{1/6}y$ (0-3) **12.** $x + y$ (0-2, 0-7)

13. $2x^2y^3\sqrt[3]{3x^2}$ (0-4) **14.** $3a^3b^2\sqrt[3]{49a^2b}$ (0-4) **15.** $\left(3a + 6\sqrt{ab}\right)/(a - 4b)$ (0-4, 0-5) **16.** $\frac{5}{2}\sqrt[3]{2}$ (0-4)

17. $\sqrt{2x}$ (0-4) **18.** $2x^3y^2\sqrt[6]{2xy}$ (0-4) **19.** 3×10, or 30 (0-2) **20.** (A) $2x$; (B) $8x$ (0-4)

Chapter 1   Exercise 1-1

1. T **3.** T **5.** T **7.** T **9.** $\{3, 4, 5, 6, 7\}$ **11.** $\{5\}$ **13.** $\varnothing$ **15.** $\{8\}$ **17.** $\{-7, 7\}$

19. $\{2, 3, 5, 7\}$ **21.** $\{-2, 1, 2\}$ **23.** 100 **25.** 61 **27.** 80 **29.** 41 **31.** 89 **33.** 20

35. (A) $\{2, 3, 4, 5, 6, 7\}$; (B) $\{2, 3, 4, 5, 6, 7\}$ **37.** $\{1, 2, 3, 4, 6\}$ **39.** Yes **41.** No **43.** No

45. They are all different: $\varnothing$ is the empty set, $\{0\}$ has one element 0, and $\{\varnothing\}$ has one element $\varnothing$. **47.** 850 **49.** 150

51. 130 **53.** 850 **55.** 150 **57.** 130 **59.** $AB-, AB+$ **61.** $A-, AB-, B-, A+, AB+, B+$ **63.** $O-$

65. $A-, AB-$

Exercise 1-2

1. 18 **3.** 9 **5.** 6 **7.** 9 **9.** $x = \frac{11}{2}$, or 5.5 **11.** 10 **13.** 8 **15.** $\frac{31}{24}$ **17.** 3 **19.** No solution

21. $\frac{8}{5}$ **23.** 2 **25.** No solution **27.** -4 **29.** $\frac{2}{3}$ **31.** $d = (a_n - a_1)/(n - 1)$ **33.** $f = d_1d_2/(d_1 + d_2)$

35. $a = (A - 2bc)/(2b + 2c)$ **37.** $x = (5y + 3)/(2 - 3y)$ **39.** 4 **41.** $x = (by + cy - ac)/(a - y)$

Exercise 1-3

1. 8, 10, 12, 14 **3.** 17 by 10 meters **5.** $90 **7.** 4,000 records **9.** $2,000 per month **11.** 337 feet

13. 55,000 feet **15.** 2,500 deer **17.** 30 liters of 20% solution and 70 liters of 80% solution **19.** 14,080 feet

21. (A) 216 miles; (B) 225 miles **23.** 264 hertz, 330 hertz **25.** 141.2 centimeters **27.** 150 feet

29. $5\frac{5}{11}$ minutes after 1 P.M.

Exercise 1-4

1. $-8 \le x \le 7$ **3.** $-6 \le x < 6$

5. $x \geq -6$ x **7.** $(-2, 6]$ x

9. $(-7, 8)$ x **11.** $(-\infty, -2]$ x

13. $[-7, 2)$; $-7 \leq x < 2$ **15.** $(-\infty, 0]$; $x \leq 0$ **17.** $x < 5$ or $(-\infty, 5)$ x

19. $x \geq 3$ or $[3, \infty)$ x **21.** $N < -8$ or $(-\infty, -8)$ N

23. $t > 2$ or $(2, \infty)$ t **25.** $m > 3$ or $(3, \infty)$ m

27. $B \geq -4$ or $[-4, \infty)$ B **29.** $-2 < t \leq 3$ or $(-2, 3]$ t

31. $\{-4, -3, -2, -1, 0, 1, 2\}$ **33.** $\{-3, -2, -1, 0, 1\}$ **35.** $q < -14$ or $(-\infty, -14)$ q

37. $x \geq 4.5$ or $[4.5, \infty)$ x **39.** $-20 \leq x \leq 20$ or $[-20, 20]$ x

41. $-30 \leq x < 18$ or $[-30, 18)$ x **43.** $-8 \leq x < -3$ or $[-8, -3)$ x

45. $-14 < x \leq 11$ or $(-14, 11]$ x **47.** Positive   **49.** (A) F; (B) T; (C) T

51. $8{,}000 \leq h \leq 20{,}000$ or $[8{,}000, 20{,}000]$ **53.** $x > 600$

Exercise 1-5

1. $\sqrt{5}$ **3.** 4   **5.** $5 - \sqrt{5}$ **7.** $5 - \sqrt{5}$ **9.** 12   **11.** 12   **13.** 9   **15.** 4   **17.** 4   **19.** 9

21. $x = \pm 7$ x **23.** $-7 \leq x \leq 7$ x

25. $x \leq -7$ or $x \geq 7$ x **27.** $y = 2$ or 8 y

29. $2 < y < 8$ y **31.** $y < 2$ or $y > 8$ y

33. $u = -11$ or -5 u **35.** $-11 \leq u \leq -5$ u

37. $u \leq -11$ or $u \geq -5$ u **39.** $x = -4, \frac{4}{3}$ **41.** $-\frac{9}{5} \leq x \leq 3$

43. $y < 3$ or $y > 5$ **45.** $t = -\frac{4}{5}, \frac{18}{5}$ **47.** $-\frac{5}{7} < u < \frac{23}{7}$ **49.** $x \leq -6$ or $x \geq 9$ **51.** $-35 < C < -\frac{5}{9}$

53. $x \geq 5$ **55.** $x \leq -8$ **57.** $x \geq -\frac{3}{4}$ **59.** $x \leq \frac{2}{5}$

61. Case 1: $a = b$; $|b - a| = |0| = 0$; $|a - b| = |0| = 0$

Case 2: $a > b$; $|b - a| = -(b - a) = a - b$

                    $|a - b| = a - b$

Case 3: $b > a$; $|b - a| = b - a$

                    $|a - b| = -(a - b) = b - a$

Exercise 1-6

1. $-3 < x < 4$
$(-3, 4)$
x

3. $x \leq -3$ or $x \geq 4$
$(-\infty, -3] \cup [4, \infty)$
x

5. $-5 < x < 2$
$(-5, 2)$
x

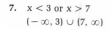

7. $x < 3$ or $x > 7$
$(-\infty, 3) \cup (7, \infty)$

9. $0 \le x \le 8$
$[0, 8]$

11. $-5 \le x \le 0$
$[-5, 0]$

13. $x < -2$ or $x > 2$
$(-\infty, -2) \cup (2, \infty)$

15. True for all real numbers.
Graph: Whole real number line

17. $-1 \le x \le 1$ or $x \ge 5$
$[-1, 1] \cup [5, \infty]$

19. $x < -5$ or $3 < x < 5$
$(-\infty, -5) \cup (3, 5)$

21. $-4 < x \le 2$
$(-4, 2]$

23. $-5 \le x \le 0$ or $x > 3$
$[-5, 0] \cup (3, \infty)$

25. $x \le -4$ or $x > 1$
$(-\infty, -4] \cup (1, \infty)$

27. $x < 0$ or $x > \frac{1}{4}$
$(-\infty, 0) \cup (\frac{1}{4}, \infty)$

29. $x < -3$ or $x \ge 3$
$(-\infty, -3) \cup [3, \infty)$

31. $-4 < x \le \frac{3}{2}$
$(-4, \frac{3}{2}]$

33. $-1 < x < 2$ or $x \ge 5$
$(-1, 2) \cup [5, \infty)$

35. No solutions

37. $x > 4$
$(4, \infty)$

39. $-2 \le x \le -\frac{1}{2}$ or $\frac{1}{2} \le x \le 2$
$[-2, -\frac{1}{2}] \cup [\frac{1}{2}, 2]$

41. $-2 \le x \le 2$
$[-2, 2]$

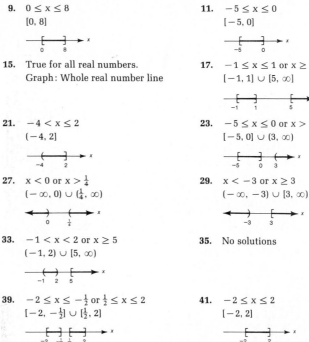

Exercise 1-7

1. $7 + 5i$ **3.** $5 + 3i$ **5.** $2 + 4i$ **7.** $5 + 9i$ **9.** $4 - 3i$ **11.** -24 or $-24 + 0i$ **13.** $-12 - 6i$
15. $15 - 3i$ **17.** $-4 - 33i$ **19.** 65 or $65 + 0i$ **21.** $\frac{2}{5} - \frac{1}{5}i$ **23.** $\frac{3}{13} + \frac{11}{13}i$ **25.** $5 + 3i$ **27.** $7 - 5i$
29. $-3 + 2i$ **31.** $8 + 25i$ **33.** $\frac{5}{7} - \frac{2}{7}i$ **35.** $\frac{2}{13} + \frac{3}{13}i$ **37.** $-\frac{2}{5}i$ or $0 - \frac{2}{5}i$ **39.** $\frac{3}{2} - \frac{1}{2}i$ **41.** $-6i$ or $0 - 6i$
43. 0 or $0 + 0i$ **45.** $-1, -i, 1, i, -1, -i, 1$ **47.** $x = 3, y = -2$ **49.** $(a + c) + (b + d)i$
51. $a^2 + b^2$ or $(a^2 + b^2) + 0i$ **53.** $(ac - bd) + (ad + bc)i$ **55.** $i^{4k} = (i^4)^k = (i^2 \cdot i^2)^k = [(-1)(-1)]^k = 1^k = 1$
57. (1) Def. addition, (2) property of real numbers, (3) def. addition

Exercise 1-8

1. $0, 2$ **3.** $-8, \frac{1}{2}$ **5.** $\frac{3}{2}, 4$ **7.** $-\frac{4}{3}, \frac{1}{2}$ **9.** ± 5 **11.** $\pm 5i$ **13.** $\pm 2\sqrt{3}$ **15.** $\pm \frac{4}{3}$ **17.** $\pm \frac{5}{2}i$
19. $-2, -8$ **21.** $3 \pm 2i$ **23.** $5 \pm 2\sqrt{7}$ **25.** $(-1 \pm \sqrt{5})/2$ **27.** $2 \pm 2i$ **29.** $(2 + \sqrt{2})/2$
31. $(-3 \pm \sqrt{17})/4$ **33.** $\frac{1}{5} \pm \frac{3}{5}i$ **35.** $3 \pm 2\sqrt{3}$ **37.** $(3 \pm \sqrt{3})/2$ **39.** $(1 \pm \sqrt{7})/3$ **41.** $(-m \pm \sqrt{m^2 - 4n})/2$
43. $-\frac{5}{4}, \frac{2}{3}$ **45.** $(3 \pm \sqrt{5})/2$ **47.** $(3 \pm \sqrt{13})/2$ **49.** $-\frac{4}{7}, 0$ **51.** $-\frac{1}{2}, 2$ **53.** $2 \pm 2i$ **55.** $-50, 2$
57. $(-5 \pm \sqrt{57})/2$ **59.** $(-3 \pm \sqrt{57})/4$ **61.** $t = \sqrt{2s/g}$ **63.** $I = (E + \sqrt{E^2 - 4RP})/2R$
65. $[(-b + \sqrt{b^2 - 4ac})/2a] \times [(-b - \sqrt{b^2 - 4ac})/2a] = [b^2 - (b^2 - 4ac)]/4a^2 = c/a$
67. Substitute $r_1 = c/r_2 a$ into $r_1 + r_2 = -b/a$ to obtain $c/r_2 a + r_2 = -b/a$ or $ar_2^2 + br_2 + c = 0$. Similarly, substitute $r_2 = c/r_1 a$
into $r_1 + r_2 = -b/a$ to obtain after simplification $ar_1^2 + br_1 + c = 0$.
69. The $\pm$ in front still yields the same two numbers even if a is negative. **71.** $1.35, 0.48$ **73.** $-1.05, 0.63$
75. Has real solutions, since discriminant is positive **77.** Has no real solutions, since discriminant is negative

Exercise 1-9

1. 8, 13   **3.** 12, 14   **5.** 5.12 by 3.12 inches   **7.** 179 miles   **9.** 20%   **11.** 5 and 12 miles/hour
13. 13.09 and 8.09 hours   **15.** 70 miles/hour   **17.** 50 miles/hour

Exercise 1-10

1. 22   **3.** 8   **5.** No solution   **7.** 0, 4   **9.** $\pm 2, \pm\sqrt{2}i$ **11.** $-\sqrt[5]{5}, \sqrt[5]{2}$ **13.** $\frac{1}{8}, -8$
15. $-2, 3, \frac{1}{2} \pm \sqrt{7}/2i$ **17.** No solution   **19.** 1   **21.** 2   **23.** $-\frac{3}{4}, \frac{1}{5}$ **25.** $\pm 1, \pm 3$ **27.** 1, 16
29. 2, 3, 7, 8   **31.** -2 **33.** 9, 16   **35.** 4, 81

Exercise 1-11   Chapter Review

1. (A) $\{1, 3, 4, 5\}$;   (B) $\{3, 5\}$;   (C) $\varnothing$;   (D) $\{3, 4, 5\}$;   (E) $\{2, 4, 6\}$ *(1-1)*
2. (A) F;   (B) T;   (C) T;   (D) T;   (E) T;   (F) T   *(1-1)*   **3.** $\{-\frac{2}{3}, \frac{5}{2}\}$ *(1-1, 1-8)*
4. Substitution property   *(1-2)*   **5.** 21   *(1-2)*   **6.** $\frac{30}{11}$ *(1-2)*   **7.** $x \geq 1$ *(1-4)*
$[1, \infty)$

8. $-14 < y < -4$ *(1-5)*   **9.** $-1 \leq x \leq 4$ *(1-5)*   **10.** $-5 < x < 4$ *(1-6)*
$(-14, -4)$ $[-1, 4]$ $(-5, 4)$

11. $x \leq -3$ or $x \geq 7$ *(1-6)*   **12.** (A) $3 - 6i$;   (B) $15 + 3i$;   (C) $2 + i$ *(1-7)*   **13.** $\pm\sqrt{\frac{7}{2}}$ or $\pm\sqrt{\frac{14}{2}}$ *(1-8)*
$(-\infty, -3] \cup [7, \infty)$

14. 0, 2   *(1-8)*   **15.** $\frac{1}{2}, 3$ *(1-8)*   **16.** $-\frac{1}{2} \pm \left(\sqrt{3}/2\right)i$ *(1-8)*   **17.** $\left(3 \pm \sqrt{33}\right)/4$ *(1-8)*   **18.** 2, 3   *(1-10)*
19. (A) $H = \{x \mid 10x + 11 = 6/x\}$;   (B) $H = \{-\frac{3}{2}, \frac{2}{5}\}$ *(1-1, 1-8)*
20. (A) $\{-1, 0, 1, 2, 3\}$;   (B) $\{1\}$;   (C) No;   (D) Yes;   (E) No;   (F) No   *(1-1, 1-4)*
21. $\varnothing$ (each of the others is a set containing one element)   *(1-1)*   **22.** -15 *(1-2)*   **23.** No solution   *(1-2)*
24. $x \geq -19$ *(1-4)*   **25.** $x < 2$ or $x > \frac{10}{3}$ *(1-5)*   **26.** $x < 0$ or $x > \frac{1}{2}$ *(1-6)*
$[-19, \infty)$ $(-\infty, 2) \cup (\frac{10}{3}, \infty)$ $(-\infty, 0) \cup (\frac{1}{2}, \infty)$

27. $x \leq 1$ or $3 < x < 4$ *(1-6)*   **28.** (A) 6;   (B) 6   *(1-5)*   **29.** (A) $5 + 4i$;   (B) $-i$ *(1-7)*
$(-\infty, 1] \cup (3, 4)$

30. (A) $-1 + i$;   (B) $\frac{4}{13} - \frac{7}{13}i$;   (C) $\frac{5}{2} - 2i$ *(1-7)*   **31.** $\left(-5 \pm \sqrt{5}\right)/2$ *(1-8)*   **32.** $1 \pm i\sqrt{2}$ *(1-8)*
33. $\left(1 \pm \sqrt{43}\right)/3$ *(1-8)*   **34.** $-\frac{27}{8}, 64$ *(1-8)*   **35.** $\pm 2, \pm 3i$ *(1-10)*   **36.** $\frac{9}{4}, 3$ *(1-10)*   **37.** $M = P/(1 - dt)$ *(1-10)*
38. $1 = \left(E \pm \sqrt{E^2 - 4PR}\right)/2R$ *(1-10)*   **39.** $A \cap B$ *(1-1)*   **40.** (A) T;   (B) F;   (C) T   *(1-1)*   **41.** 1   *(1-7)*
42. No solution   *(1-6)*   **43.** Set of all real numbers   *(1-8)*
44. $x \leq -4$ or $-2 \leq x < 0$ or $0 < x \leq 2$ or $x \geq 4$; $(-\infty, -4] \cup [-2, 0) \cup (0, 2] \cup [4, \infty)$ *(1-5)*   **45.** 9, 25   *(1-10)*
46. 450 milliliters   *(1-3)*   **47.** (A) 925;   (B) 75   *(1-1)*   **48.** $\frac{5}{3}$ or $-\frac{3}{5}$ *(1-9)*
49. (A) 2,000 and 8,000;   (B) 5,000   *(1-10)*
50. $x = \left(13 \pm \sqrt{45}\right)/2$ thousand, or approximately 3,146 and 9,854   *(1-10)*

Practice Test: Chapter 1

1. B *(1-1)*   **2.** C *(1-1)*   **3.** $\frac{8}{5} - \frac{4}{5}i$ *(1-7)*   **4.** $x \le 18$ *(1-4)*
$(-\infty, 18]$

5. $-1 < x < 4$ *(1-5)*   **6.** $x \le -1$ or $x \ge 7$ *(1-5)*   **7.** $0 \le x \le 3$ or $x > 4$ *(1-6)*
$(-1, 4)$ $(-\infty, -1] \cup [7, \infty)$ $[0, 3] \cup (4, \infty)$

8. No solution *(1-2)*   **9.** $2 \pm i\sqrt{3}$ *(1-8)*   **10.** $-3 \pm \sqrt{2}$ *(1-8)*   **11.** $-1, 32$ *(1-10)*
12. $\pm 2i, \pm\sqrt{1/2}$ or $\pm\sqrt{2}/2$ *(1-10)*   **13.** 5 (0 is extraneous) *(1-10)*   **14.** 2 miles/hour *(1-9)*
15. 20 milliliters of 30% solution, 30 milliliters of 80% solution   *(1-3)*

Chapter 2   Exercise 2-1

1.

3. Symmetric with respect to the y axis

5. Symmetric with respect to the x axis   **7.** Symmetric with respect to the x axis, y axis, and origin   **9.** $\sqrt{145}$

11. $\sqrt{68}$ **13.** $x^2 + y^2 = 49$ **15.** $(x - 2)^2 + (y - 3)^2 = 36$ **17.** $(x + 4)^2 + (y - 1)^2 = 7$
19. $(x + 3)^2 + (y + 4)^2 = 2$ **21.** Symmetric with respect to the x axis   **23.** Symmetric with respect to the y axis

25. Symmetric with respect to the x axis, y axis, and origin   **27.** Symmetric with respect to the x axis, y axis, and origin

29. Symmetric with respect to the origin   **31.** A right triangle   **33.** $x = -3, 7$ **35.** Center: $(3, 5)$; radius $= 7$

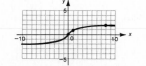

37. Center: $(-4, 2)$; radius $= \sqrt{7}$ **39.** $(x - 3)^2 + (y - 2)^2 = 49$, $C(3, 2)$, $r = 7$
41. $(x + 4)^2 + (y - 3)^2 = 17$, $C(-4, 3)$, $r = \sqrt{17}$ **43.**   Symmetric with respect to the y axis

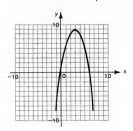

45. Symmetric with respect to the origin   **47.**   **49.** $5x + 3y = -2$

51. $(x - 4)^2 + (y - 2)^2 = 34$ **53.**   18.11   **55.**   Symmetric with respect to the y axis

57. Symmetric with respect to the y axis   **59.** Symmetric with respect to the y axis

Exercise 2-2

1. A function   **3.** Not a function   **5.** A function   **7.** A function   **9.** Not a function   **11.** Not a function
13. -8 **15.** -6 **17.** 7   **19.** 26   **21.** $-\frac{10}{9}$ **23.** 3   **25.** A function   **27.** A function
29. Not a function   **31.** A function   **33.** Not a function   **35.** A function   **37.** A function   **39.** A function
41. A function   **43.** Domain $= \{2, 4\}$; range $= \{-2, 0, 2, 4\}$; not a function
45. Domain $= \{-4, 0, 4\}$; range $= \{1\}$; a function   **47.** Domain $= \{0, 1, 4\}$; range $= \{-2, -1, 0, 1, 2\}$; not a function
49. Domain $= \{-2, 2\}$; range $= \{2, 6\}$; a function   **51.** Domain $= R$ **53.** Domain $= R$ **55.** Domain: $x \geq -2$
57. Domain: $-9 \leq x \leq 9$ **59.** Domain: All R except 2   **61.** Domain: $x \leq -3$ or $x > 2$ **63.** $-6 - h$ **65.** h
67. $2a + h$ **69.** $2x + h$ **71.** $3x^2 + 3xh + h^2$ **73.** $C(x) = 10 + 0.12x$ **75.** $A(r) = \pi r^2$
77. $V(x) = x(8 - 2x)(12 - 2x)$; domain: $0 < x < 4$
79. (A)   0, 16, 64, 144;   (B) $64 + 16h$;   (C)   64; this number appears to be the speed of the object at the end of 2 seconds.

Exercise 2-3

1. f **3.** f **5.** (A) None;   (B) $[0, \infty)$;   (C) $(-\infty, 0]$
7. (A) $(-\infty, -2)$;   (B) $[-2, -1]$, $[1, \infty)$;   (C) $[-1, 1)$ **9.** f, g, and p
11. q; discontinuous at $x = -2$ and $x = 1$ **13.** Odd   **15.** Even   **17.** Neither   **19.** Even   **21.** Neither

23.

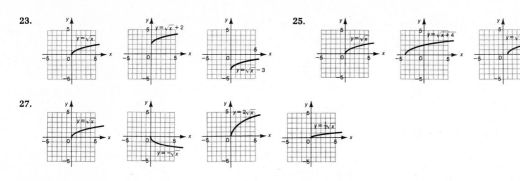

25.

27.

29. It is the same as the graph of $y = x^2$ reflected with respect to the x axis and shifted to the left two units.

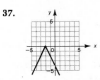

31. It is the same as the graph of $y = |x|$ reflected with respect to the x axis and shifted to the left two units.

33. It is the same as the graph of $y = -\sqrt{x}$ reflected with respect to the x axis and shifted to the right one unit.

35. Even

37.

39.

41. Even; discontinuous at $x = 0$

43. Odd; discontinuous at $x = 0$

45. Discontinuous at $x = 2$

47. Discontinuous at $x = 0$

Answers

49. Discontinuous at the integers

51. Discontinuous at the integers

53. Even

55. Odd

57. Discontinuous at the integers

Exercise 2-4

1. Slope $= 2$

3. Slope $= -\frac{3}{5}$

5. Slope $= -\frac{3}{4}$

7. Slope $= \frac{2}{3}$

9. Slope $= \frac{4}{5}$

11. Slope $= \frac{3}{2}$

13. Slope $= 2$

15. Slope not defined

17. Slope $= 0$

19. $y = -3x + 4$ **21.** $y = -\frac{2}{5}x + 2$ **23.** $y = 5$ **25.** $y = -2x + 8$ **27.** $y = -\frac{4}{3}x + \frac{8}{3}$ **29.** $y = 4$
31. $x = 4$ **33.** $y = -\frac{1}{3}x + 2$ **35.** $y = \frac{3}{4}x + 3$ **37.** $3x - y = -13$ **39.** $3x - y = 9$ **41.** $x = 2$

43. $x = 3$ **45.** $3x - 2y = 15$ **47.** $3x - y = 4$ **49.**

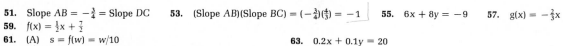

51. Slope $AB = -\frac{3}{4}$ = Slope DC **53.** (Slope AB)(Slope BC) = $(-\frac{3}{4})(\frac{4}{3}) = -1$ **55.** $6x + 8y = -9$ **57.** $g(x) = -\frac{2}{3}x$

59. $f(x) = \frac{1}{2}x + \frac{7}{2}$

61. (A) $s = f(w) = w/10$

(B) $f(15) = 1.5$ inches, $f(30) = 3$ inches

(C)   Slope $= \frac{1}{10}$

(D)

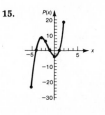

63. $0.2x + 0.1y = 20$

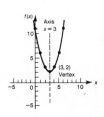

Exercise 2-5

1.

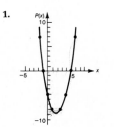

3.

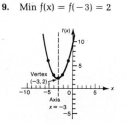

5.   Min $f(x) = f(3) = 2$

7.   Max $f(x) = f(-3) = -2$

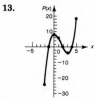

9.   Min $f(x) = f(-3) = 2$

11.   Max $f(x) = f(3) = 3$

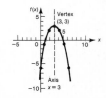

13.

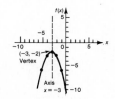

15.

17.   Min $f(x) = f(-2) = -2$

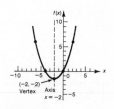

19. Max $f(x) = f(-2) = 6$

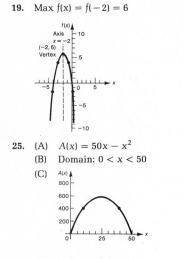

21.

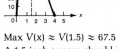

23.

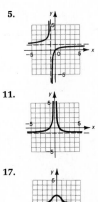

25. (A) $A(x) = 50x - x^2$

(B) Domain: $0 < x < 50$

(C)

(D) 25 by 25 feet

27. (A) $V(x) = (12 - 2x)(8 - 2x)x$
$= 4x^3 - 40x^2 + 96x$

(B) Domain: $0 < x < 4$. [Note: At $x = 0$ and $x = 4$, we have zero volume.]

(C)

(D) Max $V(x) \approx V(1.5) \approx 67.5$ cubic inches
A 1.5-inch square should be cut from each corner.

Exercise 2-6

1.

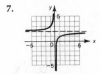

3.

5.

7.

9.

11.

13.

15.

17.

19.

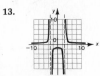

21.

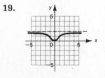

Exercise 2-7

1. $(f \circ g)(x) = (x^2 - x + 1)^7$, $(g \circ f)(x) = x^{14} - x^7 + 1$ **3.** $(f \circ g)(x) = \sqrt{2x + 5}$, $(g \circ f)(x) = 2\sqrt{x} + 5$
5. $(f \circ g)(x) = (1 - x^2)^{1/2}$, $(g \circ f)(x) = 1 - x$ **7.** $(f \circ g)(x) = |3x - 2|$, $(g \circ f)(x) = 3|x| - 2$
9. $(f \circ g)(x) = (x^3 - 4)^{2/3}$, $(g \circ f)(x) = x^2 - 4$ **11.** $(f \circ g)(x) = 7$, $(g \circ f)(x) = -7$
13. Both H and H^{-1} are functions.   **15.** Both g and g^{-1} are functions.   **17.** p is a function

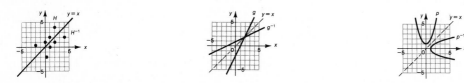

19. (A) Domain of $f = R = $ Range of f^{-1}, Range of $f = R = $ Domain of f^{-1}; (B) $(x + 2)/2$; (C) 3; (D) 2; (E) x
21. (A) Domain of $f = R = $ Range of f^{-1}, Range of $f = R = $ Domain of f^{-1}; (B) $(x + 5)/3$; (C) 3; (D) 2; (E) x
23. Not one to one
25. (A) Domain of $f = [0, \infty) = $ Range of f^{-1}, Range of $f = [1, \infty) = $ Domain of f^{-1}; (B) $\sqrt{x - 1}$; (C) $\sqrt{3}$; (D) 2;
(E) x   **27.** Not one to one
29. (A) Domain of $f = [1, \infty) = $ Range of f^{-1}, Range of $f = [0, \infty) = $ Domain of f^{-1}; (B) $\sqrt{x} + 1$; (C) 3; (D) 2; (E) x
31. Domain of $f \circ g = R$, Domain of $g \circ f = R$ **33.** Domain of $f \circ g = [-\frac{5}{2}, \infty)$, Domain of $g \circ f = [0, \infty)$
35. Domain of $f \circ g = [-1, 1]$, Domain of $g \circ f = [0, \infty)$ **37.** Domain of $f \circ g = R$, Domain of $g \circ f = R$
39. Domain of $f \circ g = R$, Domain of $g \circ f = R$ **41.** Domain of $f \circ g = R$, Domain of $g \circ f = R$
43. (A) x; (B) $4 - x$; (C) x; (D) $f^{-1}(x) = f(x)$ **45.** (A) x;   (B) $1/x$; (C) x; (D) $f^{-1}(x) = f(x)$
47. $g(x) = (2x + 1)/(x - 1)$ [Note: $g = f^{-1}$.]   **49.** $(f^{-1} \circ f)(x) = x = I(x)$ for all x in X
51. $(f \circ g)(R) = 2\pi R$, a formula for the circumference of a circle in terms of R

Exercise 2-8

1. $F = kv^2$ **3.** $f = k\sqrt{T}$ **5.** $y = k/\sqrt{x}$ **7.** $t = k/T$ **9.** $R = kSTV$ **11.** $V = khr^2$ **13.** 4   **15.** $9\sqrt{3}$
17. $U = k(ab/c^3)$ **19.** $L = k(wh^2/l)$ **21.** -12 **23.** 83 pounds   **25.** 20 amperes
27. The new horsepower must be eight times the old.   **29.** No effect   **31.** $t^2 = kd^3$ **33.** 1.47 hours (approx.)
35. 20 days   **37.** Quadrupled   **39.** 540 pounds   **41.** (A) $\Delta S = kS$; (B) 10 ounces; (C) 8 candlepower
43. 32 times per second   **45.** $N = k(F/d)$ **47.** 1.2 miles/second   **49.** 20 days
51. The volume is increased by a factor of 8.

Exercise 2-9   Chapter Review

1. (A) Reflected across x axis; (B) Shifted down three units; (C) Shifted left three units (2-3)
2. (A) $\sqrt{45}$; (B) $-\frac{1}{2}$; (C) 2 (2-1, 2-4)   **3.** Vertical: $x = -3$, slope not defined; horizontal: $y = 4$, slope $= 0$ (2-4)
4. 16 (2-2)   **5.** 1 (2-2)   **6.** 3 (2-2)   **7.** $-2a - h$ (2-2)
8. $(f \circ g)(x) = 17 - 3x^2$; $(g \circ f)(x) = 4 - (3x + 5)^2 = -21 - 30x - 9x^2$ (2-7)   **9.** (A) $(x + 1)/4$; (B) 2; (C) x (2-7)
10. Slope $= -\frac{3}{2}$ (2-4)   **11.** $2x + 3y = 12$ (2-4)   **12.** (2-6) SKIP

13. (A) $y = k(x/z)$; (B) $\frac{4}{3}$ (2-8)   **14.** $x - 2y = 2$ (2-4)   **15.** $(x - 3)^2 + (y + 2)^2 = 4$ (2-1)
16. Decreasing (2-3, 2-4)   **17.** It is symmetric with respect to all three. (2-1)   **18.** (A) f; (B) g; (C) h (2-3)

19. (2-3)

20. (A) $y = -2x - 3$; (B) $y = \frac{1}{2}x + 2$ (2-4)   **21.** Min $f(x) = f(3) = -4$ (2-5)

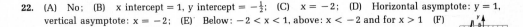

22. (A) No; (B) x intercept $= 1$, y intercept $= -\frac{1}{2}$; (C) $x = -2$; (D) Horizontal asymptote: $y = 1$, vertical asymptote: $x = -2$; (E) Below: $-2 < x < 1$, above: $x < -2$ and for $x > 1$ (F)   (2-6)

23. Center: $(3, -4)$, radius $= 5$ (2-1)   **24.** $P(x) = [(x - 2)x - 5]x + 6$ (2-5)   **25.** All but B (2-7)

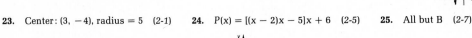

26. (A) Domain of $f = \{-2, -1, 0, 1, 2\} = $ Range of f^{-1}, Range of $f = \{\frac{1}{4}, \frac{1}{2}, 1, 2, 4\} = $ Domain of f^{-1}

     (B)

     (C) Both f and f^{-1} are functions. (2-2, 2-3, 2-7)

27. (A) Domain of $f = [0, \infty) = $ Range of f^{-1}; Range of $f = [-1, \infty) = $ Domain of f^{-1}; (B) $\sqrt{x + 1}$; (C) 2; (D) 4;
     (E) x (2-7)   **28.** $t = kwd/p$ (2-8)

29. (A) Horizontal: $y = \frac{3}{2}$, vertical: $x = -\frac{3}{2}$; (B) horizontal: $y = 0$, vertical: $x = -2$, $x = 3$ (2-6)

30. f is discontinuous at $x = -\frac{3}{2}$ and g is discontinuous at $x = -2$ and $x = 3$. (2-6)

31. (A) $(3x + 2)/(x - 1)$; (B) $\frac{11}{2}$; (C) x (2-7)   **32.** $x - y = 3$; a line (2-1, 2-4)

33. Above for $(-\infty, -2)$, below for $(-2, \infty)$ (2-6)   **34.** The force is doubled. (2-8)

Practice Test: Chapter 2

1. (A) $3x + 2y = -6$; (B) $\sqrt{52}$ (2-1, 2-4)   **2.** (A) $y = -\frac{3}{4}x + 1$; (B) $y = \frac{4}{3}x - \frac{22}{3}$ (2-4)

3. Center: $(-2, 3)$, radius $= 4$ (2-1)   **4.** (A) -1; (B) $-4a - 2h$ (2-2)

5. (A) g and h; (B) f; (C) f (2-3, 2-7)   **6.** Min $f(x) = f(3) = 2$, vertex: $(3, 2)$ (2-5)

7. (A) No; (B) At $x = -1$; (C) x intercept $= 1$, y intercept $= -\frac{1}{2}$;
     (D) Horizontal asymptote: $y = \frac{1}{2}$, vertical asymptote: $x = -1$ (2-6)

8. Above for $x < -1$ and for $x > 1$, below for $-1 < x < 1$ (2-6)   **9.** (2-6)

10. (A) $(f \circ g)(x) = \sqrt{|x|} - 8$, $(g \circ f)(x) = |\sqrt{x} - 8|$; (B) Domain of $f \circ g = R$, Domain of $g \circ f = [0, \infty)$ (2-7)
11. (A) $(x + 7)/3$; (B) 4; (C) x; (D) Increasing (2-3, 2-7)
12. (A) Domain of $f = [1, \infty) = $ Range of f^{-1}, Range of $f = [0, \infty) = $ Domain of f^{-1}
    (B) Both f and f^{-1} are functions.

    (C) $f^{-1}(x) = x^2 + 1, x \geq 0$ (2-7)
13. (A) $H = K(n/m^2)$; (B) 18; (C) H is cut in half. (2-8)
14. It is the same as the graph of g shifted to the right two units and down one unit; then turned upside down. (2-3)
15. (A) $f(x) = [(x - 3)x - 1]x + 3$
    (B)   (2-5)

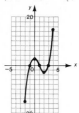

Chapter 3   Exercise 3-2

1. $2m + 1$ **3.** $4x - 5, R = 11$ **5.** $x^2 + x + 1$ **7.** $2y^2 - 5y + 13, R = -27$
9. $(x^2 + 3x - 7)/(x - 2) = x + 5 + [3/(x - 2)]$ **11.** $(4x^2 + 10x - 9)/(x + 3) = 4x - 2 - [3/(x + 3)]$
13. $(2x^3 - 3x + 1)/(x - 2) = 2x^2 + 4x + 5 + [11/(x - 2)]$ **15.** $3x^3 - 3x^2 + 3x - 4$ **17.** $x^4 - x^3 + x^2 - x + 1$
19. $2x^2 - 2x - 3, R = -2$ **21.** $2x^3 - 3x^2 - x - 5, R = -10$ **23.** $4x^3 - 6x - 2, R = 2$ **25.** $4x^2 - 2x - 4$
27. $x^2 - 1.7x + 2.49, R = -0.253$ **29.** $3x^2 + 0.6x + 2.12, R = -3.576$ **31.** $2x^2 - 3x + 2, R = 0$
33. $2x^2 + 1.6x - 2.72, R = -2.976$ **35.** $2x^2 - 7.8x + 2.92, R = 1.912$ **37.** $x^2 + (-3 + i)x - 3i$
39. (A) In both cases, the coefficient of x is a_2, the constant term is $a_2r + a_1$, and the remainder is $(a_2r + a_1)r + a_0$;
    (B) The remainder expanded is $a_2r^2 + a_1r + a_0 = P(r)$.

Exercise 3-3

1. 4   **3.** 3   **5.** -6 **7.** $3, -5$ **9.** $-\frac{1}{2}, 8, -2$ **11.** Yes   **13.** Yes   **15.** -3 **17.** 0.427
19.   **21.**   **23.** $-4, -8, 1$ **25.** $\frac{1}{8}, -\frac{3}{5}, -4$

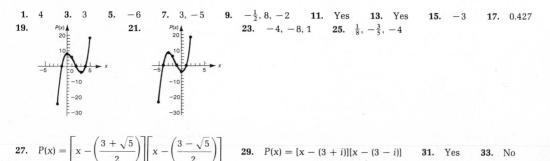

27. $P(x) = \left[x - \left(\dfrac{3 + \sqrt{5}}{2}\right)\right]\left[x - \left(\dfrac{3 - \sqrt{5}}{2}\right)\right]$ **29.** $P(x) = [x - (3 + i)][x - (3 - i)]$ **31.** Yes   **33.** No

35. 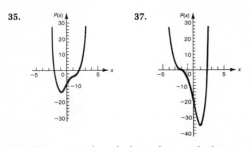   **37.**

39. $P(2) = 10$. Both methods involve exactly the same operations on the same numbers.
41. (A) $P(r) = (a_2r + a_1)r + a_0$;   (B) $P(r) = R = (a_2r + a_1)r + a_0$. Both are the same.

Exercise 3-4

1. -8 (multiplicity 3), 6 (multiplicity 2); degree of $P(x)$ is 5   **3.** -4 (multiplicity 3), 3 (multiplicity 2), -1; degree of $P(x)$ is 6
5. $P(x) = (x - 3)^2(x + 4)$, degree 3   **7.** $P(x) = (x + 7)^3(x - \frac{2}{3})(x + 5)$, degree 5
9. $P(x) = [x - (2 - 3i)][x - (2 + 3i)](x + 4)^2$, degree 4   **11.** One real and two complex, or three real
13. Six complex, two real and four complex, four real and two complex, or six real   **15.** $P(x) = (x + 4)^2(x + 1)$
17. $P(x) = (x - 1)(x + 1)(x - i)(x + i)$ **19.** $P(x) = (2x - 1)[x - (4 + 5i)][x - (4 - 5i)]$
21. Four complex, two complex and two real, or four real
23. Six complex, two real and four complex, four real and two complex, or six real   **25.** $x^2 - 8x + 41$
27. $x^2 - 2ax + (a^2 + b^2)$ **29.** $3 + i, -1$ **31.** $5i, 3$ **33.** (A) 3; (B) $-\frac{1}{2} - (\sqrt{3}/2)i, -\frac{1}{2} + (\sqrt{3}/2)i$ **35.** $n, 1$
37. No, since $P(x)$ is not a polynomial with real coefficients (the coefficient of x is the complex number $2i$).

Exercise 3-5

1. One positive, one negative   **3.** No positive, two or no negative   **5.** One or three positive, no negative
7. UB $= 2$, LB $= -1$ **9.** UB $= 2$, LB $= -3$ **11.** UB $= 1$, LB $= -2$ **13.** $P(3) = -2, P(4) = 2$
15. $P(-3) = -13, P(-2) = 3$ **17.** $Q(1) = 4, Q(2) = -1$
19. (A) Zero or two positive, one negative; (B) UB $= 3$, LB $= -3$;
    (C) 1 is a zero and one zero in each interval $(-3, -2)$ and $(2, 3)$
21. (A) One positive, zero or two negative; (B) UB $= 3$, LB $= -2$; (C) One real zero in interval $(2, 3)$
23. (A) One positive, one or three negative; (B) UB $= 2$, LB $= -5$;
    (C) Each interval contains exactly one real zero: $(-4, -3), (-2, -1) (-1, 0)$, and $(1, 2)$.
25. (A) One or three positive, zero or two negative; (B) UB $= 2$, LB $= -2$; (C) At least one real zero in interval $(1, 2)$
27. By Descartes' rule of signs $P(x)$ has one positive and one negative real zero. By the fundamental theorem of algebra, $P(x)$ has four zeros, and 0 is not a zero. Therefore, the other two zeros must be complex.
29. By Descartes' rule of signs, $P(x)$ has no positive or negative zeros. Since $P(0) = 0, P(-1) < 0$, and $P(1) > 0$, the graph crosses the x axis only at the origin.

Exercise 3-6

1. (A) $\pm 1, \pm 2, \pm 3, \pm 6$; (B) $-2, 1, 3$ **3.** (A) $\pm 1, \pm 2, \pm 4, \pm \frac{1}{3}, \pm \frac{2}{3}, \pm \frac{4}{3}$; (B) 2 (double zero), $-\frac{1}{3}$
5. (A) $\pm 1, \pm 3, \pm \frac{1}{2}, \pm \frac{3}{2}, \pm \frac{1}{3}, \pm \frac{1}{4}, \pm \frac{3}{4}, \pm \frac{1}{6}, \pm \frac{1}{12}$; (B) $-\frac{1}{2}, \frac{1}{3}, \frac{3}{2}$ **7.** (A) $\pm 1, \pm 2, \pm 4, \pm \frac{1}{3}, \pm \frac{2}{3}, \pm \frac{4}{3}$; (B) $-\frac{1}{3}$
9. (A) $\pm 1, \pm 2, \pm 3, \pm 6$; (B) No rational zeros   **11.** (A) $\pm 1, \pm 2, \pm 4, \pm 8$; (B) ± 2
13. (A) $\pm 1, \pm 2, \pm 3, \pm 6, \pm \frac{1}{3}, \pm \frac{2}{3}$; (B) $-\frac{1}{3}, 2$ **15.** $\frac{1}{2}, 1 \pm \sqrt{2}$ **17.** -2 (double root), $\pm \sqrt{5}$ **19.** $\pm 1, \frac{3}{2}, \pm i$
21. $P(x) = (x + 2)(2x - 1)(3x + 2)$ **23.** $P(x) = (x + 4)[x - (1 + \sqrt{2})][x - (1 - \sqrt{2})]$
25. $\sqrt{6}$ is a zero of $P(x) = x^2 - 6$, but $P(x)$ has no rational zeros.
27. $\sqrt[3]{5}$ is a zero of $P(x) = x^3 - 5$, but $P(x)$ has no rational zeros.

29. Inequality notation: $2 - \sqrt{3} \leq x \leq 2 + \sqrt{3}$, interval notation: $\left[2 - \sqrt{3},\ 2 + \sqrt{3}\right]$
31. Inequality notation: $-3 \leq x \leq \frac{1}{2}$ or $x \geq 2$, interval notation: $[-3, \frac{1}{2}] \cup [2, \infty)$

Exercise 3-7

1. 4.9 **3.** 0.7 **5.** $-1, 2.1$ **7.** $-\frac{1}{2}, 3, 1.9$
9. One irrational root is in each of the following intervals: $[-1, 0], [0, 1], [4, 5]$; largest ≈ 4.87
11. $P(2) = 2, P(2.5) = -0.1875, P(3) = 2$ **13.** 1.8 inches **15.** 0.4 foot

Exercise 3-8

1. $A = 2, B = 5$ **3.** $A = 7, B = -2$ **5.** $A = 1, B = 2, C = 3$ **7.** $A = 2, B = 1, C = 3$

9. $A = 0, B = 2, C = 2, D = -3$ **11.** $\dfrac{3}{x - 4} - \dfrac{4}{x + 2}$ **13.** $\dfrac{3}{3x + 4} - \dfrac{1}{2x - 3}$ **15.** $\dfrac{2}{x} - \dfrac{1}{x - 3} - \dfrac{3}{(x - 3)^2}$

17. $\dfrac{2}{x} + \dfrac{3x - 1}{x^2 + 2x + 3}$ **19.** $\dfrac{2x}{x^2 + 2} + \dfrac{3x + 5}{(x^2 + 2)^2}$ **21.** $\dfrac{x - 2 + 3}{x - 2} - \dfrac{2}{x - 3}$ **23.** $\dfrac{2}{x - 3} + \dfrac{2x + 5}{x^2 + 3x + 3}$

25. $\dfrac{2}{x - 4} - \dfrac{1}{x + 3} + \dfrac{3}{(x + 3)^2}$ **27.** $\dfrac{2}{x - 2} - \dfrac{3}{(x - 2)^2} - \dfrac{2x}{x^2 - x + 1}$ **29.** $\dfrac{x + 2 + 1}{2x - 1} - \dfrac{2}{x + 2} + \dfrac{x - 1}{2x^2 - x + 1}$

Exercise 3-9   Chapter Review

1. $2x^3 + 3x^2 - 1 = (x + 2)(2x^2 - x + 2) - 5$ (3-2) **2.** $P(3) = -8$ (3-3) **3.** $2, -4, -1$ (3-3) **4.** $1 - i$ (3-4)
5. (A) Zero or two positive, one negative; (B) Zero positive, one negative   (3-5) **6.** LB: -2 and -1, UB: 4   (3-5)
7. $P(1) = -5$ and $P(2) = 1$ are of opposite sign.   (3-5) **8.** $\pm 1, \pm 2, \pm 3, \pm 6$ (3-6) **9.** $-1, 2, 3$ (3-6)
10. $\dfrac{2}{x - 3} + \dfrac{5}{x + 2}$ (3-8) **11.** $P(x) = (x - \frac{2}{3})(3x^2 - 6x - 3) - 5$ (3-2) **12.** -4 (3-3)
13. $P(x) = \left[x - \left(1 + \sqrt{2}\right)\right]\left[x - \left(1 - \sqrt{2}\right)\right]$ (3-3) **14.** Yes, since $P(-1) = (-1)^{25} + 1 = 0$ (3-3)
15. (A) Zero or two positive, zero or two negative; (B) LB: -3, UB: 4; (C) -2 is a zero; $(-1, 0), (0, 1),$ and $(3, 4)$ each contains a
zero.   (3-3) **16.** $-2, -\frac{1}{2}, 4$ (3-6) **17.** $P(x) = 2(x + 2)(x + \frac{1}{2})(x - 4)$ (3-3) **18.** No rational zeros   (3-6)
19. $\frac{1}{2}, \dfrac{1 + \sqrt{3}i}{2}, \dfrac{1 - \sqrt{3}i}{2}$ (3-6, 3-7) **20.** $P(x) = 2(x - \frac{1}{2})\left(x - \dfrac{1 + \sqrt{3}i}{2}\right)\left(x - \dfrac{1 - \sqrt{3}i}{2}\right)$ (3-3) **21.** 1.4   (3-7)

22. $\dfrac{1}{x} - \dfrac{2}{x - 2} + \dfrac{3}{(x - 2)^2}$ (3-8) **23.** $\dfrac{3}{x} + \dfrac{2x - 1}{2x^2 - 3x + 3}$ (3-8)

24. $P(x) = [x - (1 + i)][x^2 + (1 + i)x + (3 + 2i)] + (3 + 5i)$ (3-2) **25.** $P(x) = (x + \frac{1}{2})^2(x + 3)(x - 1)^3$, degree 6   (3-3)

26. $P(x) = (x + 5)[x - (2 - 3i)][x - (2 + 3i)]$, degree 3   (3-3) **27.** $\frac{1}{2}, 3.0$ (3-6, 3-7) **28.** $\dfrac{2}{x - 3} - \dfrac{3}{x} + \dfrac{x - 1}{x^2 + 1}$ (3-8)

Practice Test: Chapter 3

1. $Q(x) = 8x^3 - 12x^2 - 16x - 8, R = 5, P(\frac{1}{4}) = R = 5$ (3-3) **2.** Since $P(-1) = 0, x - (-1) = x + 1$ is a factor of $P(x)$.   (3-3)
3. (A) $\pm 1, \pm 2, \pm 3, \pm 6, \pm \frac{1}{2}, \pm \frac{3}{2}$; (B) Two or no positive real zeros and one negative real zero   (3-5, 3-6)
4. (A) $[-1, 0], [1, 2], [2, 3]$; (B) LB: -1, UB: 4   (3-5) **5.** $\frac{3}{2}, 1 \pm \sqrt{3}$ (3-6)
6. $P(x) = 2(x - \frac{3}{2})\left[x - \left(1 + \sqrt{3}\right)\right]\left[x - \left(1 - \sqrt{3}\right)\right]$ (3-3, 3-6)

7. Since $P(x) = x^3 + 3x - 5$ has one variation in sign and $P(-x)$ has no variations in sign, by Descartes' rule of sign $P(x)$ has no negative real zeros and exactly one positive real zero.   (3-5)

8. 1.2   (3-7)   9. $\dfrac{2}{x+2} - \dfrac{1}{x-1} + \dfrac{3}{(x-1)^2}$ (3-8)   10. $\dfrac{1}{x+1} + \dfrac{2x}{x^2+4}$ (3-8)

Chapter 4   Exercise 4-1

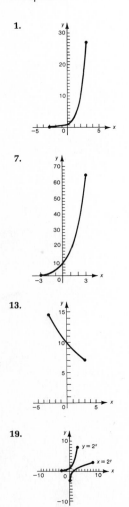

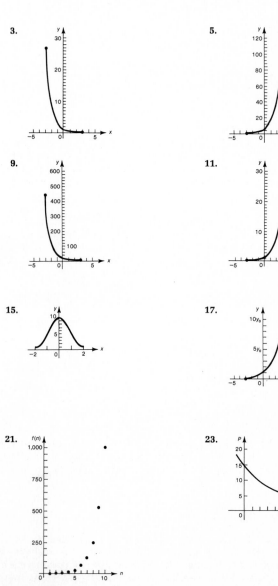

25.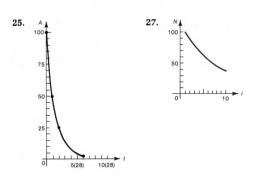

27.

Exercise 4-2

1. $9 = 3^2$ **3.** $81 = 3^4$ **5.** $1,000 = 10^3$ **7.** $1 = e^0$ **9.** $\log_8 64 = 2$ **11.** $\log_{10} 10,000 = 4$ **13.** $\log_v u = x$
15. $\log_{27} 9 = \frac{2}{3}$ **17.** 5 **19.** -4 **21.** 2 **23.** 3 **25.** $x = 4$ **27.** $y = 2$ **29.** $b = 4$
31. $0.001 = 10^{-3}$ **33.** $3 = 81^{1/4}$ **35.** $16 = (\frac{1}{2})^{-4}$ **37.** $N = a^e$ **39.** $\log_{10} 0.01 = -2$ **41.** $\log_e 1 = 0$
43. $\log_2 (\frac{1}{8}) = -3$ **45.** $\log_{81} (\frac{1}{3}) = -\frac{1}{4}$ **47.** $\log_{49} 7 = \frac{1}{2}$ **49.** u **51.** $\frac{1}{2}$ **53.** $\frac{3}{2}$ **55.** 0 **57.** $\frac{3}{2}$
59. $x = 2$ **61.** $y = -2$ **63.** $b = 100$ **65.** $b = $ Any positive real number except 1
67. Domain of f is the set of all real numbers; the range of f is $\{1\}$. The domain of f^{-1} is $\{1\}$; the range of f^{-1} is the set of all real numbers. No, f^{-1} is not.
69. (A) 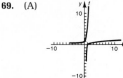   (B) Domain of f is the set of real numbers; range of f is the set of positive real numbers. The domain of f is the range of f^{-1} and the range of f is the domain of f^{-1}.
(C) f^{-1} is called the logarithmic function with base 10.

71. $f^{-1}(x) = [1 + \log_5 (x - 4)]/3$ **73.** $g^{-1}(x) = (b^{x/3} + 2)/5$

Exercise 4-3

1. $\log_b u + \log_b v$ **3.** $\log_b A - \log_b B$ **5.** $5 \log_b u$ **7.** $\frac{3}{5} \log_b N$ **9.** $\frac{1}{2} \log_b Q$ **11.** $\log_b u + \log_b v + \log_b w$
13. $\log_b AB$ **15.** $\log_b X/Y$ **17.** $\log_b wx/y$ **19.** 3.40 **21.** -0.92 **23.** 3.30 **25.** $2 \log_b u + 7 \log_b v$
27. $-\log_b a$ **29.** $\frac{1}{3} \log_b N - 2 \log_b p - 3 \log_b q$ **31.** $\frac{1}{4}(2 \log_b x + 3 \log_b y - \frac{1}{2} \log_b z)$ **33.** $\log_b x^2/y$
35. $\log_b x^3y^2/z^4$ **37.** $\log_b \sqrt[5]{x^2y^3}$ **39.** 2.02 **41.** 0.23 **43.** -0.05 **45.** 8 **47.** $y = cb^{-kt}$
49. Let $u = \log_b M$ and $v = \log_b N$; then $M = b^u$ and $N = b^v$. Thus, $\log_b M/N = \log_b b^u/b^v = \log_b b^{u-v} = u - v = \log_b M - \log_b N$.
51. $MN = b^{\log_b M} b^{\log_b N} = b^{\log_b M + \log_b N}$; hence, by definition of logarithm $\log_b MN = \log_b M + \log_b N$.

Exercise 4-4

1. 4.9177 **3.** -2.8419 **5.** 3.7623 **7.** -2.5128 **9.** 0.8627 **11.** 3.3096 **13.** -1.3840 **15.** $200,800$
17. $0.000\ 6648$ **19.** 47.73 **21.** 0.6760 **23.** 5.4843 **25.** -2.3215 **27.** 2.74×10^7 **29.** 1.58×10^{-5}
31. 4.959 **33.** 7.861 **35.** 3.301 **37.** 3.6776 **39.** -1.6094 **41.** -1.7372 **43.** 0.8544 **45.** 13.3114
47. -5.7541 **49.** 12.725 **51.** -25.715 **53.** 1.1709×10^{32} **55.** 4.2672×10^{-7}

Exercise 4-5

1. 1.46 **3.** 0.321 **5.** 1.29 **7.** 3.50 **9.** 1.80 **11.** 2.07 **13.** 20 **15.** 5 **17.** 14.2 **19.** −1.83
21. 11.7 **23.** 5 **25.** 1, e^2, e^{-2} **27.** $x = e^e$ **29.** 100, 0.1 **31.** $x = -(1/k) \ln (I/I_0)$ **33.** $I = I_0 10^{N/10}$
35. $t = (-L/R) \ln [1 - (RI/E)]$
37. Inequality sign should have been reversed when both sides were multiplied by $\log \frac{1}{2}$, a negative quantity.
39. 5 years to the nearest year **41.** Approx. 3.8 hours **43.** Approx. 35 years **45.** Approx. 28 years
47. Divide both sides by I_0, take logs of both sides, and then multiply both sides by 10. **49.** 95 feet, 489 feet

Exercise 4-6   Chapter Review

1. $n = \log_{10} m$ (4-2) **2.** $x = 10^y$ (4-2) **3.** 8 (4-2) **4.** 5 (4-2) **5.** 3 (4-2) **6.** 1.24 (4-5)
7. 11.9 (4-5) **8.** 900 (4-3, 4-5) **9.** 5 (4-3, 4-5) **10.** $y = e^x$ (4-2) **11.** $y = \ln x$ (4-2) **12.** −2 (4-2)
13. $\frac{1}{3}$ (4-2) **14.** 64 (4-2) **15.** e (4-2) **16.** 33 (4-2) **17.** 1 (4-2) **18.** 2.32 (4-5) **19.** 3.92 (4-5)
20. 92.1 (4-5) **21.** 300 (4-3, 4-5) **22.** 2 (4-3, 4-5) **23.** 1, 10^3, 10^{-3} (4-3, 4-5) **24.** 10^e (4-3, 4-5)
25. 1.95 (4-4) **26.** $y = ce^{-5t}$ (4-3, 4-5) **27.** Domain $f = (0, \infty) = $ Range f^{-1}, Range $f = R = $ Domain f^{-1} (4-2)

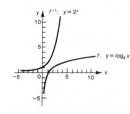

28. If $\log_1 x = y$, then we would have to have $1^y = x$; that is, $1 = x$ for arbitrary positive x, which is impossible. (4-2)
29. Let $u = \log_b M$ and $v = \log_b N$; then $M = b^u$ and $N = b^v$. Thus, $\log_b (M/N) = \log_b (b^u/b^v) = \log_b b^{u-v} = u - v = \log_b M - \log_b N$. (4-3)
30. 23.4 years (4-5) **31.** 23.1 years (4-5) **32.** 37,100 years (4-5) **33.** $I = I_0 e^{-kx}$ (4-3, 4-5)
34. $n = -\log [1 - (Pi/R)]/\log (1 + i)$ (4-3, 4-5)

Practice Test: Chapter 4

1. $y = e^{-x^2}$ (4-2) **2.** $\log_b [(x^{1/2}y^3)/z^2]$ (4-3) **3.** 1.63 (4-5) **4.** 2.11 (4-5) **5.** 11,200 (4-5)
6. −3 (4-2) **7.** 8 (4-2) **8.** $\frac{1}{10}$ (4-2) **9.** 6 (4-3, 4-5) **10.** e^{10} (4-3, 4-5)
11. $f^{-1}(x) = \ln x$, Domain of $f = R = $ Range of f^{-1}, Range of $f = (0, \infty) = $ Domain of f^{-1} (4-2) **12.** 0.837 (4-4)
13. 9.24 years (4-5) **14.** $r = (1/t) \ln (A/P)$ (4-3, 4-5) **15.** $n = -\log [1 - (Pi/R)]/\log (1 + i)$ (4-3, 4-5)

Chapter 5   Exercise 5-1

1. (2, −3) **3.** No solution (parallel lines) **5.** (6, 2) **7.** (2, −1) **9.** (2, −1) **11.** (5, 2)
13. Infinitely many solutions (dependent) **15.** $(1, -\frac{2}{3})$ **17.** No solution (inconsistent) **19.** (2,500, 200)
21. (1, 0.2) **23.** (3, −1, −2) **25.** (1, −1, 2) **27.** (−2, 1, 0) **29.** Infinitely many solutions (dependent)
31. 35 20¢ stamps, 12 15¢ stamps **33.** 40 milliliters of 50% solution and 60 milliliters of 80% solution
35. $3\frac{1}{3}$ grams of 18-carat and $6\frac{2}{3}$ grams of 12-carat **37.** 80 grams of mix A, 60 grams of mix B
39. 40 seconds; 24 seconds; 120 miles **41.** 1,200 style A; 800 style B; 2,000 style C
43. 60 grams of mix A, 50 grams of mix B, 40 grams of mix C

Exercise 5-2

1. $\begin{bmatrix} 4 & -6 & | & -8 \\ 1 & -3 & | & 2 \end{bmatrix}$ **3.** $\begin{bmatrix} -4 & 12 & | & -8 \\ 4 & -6 & | & -8 \end{bmatrix}$ **5.** $\begin{bmatrix} 1 & -3 & | & 2 \\ 8 & -12 & | & -16 \end{bmatrix}$ **7.** $\begin{bmatrix} 1 & -3 & | & 2 \\ 0 & 6 & | & -16 \end{bmatrix}$

9. $\begin{bmatrix} 1 & -3 & \bigm| & 2 \\ 2 & 0 & \bigm| & -12 \end{bmatrix}$ **11.** $\begin{bmatrix} 1 & -3 & \bigm| & 2 \\ 3 & -3 & \bigm| & -10 \end{bmatrix}$ **13.** $x_1 = 3, x_2 = 2$ **15.** $x_1 = 3, x_2 = 1$ **17.** $x_1 = 2, x_2 = 1$

19. $x_1 = 2, x_2 = 4$ **21.** No solution   **23.** $x_1 = 1, x_2 = 4$

25. Infinitely many solutions: $x_2 = s, x_1 = 2s - 3$ for any real number s

27. Infinitely many solutions: $x_2 = s, x_1 = \frac{1}{2}s + \frac{1}{2}$ for any real number s **29.** $x_1 = 2, x_2 = -1$ **31.** $x_1 = 2, x_2 = -1$

33. $x_1 = 1.1, x_2 = 0.3$

Exercise 5-3

1. Yes   **3.** No   **5.** No   **7.** Yes   **9.** $x_1 = -2, x_2 = 3, x_3 = 0$ **11.** $x_1 = 2t + 3$ **13.** No solution

$x_2 = -t - 5$

$x_3 = t$

t any real number

15. $x_1 = 2s + 3t - 5$ **17.** $\begin{bmatrix} 1 & 0 & \bigm| & -7 \\ 0 & 1 & \bigm| & 3 \end{bmatrix}$ **19.** $\begin{bmatrix} 1 & 0 & 0 & \bigm| & -5 \\ 0 & 1 & 0 & \bigm| & 4 \\ 0 & 0 & 1 & \bigm| & -2 \end{bmatrix}$ **21.** $\begin{bmatrix} 1 & 0 & 2 & \bigm| & -\frac{5}{3} \\ 0 & 1 & -2 & \bigm| & \frac{1}{3} \\ 0 & 0 & 0 & \bigm| & 0 \end{bmatrix}$

$x_2 = s$

$x_3 = -3t + 2$

$x_4 = t$

s and t any real numbers

23. $x_1 = -2, x_2 = 3, x_3 = 1$ **25.** $x_1 = 0, x_2 = -2, x_3 = 2$ **27.** $\begin{cases} x_1 = 2t + 3 \\ x_2 = t - 2 \\ x_3 = t \\ t \text{ any real number} \end{cases}$ **29.** $x_1 = (-4t - 4)/7$

$x_2 = (5t + 5)/7$

$x_3 = t$

t any real number

31. $x_1 = -1, x_2 = 2$ **33.** No solution   **35.** No solution   **37.** $x_1 = 0, x_2 = 2, x_3 = -3$

39. $x_1 = 1, x_2 = -2, x_3 = 1$ **41.** $x_1 = 2s - 3t + 3$

$x_2 = s + 2t + 2$

$x_3 = s$

$x_4 = t$

s and t any real numbers

43. 20 one-person boats, 220 two-person boats, 100 four-person boats

45. $(t - 80)$ one-person boats, $(-2t + 420)$ two-person boats, t four-person boats, $80 \le t \le 210$, t an integer

47. No solution; no production schedule will use all the work-hours in all departments.

49. 8 ounces food A, 2 ounces food B, 4 ounces food C   **51.** No solution

53. 8 ounces food A, $(-2t + 10)$ ounces food B, t ounces food C, $0 \le t \le 5$

55. Company A: 10 hours; company B: 15 hours

Exercise 5-4

1. $(-12, 5), (-12, -5)$ **3.** $(2, 4), (-2, -4)$ **5.** $(5i, -5i), (-5i, 5i)$ **7.** $(3 + 4i, -1 + 2i), (3 - 4i, -1 - 2i)$

9. $(2, 4), (2, -4), (-2, 4), (-2, -4)$ **11.** $(1, 3), (1, -3), (-1, 3), (-1, -3)$

13. $\left(-1 + \sqrt{3}i, 1 + \sqrt{3}i\right), \left(-1 - \sqrt{3}i, 1 - \sqrt{3}i\right)$ **15.** $(0, -1), (-4, -3)$ **17.** $(2, 2i), (2, -2i), (-2, 2i), (-2, -2i)$

19. $\left(2, \sqrt{2}\right), \left(2, -\sqrt{2}\right), (-1, i), (-1, -i)$ **21.** $(2, 1), (-2, 1), (2i, 3), (-2i, 3)$ **23.** $(2, 1), (-2, -1), (i, -2i), (-i, 2i)$

25. $(2, 2), (-2, -2), \left(\sqrt{2}, -\sqrt{2}\right), \left(-\sqrt{2}, \sqrt{2}\right)$ **27.** $(-3, 1), (3, -1), (-i, i), (i, -i)$

29. $\left(\frac{1}{2} + \left(\sqrt{3}/2\right)i, \frac{1}{2} - \left(\sqrt{3}/2\right)i\right), \left(\frac{1}{2} - \left(\sqrt{3}/2\right)i, \frac{1}{2} + \left(\sqrt{3}/2\right)i\right)$ **31.** 12 by 5 inches   **33.** 20¢

Exercise 5-5

1.   **3.**   **5.**

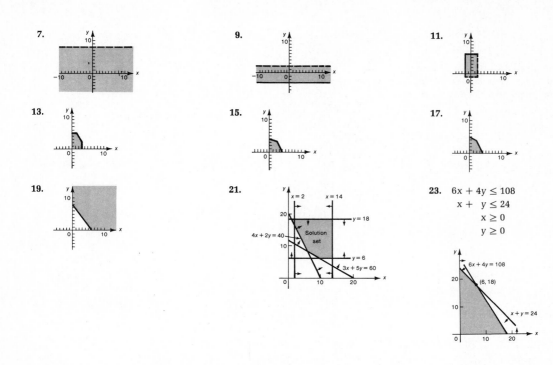

7.   **9.**   **11.**

13.   **15.**   **17.**

19.   **21.**   **23.** $6x + 4y \le 108$
$\quad\quad\quad\quad\quad\quad\quad\quad\quad\quad x + \ \ y \le 24$
$\quad\quad\quad\quad\quad\quad\quad\quad\quad\quad x \ge 0$
$\quad\quad\quad\quad\quad\quad\quad\quad\quad\quad y \ge 0$

Exercise 5-6   Chapter Review

1. $(2, 3)$ (5-1)   **2.** No solution (inconsistent)   (5-1)

3. Infinitely many solutions: $(x, (4x + 8)/3)$ for any real number x   (5-1)   **4.** $(2, 1, -1)$ (5-1)

5. $(1, -1), (\frac{7}{5}, -\frac{1}{5})$ (5-4)   **6.** $(1, 3), (1, -3), (-1, 3), (-1, -3)$ (5-4)

7. (5-1)   **8.** (5-5)

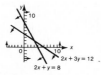

9. $x_1 = -1, x_2 = 3$ (5-2, 5-3)   **10.** $x_1 = -1, x_2 = 2, x_3 = 1$ (5-3)   **11.** $x_1 = 2, x_2 = 1, x_3 = -1$ (5-3)

12. Infinitely many solutions: $x_1 = -5t - 12, x_2 = 3t + 7, x_3 = t$, t any real number   (5-3)   **13.** No solution   (5-3)

14. Infinitely many solutions: $x_1 = -\frac{3}{7}t - \frac{4}{7}, x_2 = \frac{5}{7}t + \frac{9}{7}, x_3 = t$, t any real number   (5-3)

15. $\left(2, \sqrt{2}\right), \left(2, -\sqrt{2}\right), (-1, i), (-1, -i)$ (5-4)   **16.** $(2, 2), (-2, -2), \left(\sqrt{2}, -\sqrt{2}\right), \left(-\sqrt{2}, \sqrt{2}\right)$ (5-4)

17. (5-5)   **18.** $x_1 = 1,000, x_2 = 4,000, x_3 = 2,000$ (5-3)

19. $(2, 2), (-2, -2), \left(\frac{4}{7}\sqrt{7}, -\frac{2}{7}\sqrt{7}\right), \left(-\frac{4}{7}\sqrt{7}, \frac{2}{7}\sqrt{7}\right)$ (5-4)   **20.** $48\frac{1}{2}$-pound packages, $72\frac{1}{3}$-pound packages   (5-1, 5-3)

21. 6 by 8 meters   (5-1, 5-3)   **22.** 40 grams mix A, 60 grams mix B, 30 grams mix C   (5-1, 5-3)

Practice Test: Chapter 5

1. (5-1)

2. (5-5)

3. (5-5)

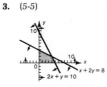

4. $(1 + i, 2i)$ $(1 - i, -2i)$ (5-4)   **5.** $(1, 3), (1, -3), (-1, 3), (-1, -3)$ (5-4)
6. $\left(0, 2\sqrt{2}\right), \left(0, -2\sqrt{2}\right), \left(\sqrt{2}, \sqrt{2}\right), \left(-\sqrt{2}, -\sqrt{2}\right)$ (5-4)   **7.** $x_1 = 3, x_2 = -1$ (5-2, 5-3)
8. $x_1 = 2, x_2 = -2, x_3 = 1$ (5-3)   **9.** No solution   (5-3)
10. $x_1 = 2t - 3, x_2 = -5t + 1, x_3 = t$, t any real number   (5-3)

Chapter 6   Exercise 6-1

1. $2 \times 2, 1 \times 4$ **3.** 2   **5.** $\begin{bmatrix} 0 & 0 \\ 0 & 0 \end{bmatrix}$ **7.** C, D **9.** A, B **11.** $\begin{bmatrix} -1 & 0 \\ 5 & -3 \end{bmatrix}$ **13.** $\begin{bmatrix} -2 \\ 3 \\ 0 \end{bmatrix}$ **15.** $\begin{bmatrix} -1 \\ 6 \\ 5 \end{bmatrix}$

17. $\begin{bmatrix} -15 & 5 \\ 10 & -15 \end{bmatrix}$ **19.** $[0 \quad 1 \quad 4]$ **21.** $\begin{bmatrix} 250 & 360 \\ 40 & 350 \end{bmatrix}$ **23.** $\begin{bmatrix} -1 & 0 \\ 0 & 6 \\ -1 & -1 \end{bmatrix}$ **25.** $a = -1, b = 1, c = 3, d = -5$

27. Guitar Banjo
$\begin{bmatrix} \$33 & \$26 \\ \$57 & \$77 \end{bmatrix}$ Materials
  Labor

29. $\begin{bmatrix} 135 & 282 & 50 \\ 55 & 258 & 155 \end{bmatrix}, \begin{bmatrix} 0.14 & 0.30 & 0.05 \\ 0.06 & 0.28 & 0.17 \end{bmatrix}$

Exercise 6-2

1. 10   **3.** -1 **5.** $[12 \quad 13]$ **7.** $\begin{bmatrix} 5 \\ -3 \end{bmatrix}$ **9.** $\begin{bmatrix} 2 & 4 \\ 1 & -5 \end{bmatrix}$ **11.** $\begin{bmatrix} 10 & 1 \\ -2 & 10 \end{bmatrix}$ **13.** 6   **15.** 15

17. $\begin{bmatrix} 0 & 9 \\ 5 & -4 \end{bmatrix}$ **19.** $\begin{bmatrix} 5 & 8 & -5 \\ -1 & -3 & 2 \\ -2 & 8 & -6 \end{bmatrix}$ **21.** $[11]$ **23.** $\begin{bmatrix} 3 & -2 & -4 \\ 6 & -4 & -8 \\ -9 & 6 & 12 \end{bmatrix}$ **25.** $AB = \begin{bmatrix} 5 & 7 \\ 2 & 3 \end{bmatrix}, BA = \begin{bmatrix} 1 & 3 \\ 2 & 7 \end{bmatrix}$

27. Both sides equal $\begin{bmatrix} 0 & 12 \\ 1 & 5 \end{bmatrix}$ **29.** (A) \$9 per boat   (B) $[1.5 \quad 1.2 \quad 0.4] \cdot \begin{bmatrix} 7 \\ 10 \\ 4 \end{bmatrix} = \$24.10$ (C) 3×2

(D)
	I	II	
	\$9.00	\$11.00	One-person
	\$14.10	\$17.20	Two-person
	\$19.80	\$24.10	Four-person

Labor costs per boat
at each plant

31. (A) \$2,025   (B) $[2,000 \quad 800 \quad 8,000] \cdot \begin{bmatrix} \$0.40 \\ \$0.75 \\ \$0.25 \end{bmatrix} = \$3,400$ (C) $\begin{bmatrix} \$2,025 \\ \$3,400 \end{bmatrix}$ Berkeley
Oakland
Cost per
town

(D)
Telephone	House	Letter
[3,000	1,300	13,000]

Number of each type of contact made

Exercise 6-3

1. $\begin{bmatrix} 2 & -3 \\ 4 & 5 \end{bmatrix}$ **3.** $\begin{bmatrix} -2 & 1 & 3 \\ 2 & 4 & -2 \\ 5 & 1 & 0 \end{bmatrix}$ **9.** $x_1 = -8, x_2 = 2$ **11.** $x_1 = 0, x_2 = 4$ **13.** $\begin{bmatrix} 3 & -2 \\ -1 & 1 \end{bmatrix}$

15. $\begin{bmatrix} 7 & -3 \\ -2 & 1 \end{bmatrix}$ **17.** $\begin{bmatrix} 7 & 6 & -3 \\ 2 & 2 & -1 \\ -6 & -5 & 3 \end{bmatrix}$ **19.** $\frac{1}{2}\begin{bmatrix} 3 & -1 & -1 \\ -1 & 1 & 1 \\ -3 & 1 & 3 \end{bmatrix}$ **21.** (A) $x_1 = -3, x_2 = 2$
(B) $x_1 = -1, x_2 = 2$
(C) $x_1 = -8, x_2 = 3$

23. (A) $x_1 = 17, x_2 = -5$ **25.** (A) $x_1 = 1, x_2 = 0, x_3 = 0$ **27.** (A) $x_1 = 1, x_2 = 1, x_3 = 3$
(B) $x_1 = 7, x_2 = -2$ (B) $x_1 = -1, x_2 = 0, x_3 = 1$ (B) $x_1 = -1, x_2 = 1, x_3 = -1$
(C) $x_1 = 24, x_2 = -7$ (C) $x_1 = -1, x_2 = -1, x_3 = 1$ (C) $x_1 = 5, x_2 = -1, x_3 = -5$

35. Concert 1: 6,000 \$4 tickets and 4,000 \$8 tickets; concert 2: 5,000 \$4 tickets and 5,000 \$8 tickets; concert 3: 3,000 \$4 tickets and 7,000 \$8 tickets

37. Diet 1: 60 ounces mix A and 80 ounces mix B; diet 2: 20 ounces mix A and 60 ounces mix B; diet 3: 0 ounces mix A and 100 ounces mix B

Exercise 6-4

1. 8 **3.** -20 **5.** -0.88 **7.** $\begin{vmatrix} a_{22} & a_{23} \\ a_{32} & a_{33} \end{vmatrix}$ **9.** $\begin{vmatrix} a_{11} & a_{12} \\ a_{31} & a_{32} \end{vmatrix}$ **11.** $(-1)^{1+1}\begin{vmatrix} a_{22} & a_{23} \\ a_{32} & a_{33} \end{vmatrix}$ **13.** $(-1)^{2+3}\begin{vmatrix} a_{11} & a_{12} \\ a_{31} & a_{32} \end{vmatrix}$

15. $\begin{vmatrix} 1 & -2 \\ -4 & 8 \end{vmatrix}$ **17.** $\begin{vmatrix} -2 & 0 \\ 5 & -2 \end{vmatrix}$ **19.** $(-1)^{1+1}\begin{vmatrix} 1 & -2 \\ -4 & 8 \end{vmatrix} = 0$ **21.** $(-1)^{3+2}\begin{vmatrix} -2 & 0 \\ 5 & -2 \end{vmatrix} = -4$ **23.** 10

25. -21 **27.** -40 **29.** $(-1)^{1+1}\begin{vmatrix} a_{22} & a_{23} & a_{24} \\ a_{32} & a_{33} & a_{34} \\ a_{42} & a_{43} & a_{44} \end{vmatrix}$ **31.** $(-1)^{4+3}\begin{vmatrix} a_{11} & a_{12} & a_{14} \\ a_{21} & a_{22} & a_{24} \\ a_{31} & a_{32} & a_{34} \end{vmatrix}$ **33.** 22 **35.** -12

37. 0 **39.** 6 **41.** 60 **43.** $\begin{vmatrix} a & b \\ ka & kb \end{vmatrix} = akb - kab = 0$ **45.** $\begin{vmatrix} a & b \\ c & d \end{vmatrix} = ad - cb = ad - bc = \begin{vmatrix} a & c \\ b & d \end{vmatrix}$

49. $49 = (-7)(-7)$

Exercise 6-5

1. Theorem 2 **3.** Theorem 2 **5.** Theorem 3 **7.** Theorem 4 **9.** Theorem 6 **11.** $x = 0$ **13.** $x = 5$
15. 25 **17.** -12 **19.** Theorem 2 **21.** Theorem 3 **23.** Theorem 6 **25.** $x = 5, y = 0$
27. $x = -3, y = 10$ **29.** -28 **31.** 106 **33.** 0 **35.** 6 **37.** 14
39. Expand the left member of the equation using minors. **41.** Expand both members of the equation and compare.
43. This follows from Theorem 5.
45. Expand the determinant about the first row to obtain $(y_1 - y_2)x - (x_1 - x_2)y + (x_1y_2 - x_2y_1) = 0$; then show that the two points satisfy this linear equation.
47. If the determinant is 0, then the area of the triangle formed by the three points is 0. The only way that this can happen is if the three points are on the same line; that is, the points are collinear.

Exercise 6-6

1. $x = 5, y = -2$ **3.** $x = 1, y = -1$ **5.** $x = -1, y = 1$ **7.** $x = 2, y = -2, z = -1$ **9.** $x = 2, y = -1, z = 2$
11. $x = 2, y = -3, z = -1$ **13.** $x = 1, y = -1, z = 2$
15. Since $D = 0$, the system has either no solution or infinitely many. Since $x = 0, y = 0, z = 0$ is a solution, the second case must hold.

Exercise 6-7   Chapter Review

1. $\begin{bmatrix} 3 & 3 \\ 4 & 2 \end{bmatrix}$ (6-1) **2.** Not defined (6-1) **3.** $\begin{bmatrix} -3 & 0 \\ 1 & -1 \end{bmatrix}$ (6-1) **4.** $\begin{bmatrix} 4 & 3 \\ 7 & 4 \end{bmatrix}$ (6-2) **5.** Not defined (6-2)

6. $\begin{bmatrix} 5 \\ 5 \end{bmatrix}$ (6-2) **7.** $\begin{bmatrix} 2 & 3 \\ 4 & 6 \end{bmatrix}$ (6-2) **8.** 8 (a real number) (6-2) **9.** Not defined (6-2) **10.** $\begin{bmatrix} 3 & -2 \\ -4 & 3 \end{bmatrix}$ (6-3)

11. (A) $x_1 = -1, x_2 = 3$; (B) $x_1 = 1, x_2 = 2$; (C) $x_1 = 8, x_2 = -10$ (6-3) **12.** -17 (6-4) **13.** 0 (6-4)

14. $x = 2, y = -1$ (6-6) **15.** Not defined (6-1) **16.** $\begin{bmatrix} 10 & -8 \\ 4 & 6 \end{bmatrix}$ (6-1, 6-2) **17.** $\begin{bmatrix} -2 & 8 \\ 8 & 6 \end{bmatrix}$ (6-1, 6-2)

18. 9 (a real number) (6-2) **19.** [9] (a matrix) (6-2) **20.** $\begin{bmatrix} 10 & -5 & 1 \\ -1 & -4 & -5 \\ 1 & -7 & -2 \end{bmatrix}$ (6-1, 6-2)

21. $\begin{bmatrix} -\frac{5}{2} & 2 & -\frac{1}{2} \\ 1 & -1 & 1 \\ \frac{1}{2} & 0 & -\frac{1}{2} \end{bmatrix}$ or $\frac{1}{2}\begin{bmatrix} -5 & 4 & -1 \\ 2 & -2 & 2 \\ 1 & 0 & -1 \end{bmatrix}$ (6-3)

22. (A) $x_1 = 2, x_2 = 1, x_3 = -1$; (B) $x_1 = 1, x_2 = -2, x_3 = 1$; (C) $x_1 = -1, x_2 = 2, x_3 = -2$ (6-3)

23. $-\frac{11}{12}$ (6-4) **24.** 35 (6-4, 6-5) **25.** $y = \frac{10}{5} = 2$ (6-6) **26.** $\begin{bmatrix} -\frac{11}{12} & -\frac{1}{12} & 5 \\ \frac{10}{12} & \frac{2}{12} & -4 \\ \frac{1}{12} & -\frac{1}{12} & 0 \end{bmatrix}$ or $\frac{1}{12}\begin{bmatrix} -11 & -1 & 60 \\ 10 & 2 & -48 \\ 1 & -1 & 0 \end{bmatrix}$ (6-3)

27. $x_1 = 1,000, x_2 = 4,000, x_3 = 2,000$ (6-3) **28.** 42 (6-4, 6-5)

29. $\begin{vmatrix} u + kv & v \\ w + kx & x \end{vmatrix} = (u + kv)x - (w + kx)v = ux + kvx - wv - kvx = ux - wv = \begin{vmatrix} u & v \\ w & x \end{vmatrix}$ (6-4, 6-5)

Practice Test: Chapter 6

1. 0 (a real number) (6-2) **2.** $\begin{bmatrix} 7 & -8 & 6 \\ -4 & 5 & -3 \end{bmatrix}$ (6-2) **3.** Not defined (6-2) **4.** $\begin{bmatrix} 6 & -12 & -12 \\ 5 & 4 & 7 \\ 0 & -4 & -4 \end{bmatrix}$ (6-1, 6-2)

5. Not defined (6-1) **6.** $A^{-1} = \begin{bmatrix} -4 & 3 \\ -3 & 2 \end{bmatrix}$ (6-3) **7.** (A) $x_1 = -17, x_2 = -12$; (B) $x_1 = 11, x_2 = 8$ (6-3)

8. 20 (6-4, 6-5) **9.** $z = \frac{70}{14} = 5$ (6-6)

10. Since $D = 0$, we conclude (from Cramer's rule) that the system has either no solution or infinitely many. Since (0, 0, 0) is a solution, the first case does not hold; therefore, there must be infinitely many solutions. (6-6)

Chapter 7   Exercise 7-1

1. $-1, 0, 1, 2$ **3.** $0, \frac{1}{3}, \frac{1}{2}, \frac{3}{5}$ **5.** $4, -8, 16, -32$ **7.** 6 **9.** $\frac{99}{101}$ **11.** $S_5 = 1 + 2 + 3 + 4 + 5$

13. $S_3 = \frac{1}{10} + \frac{1}{100} + \frac{1}{1,000}$ **15.** $S_4 = -1 + 1 - 1 + 1$ **17.** $1, -4, 9, -16, 25$ **19.** $0.3, 0.33, 0.333, 0.3333, 0.333\ 33$

21. $1, -\frac{1}{2}, \frac{1}{4}, -\frac{1}{8}, \frac{1}{16}$ **23.** $7, 3, -1, -5, -9$ **25.** $4, 1, \frac{1}{4}, \frac{1}{16}, \frac{1}{64}$ **27.** $a_n = n + 3$ **29.** $a_n = 3n$

31. $a_n = n/(n + 1)$ **33.** $a_n = (-1)^{n+1}$ **35.** $a_n = (-2)^n$ **37.** $a_n = x^n/n$ **39.** $S_4 = \frac{4}{1} - \frac{8}{2} + \frac{16}{3} - \frac{32}{4}$

41. $S_3 = x^2 + (x^3/2) + (x^4/3)$ **43.** $S_5 = x - (x^2/2) + (x^3/3) - (x^4/4) + (x^5/5)$ **45.** $S_4 = \sum_{k=1}^{4} k^2$ **47.** $S_5 = \sum_{k=1}^{5} \frac{1}{2^k}$

49. $S_n = \sum_{k=1}^{n} \frac{1}{k^2}$ **51.** $S_n = \sum_{k=1}^{n} (-1)^{k+1} k^2$ **55.** (A) $3, 1.83, 1.46, 1.415$; (B) $\sqrt{2} = 1.414$; (C) $1, 1.5, 1.417, 1.414$

57. Series approx. of $e^{0.2} = 1.221\ 400\ 0$, calculator value of $e^{0.2} = 1.221\ 402\ 8$

Exercise 7-2

1. Fails at $n = 2$ **3.** Fails at $n = 3$ **5.** $P_1: 2 = 2 \cdot 1^2$; $P_2: 2 + 6 = 2 \cdot 2^2$; $P_3: 2 + 6 + 10 = 2 \cdot 3^2$

7. $P_1: a^5 a = a^{5+1}$; $P_2: a^5 a^2 = a^5(a^1 a) = (a^5 a)a = a^6 a = a^7 = a^{5+2}$; $P_3: a^5 a^3 = a^5(a^2 a) = a^5(a^1 a)a = [(a^5 a)a]a = a^8 = a^{5+3}$

9. $P_1: 9^1 - 1 = 8$ is divisible by 4; $P_2: 9^2 - 1 = 80$ is divisible by 4; $P_3: 9^3 - 1 = 728$ is divisible by 4

11. $P_k: 2 + 6 + 10 + \cdots + (4k - 2) = 2k^2$; $P_{k+1}: 2 + 6 + 10 + \cdots + (4k - 2) + (4k + 2) = 2(k + 1)^2$

13. $P_k: a^5 a^k = a^{5+k}$; $P_{k+1}: a^5 a^{k+1} = a^{5+k+1}$ **15.** $P_k: 9^k - 1 = 4r$; $P_{k+1}: 9^{k+1} - 1 = 4s$; $r, s \in N$

39. Formula: $2 + 4 + 6 + \cdots + 2n = n(n + 1)$ **41.** $1 + 2 + 3 + \cdots + (n - 1) = n(n - 1)/2, n \geq 2$

47. $3^4 + 4^4 + 5^4 + 6^4 \neq 7^4$

Exercise 7-3

1. (B) $d = -0.5$; $5.5, 5$; (C) $d = -5$; $-26, -31$ **3.** $a_2 = -1, a_3 = 3, a_4 = 7$ **5.** $a_{15} = 67, S_{11} = 242$

7. $S_{21} = 861$ **9.** $a_{15} = -21$ **11.** $d = 6, a_{101} = 603$ **13.** $S_{40} = 200$ **15.** $a_{11} = 2, S_{11} = \frac{77}{6}$ **17.** $a_1 = 1$

19. $S_{51} = 4{,}131$ **21.** $-1{,}071$ **23.** $4{,}446$ **27.** $a_n = -3 + (n - 1)3$

29. (A) 336 feet; (B) 1,936 feet; (C) $16t^2$ **33.** Hint: $y = x + d, z = x + 2d$ **35.** $x = -1, y = 2$

Exercise 7-4

1. (A) $r = -2$; $-16, 32$; (D) $r = \frac{1}{3}$; $\frac{1}{54}, \frac{1}{162}$ **3.** $a_2 = 3, a_3 = -\frac{3}{2}, a_4 = \frac{3}{4}$ **5.** $a_{10} = \frac{1}{243}$ **7.** $S_7 = 3{,}279$

9. $r = 0.398$ **11.** $S_{10} = -1{,}705$ **13.** $a_2 = 6, a_3 = 4$ **15.** $S_7 = 547$ **17.** $\frac{1{,}023}{1{,}024}$ **19.** $x = 2\sqrt{3}$

21. $S_\infty = \frac{9}{2}$ **23.** No sum   **25.** $S_\infty = \frac{8}{5}$ **27.** $\frac{7}{9}$ **29.** $\frac{6}{11}$ **31.** $3\frac{8}{37}$ or $\frac{119}{37}$

33. $A = P(1 + r)^n$; approx. 12 years   **35.** 900   **37.** $a_n = (-2)(-3)^{n-1}$ **41.** \$4,000,000

Exercise 7-5

1. \$700 per year; \$115,500   **3.** 0.0015 pound/square inch   **5.** 1,250,000   **7.** $A = A_0 2^{2t}$ **9.** $r = 10^{-0.4} = 0.398$

11. 2   **13.** $\$9.223 \times 10^{16}$; $\$1.845 \times 10^{17}$

Exercise 7-6

1. 720   **3.** 20   **5.** 720   **7.** 15   **9.** 1   **11.** 28   **13.** $9!/8!$ **15.** $8!/5!$ **17.** 126   **19.** 6   **21.** 1

23. 2,380   **25.** $m^3 + 3m^2 n + 3mn^2 + n^3$ **27.** $8x^3 - 36x^2 y + 54xy^2 - 27y^3$ **29.** $x^4 - 8x^3 + 24x^2 - 32x + 16$

31. $m^4 + 12m^3 n + 54m^2 n^2 + 108mn^3 + 81n^4$ **33.** $32x^5 - 80x^4 y + 80x^3 y^2 - 40x^2 y^3 + 10xy^4 - y^5$

35. $m^6 + 12m^5 n + 60m^4 n^2 + 160m^3 n^3 + 240m^2 n^4 + 192mn^5 + 64n^6$ **37.** $5{,}005u^9 v^6$ **39.** $264m^2 n^{10}$ **41.** $924w^6$

43. $-48{,}384x^3 y^5$ **45.** 1.1046   **47.** $\binom{n}{r} = \dfrac{n!}{r!(n - r)!} = \dfrac{n!}{(n - r)![n - (n - r)]!} = \binom{n}{n - r}$

49. $\binom{k}{r - 1} + \binom{k}{r} = \dfrac{k!}{(r - 1)!(k - r + 1)!} + \dfrac{k!}{r!(k - r)!} = \dfrac{rk! + (k + 1 - r)k!}{r!(k + 1 - r)!} = \dfrac{(k + 1)!}{r!(k + 1 - r)!} = \binom{k + 1}{r}$

51. $\binom{k}{k} = \dfrac{k!}{k!(k - k)!} = 1 = \dfrac{(k + 1)!}{(k + 1)![(k + 1) - (k + 1)]!} = \binom{k + 1}{k + 1}$

Exercise 7-7   Chapter Review

1. (A) Geometric; (B) Arithmetic; (C) Arithmetic; (D) Neither; (E) Geometric (7-3, 7-4)

2. (A) $5, 7, 9, 11$; (B) $a_{10} = 23$; (C) $S_{10} = 140$ (7-1, 7-3, 7-4)

3. (A) $16, 8, 4, 2$; (B) $a_{10} = \frac{1}{32}$; (C) $S_{10} = 31\frac{31}{32}$ (7-1, 7-3, 7-4)

4. (A) $-8, -5, -2, 1$; (B) $a_{10} = 19$; (C) $S_{10} = 55$ (7-1, 7-3, 7-4)

5. (A) $-1, 2, -4, 8$; (B) $a_{10} = 512$; (C) $S_{10} = 341$ (7-1, 7-3, 7-4)   6. $S_{\infty} = 32$ (7-4)   7. 720   (7-6)

8. $20 \cdot 21 \cdot 22 = 9{,}240$ (7-6)   9. 21   (7-6)

10. $P_1 : 5 = 1^2 + 4 \cdot 1$; $P_2 : 5 + 7 = 2^2 + 4 \cdot 2$; $P_3 : 5 + 7 + 9 = 3^2 + 4 \cdot 3$ (7-2)

11. $P_1 : 2 = 2^{1+1} - 2$; $P_2 : 2 + 4 = 2^{2+1} - 2$; $P_3 : 2 + 4 + 8 = 2^{3+1} - 2$ (7-2)

12. $P_1 : 49^1 - 1 = 48$ is divisible by 6; $P_2 : 49^2 - 1 = 2{,}400$ is divisible by 6; $P_3 : 49^3 - 1 = 117{,}648$ is divisible by 6   (7-2)

13. $P_k : 5 + 7 + 9 + \cdots + (2k + 3) = k^2 + 4k$; $P_{k+1} : 5 + 7 + 9 + \cdots + (2k + 3) + (2k + 5) = (k + 1)^2 + 4(k + 1)$ (7-2)

14. $P_k : 2 + 4 + 8 + \cdots + 2^k = 2^{k+1} - 2$; $P_{k+1} : 2 + 4 + 8 + \cdots + 2^k + 2^{k+1} = 2^{k+2} - 2$ (7-2)

15. $P_k : 49^k - 1 = 6r$ for some integer r; $P_{k+1} : 49^{k+1} - 1 = 6s$ for some integer s (7-2)

16. $S_{10} = -6 - 4 - 2 + 0 + 2 + 4 + 6 + 8 + 10 + 12 = 30$ (7-1, 7-3, 7-4)

17. $S_7 = 8 + 4 + 2 + 1 + \frac{1}{2} + \frac{1}{4} + \frac{1}{8} = 15\frac{7}{8}$ (7-1, 7-3, 7-4)   18. $S = \frac{81}{5}$ (7-4)   19. $S_n = \sum\limits_{k=1}^{n} \dfrac{(-1)^{k+1}}{3^k}$, $S_{\infty} = \frac{1}{4}$ (7-4)

20. $d = 3, a_5 = 25$ (7-3)   21. $\frac{8}{11}$ (7-4)   22. 190   (7-6)   23. 1,820   (7-6)   24. 1   (7-6)

25. $x^5 - 5x^4 y + 10x^3 y^2 - 10x^2 y^3 + 5xy^4 - y^5$ (7-6)   26. $-1{,}760 x^3 y^9$ (7-6)   30. 49g/2 feet; 625g/2 feet   (7-3)

31. $x^6 + 6ix^5 - 15x^4 - 20ix^3 + 15x^2 + 6ix - 1$ (7-6)

Practice Test: Chapter 7

1. (A) Neither; (B) Geometric, $r = -\frac{1}{4}$; (C) Arithmetic, $d = -4$ (7-3, 7-4)

2. $64, 60, 56, 52$; $a_{51} = -136$; $S_{31} = 124$ (7-3)   3. $32, 16, 8, 4$; $a_{10} = \frac{1}{16}$; $S_{10} = \frac{1{,}023}{16}$ (7-4)

4. $x^2 + 4x^5 + 9x^8 + 16x^{11}$ (7-1)   5. $\sum\limits_{k=1}^{4} (-1)^{k+1} (\frac{2}{3})^k$ (7-1)   6. $\frac{2}{111}$ (7-4)

7. Write $P_n : 3 + 3^2 + 3^3 + \cdots + 3^n = (3^{n+1} - 3)/2$

$S = \{n \in N \,|\, P_n \text{ is true}\}$

Part 1. Show $1 \in S$. $3 = (3^2 - 3)/2$ $\therefore 1 \in S$

Part 2. Show that S is inductive.

$$3 + 3^2 + 3^3 + \cdots + 3^k = (3^{k+1} - 3)/2$$
$$3 + 3^2 + 3^3 + \cdots + 3^k + 3^{k+1} = (3^{k+1} - 3)/2 + 3^{k+1}$$
$$= (3 \cdot 3^{k+1} - 3)/2 = (3^{k+2} - 3)/2$$

Thus, $k \in S \Rightarrow k + 1 \in S$, and S is inductive. $\therefore S = N$ (7-2)

8. (A) 455; (B) 253   (7-6)   9. $81x^4 - 108x^3 y + 54x^2 y^2 - 12xy^3 + y^4$ (7-6)   10. $-4{,}032 x^4$ (7-6)

Chapter 8   Exercise 8-2

1. 4   3. 8   5. 4   7. 8   9. 36   11. 360, 1,296   13. 19   15. 56   17. $26^3 \cdot 10^3 = 17{,}576{,}000$

19. 720, 6,840, $n(n-1)(n-2)$ 21. (A) $26 \cdot 25 \cdot 24 = 15{,}600$; (B) $26^3 = 17{,}576$; (C) $26 \cdot 25 \cdot 25 = 16{,}250$

23. 5!, 10!, n!   25. $12! \, 14! \approx 4.176 \times 10^{19}$

27. (A) 12; (B) Let N be the number of distinguishable arrangements. Then, $N \cdot 2! = 4!$ and $N = 12$.   29. 2^5

Exercise 8-3

1. 12   3. 2,652   5. 6   7. 1,326   9. 56   11. 360   13. $P_{6,3} = 120$ 15. $C_{25,2} = 300$ 17. $6! = 720$

19. $P_{5,3} = 60$, $P_{5,5} = 5! = 120$ 21. $C_{52,13} = 635{,}013{,}559{,}600$ 23. $\dbinom{52}{13, 13, 13, 13} \approx 5.364 \times 10^{28}$

25. $\dbinom{5}{2, 3} = 10$

27. $C_{60,5}C_{40,5} \approx 3.594 \times 10^{12}$　　**29.** $C_{n,r} = \dfrac{n!}{r!(n-r)!} = \begin{pmatrix} n \\ r, n-r \end{pmatrix}$　　**31.** $2! = 2$　　**33.** $\begin{pmatrix} 8 \\ 1, 1, 2, 2, 2 \end{pmatrix} = 5,040$

35. Terms involving $a^{n-k}b^k$ can be obtained by selecting the b's from k different factors and the a's from the rest. This can be done in $C_{n,k}$ ways.

Exercise 8-4

1. $S = \{1, 2, 3, 4, 5, 6, 7, 8\}$　　**3.** $\frac{1}{8}$　　**5.** $E = \{1, 3, 5, 7\}$, $n(E)/n(S) = \frac{4}{8} = \frac{1}{2}$　　**7.** $\frac{3}{4}$　　**9.** The event E cannot happen.
11. $S = \{HHH, HHT, HTH, HTT, THH, THT, TTH, TTT\}$　　**13.** $E = \{HTT, THT, TTH, TTT\}$, $P(E) = \frac{1}{2}$
15. (A) because $P(B)$ is negative and $P(R) > 1$, (B) because $P(Y) + P(B) + P(R) \neq 1$　　**17.** .57　　**19.** $\frac{1}{6}$　　**21.** $\frac{1}{18}$
23. $\frac{15}{36}$　　**25.** $\frac{35}{36}$　　**27.** $C_{16,5}/C_{52,5} \approx .001\ 68$　　**29.** $48/C_{52,5} \approx .000\ 0185$　　**31.** $4/C_{52,5} \approx .000\ 0015$
33. $C_{4,2}C_{4,3}/C_{52,5} \approx .000\ 009$　　**35.** $1/C_{6,2} = \frac{1}{15} \approx .067$　　**37.** $1/P_{6,3} = \frac{1}{120} \approx .008$

Exercise 8-5

1. .352　　**3.** .764　　**5.** P(point down) = .392; P(point up) = .608; no
7. (A)　P(3 tails) $\approx$ .128, P(2 tails) $\approx$ .371, P(1 tail) $\approx$ .380, P(0 tails) $\approx$ .121;
　　(B)　P(3 tails) = .125, P(2 tails) = .375, P(1 tail) = .375, P(0 tails) = .125;
　　(C)　3 tails: 125, 2 tails: 375, 1 tail: 375, 0 tails: 125

Exercise 8-6　　Chapter Review

1. (A)　Twelve combined outcomes;　(B)　$2 \cdot 6 = 12$　(8-2)　　**2.** 35　(8-3)　　**3.** 210　(8-3)　　**4.** $C_{13,3} = 286$　(8-3)

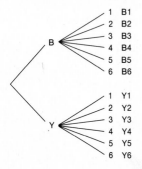

5. $P_{8,2} = 56$　(8-3)　　**6.** $P_{7,7} = 7! = 5,040$　(8-3)　　**7.** $\frac{1}{2}, \frac{2}{3}$　(8-4)　　**8.** .75　(8-5)　　**9.** 14　(8-2)　　**10.** 72　(8-2)
11. (A)　$C_{10,5} = 252$;　(B)　$P_{10,5} = 30,240$　(8-3)　　**12.** No, since $P(e_i)$ must be nonnegative　(8-4)

13. $E = \{(1, 3), (2, 2), (3, 1)\}$, $P(E) = \frac{3}{36} = \frac{1}{12}$　(8-4)　　**14.** $\dfrac{1}{P_{12,2}} = \dfrac{1}{132} \approx .007\ 58$　(8-4)　　**15.** $\dfrac{1}{C_{12,2}} = \dfrac{1}{66} \approx .0152$　(8-4)

18. $C_{14,2}C_{12,2} = 6,006$　(8-2, 8-3)　　**19.** $\dfrac{C_{13,2}C_{13,3}}{C_{52,5}} \approx .0086$　(8-4)　　**20.** 56　(8-3)　　**21.** Both equal $\dfrac{n!}{r!(n-r)!}$　(8-3)

22. $\begin{pmatrix} 52 \\ 13, 13, 13, 13 \end{pmatrix} \approx 5.36 \times 10^{28}$　(8-3)　　**23.** $\begin{pmatrix} 8 \\ 3, 2, 2, 1 \end{pmatrix} = 1,680$　(8-3)

Practice Test: Chapter 8

1. (A) $\frac{1}{8}$; (B) $\frac{5}{8}$ (8-4)   **2.** 120 (8-3)   **3.** 132 (8-3)   **4.** 66 (8-3)   **5.** (A) 210; (B) 343 (8-2, 8-3)

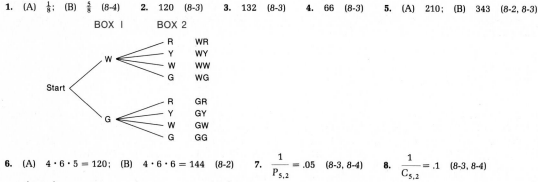

6. (A) $4 \cdot 6 \cdot 5 = 120$; (B) $4 \cdot 6 \cdot 6 = 144$ (8-2)   **7.** $\dfrac{1}{P_{5,2}} = .05$ (8-3, 8-4)   **8.** $\dfrac{1}{C_{5,2}} = .1$ (8-3, 8-4)

9. $\begin{pmatrix} 9 \\ 2, 3, 4 \end{pmatrix} = \dfrac{9!}{2!\,3!\,4!} = 1{,}260$ (8-3)   **10.** $\begin{pmatrix} 7 \\ 4, 3 \end{pmatrix} = 35$ (8-3)   **11.** $\dfrac{C_{6,2}}{C_{10,2}} \approx .33$ (8-4)

12. (A) Red, .13; green, .29; yellow, .58;   (B) Red, .10; green, .30, yellow, .60   (8-4, 8-5)

Index

Chapter-Opening Photo Sources

Chapter 0 Dragon fly
Photo © by P. J. Bryant, University
of California, Irvine/BPS.

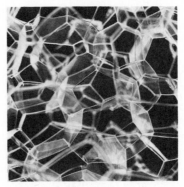

Chapter 1 Soap bubbles
Photo © by Peter Pearce.

Chapter 2 Feather
Photo © by Anne Monk.

Chapter 3 Fractured rock
Photo © by William C. Ferguson.

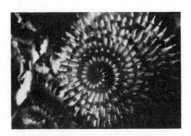

Chapter 4 Christmas tree worm
Photo © by S. K. Webster, Monterey Bay Aquarium/BPS.

Chapter 5 Leaf with rain drops
Photo © by Anne Monk.

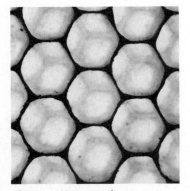

Chapter 6 Honeycomb
Photo © by Peter Pearce.

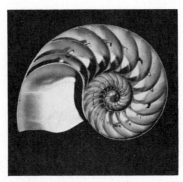

Chapter 7 Shell
Photo © by Anne Monk.

Chapter 8 Cracked mud
Photo © by Peter Pearce.